Protein Structure
and Engineering

NATO ASI Series

Advanced Science Institutes Series

A series presenting the results of activities sponsored by the NATO Science Committee, which aims at the dissemination of advanced scientific and technological knowledge, with a view to strengthening links between scientific communities.

The series is published by an international board of publishers in conjunction with the NATO Scientific Affairs Division

A	**Life Sciences**	Plenum Publishing Corporation
B	**Physics**	New York and London
C	**Mathematical**	Kluwer Academic Publishers
	and Physical Sciences	Dordrecht, Boston, and London
D	**Behavioral and Social Sciences**	
E	**Applied Sciences**	
F	**Computer and Systems Sciences**	Springer-Verlag
G	**Ecological Sciences**	Berlin, Heidelberg, New York, London,
H	**Cell Biology**	Paris, and Tokyo

Recent Volumes in this Series

Volume 177—Prostanoids and Drugs
edited by B. Samuelsson, F. Berti, G. C. Folco,
and G. P. Velo

Volume 178—The Enzyme Catalysis Process: Energetics, Mechanism,
and Dynamics
edited by Alan Cooper, Julien L. Houben, and Lisa C. Chien

Volume 179—Immunological Adjuvants and Vaccines
edited by Gregory Gregoriadis, Anthony C. Allison,
and George Poste

Volume 180—European Neogene Mammal Chronology
edited by Everett H. Lindsay, Volker Fahlbusch,
and Pierre Mein

Volume 181—Skin Pharmacology and Toxicology: Recent Advances
edited by Corrado L. Galli, Christopher N. Hensby,
and Marina Marinovich

Volume 182—DNA Repair Mechanisms and their Biological Implications
in Mammalian Cells
edited by Muriel W. Lambert and Jacques Laval

Volume 183—Protein Structure and Engineering
edited by Oleg Jardetzky

Series A: Life Sciences

Protein Structure and Engineering

Edited by
Oleg Jardetzky
Stanford University
Stanford, California

Assistant Editor

Robin Holbrook
Stanford University
Stanford, California

Plenum Press
New York and London
Published in cooperation with NATO Scientific Affairs Division

Proceedings of a NATO Advanced Study Institute
and Tenth Course of the International School of
Pure and Applied Biostructure
on Protein Structure and Engineering,
held June 19–30, 1989,
in Erice, Sicily, Italy

Library of Congress Cataloging-in-Publication Data

Protein structure and engineering / edited by Oleg Jardetzky ;
  assistant editor, Robin Holbrook.
        p.   cm. -- (NATO ASI series. Series A, Life sciences ; vol.
  183)
      "Proceedings of a NATO Advanced Study Institute and Tenth Course
  of the International School of Pure and Applied Biostructure on
  protein structure and engineering, held June 19-30, 1989, in Erice,
  Sicily, Italy"--T.p. verso.
      "Published in cooperation with NATO Scientific Affairs Division."
    Includes bibliographical references.
    ISBN 0-306-43484-9
      1. Protein engineering--Congresses. 2. Proteins--Structure-
  -Congresses. I. Jardetzky, Oleg. II. Holbrook, Robin.
  III. International School of Pure and Applied Biostructure. Course
  (10th : 1989 : Erice, Sicly) IV. North Atlantic Treaty
  Organization. Scientific Affairs Division. V. Series: NATO ASI
  series. Series A, Life sciences ; v. 183.
  TP248.65.P76P77   1990
  660'.63--dc20   89-49339
                                                                  CIP

© 1989 Plenum Press, New York
A Division of Plenum Publishing Corporation
233 Spring Street, New York, N.Y. 10013

Printed in the United States of America

ACKNOWLEDGEMENTS

The initiative for the 10th Course of the International School of Pure and Applied Biostructure, "PROTEIN STRUCTURE and ENGINEERING", came from Professor Claudio Nicolini, Director of the International School of Pure and Applied Biostructure, who also took the responsibility for obtaining funding and served as Co-Director of the Course. The course was held at the Ettore Majorana Center for Scientific Culture in Erice, Italy 19 - 30 June 1989 as a NATO Advanced Study Institute. The organizers and participants are also indebted to Professor A. Zichichi, Director of the Ettore Majorana Centre for Scientific Culture, for his support and to the staff of the Centre, notably Dr. Gabriele, Dr. J. Pilarski and Dr. P. Sevelli for their outstanding management of the course administration and all practical arrangements.

CONTENTS

Introduction.. 1
    Oleg Jardetzky

X-RAY

Exploitation of Geometric Redundancies as a Source of Phase Information in
    X-ray Structure Analysis of Symmetric Protein Assemblies 3
    Rudolf Ladenstein and Adelbert Bacher

Escherichia Coli Aspartate Transcarbamylase:
    the Relationship Between Structure and Function 35
    Evan R. Kantrowitz and William N. Lipscomb

Fundamentals of Neutron Diffraction 49
    V. Ramakrishnan

Purification of Complexes Between Peptide Antigens and Class II Major
    Histocompatibility Complex Antigens Using Biotinylated Peptides 61
    Theodore Jardetzky, Joan Gorga, Robert Busch, Jonathan Rothbard,
    Jack L. Strominger and Don Wiley

NUCLEAR MAGNETIC RESONANCE

NMR Method for Protein Structure Determination in Solution.................................... 69
    Kurt Wüthrich

Determination of Structural Uncertainty from NMR and Other Data:
    The *Lac* Repressor Headpiece.. 79
    Russ B. Altman, Ruth Pachter and Oleg Jardetzky

The Generation of Three-Dimensional Structures from NMR-Derived Constraints 97
    Denise D. Beusen and Garland R. Marshall

2D-NMR for 3D-Structure of Membrane Spanning Polypeptides:
    Gramacidin A and Fragments of Bacteriorhodopsin 111
    V. F. Bystrov, A. S. Arseniev, I. L. Barsukov, A. L. Lomize,
      G. V. Abdulaeva, A. G. Sobol, I. V. Maslennikov, and A. P. Golovanov

The Dynamics of Oligonucleotides and Peptides Determined by Proton NMR 139
    Patrice Koehl, Bruno Kieffer and Jean-François Lefèvre

Methods of Stable-Isotope-Assisted Protein NMR Spectroscopy in Solution.................. 155
    Brian J. Stockman and John L. Markley

NMR Studies of Protein Dynamics and Folding... 193
    Christopher M. Dobson

Understanding the Specificity of the Dihydrofolate Reductase
    Binding Site 209
    Gordon C. K. Roberts

Histone H1 Solution Structure and the Sealing of Mammalian Nucleosome 221
    Claudio Nicolini, Paolo Catasti, Mario Nizzari and Enrico Carrara

^{1}H NMR Studies of Genetic Variants and Point Mutants of Myoglobin:
    Modulation of Distal Steric Tilt of Bound Cyanide Ligand 243
    Gerd N. La Mar, S. Donald Emerson, Krishnakumar Rajarathnam,
    Liping P. Yu, Mark Chiu and Stephen A. Sligar

MOLECULAR DYNAMICS

Spectroscopy of Molecular Structure and Dynamics 257
    Rudolf Rigler

Molecular Dynamics:
    Applications to Proteins.. 269
    Martin Karplus

SITE-DIRECTED MUTAGENESIS

Protein Engineering and Biophysical Studies of Metal Binding Proteins 291
    Sture Forsén

Growth Hormones:
    Expression, Structure and Protein Engineering 309
    K. G. Skryabin, P. M. Rubtsov, V. G. Gorbulev, A. A. Schulga,
    A. Sh.Parsadanian, M. P. Kirpichnikov, A. A. Bayev,
    A. G. Pavlovskii, S. N. Borisova, B. K. Vainstein and A. A. Bulatov

Cloning, Sequencing and Expression of a New β-Galactosidase from
    the Extreme Thermophilic Sulfolobus Solfataricus 325
    Mosè Rossi, Maria Vittoria Cubellis, Carla Rozzo, Marco Moracci
    and Rocco Rella

Site-Directed Mutagenesis and the Mechanism of Flavoprotein Disulphide
    Oxidoreductases... 333
    Richard N. Perham, Alan Berry, Nigel S. Scrutton and Mahendra P. Deonarain

Resonance Raman and Site-Directed Mutagenesis Studies of Myoglobin Dynamics.......... 347
    Paul M. Champion

Comparison of the Secondary Structures of Human Class I and Class II
    MHC Antigens by FTIR and CD Spectroscopy... 355
    Joan C. Gorga, Aichun Dong, Mark C. Manning, Robert W. Woody,
    Winslow S. Caughey and Jack L. Strominger

Specificities of Germ Line Antibodies... 367
    Thomas P. Theriault, Gordon S. Rule and Harden M. McConnell

Contributors... 377

Index   379

INTRODUCTION

The development of molecular biology over the past thirty years has lead to an explosive growth of knowledge of protein structure and of methods for the design and manufacture of proteins of almost any desired sequence. While this seemingly opens limitless possibilities for the engineering of proteins with novel structures and functions, progress is in reality limited by our inability to predict function from structure, or even structure from sequence. Some progress in the understanding of protein folding and of the determinants of biological specificity has been made, but in essence these remain among the major unsolved problems of molecular biology. Therefore, no contemporary discussion of the subject can give the newcomer to the field a recipe for making a protein that will do exactly what he wants it to do. The best one can hope for is to survey the limits of the known in order to chart a few new inroads into the unknown.

The 10th Course of the International School of Pure and Applied Biostructure "PROTEIN STRUCTURE and ENGINEERING", held at the Ettore Majorana Center in Erice, Italy 19 - 30 June 1989 as a NATO Advanced Study Institute, was organized around three sets of issues:

(1) How is one to know what structures to make?
(2) How can one make them?
(3) What potential applications can be expected?

One might note that at this time only a partial answer to the first question can be given; it is possible to describe the methods available for the study of structures and their behavior - these include X-ray diffraction, High Resolution NMR, as well spectroscopic and theoretical methods for the study of molecular dynamics; it is also possible to cite specific examples in which site-directed mutants, designed from prior knowledge of the parent structure, have been used to test various hypotheses about the structure, dynamics, folding or function of a particular protein. For reasons noted above, it is not yet possible to define any kind of general design or architectural principles that would be useful in making decisions for the design of proteins with novel functions at will.

Techniques of genetic engineering have provided many specific answers to the second question. It would be only a slight overstatement to say that, the limitations of existing methods notwithstanding, almost any desired protein sequence could be produced at this time, using existing technology. If the published volume of the proceedings were to accurately reflect the state of knowledge in the field, description of such techniques and their known use would be the by far dominant section. The decision of the organizers was however to focus attention on answering the as yet unanswered questions of rational design and on structural studies that can allow it. For this reason only a sample of the available preparative methods have been included.

Knowledge of protein design and architectural principles will be necessary to transcend the narrow limits within which the third question can now be answered. The use of genetic engineering methods in the manufacture of known proteins for a variety of pharmacological, agricultural and industrial uses is a reality, but hardly involves any protein design. For the present, realistic answers to the question of new applications are limited to modification of known proteins to endow them with modified and, occasionally, new functions. The recent work on catalytic antibodies may serve as a prime example. Several other examples can be found in the present volume. The more bold and speculative suggestions made for stimulating conversation at the meeting, but did not crystallize in a form fit to print.

While the modification of known proteins may seem as a modest framework, compared to the limitless opportunities one can imagine, it must be noted that it has a very important role to play. It can serve to answer the key questions that need to be answered before one could speak of protein architectural principles and embark on a grand scheme of new protein design: What contribution does a particular amino acid at a particular point in the sequence make to the stability of a folded structure - or to the folding pathway? What is its contribution to a specific interaction with a ligand (be it a substrate, inhibitor or regulator)? The number of combinations to be considered in any attempt to answer these and related questions is staggering and, despite the rapidly growing literature, the surface of the problem has barely been scratched.

No single conference and no single volume can aspire to present an encyclopedic review of the large number of current studies relevant to the subject. At best one can hope to present a small selection representative of the cutting edge of the field. It is in this spirit that this volume is offered to the reader.

Oleg Jardetzky
September 29, 1989

EXPLOITATION OF GEOMETRIC REDUNDANCIES AS A SOURCE OF PHASE INFORMATION IN X-RAY STRUCTURE ANALYSIS OF SYMMETRIC PROTEIN ASSEMBLIES - Including a worked example: the three-dimensional structure of the icosahedral β_{60} capsid of heavy riboflavin synthase from *Bacillus subtilis*

Rudolf Ladenstein* and Adelbert Bacher**

*Max-Planck-Institut für Biochemie, D-8033 Martinsried & **Institut für Organische Chemie und Biochemie der Technischen Universität München D-8046 Garching, West Germany

INTRODUCTION

Scope of the Work

The construction of symmetric structures from asymmetric building blocks represents an important feature of nature and is studied by several disciplines of science from different viewpoints. In the field of molecular biology the symmetries of complex macromolecules are of special interest. They constitute the basis for structural organization and biological function in many cases. Maximum stability in oligomeric macromolecules is usually achieved by arranging the subunits in a symmetrical manner such that all of the subunits can form equivalent contacts.

During the past decades of research on biological macromolecules evidence has accummulated that icosahedral symmetry is an important feature which governs the self-organization of protein monomers in the formation of highly symmetric oligomeric complexes. The crystallographic work on virus structures is presently revealing the beauty, complexity and functionality of large macromolecular assemblies (Harrison, 1984; Liljas, 1986). The structure analytic study on heavy riboflavin synthase described in this paper will show that icosahedral symmetry may also be of importance for the structural organization of a bifunctional enzyme complex.

The known symmetries of a complex protein oligomer may very much benefit the determination of its three-dimensional structure. The well-known Patterson search methods (Huber, 1985) represent efficient correlation procedures which enable the crystallographer to extract the symmetry relations among the subunits of a crystalline oligomeric macromolecule from the crystallographic intensity data alone without prior conditions. The knowledge of these symmmetries in turn allows one to use the geometric redundancy in the intensity data set of a crystalline macromolecule in order to derive new structural information by averaging (Bricogne, 1976) of electron density maps in real space. The applicability of these methods has profitted from the development of efficient computers with high storage capacity which have made it possible to treat the vast experimental data in short time and with high accuracy.

The complex structures of highly symmetric protein molecules are fascinating in that they provide a picture of the immense potential for self-organization inherently present in matter.

3

Furthermore, the detailed knowledge of the structure of a macromolecular system serves as an important basis for the deeper understanding of its function. In most cases it will allow the intelligent planning of investigations with complementary methods, which may provide, together with the structure data, an insight into the complicated structure-function relations of a macromolecular assembly.

Heavy Riboflavin Synthase and Related Macromolecules

Heavy riboflavin synthase (HRS) from *Bacillus subtilis* is a bifunctional enzyme complex with a molecular weight of 10^6 Daltons. It is composed of 60 identical β subunits ($M_\beta = 16200$) which form an icosahedral capsid that encloses a trimer of α subunits ($M_\alpha = 23500$, Bacher et al. 1980). It has been shown by immunochemical methods that the immunological determinants of the α_3 trimers are not accessible for specific antibodies in the native complex $\alpha_3\beta_{60}$ (Bacher et al., 1980; Bacher et al., 1986). On the basis of electron microscopic data (Bacher et al., 1980) and X-ray small angle scattering (Ladenstein et al., 1986) a particle diameter of approximately 150 Å has been derived. The complex $\alpha_3\beta_{60}$ is stable only in a rather narrow pH region around pH 7. Dependent on pH and the concentration of specific substrate- and product analogous ligands (see Fig. 2) disaggregation of the native complex but also reaggregation to stable β_{60} aggregates (26 S), which are characterized by a hollow sphere shape, can occur (Bacher et al., 1986). In the absence of the stabilizing ligands polydisperse mixtures of large β aggregates are formed. The dominating species is characterized by an approximate particle diameter of 290 Å. Its architecture presumably follows the construction principles of truncated icosahedrons (Bacher et al., 1986). The well characterized reactions leading to related β subunit assemblies are shown in Figure 1.

Thus the complex $\alpha_3\beta_{60}$ represents an ideal system for the study of protein-protein and protein-ligand interactions and the self-assembly of macromolecular systems with icosahedral symmetry.

The Catalytic Reaction

The bifunctional complex $\alpha_3\beta_{60}$ catalyzes the final reactions in the biosynthesis of riboflavin (vitamin B_2). Briefly, the β subunits catalyze the condensation of a 3,4 - dihydroxy - butanone 4 - phosphate (**1**) with 5-amino-6-ribitylamino-2,4 (1H,3H) - pyrimidinedione (**2**) yielding 6,7-dimethyl-8-ribityllumazine (**3**) (Bacher et al., 1978; Neuberger et al., 1986). The subsequent dismutation of **3** is catalyzed by the α subunits yielding riboflavin (**4**) and the pyrimidinedione **2** which can be reutilized by the β subunits (Bacher et al.), (Figure 2). The kinetics of the catalytic steps are incompletely understood.

Crystals of Heavy Riboflavin Synthase

The complex $\alpha_3\beta_{60}$ could be crystallized from 1.35 M phosphate buffer pH 8.7 in the presence of 0.5 mM ligand **5** (Ladenstein et al., 1983). The increased pH stability under the influence of the substrate analogous ligand turned out to be crucial for successful crystallization. The crystals diffract X-rays to a resolution of 3.3 Å and belong to space group $P6_322$ of the hexagonal system. The unit cell dimensions are $\mathbf{a} = \mathbf{b} = 156.4$ Å, $\mathbf{c} = 298.5$ Å, $\alpha = \beta = 90°$, $\gamma = 120°$. As a consequence of space group symmetry, particle dimensions and threefold particle symmetry, the particle centers must sit on points with symmetry [32]. Thus the crystalline packing may be described either by hexagonal densest packing or by packing in hexagonal layers (Ladenstein et al., 1983). In Figure 3 these two possibilities are shown. By electron microscopic investigation of freeze-etched 3D-crystals the packing in hexagonal layers could be verified (Ladenstein et al., 1986).

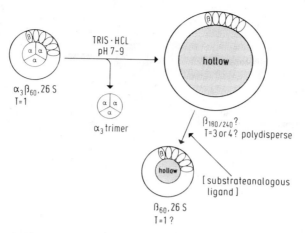

Figure 1. Disaggregation and reaggregation of heavy riboflavin synthase ($\alpha_3\beta_{60}$); formation of hollow β_{60} (26 S) particles.

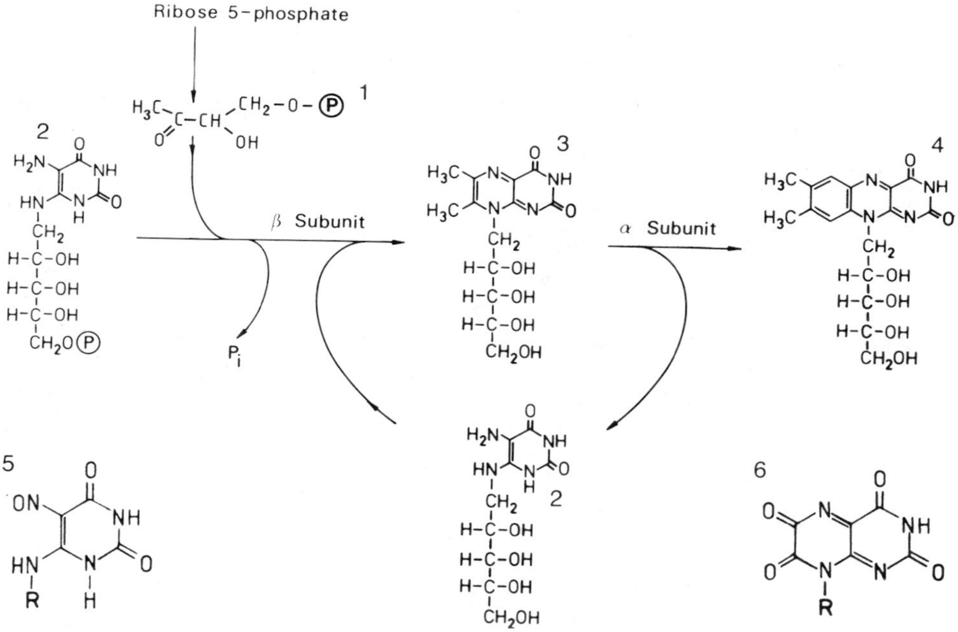

Figure 2. Biosynthesis of riboflavin: **1** = 3,4 - dihydroxy - 2 - butanone 4 - phosphate; **2** = 5-amino-6-(D-ribitylamino)-2,4(1H,3H)-pyrimidinedione; **3** = 6,7-dimethyl-8-(-ribityl)-lumazine; **4** = riboflavin.
Inhibitors of heavy riboflavin synthase: **5** = 5-nitroso-6-ribitylamino-2,4(1H,3H)-pyrimidinedione; **6** = 6,7-dioxo-8-ribityl-5,6,7,8-tetrahydrolumazine.

5

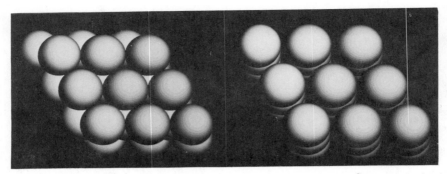

Figure 3. Sphere models showing possible crystalline packing of $\alpha_3\beta_{60}$ particles in the hexagonal unit cell. a) densest hexagonal packing; b) layer packing.

CRYSTALLOGRAPHIC FUNDAMENTALS

Diffraction of X-Rays by a Periodic Object

When a plane wave is scattered by an object, the scattered radiation may be described by the equation

$$F(h) = \int \rho(x) \cdot \exp 2\Pi ihx \, dx \qquad [1]$$

where F is a complex number which represents the amplitude and phase of the scattered radiation in a direction determined by the vector h, $\rho(x)$ is the scattering function at a position x in the object, and the integral is taken over the volume of the 3D object. For X-ray diffraction, $\rho(x)$ is the electron density at position x (excluding the effects of anomalous scattering). Thus the scattered radiation is described by the Fourier transform of the object as seen by the incident radiation. By taking the inverse Fourier transform, we get

$$\rho(x) = \int F(h) \cdot \exp -2\Pi ihx \, dh \qquad [2]$$

The integral is taken over the volume V^* of the space spanned by the vector h.

We are interested in the special case in which the scattering object is a 3D crystal. The fundamental property of a crystal is that $\rho(x)$ is periodic in all three dimensions of space. It is known that the Fourier transform of a periodic function is zero, except when h is an integer multiple of the periodicity. Thus the structure factors $F(h)$ are zero except on a three-dimensional lattice, the so-called reciprocal lattice.

The natural coordinate system for a crystal is

$$x = x \cdot a + y \cdot b + z \cdot c \qquad [3]$$

where a, b, c represent the basis vectors of the unit cell of the crystal. These vectors are not necessarily orthogonal but they define a three-dimensional space, which is called **direct space**. The dimension of direct space is $length^3$ (L^3). In turn we may define the space of the structure factor lattice, which is related to the recorded diffraction pattern, by

$$h = h \cdot a^* + k \cdot b^* + l \cdot c^* \qquad [4]$$

with the lengths of the basis vectors **a*,b*,c*** inversely proportional to the lengths of the basis vectors **a,b,c** of the unit cell. The space in which the structure factors are defined is generally called **reciprocal space**; its dimension is length^{-3} (L^{-3}).

In evaluation of diffraction experiments it is often necessary to transform a function defined in reciprocal space into direct space and vice versa. As we have seen these operations can be performed by Fourier transformation (F) and inverse Fourier transformation (F^{-1}) as

$$F[\rho(x)] = F(h) \qquad\qquad [5]$$
$$F^{-1}[F(h)] = \rho(x).$$

The Electron Density Function

If we choose **a***, **b*** and **c*** to obey the Laue relations

$$
\begin{array}{lll}
a \cdot a^* = 1 & a \cdot b^* = 0 & a \cdot c^* = 0 \\
b \cdot a^* = 0 & b \cdot b^* = 1 & c \cdot c^* = 0 \\
c \cdot a^* = 0 & c \cdot b^* = 0 & c \cdot c^* = 1
\end{array} \qquad [6]
$$

the vector products in the integrals (1) and (2) simplify to $h \cdot x = hx + ky + lz$. With these definitions we are able to normalize (1) and (2) to reflect the contents of one unit cell; we get

$$F(h) = V \iiint \rho(x) \cdot \exp[2\Pi i(hx + ky + lz)]dxdydz \qquad [7]$$
$$\rho(x) = 1/V \iiint F(h) \cdot \exp[-2\Pi i(hx + ky + lz)]dhdkdl \qquad [8]$$

The discrete nature of **F(h)** allows the conversion from an integral to a sum in Eq. [8]

$$\rho(x) = 1/V \sum_h \sum_k \sum_l F(h) \cdot \exp[-2\Pi i(hx + ky + lz)] \qquad [9]$$

Equation [9] represents the well-known electron density equation which can be calculated by inverse Fourier transformation of the scattered waves, described by the structure factors **F(h)**. The structure factor **F(h)** is characterized by an amplitude |F(h)| and a phase $\alpha(h)$ according to

$$F(h) = |F(h)| \cdot \exp i\alpha(h) \qquad [10]$$

The phase information $\alpha(h)$ is lost in the diffraction experiment; phases are generally determined by methods such as single (SIR) and multiple (MIR) isomorphous replacement and phase extension. The amplitudes $|F(h)| = const. \cdot \sqrt{I(h)}$ are obtained from measurement of the crystallographic intensities in diffraction patterns (e.g. photographic rotation method (Arndt et al., 1977), area detectors (Messerschmidt et al., 1987) and diffractometers (Blundell et al., 1976). The set of all symmetry independent I(h) represents the unique intensity data set of a crystal. Crystallographic intensity data of native and derivative crystals of heavy ribo-flavin synthase are shown in Table 1.

Formal Description of Symmetries

By definition the crystallographic symmetries represent the set of symmetry elements valid in a crystal; they relate the asymmetric units of the crystal cell. The noncrystallographic or local symmetry elements are confined to the asymmetric unit and can be described by the set of symmetry operations which are valid in the asymmetric unit. In the case of an oligomeric protein the asymmetric unit of a crystal cell may contain more than one copy of a subunit. The positions of these subunits in direct space are defined by the symmetry operations of the assymetric unit.

7

Table 1. Intensity data statistics ($F^2 > 1.0\ \sigma$).

Derivative	Resolution	Measurements	Independent Reflections	Measured/Possible Reflections	$R_{merge}[\%]$
ITAN	∞-3.3	75100	27700	0.850 to 3.3Å	13.1
AuCN	∞-3.2	87250	32200	0.673 to 3.2Å	12.8
CMAA	∞-3.6	71300	24800	0.767 to 3.6Å	13.9
WAC	∞-3.4	38200	23000	0.741 to 3.4Å	12.8
WP	∞-3.6	19000	13700	0.434 to 3.6Å	9.4
LUMO	∞-3.6	32500	18300	0.684 to 3.6Å	11.0

ITAN, native crystals; $R_{merge} = \Sigma\Sigma|<I_h>-I_{hi}|/\Sigma N_h<I_h>$, where $<I_h>$ is the average intensity of N_h measurements and I_{hi} is the individual intensity of a reflection h.
LUMO, functional derivative obtained by soaking the native crystals with 1mM of the dioxolumazine (Ligand **6**) in 1.75 M potassium phosphate buffer, pH = 8.7, at 20°C.

Generally we may define a symmetry operation in 3D space by way of a linear transformation including a 3x3 Matrix **R** and a translation vector **t** as

$$\mathbf{x}' = \mathbf{R} \cdot \mathbf{x} + \mathbf{t}$$ [12]

and $$\rho(\mathbf{x}') = \rho(\mathbf{x})$$ [13]

for all positions **x** within a crystal or its asymmetric unit.

The Symmetry of Icosahedral Polyhedrons

The ancient Greek mathematicians already knew that only five regular polyhedrons can exist, the so-called platonic solids. These are tetrahedron, octahedron, icosahedron, cube and dodecahedron. Nowadays it is well established that geometry and symmetry of these bodies represent an important principle which governs the self-assembly of protein subunits in quite a large number of cases. A regular icosahedron (Figure 4) is constructed from 20 equilateral triangles and possesses 6 fivefold (n = 72°), 10 threefold (n = 120°) and 15 twofold (n = 180°) rotation symmetry axes. All of these axes intersect at a common point which represents the center (origin) of the particle.

A regular icosahedron is characterized by 60 asymmetric units. This number represents the maximum number of identical units which may be arranged on a closed symmetrical shell. Thus

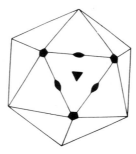

Figure 4. Icosahedron with all symmetry elements of one triangular face indicated: (●) twofold axis, (▼) threefold axis, (◆) fivefold axis. All of the symmetry axes intersect at the center of the particle: an icosahedral asymmetric unit is defined as the triangular region in between two fivefold and one threefold axis.

each of the 60 units takes an exactly equivalent position. The biological advantages of this mode of aggregation of protein subunits are obvious (e.g. energetic stabilization, macromolecules within the shell are protected from degradation).

Geometric considerations show that icosahedral polyhedra can be described by a triangulation number $T = h^2 + hk + k^2$ (h,k are integers), which represents the number of triangles of a hexagonal net on one triangular icosahedral face. T can assume only certain values, the smallest ones being $T = 1, 3, 4, 7$.

Most of the known spherical viruses contain more than 60 subunits in their capsids and can be characterized by icosahedral symmetries. In a now classic paper (Caspar & Klug, 1962), a theory was developed describing how multiples of 60 identical units could be arranged in a quasi-equivalent manner to give stable protein complexes. These arrangements can be described as truncated icosahedrons, characterized by triangulation numbers $T > 1$ and can be constructed from $60 \cdot T$ subunits. Among spherical viruses quasi-equivalence was detected in several cases (Harrison et al., 1978; Rossmann et al., 1985; Hogle et al., 1985).

The Patterson Function

The Patterson function represents an important tool of protein crystallography (Huber, 1985; Dodson, 1985). Together with suitable correlation functions, it may be used for the determination of heavy atom positions as well as molecular orientation and translation parameters (Rossmann et al., 1963). The Patterson function is defined by a convolution integral as follows

$$P(u) = V \cdot \int \rho(x) \cdot \rho(x + u) dx \qquad [14]$$

The vector $u = (u,v,w)$ spans Patterson space. The function values $P(u)$ may be calculated by taking the product of the electron density at position x with the electron density at position $x + u$. By substitution of the electron densities, $\rho(x)$ and $\rho(x + u)$, for those defined by [9], it may be shown that

$$P(u) = 1/V \sum_u \sum_v \sum_w I(h) \cdot exp2\Pi i h \cdot u \qquad [15]$$

It is obvious that $P(u)$ can be calculated without prior conditions directly by Fourier transformation of the crystallographic intensities $I(h) \propto |F(h)|^2$. If the positions x and $x + u$ are identical with atomic positions in the crystal, the function $P(u)$ shows maximum values. The vector u then represents an interatomic vector. Thus the function $P(u)$ may be characterized as a distribution of all interatomic vectors of the unit cell in space. $P(u)$ possesses characteristic symmetry properties:
a) the vector distribution is centrosymmetric,
b) $P(u)$ contains the same rotation symmetries as the particle in direct space,
c) there are no translation symmetries in $P(u)$.

STRUCTURE ANALYTIC METHODS

Definition of the Problem

Protein molecules frequently crystallize with several identical subunits in the asymmetric unit, or in several crystal forms which contain the same molecule in different arrangements. Rossmann & Blow recognized that intensity data collected from such structures are redundant

(Rossmann et al., 1963), and that their redundancy could be a source of phase information. Very efficient symmetry averaging techniques (Bricogne, 1976) have been developed which use the geometric redundancy in the intensity data for the refinement of phase angles.

In heavy riboflavin synthase, the subject of this study, the icosahedral symmetry of the β_{60} capsid gives rise to local symmetries in the crystal asymmetric unit. The asymmetric unit of the hexagonal unit cell (space group P6$_3$22) contains 10 β subunits (Ladenstein et al., 1983 ; Ladenstein et al., 1986). If there is more than one copy of a subunit of an oligomeric protein in the crystallographic asymmetric unit, the basic problem is to find the mathematical operation which superimposes one subunit on the other. This is equivalent to the determination of six variables which specify rotation ($\mathbf{R}_i$) and translation ($\mathbf{t}_i$) of the structure motif (ie. one subunit) in 3D space. By reference to Figure 5, the following parameters must be known as a prerequisite for symmetry averaging of electron density maps:
a) orientation and foldedness of the particle symmetry axes P,Q,R relative to the crystal axes x,y,z (rotation matrices $\mathbf{R}_i$),
b) translation in between the origin of the crystal system x,y,z and the origin of the molecular coordinate system P,Q,R (translation vectors $\mathbf{t}_i$).

In addition, an envelope enclosing the 3D region occupied by one subunit or an appropriate subunit aggregate must be defined. In the following sections solutions to the rotation, translation, starting phase and envelope problems will be discussed.

The Rotation Problem

The Patterson function contains information on the local rotational symmetries of the subunits within an asymmetric unit. The orientation as well as the foldedness of rotation symmetry axes and thus the particle orientation may be extracted from this information by suitable correlation procedures. The correlation of two 3D structures $P_1(x)$ and $P_2(x)$ in Patterson space is given as a function of their relative orientation by (Rossmann et al., 1962)

$$R(\Psi,\Phi,X) = \int P_1(x)\cdot P_2[C(\Psi,\Phi,X)x]dV \qquad [16]$$

In determining the orientation of a symmetric protein molecule P_1 and P_2 represent the same

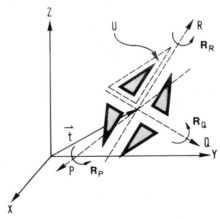

Figure 5. Determination of molecular rotation, translation parameters and of a molecular envelope: x, y, z - crystallographic coordinates, P, Q, R - molecular coordinates, **R** - rotational operator, **t** - translation vector, U - envelope.

Patterson structures, i.e. $P_1 = P_2$, and R is an auto-correlation function, which shows an origin peak. R may be easily calculated in a 3D grid by summation of the products $\Sigma P_1(\mathbf{x}) \cdot P_2(\mathbf{C} \cdot \mathbf{x})$; $\mathbf{C} = (\mathbf{D} \cdot \mathbf{T}(\Psi, \Phi, X) \cdot \mathbf{O})$ is a matrix operator which is dependent on the angular variables Ψ, Φ, X; $\mathbf{O}$ and $\mathbf{D}$ are orthogonalization and deorthogonalization matrices which allow transformation between a Cartesian and the crystal coordinate system. The elements of $\mathbf{T}$ may be calculated from trigonometric functions of Ψ, Φ, X (Patterson, 1959).

Several suitable coordinate systems may be chosen for these calculations (Rossmann et al., 1962). Here we refer to spherical polar coordinates (Figure 6). The Cartesian coordinates x,y,z of a point P lying on a rotation axis are then given by

$$x = r \cdot \sin \Psi \cdot \cos \Phi$$
$$y = r \cdot \cos \Psi$$
$$z = -r \cdot \sin \Psi \cdot \sin \Phi$$

[17]

The results of the self-correlation search for rotational symmetries may be easily represented by two-dimensional stereogram plots (Figure 7).

The hexagonal unit cell of HRS contains two complete icosahedral particles $\alpha_3 \beta_{60}$ (Ladenstein et al., 1983). The concept of the rotation function is based on determining the rotational relationship between the self-vectors within a molecule and, as far as possible, omitting the (generally) longer crossvectors between adjacent particles, since crossvectors are related to packing symmetries. This implies that the solutions $R(\Psi, \Phi, X)$ are dependent on the integration radius. For this reason a radius of integration around the Patterson origin must be chosen which is less than the diameter of the particle. In this investigation (Ladenstein et al., 1986) the rotation function was explored in 5° intervals. The clearest result for the icosahedral $\alpha_3 \beta_{60}$ particle was obtained with an integration radius of 50 Å ($\approx$ 1/3 particle diameter). For the calculation of [16] we have used intensity data $I(\mathbf{h}) > 1\sigma(I)$ between ∞ and 5 Å resolution. The results of the correlation calculations in Patterson space show a set of maxima for 2-fold, 3-fold and 5-fold local axes, accurately consistent with icosahedral symmetry; the 2D stereograms derived from the rotation function sections $R[\Psi, \Phi(X = 180°, 120°, 72°)]$ are shown in Figure 7.

The orientations of the icosahedral particle axes in the spherical polar coordinate system (Figure 6) are given by the coordinates (Ψ, Φ) of the correlation maxima, assigned the symbol (o). The particle orientation relative to the hexagonal crystal cell axes $\mathbf{a}, \mathbf{b}, \mathbf{c}$ could be derived from a comparison with two possible models $\mathbf{A}, \mathbf{B}$ as shown in Figure 8. The results of the correlation calculations for the icosahedral heavy riboflavin synthase particle are compatible only with orientation $\mathbf{A}$. The rotation operations defined by the polar angle triplets $(\Psi, \Phi, X)_i$ can easily be described by conversion to 3x3 rotation matrices.

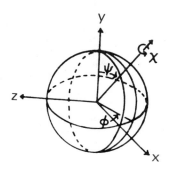

Figure 6. Spherical polar coordinate system. x, y, z - orthogonal coordinates, (Ψ, Φ) - orientation of a rotation axis, (χ) - foldedness.

11

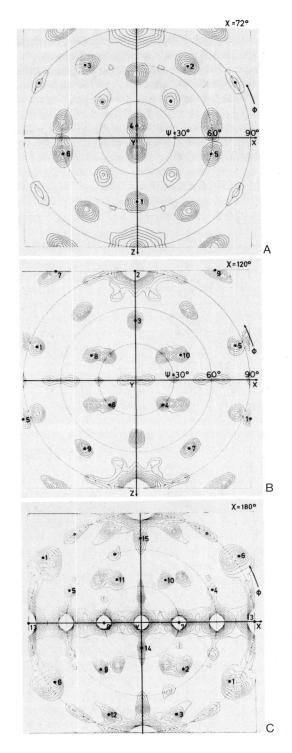

Figure 7. Stereogram plots for two-, three- and fivefold rotation axes, derived from correlation calculations in Patterson space: (A) $\chi = 72°$, (B) $\chi = 120°$, (C) $\chi = 180°$, from the

The Translation Problem

Starting from an unknown molecular model, the translation problem may be solved by two approaches. The translation vector(s) t_i, pointing from the origin of the crystal unit cell to the origin(s) of the molecular coordinate system(s) can be found from:
a) the center of gravity of heavy atom positions
b) analysis of molecular packing in the crystal unit cell.

Method a) requires the introduction of heavy atoms into the protein oligomer under investigation (single isomorphous replacement, SIR (Blundell et al., 1976). The subsequent determination of the heavy atom coordinates by difference Patterson methods is, depending on the complexity of the problem, not always straightforward.

Method b) is the method of choice for large macromolecules which may be observed easily by electron microscopy. The method requires the availability of microcrystals of the protein under investigation. Molecular packing in the crystal unit cell can then be directly deduced from electron micrographs obtained by the freeze etching technique (Ludwig et al.,1982, Ladenstein et al., 1986).

The crystalline packing of the two icosahedral HRS molecules can be described by two equally plausible packing models as has been mentioned previously: densest hexagonal packing or packing in hexagonal layers. By electron microscopy of freeze etched HRS crystals (Ladenstein et al., 1986), the hexagonal layer packing has been verified (Figure 9). This kind of molecular packing is characterized by particle positions on the sixfold screw axes 6_3 of the unit cell. However, the crystal symmetries of space group $P6_322$ place some restrictions on the z-coordinates of the particle positions. The following positions (x,y,z) may be adopted:

	particle 1	particle 2	
case 1	(0,0,0)	(0,0,1/2c)	
			[18]
case 2	(0,0,1/4c)	(0,0,3/4c)	

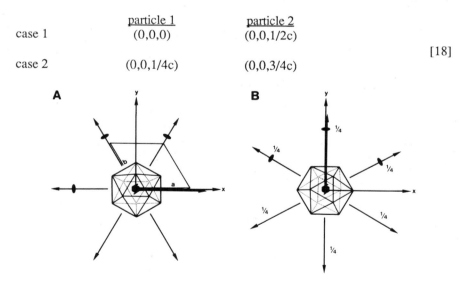

Figure 8. Orientation of the icosahedral particles relative to the hexagonal crystal cell, defined by **a,b,c** and to the orthogonal system x,y,z. (●) - twofold symmetry axes in the **ab** plane; two orientations corresponding to models **A, B** are shown.

positions of the maxima (•) the orientations (ψ,ϕ) of icosahedral symmetry axes can be deduced. Peak height = 100 for crystallographic twofolds, mean value 48, contour lines drawn from 50 - 60 in steps of 1 arbitrary unit, $\sigma = 0.3$.

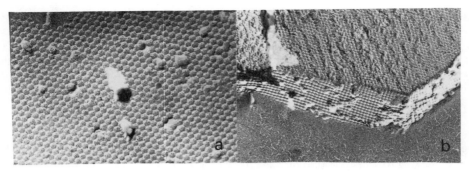

Figure 9. Electron micrographs of freeze-etched heavy riboflavin synthase micro crystals: (a) ab plane, (b) ac plane showing hexagonal layer packing. Distance of particle centers - 152 Å.

The orientation of the particles in the two cases differs by a particle rotation around the z-axis of $\Delta\Phi_z = 30°$ (see Figs 7,9). The rotation function analysis described in the previous chapter is only compatible with the particle positions of case 1 in which a certain subset of local twofold icosahedral axes coincides with crystallographic twofold axes. For simplicity, the particle at the crystallographic origin (translation vector $\mathbf{t} = (0,0,0)$, zero vector) was chosen as a reference particle for the symmetry averaging calculations. In Figure 10 a stereodiagram of the molecular packing of HRS particles in the unit cell is shown.

The Problem of Getting Starting Phases

The problem can be characterized by the minimum phase information (see equations 9, 10) which is necessary to solve an unknown protein structure by direct-space averaging of electron density maps. Different approaches may be chosen to solve the problem:

a) Because geometrically redundant intensities and phases are mathematically related (Rossmann et al., 1963; Crowther, 1967; Crowther, 1969), the phases could in general be calculated *ab initio*. However a possible snag lies in enantiomorph ambiguity. Crowther (Crowther, 1969) showed that if the arrangement of envelopes was centrosymmetric (this is true for all spherical envelopes), only the real parts A of the complex structure factors $\mathbf{F}(\mathbf{h}) = A + iB$ could be determined by exploitation of the noncrystallographic symmetry alone. Therefore phases

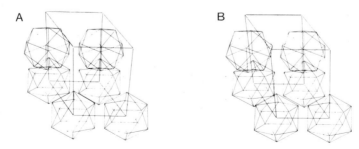

Figure 10. Crystal packing of heavy riboflavin synthase molecules drawn to scale. For clarity, 4 particles at z = O and only 2 particles at z = 1/2 are shown (A & B - stereo pair of Riboflavinsynthase crystal packing); drawn by a computer program written by Lesk & Hardman (1982).

14

generated from such information may not even be immediately useful to locate heavy atoms in isomorphous derivatives.

b) In a more promising approach a structurally similar protein is used as a source of preliminary phase information. As an example, Luo et al.(Luo et al., 1987) have determined the structure of the icosahedral mengo virus, a member of the picorna virus group, with phase information from the structurally related rhinovirus HRV 14.

c) The availability of one heavy atom derivative obtained by the method of single isomorphous replacement (SIR) (Blundell et al., 1976) will usually provide a straightforward solution to the problem. As already mentioned, the translations may be determined from the heavy atom positions, either directly (for proper symmetries) or after correlation search in the SIR electron density map (for improper symmetries). Averaging or solvent-flattening (Wang, 1985) of the SIR map further will provide a reasonably accurate envelope. Refinement of the SIR phases by symmetry averaging and phase combination (Hendrickson et al., 1970) will usually solve the phase problem. Since geometric redundancies also allow one to extend the phases to higher resolution by cyclic averaging, low resolution SIR phases (5 - 10 Å resolution) are generally sufficient.

A specialized SIR method which seems to be extremely suitable for icosahedral macromolecules (e.g. spherical viruses) and other symmetric protein complexes was found by using symmetrical heavy atom cluster compounds (Ladenstein et al., 1987) which match the symmetry of their binding sites on the protein. Due to clustering of heavy atoms on defined positions on the symmetry axes, these complexes behave as 'superscatterers', which may easily be identified in difference Patterson maps. This method may significantly facilitate phase determination in structure analysis of symmetrical protein complexes and spherical viruses. Heavy atom derivatives are generally obtained by soaking the native crystals in suitable heavy atom solutions. The conditions and heavy atom reagents used to prepare isomorphous derivatives of heavy riboflavin synthase crystals are given in Table 2.

Table 2. Heavy atom derivatives of heavy riboflavin synthase

Reagent	Abbreviation	Concentration	Treatment*
$K[Au(CN)_2]$	AuCN	1mM	1 week
$Cl\text{-}Hg\text{-}CH_2\text{-}CHO$	CMAA	1mM	24 hours
$[W_3O_2(O_2CCH_3)_6]^{2+}$	WAC	1mM	3 weeks
$[NaP_5W_{30}O_{110}]^{14-}$	WP	<4mM**	3 weeks

* in 1.75 M phosphate buffer, pH 8.7, 0.5 mM Ligand **5**
** saturated solution of WP in *, 20°C

The vector set corresponding to the heavy atom structure is represented by a difference Patterson synthesis

$$\Delta P(u) = 1/V \sum_u \sum_v \sum_w (|F_D| - |F_p|)^2 \cdot exp2\Pi ih \cdot u \qquad [19]$$

$|F_D|$ and $|F_p|$ represent the structure factor amplitudes of derivative and native protein, respectively. In principle, the heavy atom structure may be derived from [19] but with highly symmetric proteins showing many heavy atom positions within an asymmetric unit, its

15

interpretation is usually difficult. If the heavy atom positions obey the local symmetries the problem may be solved more easily. Computerized vector search procedures allow one to explore a difference Patterson map systematically by using the known local and crystallographic symmetry operations.

In the work on heavy riboflavin synthase difference Patterson maps were calculated with intensity data $|F|^2 > 1\sigma$ from 30-4.0 Å resolution on a 1.75 Å grid. The vector search program is a part of the real space search routines of the PROTEIN system (Steigemann, 1974). The program was modified to allow in addition to crystallographic symmetries, the inclusion of local symmetry operators, which can be defined in the input stream either by rotation matrices or by angle triplets of different angle systems together with the appropriate translation vectors t_i. Scan parameters are the x,y,z coordinates of the heavy atom site. From scan point $r_i = (x,y,z)$ all symmetrical points defined by local and crystallographic operators S_k ($k = n \cdot m$, if n is the number of crystal symmetry operators and m is the number of local symmetry operators) and therefore all Patterson difference vectors $d_{ik} = (r_i - S_k r_i)$ are computed. Because we used a product function

$$C = [\Pi \, \Delta P_i(u)]^{1/k} \qquad\qquad [20]$$
$$i$$

as a correlation function, the function values were scaled by taking the kth root. This function turned out to represent a very strict measure for the correlation between vector densities and positions. For the successful use of vector search methods it is necessary to carefully define the minimum scan region. The scan region must include at least one complete asymmetric unit, which for an icosahedron is defined as the region in between two fivefold and one threefold local rotation axes. For our puposes a scan region defined by the grid values {0<x<45, -8<y<30, 0<z<19} was appropriate. This region consisted of 34,086 scan points. For an independent scan with 10 local and 12 crystallographic symmetry operators S_k roughly 100 hours computing time on a VAX 11/782 computer were necessary. The vector search applied to the derivative AUCN clearly showed a single site at $(x,y,z)_{grid} = (35, 5, 4)$, close to a local twofold axis of the particle. A preliminary interpretation in terms of this site was the basis for a successful initiation of phase calculation. The heavy atom positions of all other derivatives were subsequently determined from difference electron density maps calculated with SIR phases from the derivative AUCN in the following way

$$\Delta\rho(x) = 1/V \sum\sum\sum \Delta F \cdot exp\text{-}2\Pi ih \cdot x \qquad\qquad [21]$$
$$ h\,k\,l$$

$$\Delta F = |F_h(DERI)| - |F_h(NATI)| \cdot exp i\alpha_{SIR}$$

The heavy atom positions were refined (Dickerson et al., 1961; Steigemann, 1974) and phases calculated (Blow et al., 1959) in the usual way; Table 3 shows the heavy atom parameters used for phase calculation.

The Envelope Problem

The envelope (see Figure 5) represents a closed surface in 3D space which encloses one subunit M or an appropriate subunit aggregate M_n of a symmetric protein molecule. It should reflect the subunit boundaries as exactly as possible since the direct space averaging calculations (see section on Cyclic Averaging of Electron Densities and Phase Extension, following) hardly converge with an incorrectly defined envelope.

The problem of envelope definition is intimately connected with the phase problem because preliminary phase information is required, i.e. phase information which allows the calculation of a low resolution electron density map (Figure 12).

Table 3. Heavy atom parameters used for isomorphous replacement phasing

Derivative	Site Number	Fractional cell coordinates			Occupancy Z,σ(Z)	Temperature factor B,σ(B)
		x	y	z		
AuCN	1	0.39093	0.05415	0.02123	66.42,7.28	39.93,18.58
	2	0.25025	-0.12032	-0.00469	18.30,3.16	30.49,11.25
CMAA	1	0.38928	0.05361	0.02164	63.78,4.53	52.37, 7.34
	2	0.26069	-0.07788	-0.00152	14.77,3.70	59.39,10.20
	3	0.35542	0.11007	0.04134	9.13,4.80	40.89, 8.72
WAC	1	0.01277	0.00707	0.14352	19.33,7.47	22.88, 7.85

Fractional cell coordinates of the heavy atom sites within a given isocahedral asymmetric unit are listed.
Z - Mean value of the site occupancy in electrons over all 10 sites of a crystallographic asymmetric unit; $\sigma(Z)$ - standard deviation.
B - mean value of the temperature factor over all 10 sites of a crystallographic asymmetric unit; $\sigma(B)$ - standard deviation.

In the $P6_322$ modification of riboflavin synthase crystals (Ladenstein et al., 1983) one crystallographic asymmetric unit contains 10 icosahedral asymmetric units or, in other words, one sixth of a complete enzyme particle. Proper averaging of an electron density map about local (non-crystallographic) symmetry elements requires truncation of the map by an envelope that contains those portions obeying the local symmetry. In the present case the envelope, which we shall term UP6322, has to cover the 1/6 particle described above.

Envelopes were defined by use of the program system ENVELOPE, developed by G.Remington (unpublished work). The program was run on a GENISCO raster graphics system linked to a VAX 11/782 computer. Sections of a closed envelope (-1/4<z<1/4) were traced on a position-sensitive magnetic tablet on a tenfold averaged MIR electron density map at 5 Å resolution (Figure 12). This map had been calculated on a 1.75 Å grid and allowed a clear determination of the molecular boundaries of the β_{60} capsid. The envelope contains 1/6 of the particle centered about $(x,y,z) = (0,0,0)$, but includes densities from two different crystallographic asymmetric units, denoted (1), (2) in Figure 11. The corresponding points of a.u.(2) were reduced to a.u.(1) by space group specific subroutines (SPGRP1, SPGRP2) called by the GENERATE main program (Bricogne, 1976). From the file containing the envelope z-sections a mask file was generated containing the logical variables .TRUE./.FALSE. at positions inside/outside of the envelope, respectively. At later stages of the calculations the envelope was redefined several times in order to give a tight fit on the one hand and to avoid cut-off of electron densities which belong to the capsid on the other hand.

Cyclic Averaging of Electron Densities and Phase Extension

Crystallographic intensity data collected from highly symmetric protein structures, which frequently crystallize with several identical subunits in the asymmetric unit are redundant; their redundancy could be a source of new phase information. The phase constraints implied by the consistency of geometrically redundant intensities were generalized by Rossmann & Blow (1963).

Let us consider now the case of a highly symmetric, oligomeric protein structure, whose crystal asymmetric unit contains n copies of the subunit M; in addition, the following information is available:
a) local symmetries $(\mathbf{R_i}, \mathbf{t_i})$
b) SIR starting phases $(\alpha(\mathbf{h})_{SIR})$

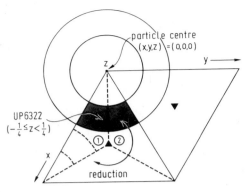

Figure 11. Definition of the closed envelope (UP6322). The shaded region $(-1/4 \leq z \leq 1/4)$ contains 10 β subunits which are related by local symmetries. During the averaging calculations the portion within crystallographic asymmetric unit (2) was reduced to crystallographic asymmetric unit (1) by space group specific subroutines.

The solution of this structure is performed, in outline, as follows (see also Figure 13):

1. Calculate an SIR electron density map, $(\alpha(\mathbf{h})_{SIR})$
2. Define an envelope U of the basic aggregate M or M_n
3. Iterate the following procedure until the crystallographic R-factor, defined as

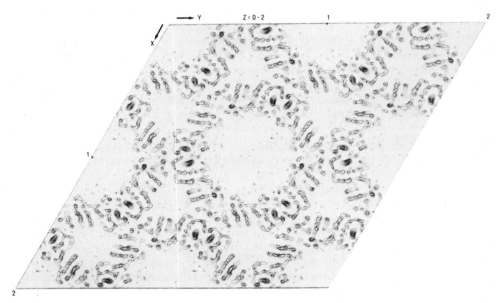

Figure 12. Icosahedrally averaged electron density map at 5.0 Å resolution, calculated with SIR phases from the derivative AUCN. The contour lines of three z-sections $(0 \leq z \leq 2)$ are overlaid; grid: x,y,z = 90, 90, 180; contour lines $\geq 1.0\ \sigma(\rho(x))$ in steps of 1.5 σ; in the directions x, y two full unit cell lengths are drawn; the ring-shaped density of the β_{60}-capsid is clearly visible.

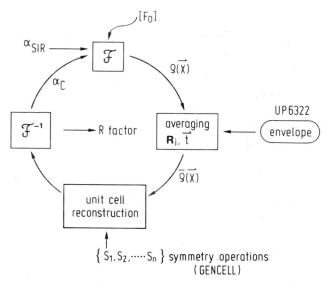

Figure 13. Cyclic averaging of electron density maps. F - Fourier transformation, F^{-1} - inverse Fourier transformation, ρ (x) - electron density, α_{SIR} - starting phases, α_c - calculated phases, R_i, t_i - rotation and translation operators.

$$R = \sum_h ||F_o(\mathbf{h})| - |F_c(\mathbf{h})|| / \sum |F_o(\mathbf{h})| \qquad [22]$$

(F_o and F_c representing observed and calculated structure factors, respectively)

converges:

a) Calculate a starting electron density map

$$\rho_{init}(\mathbf{x}) = F^{-1}[|F_o| \cdot \exp i\alpha(\mathbf{h})_{SIR}] \qquad [23]$$

b) Average the n independent copies of M within U (local symmetries R_i, t_i) and calculate average density in solvent regions, $\rho_o(\mathbf{x})$.

c) Rebuild the asymmetric unit from the averaged aggregate and reconstruct the complete unit cell by using crystallographic symmetries $S_1, S_2 ... S_m$.

d) Compute structure factors from this averaged map, $\rho_{av}(\mathbf{x})$, by Fourier back-transformation

$$F[\rho_{av}(\mathbf{x})] = F_c(\mathbf{h}) = |F_c(\mathbf{h})| \cdot \exp i\alpha_c(\mathbf{h}) \qquad [24]$$

e) Calculate an improved electron density map by using the calculate phases $\alpha_c(\mathbf{h})$ and the observed structure amplitudes $|F_o(\mathbf{h})|$

$$\rho_{new}(\mathbf{x}) = F^{-1}[|F_o(\mathbf{h})| \cdot \exp i\alpha_c(\mathbf{h})] \qquad [25]$$

The averaging operations 3 b,c pose serious computational problems as soon as the elecron density map becomes too large to fit into the core memory of the computer. In order to calculate the average density at a given point under a symmetry operation of order n, densities have to be fetched at n - 1 other points, and these can be anywhere in an oversampled map which may

19

contain several millions of points. Thus the problem is basically an adressing problem. The approach to this problem provided by Bricogne eliminates the initial random access requirement by means of two sorting operations (double sorting technique, Bricogne et al., 1976).

An appropriately modified version of the program system GENERATE (Bricogne, 1976; Ladenstein et al., 1987) was used to perform the averaging calculations on the electron density maps of the heavy riboflavin synthase β_{60} capsid. The programs were almost exclusively used on a Digital MicroVax II computer with 9 Mbyte core memory and 500 Mbyte magnetic discs. Space group specific subroutines (SPGRP1, SPGRP2) for $P6_322$ had to be written and were linked to the GENERATE main program (Bricogne, 1986). One complete cycle required about 8 hours computing time on the MicroVax II. The program GENCELL is a stand-alone version that generates the electron density of a complete unit cell from the density within an asymmetric unit by application of the corresponding crystallographic symmetries. Since the AUCN data was much more complete than the native data and extended to higher resolution as a consequence of better crystalline order (possibly introduced by the treatment with $K[Au(CN_2)]$), we treated the derivative AUCN as the native compound. An electron density map was first calculated to 5.0 Å resolution with coefficients $|F_o(AUCN)| \cdot \text{expi}(\alpha(h)_{MIR})$. This map served as a starting density for the cyclic calculations.

Since multiple isomorphous replacement (MIR) showed reasonable heavy atom contributions to the phase calculations only to about 4 Å resolution, we extended the phases by cyclic averaging to 3.3 Å, the maximum resolution available in the AUCN data set. The Fourier inversion step in the cyclic procedure (Fig. 13) can be used to calculate phases to any desired resolution independent of the limiting resolution of the phases used for calculation of the initial map. Several authors have shown that local symmetry leads to phase relationships between reflections of similar resolution (Rossmann & Blow, 1963; Main & Rossmann, 1966). Phases were extended from 4.5 to 3.3 Å in resolution shells with radii increased by a constant increment, ranging from $\Delta d^* = 0.1$ to 0.06 Å and containing roughly 2000 reflections for which new phases had to be generated. Each step consisted of several cycles: one expansion cycle from d^* to $d^* + \Delta d^*$, which was followed by 3 - 10 averaging cycles during which all reflections up to $d^* + \Delta d^*$ obtained a new phase (Figure 14). Some statistics showing the convergence of the final phase refinement as a function of resolution are given elsewhere (Ladenstein et al., 1987). For the total set of 30,717 reflections to 3.3 Å resolution an overall R-value, $R_{lin} = \Sigma |F_o| - |F_c| / \Sigma |F_o| = 24.4$ %, was obtained.

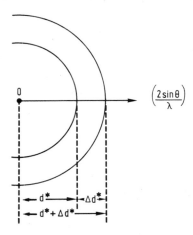

Figure 14.   Phase extension in concentric shells of reciprocal space.
Resolution: $1/d_{hkl} = 2 \sin \theta / \lambda$; O, origin.

THE STRUCTURE OF THE ICOSAHEDRAL β SUBUNIT CAPSID OF HEAVY RIBOFLAVIN SYNTHASE

Interpretation of the Final Electron Density Map

The main peptide chain of the β subunit was traced on z-sections of a 5 Å minimap drawn onto plexiglass sheets (scale 4.3 Å/cm). A complete atomic model of the β-subunit was built by fitting the amino acid sequence, obtained from chemical sequencing (Ludwig et al., 1987) to the averaged electron density map. All of the necessary steps were performed on an Evans & Sutherland PS 330 vector graphics system connected to a Digital MicroVax II computer. A modified version of the model building program system FRODO (Jones, 1978) was used.

The clarity of the improved electron density map can be seen in Figure 15. Several density sections (z > 0) of the enzyme particle at x,y,z = (0,0,0) are shown. With increasing z the diameter of the ring-shaped density representing the β capsid decreases. The assignment of the boundaries of single subunits at first proved difficult, since the interfaces, chiefly those belonging to the pentamers, appeared rather tight. Density features which stem from a central β sheet structure with neighbouring α helices were visible already in the initial averaged 5.0 Å density map. Although the electron density is at a nominal resolution of only 3.3 Å, the path of the backbone was clear, with bifurcations conceivable at only two points. We were faced with some difficulties at the N-terminal β-strand, which at first was assigned to a neighbouring subunit. Two connectivities at loops near to the substrate binding site were initially unclear, because the density representing the bound substrate analogue (ligand 5) interfered with main- and side-chain densities of the protein. There are only two termini and the direction of the chain was evident from the four α-helices. The fit of the side chains to the electron density is generally acceptable, with one exception: the region from residues 129 to 136, containing several glycines, was difficult to fit. Correct placement of the amino acid sequence within the density map was confirmed by the following points:

(1) The positions of two cysteine residues agree with the crystallographically determined locations labeled by binding of gold atoms (derivative AUCN = K(Au(CN)$_2$)).
(2) The positions of the three methionines and of several aromatic residues, especially Trp, correspond with well-defined side chain densities.
(3) The absence of density for glycine was a good positional indicator.
(4) Hydrophilic residues were predominantly located in regions exposed to solvent, and hydrophobic residues in buried regions.
(5) Most α helices and β strands predicted by a modified Chou-Fasman/Robson-Algorithm (Ludwig et al, 1987) match observed secondary structure elements surprisingly well.

The final model of the HRS β subunit is characterized by an R-Value of 39.9 % for 30,717 independent reflections in the range of 20.0 - 3.3 Å resolution.

Folding of the Main Chain of a β Subunit

The 154 residues comprising an entire β subunit of HRS are folded into an irregular ellipsoid of dimensions 47x29x18 Å. The N-terminal strand β_1 (Met1-AsN7) protrudes from the core of the subunit like a tail, approximately 17 Å in length. The main chain folds into α helix/β sheet motifs which are repeated four times (Figure 16). The respective β segments organize into a parallel four-stranded β sheet structure flanked on both sides by two α-helical segments. Two helices, α_2 and α_3, on one side of the sheet structure run fairly parallel, whereas the other pair, helices α_1 and α_4, are skew, with an angle of roughly 45 degrees between their helix axes. The C-terminal helix α_4 shows a bend near its N-terminal end due to three glycines. The central β-

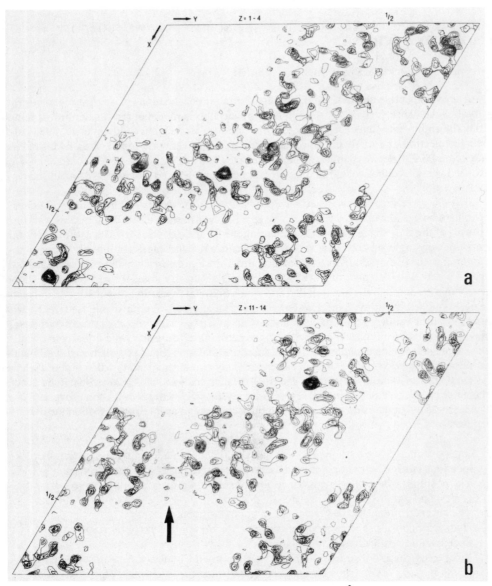

Figure 15. Two sections of the final electron density map at 3.3 Å resolution obtained by icosahedral averaging and phase extension (derivative AUCN). The cell axes x,y are not completely drawn; the contour lines of several z-sections are overlaid, (A) z = 1-4, (B) z = 11-14; x,y,z = 150, 150, 300; contour levels: ≥ 1.4 $\sigma(\rho(x))$ in steps of 1.0 σ; the arrow (↑) indicates a pentamer channel crossing the capsid wall; high density maxima stem from heavy atom binding sites of AUCN.

sheet structure shows a right-handed twist of about 50 degrees. N- and C-termini are found in the same region of the monomer, at a distance of approximately 15 Å. The cysteines (Cys93, Cys139) are part of the helix segments α_3 and α_4, respectively. They are about 16 Å apart and are involved in the binding of the heavy atom compounds AUCN and CMAA.

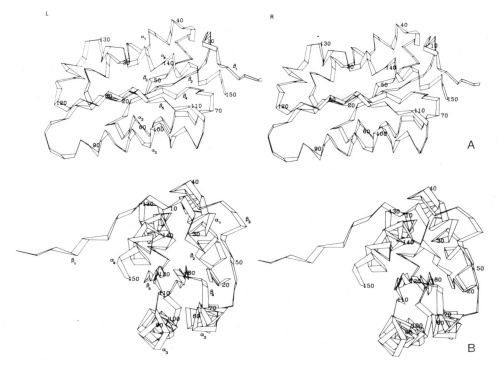

Figure 16. Main chain folding of a heavy riboflavin synthase β subunit (stereo pairs); only C_α carbon positions are shown; direction of projection tangential (A) and perpendicular (B) with respect to the capsid surface.

A large proportion of the amino acids of the β-subunit are involved in well-defined secondary structure (~51% α-helix, ~25% β-sheet, ~10% β-turns), providing strong stabilization by intramolecular hydrogen bonding. A substantial amount of α-helix and β-structure was properly located by secondary structure predictions with modified Chou-Fasman- and Robson procedures, based on the chemically determined amino acid sequence (Ludwig et al, 1987). The results were of some value in the assignment of uncertain connectivities and in proper alignment of the sequence. In total, 60% of the residues involved in the formation of α-helix and 72% of β-structure were correctly predicted.

Resemblance with the Flavodoxin Fold

The folding of the HRS β subunit shows some resemblance with flavodoxins whose structures have been studied extensively (Ludwig et al., 1982). The flavodoxins represent a class of microbial electron carrier proteins with one non-covalently bound riboflavin- 5´-phosphate (FMN) prosthetic group per molecule and molecular weights in the range of 15,000 Daltons. The three-dimensional structures are generally organized about a five-stranded parallel sheet exhibiting a typical twist. Pairs of α-helices are found on either side of the sheet. Upon formation of β-subunit pentamers, a core consisting of a five-stranded parallel β-sheet is built up. Four parallel strands are contributed by each subunit and one strand, the N-terminal extension $β_1$, by its neighbour. This topology resembles closely that of the five-stranded parallel sheet of the flavodoxins. The structures of the flavodoxin from Clostridium MP and the HRS β subunit deviate by 5.9 Å (mean deviation) after optimal superposition of eight C_α positions.

The topologies and connectivities of the helices α_2, α_3, α_4 and the strands β_4, β_5 are similar in both proteins under consideration. However, the connectivities of the remaining secondary structure elements show some significant differences (Figure 17). Thus, we find the N-terminal strand β_1 of the HRS β-subunit directly connected with strand β_2 and then with helix α_1. In the flavodoxins the N-terminal strand β_1 is followed by helix α_1 and strand β_2 is then connected to strand β_3 by an extended loop, which contains a one-turn helix. A sequence comparison of different flavodoxins with the HRS β subunit did not reveal any significant homology. In summary, there is presently no evidence for an evolutionary relatedness.

The Icosahedral β_{60} Capsid

Due to the symmetry relations in an icosahedral protein capsid, intersubunit contacts are those of dimers, trimers and pentamers. The present qualitative description of these contacts treats only the major features. A more detailed analysis will be based on a refined structure. It should be emphasized that the oligomeric structures which will be discussed in the following section represent partial structures of the β_{60} capsid. These substructures must not be mistaken for actual molecular entities. Although one would intuitively assume that the assembly of the capsid from protomers involves the intermediate formation of small oligomers, these have so far resisted a detailed elucidation. However, the discussion of symmetry-related substructures appears helpful in a description of the complex interactions in the icosahedral shell.

The contacts observed within dimers appear to be hydrophobic as well as hydrophilic in nature. They are confined to a rather small region near to the outer capsid wall.

Trimer contacts are formed primarily by several segments at the inner capsid wall (residues 21-29, 81-85, 120-134). A conspicuous feature of the contact area is a cluster consisting of three glutamic acids (Glu126, Glu126', Glu126") sharing no other ligands which could compensate their negative charges. The positively charged tungsten complex compound, WAC, which was used for MIR phase calculation, was detected at this site (Ladenstein et al., 1987). In the native complex $\alpha_3\beta_{60}$ this site may provide ionic groups for the binding of complementarily charged side chains of the α_3 trimer.

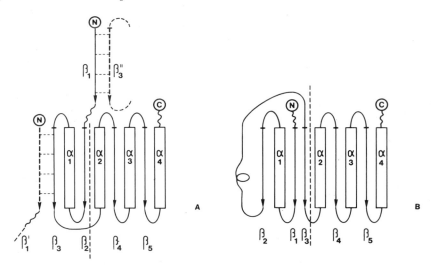

Figure 17. Comparison of the folding topology of a heavy riboflavin synthase β subunit (A) and the flavodoxins (B); α_i, α-helix, β_i,-strand, β_1', β_3", β-strands of fivefold related subunits.

As compared with dimer and trimer contact areas, the pentamer contacts exhibit by far the most extended interfaces, comprising all possible types of non-covalent bonds. An estimated area of more than half of the total accessible surface area of the monomer is involved in the formation of pentamer contacts. If the reduction in accessible surface area occurring upon subunit assembly is a rough measure of the free energy of association, as suggested by Chothia (Chothia, 1974), the assembly of pentamers should be very much preferred over the formation of dimers or trimers.

In the pentamer, the extended N-terminal segment (strand β_1) of each monomer is attached to the neighbouring monomer and comprises the fifth strand of a parallel b-sheet (Figures 17, 18a, b). A similar structural arrangement has been detected in lactate dehydrogenases from vertebrates. Structural rearrangement of the N-terminal segment β_1 may occur during assembly. On fixation of the strand β_1 to its neighbour several main chain hydrogen bonds and side chain hydrophobic contacts are newly formed. We believe tentatively, that this structural feature plays a role in the recognition of the β-subunits during assembly. Furthermore the N-terminal tail may control the proper closure of the cyclic five-fold assembly. The pentamer may be an obligatory intermediate in the assembly process and favoured energetically and kinetically over other intermediates. This is open to testing by assembly studies with truncated β subunits.

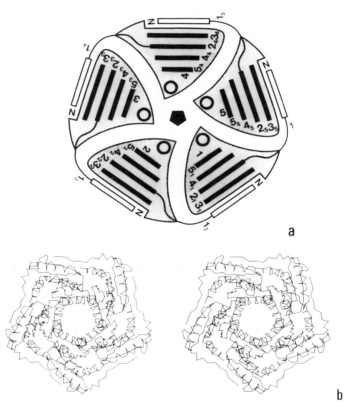

Figure 18. (a) Subunit arrangement in a β_5 pentamer, view direction along a fivefold particle axis; β strands and N-terminal segments are emphasized. (o), approximate positions of the helix segments which built up the superhelical channel (it should be noted that only selected secondary structure elements are shown); (b) C_α model of a β_5 pentamer (stereo pair); direction of projection along a fivefold particle axis.

Along the fivefold pentamer axes channels are formed by α-helix α_3 and its four symmetry related neighbours. Reminiscent of α-helix arrangement in integral membrane proteins, these a helices are nearly perpendicular to the capsid surface. They show a left-handed twist and build up a large superhelical structure motif, an α-helical coiled coil. This structural element resembles a pore, approximately 27 Å in length and 9 Å in diameter, which spans the capsid wall. α-Helical coiled coils have been found in a remarkable number of cases as a structure motif in proteins (Cohen & Parry, 1986). In coiled coil α-helices meshing of side chains allows intimate contacts. In the HRS pentamer these residues are largely apolar but some polar contacts also occur. The inner surface of the pores consists primarily of polar side chains of the α-helical segment α_3 providing a hydrophilic surface. The distance of these residues within the sequence (3-4 residues) is typical for an amphiphilic α helix. The side chains of intervening residues are mainly apolar and point into the interior of the protein (92 Val, 93 Cys, 96 Ala, 97 Ala, 99 Gly, 100 Ile, 101 Ala, 102 Gln, 104 Ala, 105 AsN, 106 Thr). We assume tentatively that the pores of the capsid are involved in substrate import to and product export from the catalytic centers which are found in the core of the particle.

The capsid of HRS is composed of 60 identical β subunits which obey the symmetry relations of a $T = 1$ icosahedron. Figure 20a shows a C_α model of the complete β_{60} assembly. The view is down an icosahedral fivefold axis. One of the twelve pores of the particle is clearly visible. The overall appearance of the capsid is spherical; triangular faces, geometrically characterizing an icosahedron, are not visible even by way of suggestion. From the cross section of the capsid shown in Figure 19b it is obvious that most of the main secondary structure elements are oriented nearly perpendicular to the capsid surface. The capsid has inner and outer diameters of $R_i = 39$ Å and $R_a = 78.5$ Å, respectively. These values derived from the atomic structure are in good agreement with those found from X-ray small angle scattering ($R_a = 83.8$ Å) and electron microscopy ($R_a = 77$ Å) (Bacher et al., 1986; Ladenstein et al. 1986). The available core space of the β_{60} capsid, ($V = 248,000$ Å^3), is large enough to accomodate an α subunit trimer of $M = 70,000$ Daltons. It has been mentioned earlier that the α subunits do not obey the crystallographic symmetry and obviously adopt different orientations within the population of particles comprising a HRS crystal (Ladenstein et al., 1983). Therefore it is not possible to determine the atomic structure of the α_3 trimers from intensity data of the crystalline complex $\alpha_3\beta_{60}$. Attempts are being made towards a separate crystallization of the α_3 trimers. By the asymmetric distribution of charges at the inner and outer capsid walls, an electrical polarity of the the capsid wall is generated. The electrostatic potential field due to the titratable groups and α-helix macrodipoles has been calculated on the basis of the Coulomb relation (Karshikov & Ladenstein, 1989). The electrostatic potential shows a radially polar distribution with a positive pole at the inner capsid wall and a negative pole outside the capsid. An interesting feature of the electrostatic field is the formation of positive potential "channels" that coincide with the channels constituted by the trimeric and pentameric β-subunit aggregates. It is supposed that the electrostatic potential field plays a role in enzyme-substrate recognition.

The charge distribution in the capsid of HRS is opposite to the distribution of charges observed in spherical animal and plant viruses. Nevertheless, a positive pole of the electrostatic potential field is found at the inner capsid wall. The virus capsids carry positive charges at their inner surface which are believed to be involved in binding interaction with phosphate groups of the nucleic acids residing in the central core (Liljas, 1986).

The Substrate Binding Site

The HRS β subunit catalyzes the formation of the lumazine **3** from the pyrimidinedione **2**; (Fig. 2). Each β subunit can bind one molecule of the substrate analogue **5** or the product analogue **6**, presumably at the active site (Bacher & Ludwig, 1982). These analogues have been

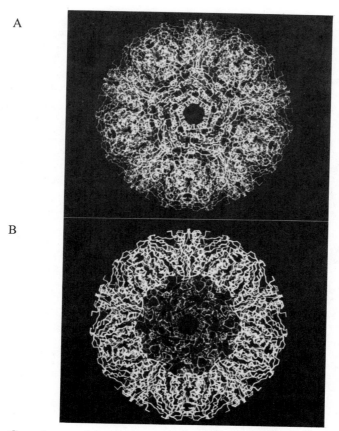

A

B

Figure 19. C_α carbon models of the heavy riboflavin synthase β_{60} capsid. (A) Direction of projection along a fivefold icosahedral axis, one of the twelve channels of the particle is clearly visible. (B) Central section of the β_{60} capsid, inner radius, $R_i = 39$ Å, outer radius, $R_a = 78.5$ Å.

used in earlier studies on ligand driven reaggregation of β_{60} particles (Bacher et al., 1986). The substrate analogue ligand **5** was a key factor in the crystallization of HRS (Ladenstein et al., 1983) and is bound at a 1:1 stoichiometry (ligand/β subunit) in the crystal forms ITAN, AUCN, CMAA, WP, WAC used in the present study. It should be noted that the analogues **5** and **6** can also bind to α subunits as demonstrated by studies with light riboflavin synthase (α_3 trimers) (Otto & Bacher, 1981).

In attempts to localize the ligand binding site we have prepared the derivative LUMO (Table 2) by soaking of AUCN crystals in 1.75 M phosphate buffer pH 8.7 containing 1 mM ligand **6**. We expected that this treatment should result in the exchange of ligand **5** bound in the crystals with the lumazine **6**. This exchange of ligands should account for a mass difference of 44 daltons (2 oxygens, 1 carbon) per β subunit. In line with this expectation, a tenfold averaged 3.7 Å difference Fourier map calculated between the derivatives LUMO and AUCN showed a difference electron density of the size we expected.

The difference electron density proved to be remarkably clear, thus demonstrating again the power of real space averaging methods. The maximum of the difference density was found directly adjacent to a drumstick-shaped excess density near to the inner capsid wall which is not accounted for by the peptide chain and had therefore been tentatively considered as a density feature which could represent the bound substrate analogue ligand **5**, present in the derivative AUCN. The difference density LUMO-AUCN which stems from the atom C7 and both oxygens at C6 and C7 of ligand **6** proved to be crucial for the correct orientation of the bound ligand. A standard group representing ligand **5** was constructed from X-ray data and fitted into the excess electron density not occupied by the β subunit model, as shown in Figure 20.

The substrate binding site of a HRS β subunit is found at the inner capsid wall pointing into the core, which accommodates a subunit trimers in $\alpha_3\beta_{60}$ particles. It is formed by several segments (β turns: 21-24, 54-58, 81-92, α helix: 127'-142'; β strand: 113'-116') which belong to two neighbouring subunits of the β_5 pentamers. The shortest distances of the binding site (reference point: N atom of the ribitylamino substituent of ligand **5**) to neighbouring fivefold, threefold and twofold particle axes are roughly 18 Å, 18.5 Å, and 25.5 Å, respectively.

The substrate binding site at the pentamer interface has the shape of an elongated pocket.

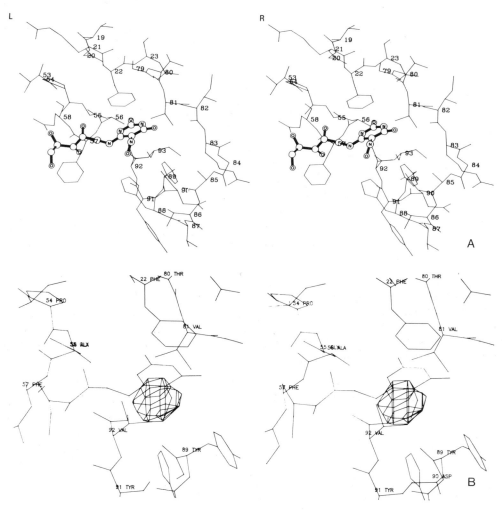

Figure 20.   (A) Substrate binding site of a β subunit of heavy riboflavin synthase, (B) the bound ligand **1** is overlaid with the icosahedrally averaged 3.7 Å difference electron density (LUMO - AUCN), contour level ≥ 5.0 σ(ρ), (stereo pairs).

The nitroso-pyrimidinedione ring is placed in a relatively hydrophobic region of the pocket formed by the side chains of Phe 22, Tyr 89, Ala 56, Val 81 and Val 92. The keto groups of the pyrimidinedione system may be hydrogen bonded to main chain groups O14-N82 and O13-N56. The oxygen atom of the nitroso group is found about 3 Å away from N_ε of His 88 and may contribute an additional hydrogen bond. His 88 is assumed to take part as a proton abstracting base in the catalytic reaction. The hydrophilic tail of the ligand molecule, consisting of the ribityl amino group is oriented such that it points to segments 127'-142' (α helix) and 113'-116' (β strand) of the neighbouring subunit. Cys 139'-S is found at a distance of $\approx$7 Å with respect to the secondary amino group of the ribityl tail and could in principle take part in the catalytic reaction, probably with the second substrate. Hydrogen bond interaction of Cys 139'-S with residues ribityl O18 and Lys 135 O (main chain) may be possible. The ribityl hydroxy groups OH 17, OH 19 and OH 20 can be hydrogen bonded to residues Gly 55 N (main chain), Glu 58 OE1 and Phe 113 O (main chain), respectively. A general property of the binding pocket is its construction from an apolar region, in which the pyrimidine ring is accommodated, and a polar region, showing a strong hydrophilic character (Lys 131, ion pair Lys 135/Asp 138, and Arg 127). The ribityl group is bound by this hydrophilic region, as might have been expected.

The second substrate taking part in the catalytic reaction, 3,4-dihydroxy-2-butanone 4-phosphate, is negatively charged due to its phosphate group. In catalysis it may be oriented such that these negative charges are stabilized by interaction with the side chains of Lys 131 and Lys 135. The binding of the substrates and products and the consequences for the catalytic reaction are currently being investigated. For a detailed picture which can describe the catalytic process, however, complementary information from NMR measurements must also be considered.

Comparison with Viral Capsid Structures

The structure of the HRS β subunit capsid obeys the same symmetry relations as the capsids of small spherical viruses (Liljas, 1986). However the folding pattern of the β subunits and the known subunit structures of spherical viruses show no apparent similarity. Thus the well-known eight-stranded β barrel motif found in spherical viruses does obviously not represent a necessary condition for the formation of an icosahedral assembly.

The central core space of the capsid occupied in icosahedral viruses by nucleic acids, hosts a trimer of enzymatically active α subunits. It should be noted in this context that the HRS β_{60} capsid is constructed from only one subunit type with a molecular weight of only 16,200 daltons. This relatively simple molecular system provides therefore an attractive model for studies on the factors responsible for stability and function of a highly symmetric protein assembly.

Stability Features of the Icosahedral Capsid

The β subunit has the capacity to form quite a variety of different molecular aggregates in spite of its relatively small size (16,200 daltons). The native enzyme present in the bacterial cells is the icosahedral complex $\alpha_3\beta_{60}$ which is the subject of the present paper. In the absence of appropriate ligands this assembly is relatively unstable (Bacher et al., 1980; Bacher et al., 1986). More specifically, it is highly sensitive to pH and buffer composition. The particle has excellent longterm stability in phosphate buffer, but complete dissaggregation can occur even under very mild conditions such as 0.1 M Tris chloride at pH 7.

The disruption of the enzyme molecule results in the liberation of α subunits, whereas the β subunits associate under formation of even larger aggregates, which appear as hollow spheres with a radius in the range of 150 Å, (see Fig.1). Hydrodynamic studies suggest the involvement of at least three different molecular species with sedimentation velocities close to 48 S. These species may be in a state of dynamic equilibrium, but details are not yet known. It is conceivable that the large hollow spheres represent empty icosahedral capsids with triangulation numbers of $T = 3$ or 4. An approximate molecular weight of 3 200,000 daltons derived from buoyant density equilibrium experiments appears consistent with this model (Bacher et al., 1986). Furthermore, the assembly state of β subunits is highly susceptible to perturbation by substrate analogues. Noncovalent binding of the ligand **5** drives the formation of spherical particles which consist of 60 β subunits. The ligand-driven reaggregation proceeds initially very fast (on the time scale of seconds) but nevertheless does not go to completion in phosphate buffer at pH 7. However, quantitative formation of hollow aggregates is possible from 6 M urea solution of β subunits by dialysis against phosphate buffer containing ligand **5** (Bacher et al., 1986).

The β_{60} aggregate has been characterized by hydrodynamic and electron microscopic studies as well as small angle X-ray scattering. All data support the conclusion that these molecules represent an empty $T = 1$ icosahedral capsid (Bacher et al., 1986; Ladenstein et al., 1986). Tentatively, we conclude that the ligand **5** increases the stability of a β_5 pentamer as compared to a β_6 hexamer. This could drive the association equilibrium in the direction of the $T = 1$ icosahedron.

The crystallographic analysis has located the binding sites for ligand **5** at the interface regions of the β_5 pentamer near the inner capsid wall (section on The Substrate Binding Site). This could well explain the observed contribution of the ligand to the stability of the $T = 1$ β_{60} capsid. It should be noted that the hollow β_{60} aggregate absolutely requires the presence of an appropriate substrate analogue such as **5** for its stability. Removal of the stabilizing ligand by dialysis induces the formation of the large hollow sphere modification.

We have already mentioned that native HRS ($\alpha_3\beta_{60}$) does not require the presence of the ligand for stability under these experimental conditions and conclude therefore that the presence of a subunit trimers contributes to the thermodynamic stability of the β_{60} icosahedron. Our data further suggest that the stabilizing contribution of α subunits and substrate analogues are additive. In the presence of ligand **5** the complex $\alpha_3\beta_{60}$ is stable at pH values up to 11.

It is apparent that the formation of the $\alpha_3\beta_{60}$ oligomer is accompanied by a reduction of the free enthalpy of the particle. This is also reflected in an increased thermostability of the α subunit trimer if its enveloped by the icosahedral capsid. The thermal inactivation of isolated α_3 trimers (light riboflavin synthase) compared with the $\alpha_3\beta_{60}$ aggregate (HRS) has shown an increased thermal stability of $\Delta T = 12°C$ for the high molecular complex (Bacher et al., 1980). It is conceivable that the increase in thermostability was a selection factor involved in the evolution of the HRS complex in Bacillaceae.

A Comment on Structure-Function Relations

Multifunctional enzymes and enzyme complexes are found in all classes of organisms. The bacterial enzyme complex HRS catalyzes two distinct reaction steps in the biosynthesis of riboflavin (Fig. 2). In the majority of multifunctional enzymes the common intermediates are non-covalently bound. This is also true for HRS and implies that the component activities, the formation of the intermediate, 6,7-dimethyl-8-ribityllumazine, and of the end product, riboflavin, can be assayed independently (Neuberger & Bacher, 1986).

The maximum velocity, V_{max}, of lumazine synthesis by the complex $\alpha_3\beta_{60}$ is increased by a factor $F = 12$ as compared to the velocity of the enzymatic reaction catalyzed by β subunit aggregates devoid of α subunits (Neuberger & Bacher, 1986). We assume tentatively that hollow β_{60} capsids possess enzymatic activity. It is not yet known whether the same is true for the large hollow aggregates.

The complex $\alpha_3\beta_{60}$ shows a Michaelis constant, K_M, with respect to the lumazine substrate which is ten times the K_M characterizing the enzymatic turnover of the lumazine substrate by α_3 trimers. Furthermore, the turnover number decreases upon complex formation from 2 s^{-1}(α_3 trimer) to 1.2 s^{-1} ($\alpha_3\beta_{60}$ complex) (Bacher et al., 1980). These kinetic data do not support any increase of the catalytic power of the complex with respect to the catalytic activities of its isolated constituents.

The substrate binding site of the β subunits is built up by fivefold contact surfaces and is located at the inner wall of the β_{60} capsid. α_3 Trimers are localized in the central core. The reaction catalyzed by the β subunits requires the binding of the substrates 1 and 2 in close proximity. Similarly the reaction catalyzed by the α subunits requires the binding of two molecules of the lumazine 3 per active center (Plaut & Harvey, 1971; Otto & Bacher, 1981). A substantial proportion of the lumazine 3 which is formed at the active site of the β subunit equilibrates with the bulk solvent (Neuberger & Bacher, 1986). The remaining lumazine molecules may diffuse to neighbouring α subunit sites. This step of the reaction might be facilitated by substrate channeling which generally is accompanied by a reduction of dimensionality in diffusion (Adam & Delbrück, 1968).

The question arises how the exchange of substrates and products between bulk solvent and the respective active sites can proceed in the light of the rather tight capsid structure. An attractive possibility would be the diffusion of substrates and products through the channels formed at the fivefold axes or alternatively the funnels at the threefold axes. In this context also the asymmetric charge distribution of the capsid wall may be of significance. However, a spacefilling model of the β_{60} capsid suggests, that even penetration of the substrate molecules through clefts in the capsid wall cannot be excluded.

ACKNOWLEDGEMENTS

We thank Prof. Robert Huber for his continuous interest and support. Drs. Johann Deisenhofer and Wolfgang Steigemann have helped with computing problems. Dipl. Biochem. Jörg Schäffer and Dr. Rudi Müller have helped with the collection of intensity data. We also thank Dipl. Chem. Karin Schott and Astrid König for help with enzyme preparation and Angelika Koehnle for the preparation of substrates and substrate analogues. This work was supported by grants from the Deutsche Forschungsgemeinschaft (to R.L. and A.B.) and by the Fonds der Chemischen Industrie (to A.B.).

REFERENCES

Adam, G. and Delbrück, 1968, in: "Structural Chemistry and Molecular Biology," A. Rich and N. Davidson, eds., Freeman, San Francisco.

Arndt, U.W. and Wonacott, A.J., 1977, The Screenless Rotation Method, in: "The Rotation Method in Crystallography," North Holland Publ. Co., Amsterdam, p. 5.

Bacher, A., Baur, R., Eggers, U., Harders, H., Otto, M.K. and Schnepple, H., 1980, Riboflavin Synthases of *Bacillus Subtilis*: Purification and Properties, *J. Biol. Chem.* 855:632.

Bacher, A. and Ludwig, H.C., 1982, Ligand-Binding Studies on Heavy Riboflavin Synthase of *Bacillus Subtilis, Eur. J. Biochem.* 127:539.

Bacher, A., Ludwig, H.C., Schnepple, H. and Ben-Shaul, Y., 1986, Heavy Riboflavin Synthase from *Bacillus Subtilis*, *J. Mol. Biol.* 187:756.

Bacher, A. and Mailänder, B., 1978, Biosynthesis of Riboflavin in *Bacillus Subtilis*: Function and Genetic Control of the Riboflavin Synthase Complex., *J. Bacteriol.* 134:476.

Bacher, A., Schnepple, H., Mailänder, B., Otto, M.K., and Ben-Shaul, Y., 1980, "Structure and Function of the Riboflavin Synthase Complex of *Bacillus Subtilis*, in: Flavins and Flavoproteins," K. Yagi & T. Yamano, eds., Japan Scientific Societies Press, Tokyo, p. 579.

Bachmann, L., Becker, R., Leupold, G., Barth, M., Guckenberger, R. and Baumeister, W., 1985, Decoration and Shadowing of Freeze-etched Catalase Crystals, *Ultramicroscopy* 16:305.

Blow, D.M. and Crick, F.H.C., 1959, The Treatment of Errors in the Isomorphous Replacement Method, *Acta Cryst.* 12:794.

Blundell, T.L. and Johnson, L.N., 1976, Isomorphous Replacement, *in:* "Protein Crystallography," Academic Press, New York, p.284.

Bricogne, G., 1976, Methods and Programs for Direct-Space Exploitation of Geometric Redundancies, *Acta Cryst.* A32:832.

Caspar, D.L.D., and Klug, A., 1962, Physical Principles in the Construction of Regular Viruses, *Cold Spring Harbour Symp. Quant. Biol.* 27:1.

Chothia, C., 1974, Hydrophobic bonding and accessible surface area in proteins, *Nature* 248:338.

Cohen, C. and Parry, D.A.D., 1986, α-Helical coiled coils - a widespread motif in proteins, *TIBS* 11:245.

Crowther, R.A., 1967, A Linear Analysis of the Non-Crystallographic Symmetry Problem, *Acta Cryst.* 22:758.

Crowther, R.A.,1969, The Use of Non-Crystallographic Symmetry for Phase Determination, *Acta Cryst.* 25:2572.

Dickerson, R.E., Kendrew, J.C. and Strandberg, B.E., 1961, The Crystal Structure of Myoglobin; Phase Determination to a Resolution of 2 Å by the Method of Isomorphous Repacement, *Acta Cryst.* 14:1188.

Dodson, E.J., 1985, Molecular Replacement: The method and its problems, *in:* "Proceedings of the Daresbury Study Weekend on Molecular Replacement," Daresbury Laboratory, p. 33.

Harrison, S.C.,1984, Multiple modes of subunit association in the structures of simple spherical viruses, *TIBS* 9:345.

Harrison, S.C., Olson, A.J., Schutt C.E., Winkler, F.K. and Bricogne, G.,1978, Tomato bushy stunt virus at 2.9 Å resolution, *Nature* 276:368.

Hendrickson, W.A. and Lattmann, E.E., 1970, Representation of Phase Probability Distributions for Simplified Combination of Independent Phase Information, *Acta Cryst.* B26:136.

Hogle, J.M., Chow, M. and Filman, D.J., 1985, Three-Dimensional Structure of Poliovirus at 2.9 Å Resolution, *Science* 229:1358.

Huber, R., 1985, Experience with the application of Patterson search techniques, *in:* "Proceedings of the Daresbury Study Weekend on Molecular Replacement," Daresbury Laboratory, p. 58.

Jones, T.A., 1978, A Graphics Model Building and Refinement System for Macromolecules, *J. Appl. Crystallogr.* 11:268.

Karshikov, A. & Ladenstein, R., 1989, Electrostatic Effects in a Large Enzyme Complex: Subunit Interactions and Electrostatic Potential Field of the Icosahedral β_{60} Capsid of Heavy Riboflavin Synthase, *Proteins* 5:248.

Ladenstein, R., Bacher, A. and Huber, R., 1987, Some Observations of a Correlation Between the Symmetry of Large Heavy-Atom Complexes and Their Binding Sites on Proteins, *J.Mol.Biol.* 195:751.

Ladenstein, R. Ludwig, H.C. and Bacher, A.,1983,Crystallization and Preliminary X-Ray Diffraction Study of Heavy Riboflavin Synthase from *Bacillus Subtilis, J. Biol. Chem.* 258:11981.

Ladenstein, R., Meyer, B., Huber, R., Labischinski, H., Bartels, K., Bartunik, H.D., Bachmann, L., Ludwig, H.C. and Bacher, A., 1986, Heavy Riboflavin Synthase from *Bacillus Subtilis*: Particle Dimensions, Crystal Packing and Molecular Symmetry, *J. Mol. Biol.* 187:870.

Ladenstein, R., Schneider, M., Huber, R., Bartunik, H.D., Wilson, K.S., Schott, K. and Bacher, A., 1988, Heavy Riboflavin Synthase from *Bacillus subtilis*: Crystal Structure Analysis of the Icosahedral β_{60} Capsid at 3.3 Å Resolution, *J. Mol. Biol.* 203:1045.

Liljas, L., 1986, The structure of spherical viruses, *Progr. Biophys. Molec. Biol.* 48:16.

Ludwig, H.C., Lottspeich, F., Henschen, A., Ladenstein, R. and Bacher, A., 1987, Heavy Riboflavin Synthase from *Bacillus subtilis*: Primary Structure of the ß Subunit, *J. Biol. Chem.* 262:1016.

Ludwig, M.L., Pattridge, K.A., Smith, W.W., Jensen, L.H., and Watenpaugh, K.D., 1982, Comparison of Flavodoxin Structures, *in:* Flavins and Flavoproteins, V.Massey and C.H.Williams, eds., Elsevier, Amsterdam, p.19.

Luo, M., Vriend, G., Kamer, G., Minor, J., Arnold, E., Rossmann, M.G., Boege, U., Scraba, D.G., Duke, G.M. and Palmenberg, A.C., 1987, The Atomic Structure of Mengo Virus at 3.0 Å Resolution, *Science* 235:182.

Main, P. and Rossmann, M.G., 1966, Relationships among Structure Factors due to Identical Molecules in Different Crystallographic Environments, *Acta Cryst.* 21:67.

Neuberger, G. and Bacher, A., 1986, Biosynthesis of Riboflavin: Enzymatic Formation of 6,7 - Dimethyl-8-Ribityllumazine by Heavy Riboflavin Synthase from *Bacillus Subtilis*, *Biochem. Biophys. Res. Commun.* 139:1111.

Otto, M.K. and Bacher, A., 1981, Ligand-Binding Studies on Light Riboflavin Synthase from *Bacillus Subtilis, Eur J. Biochem.* 115:511.

Plaut, G.W.E.and Harvey, R.A., 1971, The Enzymatic Synthesis of Riboflavin, *Methods Enzymol.* 18B:.515.

Rossmann, M.G., Arnold, E., Erickson, J.W., Frankenberger, E.A., Griffith, J.P., Hecht, H.J., Johnson, J.E., Kamer, G., Luo, M., Mosser, A.G., Rueckert, R.R., Sherry, B. and Vriend, G., 1985, Structure of a human common cold virus and functional relationship to other picorna viruses, *Nature* 317:145

Rossmann, M.G. and Blow, D.M., 1962, The Detection of Sub-Units Within the Crystallographic Asymmetric Unit, *Acta Cryst.* 15:2.

Rossmann, M.G. and Blow, D.M., 1963, Determination of Phases by the conditions of Non-Crystallographic Symmetry, *Acta Cryst.* 16:39.

Steigemann, W., 1974, Dissertation: "Die Entwicklung und Anwendung von Rechenverfahren und Rechenprogrammen zur Strukturanalyse von Proteinen," Technische Universität München, West Germany.

Wang, B.C., 1985, Resolution of Phase Ambiguity in Macromolecular Crystallography, *in:* "Methods in Enzymology," Wyckoff, Timasheff & Hirs, eds., 115:90.

ESCHERICHIA COLI ASPARTATE TRANSCARBAMYLASE: The Relationship Between Structure and Function

Evan R. Kantrowitz* and William N. Lipscomb†

*Department of Chemistry, Boston College, Chestnut Hill, MA 02167 and
†Harvard University, Gibbs Laboratory, 12 Oxford Street, Cambridge, MA 02138

INTRODUCTION

Aspartate transcarbamylase of *Escherichia coli* [EC 2.1.3.2] is a member of a special class of enzymes that not only catalyzes a cellular reaction but also controls the rate of a metabolic pathway. The enzymes in this class are usually large, each composed of more than one polypeptide chain, and catalyze a reaction at or near the beginning of a metabolic pathway. Aspartate transcarbamylase catalyzes the condensation of carbamyl phosphate with L-aspartate to produce N-carbamyl-L-aspartate and inorganic phosphate (Jones et al., 1955; Reichard & Hanshoff, 1956). This reaction is particularly important because once carbamylaspartate is formed it is committed to the biosynthesis of pyrimidines, a necessary component for nucleic acid biosynthesis. Aspartate transcarbamylase controls the rate of pyrimidine biosynthesis by altering its catalytic velocity in response to cellular levels of both pyrimidines and purines. The end product of the pyrimidine pathway, CTP, induces a decrease in catalytic velocity while, ATP, the end product of the parallel purine pathway, exerts the opposite effect, stimulating the catalytic activity (Yates & Pardee, 1956; Gerhart and Pardee, 1962 & 1963). In part, the relative amounts of purines and pyrimidines in the cell are thereby kept in balance for nucleic acid synthesis. By using X-ray crystallography and site-directed mutagenesis we have begun to delineate on the molecular level how this complex enzyme catalyzes the formation of carbamylaspartate and how it alters its catalytic activity in response to cellular metabolites.

Aspartate transcarbamylase is composed of 12 polypeptide chains of two types. Each of the 6 larger, or catalytic, chains (C1-C6) has a molecular weight of 33,000, and they are grouped together in two trimers (Figure 1). Each of the smaller, or regulatory, chains (R1-R6), has a molecular weight of 17,000 and they are organized in three dimers (Figure 1). The packing of the two catalytic trimers (catalytic subunits) and the three regulatory dimers (regulatory subunits) results in a highly symmetric molecule with D_3 symmetry.

The holoenzyme[1] can be dissociated into the catalytic and regulatory subunits which can easily be isolated. Furthermore, the holoenzyme can be reconstituted from the separate subunits

----

[1]The holoenzyme consists of the entire aspartate transcarbamylase molecule composed of two catalytic subunits (C_3) and three regulatory subunits (R_2).

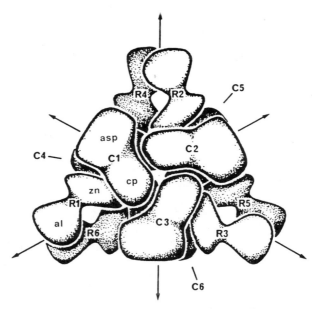

Figure 1. Schematic representation of the quaternary structure of aspartate transcarbamylase viewed along the three-fold axis. The six catalytic (C) and regulatory (R) chains are numbered. Catalytic chains C1-C2-C3 and C4-C5-C6 correspond to the two catalytic subunits while the regulatory chains R1-R6, R2-R4 and R3-R5 correspond to the three regulatory subunits. The catalytic chain is composed of the aspartate (asp) and the carbamyl phosphate (cp) domains. The regulatory chain is composed of the allosteric (al) and zinc (zn) domains. The arrows indicate the molecular two-fold axes.

under appropriate conditions. Only the isolated catalytic subunits have enzymatic activity. The isolated regulatory subunits bind the nucleotide effectors ATP and CTP. A comparison of the kinetics of the catalytic subunit and the holoenzyme reveals two important differences. First, the specific activity of the catalytic subunit is about 50 percent higher than that of the holoenzyme; and second, the substrate saturation curves change from sigmoidal for the holoenzyme to hyperbolic for the catalytic subunit.

A sigmoidal substrate saturation curve is a characteristic of the class of enzymes which exert allosteric control of biosynthetic pathways. Over a very narrow concentration range these enzymes are capable of substantial alterations in their catalytic ability. These sigmoidal saturation curves imply interacting sites, where the binding of the first substrate induces changes in substrate affinity and/or catalytic efficiency at the other, sometimes distant active sites. Both the Monod, Wyman and Changeux (1965) and the Koshland, Nemethy and Filmer (1966) models require two alternate states of the enzyme, which have often been called the T and R states. The T-state has lower affinity for substrate and lower activity than does the R-state[2].

For aspartate transcarbamylase, the R-state can be induced by the substrates as well as by

---

[2]We use T and R for the structural states. If the high-affinity (HA) or low-affinity (LA) states do not correspond to these structural states, we propose notations such as T_{HA} and R_{LA}.

certain substrate analogs such as N-phosphonacetyl-L-aspartate (PALA), which resembles the natural substrates carbamyl phosphate and aspartate (Collins & Stark, 1971) (see Figure 2). Either PALA or a combination of carbamyl phosphate and succinate, an analogue of aspartate, can convert the enzyme from the T-state to the R-state. This conversion has been monitored, for example, by ultraviolet difference spectroscopy (Collins & Stark, 1969, 1971), difference sedimentation (Kirschner & Schachman, 1971, 1973; Gerhard and Schachman, 1968), circular dichroism (Griffin et al., 1973), and X-ray solution scattering (Moody et al., 1979). It became clear from the single crystal X-ray diffraction studies (Honzatko et al., 1982; Ke et al., 1984; Honzatko & Lipscomb, 1982; Kim et al., 1987; Krause et al., 1985, 1987; Voltz et al., 1986) that contraction of the catalytic subunits when PALA binds to the holoenzyme involved movements of as much as 13 Å during closure of the active sites, and that the T to R transition involved an expansion along the molecular three-fold axis by 12 Å, as we now describe.

Structure of the Enzyme

High resolution structures are known from the unligated enzyme (Honzatko et al., 1982; Ke et al., 1984), the CTP enzyme (Honzatko et al., 1982; Kim et al., 1987) and the PALA enzyme (Krause et al., 1985, 1987; Voltz et al., 1986). One RC unit is shown in Figure 3, where the location of the active site is indicated between the carbamyl phosphate domain and the aspartate domain of the catalytic chain. The structural (non-catalytic) Zn^{+2} ion is shown in the zinc domain (black dot), and the effector binding site is indicated in the allosteric domain. As shown in Figure 1, the two catalytic trimers stack upon one another along the molecular three-fold axis in a nearly eclipsed configuration. Each regulatory dimer joins two catalytic chains which are in different C_3 units and which are rotated about 120° about the three-fold axis, e.q. C1-R1-R6-C6 (Figure 1).

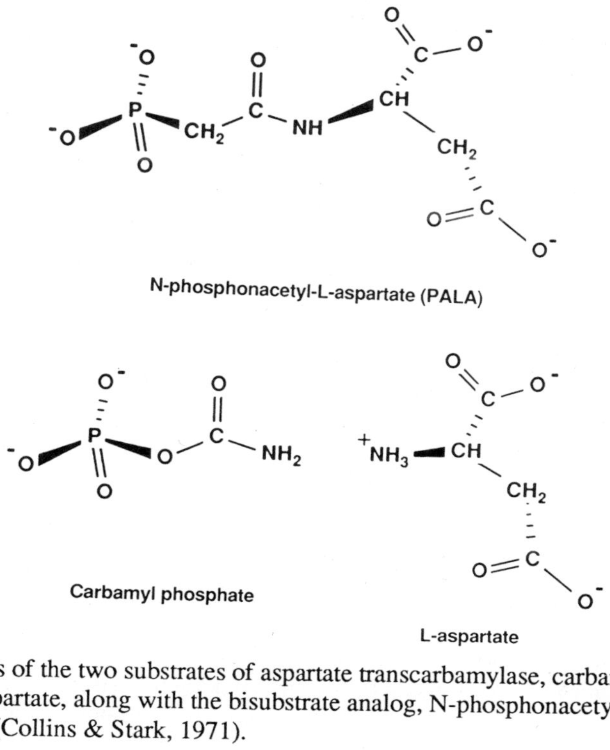

N-phosphonacetyl-L-aspartate (PALA)

Carbamyl phosphate

L-aspartate

Figure 2. Structures of the two substrates of aspartate transcarbamylase, carbamyl phosphate and L-aspartate, along with the bisubstrate analog, N-phosphonacetyl-L-aspartate (PALA) (Collins & Stark, 1971).

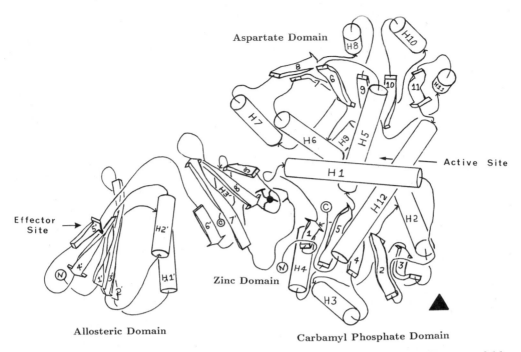

Figure 3. Secondary structure of one catalytic-regulatory pair as viewed along the three fold axis (solid triangle). The α-helices are shown as cylinders and the β-sheet as arrows. The allosteric and zinc domains of the regulatory chain are towards the left, while the aspartate and carbamyl phosphate domains of the catalytic chain are towards the right. The active site is indicated between the aspartate and carbamyl phosphate domains of the catalytic chain, and the nucleotide effector site is indicated in the allosteric domain of the regulatory chain.

Each catalytic chain contains one active site close to the interface between adjacent catalytic chains in a C_3 unit. Residues required for maximal activity are contributed by these adjacent pairs of catalytic chains. These shared active sites have recently been demonstrated using hybrid subunits in the holoenzyme (Robey & Schachman, 1985; Wente & Schachman, 1987). The six active sites face a central cavity, which is readily accessible from the outside of the molecule in both the T and R forms.

The structure of the PALA-ligated enzyme shows interactions of PALA with Ser-52, Thr-53, Arg-54, Thr-55, Arg-105, His-134, Arg-167, Arg-229, Gln-231 and Leu-267 from one catalytic chain, and Ser-80 and Lys-84 from the adjacent catalytic chain (Figure 4). Thus, major interactions occur between the negative charges of PALA and four positively charged residues of the enzyme. When PALA is displaced by carbamyl phosphate plus succinate (Gouaux & Lipscomb, 1988), the terminal oxygens of carbamyl phosphate and of the carboxylate groups of succinate are close to those in the PALA enzyme. Also the plane of the amide group of carbamyl phosphate is almost perpendicular to the plane of the peptide-like bond of PALA. In this region, interactions of main chain carbonyls of Pro-266 and Leu-267 and the side chain carbonyl of Gln-137 occur with the primary carbamyl nitrogen. Also, the carbonyl oxygen of carbamyl phosphate interacts with Thr-55, Arg-105 and His-134; and the mixed anhydride oxygen of carbamyl phosphate interacts with Arg-54. In a later section we consider the effects of site specific mutagenesis on many of these interactions. Here, we note that PALA seems to be a good analogue for carbamyl phosphate and succinate. If a primary amino group is added to

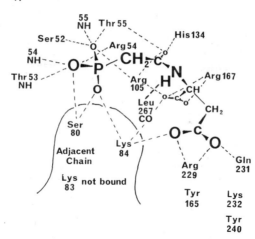

Figure 4. Schematic illustration of the PALA binding site in aspartate transcarbamylase. A very recent structure of the enzyme with carbamyl phosphate and succinate bound suggests that the terminal phosphate unit and the carboxylate anions of PALA bind to the enzyme in a similar fashion as this substrate and substrate analog pair (Gouaux & Lipscomb, 1988). The active site is composed of residues from two adjacent catalytic chains (e.g., C1-C2). In this figure all of the side chains that interact with PALA are shown. All the residues shown come from one catalytic chain except for Ser-80 and Lys-84, which come from the other catalytic chain. The hydrogen bonding interactions which stabilize PALA in the active site are also shown as dotted lines.

succinate to make aspartate, we have a model of an important step: this amino group is in a position to attack the carbonyl carbon of carbamyl phosphate perpendicular to the plane of the carbamyl group. From the purely structural view, the removal of a proton from this attacking amino group is ambiguous in this model, and could be accomplished by His-134, a phosphate oxygen, or a water molecule as an intermediate transfer agent.

The major change in tertiary structure induced by the binding of PALA or substrate analogues is the α-carbon movement of up to 8 Å of the 240s loop (residues 225-245). The accompanying closure at the active site is about 2 Å (residues 50-55). In addition, movements of about 5 Å are seen in the α-carbon movements of residues 70-75, associated with the split active site.

These changes in tertiary structure cause a large change in the quaternary structure (Figure 5). As the T form is converted to the R form by binding of substrate analogues, the two C_3 units move apart by 12 Å, and reorient about the molecular three-fold axis by $\pm 5°$ (a relative reorientation of 10°). Also, the regulatory dimers reorient about the molecular two-fold axes by 15° in a way that preserves CRRC interactions, where the two C's are in different catalytic trimers and are about 120° apart when viewed along the molecular three-fold axis.

An indication of the closure of the active site and of the changes in many of the salt links and strong hydrogen bonds is shown schematically in Figure 6, where domains of catalytic chains C1 and C4 are outlined. The extensive C1-C4 interface of the T form is greatly reduced in the R form. A triple interaction of Glu-239 of C4 (or C1) with Lys-164 and Tyr-165 of C1 (or C4) is broken as T goes to R; in this same transition an intradomain interaction between

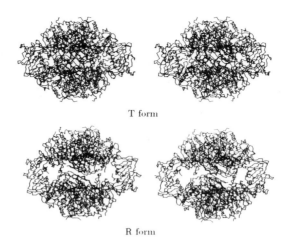

T form

R form

Figure 5. Stereoviews of the T-state (A) and R-state (B) of aspartate transcarbamylase. Data for
the T-state are from the CTP liganded structure (Honzatko & Lipscomb, 1982; Kim et
al., 1987) and for the R-state from the PALA liganded structure (Krause et al., 1985,
1987; Voltz et al., 1986). When PALA binds to aspartate transcarbamylase there is a
12 Å elongation of the molecule along the 3-fold axis which is accompanied by
opposite 5° rotations of each catalytic subunit and a 15° rotation of the regulatory
subunits around each of the two-fold axes. Additional conformational changes also
take place on the tertiary level.

Tyr-240 and Asp-271 is lost. In the R form new intrasubunit interdomain interactions are
formed among Glu-50, Arg-167 and Arg-234. Also, within the aspartate domain new
interactions occur between Arg-229 and Glu-233 and among Glu-239, Lys-164 and Tyr-165 in
the R form. There are of course, many other interactions (Ke et al., 1988) that have not been
described here. Mutagenic experiments involving most of the above mentioned interactions are
discussed below.

Function of Active Site Residues

Previous chemical modifications have implicated amino acid side chains in catalysis or
binding. For example, phenylglyoxal inactivates the enzyme by modifying a single unidentified
arginine residue per active site (Kantrowitz & Lipscomb, 1976). Bromosuccinate (Lauritzen &
Lipscomb, 1982) or trinitrobenzene sulfonate (Lahue & Schachman, 1984) cause loss of activity
by reaction with Lys-83 and Lys-84, while pyridoxylation inactivates by modifying only Lys-84
(Greenwell et al., 1973; Kempe & Stark, 1975). Photooxidation of this pyridoxylated enzyme
implicates two unidentified histidine residues (Greenwell et al., 1973). Besides the ambiguities
in identification, these chemical changes often introduce bulky substitutents which themselves
may cause the inactivation.

Site-directed mutagenesis reduces some of these ambiguities. When Lys-83 is converted to
Gln the activity is reduced slightly, but when Lys-84 is changed to Gln the activity is reduced by
a factor of 4000 relative to the wild-type (Robey et al., 1986). The replacement of Ser-52 by the
bulky Phe, which probably remains in the phosphate binding site, causes essentially a total loss
of activity (Schachman et al., 1984). Significantly, the replacement of His-134 by Ala reduces
activity by 20-fold, and substantially increases the aspartate concentration required for half the
maximal velocity (Robey et al., 1986).

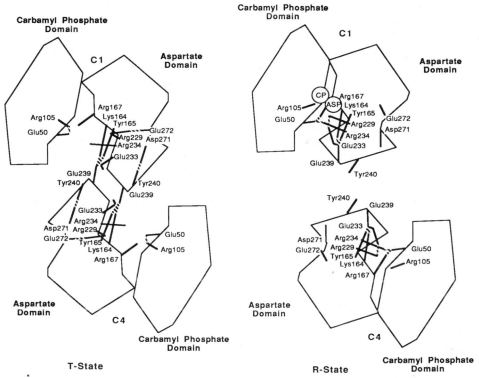

Figure 6. A model for the mechanism of homotropic cooperativity in aspartate transcarbamylase. Shown schematically are the two extreme conformations of a C1–C4 pair in the T–state (Left) and the R–state (Right), as deduced from X-ray crystallography (Honzatko & Lipscomb, 1982; Kim et al., 1987; Krause et al., 1985, 1987; Voltz et al., 1986). For clarity, only one catalytic chain from each of the upper (C1) and lower (C4) catalytic subunits is shown. Because of the molecular 3-fold axis, the various interactions shown here are repeated in the C2-C5, and C3-C6 pairs. Upon aspartate binding (in the presence of carbamyl phosphate), the aspartate domain moves towards the carbamyl phosphate domain resulting in the closure of the active site. The 240s loops of C1 and C4 undergo a large alteration in position and change from being side by side in the T–state to almost one on top of the other in the R–state. On the quaternary level, the catalytic subunits move apart resulting in an elongation of the molecule. The binding of carbamyl phosphate to the enzyme induces a local conformational change in the enzyme which enhances the binding of aspartate and perhaps causes the loss of the salt-link between Glu-50 and Arg-105, allowing Arg-105 to reorient and bind carbamyl phosphate. The binding of the substrates at one active site induces the domain closure in that catalytic chain, and requires a quaternary conformational change which allows the 240s loops of the upper and lower catalytic chains to move to their final positions. As outlined in the text, the quaternary conformational change causes the loss of a whole series of interactions which normally stabilize the constrained T-state of the enzyme and results in the formation of the high activity–high affinity R-state. The formation of the R-state, in a concerted fashion, is further stabilized by a variety of new interactions that are both interdomain and intrachain in nature. The various interactions which stabilize these two allosteric states of the enzyme have been observed in the X-ray structures and their functional importance has been deduced by site-directed mutagenesis.

41

Preliminary results on site-directed mutants in or near the active site, indicate that Arg-54, Gln-137, Arg-167 and Arg-229 are all required for activity (Figure 7). In addition, replacement of Arg-105 by histidine caused little alteration in catalytic activity, while conversion of Arg-105 to Gln reduced the activity 1000-fold. These results confirm the location of the active sites in the X-ray diffraction studies. Moreover, the three dimensional structure of the enzyme and modifications of His-134 are suggestive that this residue is important for catalysis. If, in a complex of the enzyme with carbamyl phosphate and succinate (Gouaux & Lipscomb, 1988), the succinate is replaced by aspartate as a model, the amino group of aspartate is about 4 Å from His-134. Hence, a 1 Å movement could allow His-134 to deprotonate the amino group of aspartate. A minor part of the reaction could also proceed by transfer of a proton from this amino group to the phosphate directly or through another agent such as H_2O. It is to be emphasized that these mechanistic conclusions have not been proved.

Some further proposals arise from the site-directed mutants at Gln-137, Arg-54 and Arg-167. As noted above, replacement of Gln-137 by Ala reduces activity, but also reduces the affinity of the enzyme for carbamyl phosphate and, surprisingly, for aspartate. Circular dichroism experiments indicate that the conformational change induced by the binding of carbamyl phosphate in the wild-type enzyme is greatly reduced in this mutant. Since the conformational change induced by carbamyl phosphate enhances the binding of aspartate, the reduction of the conformational change in this mutant thus affects binding of aspartate, as well as carbamyl phosphate. Turning now to Arg-54, which interacts with the anhydride oxygen of carbamyl phosphate, we speculate that this residue promotes the release of phosphate from the tetrahedral intermediate. Arg-167, which interacts with the α-carboxylate of PALA or succinate is proposed to orient aspartate correctly for the true enzymatic reaction. Further mutagenic and three-dimensional structural studies may provide hypotheses for functions of other residues in the catalytic mechanism.

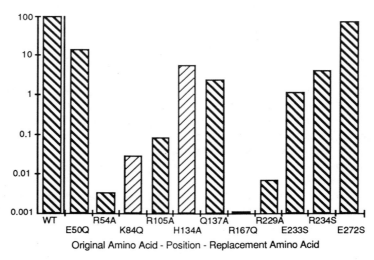

Figure 7. Histogram showing the percent maximal activity for a series of mutant versions of aspartate transcarbamylase. All the mutant enzymes except for K84Q and H134A (Robey et al., 1986) were created in the laboratory of one of the authors (ERK). A logarithmic axis is used to depict the variation in activity of these mutant enzymes, which in the worse case approaches a 10^5 reduction in activity. The one letter amino acid codes are used both to represent the original amino acid, which precedes the substitution site, as well as the new amino acid, which follows the substitution site.

Function of Residues in the Allosteric Transition

In the homotropic transition (in the absence of ATP or CTP) changes in salt links or strong hydrogen bonds occur when substrate analogues induce the T to R transition. We have concentrated site specific mutagenesis on these interactions. For example, when PALA binds, the closure of the aspartate domain towards the carbamyl phosphate domain is stabilized by interactions among Glu-50, Arg-167 and Arg-234. When Glu-50 is changed to Gln (Figure 8) the mutant has no cooperativity and greatly reduced affinity for substrates. It appears that these interdomain bridging interactions are critical for the formation of the high affinity–high activity conformation of the active site (Ladjimi et al., 1988). When Arg-234 is replaced by Ser the mutant has similar properties to the Gln-50 mutant (Figure 8).

The interdomain bridging interactions and the closure of the domains of the catalytic chain are directly related to the geometry of the active site. Using site-directed mutagenesis we have shown that both Arg-167 and Arg-229 are catalytically important residues (see Figure 7). Arg-167 forms a salt link with Glu-50 as part of the interdomain bridging interactions and is directly bound to PALA in the R-state. Furthermore, Glu-233, which seems to have no function in the T-state, is bound to Arg-229 in the R-state. The conversion of Glu-233 to Ser results in an enzyme with 80-fold lower activity and reduced affinity for aspartate, suggesting that the function of Glu-233 is to hold Arg-229 in the proper orientation for catalysis. Furthermore, the specific interaction between Arg-229 and Glu-233 can take place only after the closure of the domains of the catalytic chain. Arg-105 and Glu-50 are also involved in a similar type of interaction. In the T-state, Arg-105 is linked to Glu-50, but after the domain closure Arg-105 is found to interact with the phosphonate portion of PALA while Glu-50 is involved in the interdomain bridging interactions. Thus, the closure of the domains assists in the formation of the active site pocket.

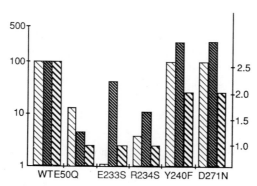

Figure 8. A comparison of the properties of a series of mutants which are important for the stabilization of the two allosteric forms of the enzyme. The interdomain bridging interaction between Glu-50 and Arg-234 stabilizes the R-state (Ladjimi et al., 1988), while the link between Tyr-240 and Asp-271 stabilizes the T-state (Middleton & Kantrowitz, 1986). The salt link between Glu-233 and Arg-229 correctly positions the latter residue to interaction with the β-carboxylate of aspartate in the R-state.

▨ Percent Maximal Activity

▨ Relative Asp Affinity

▨ Hill Coefficient

Domain closure alone is not sufficient to orient Glu-233 correctly for the stabilization of Arg-229 or Arg-234 correctly for the establishment of the interdomain bridging interactions. As the comparison of the T and R state structures has revealed, there is a large reorientation of the 240s loop. This loop movement shifts the position of the guanidinium group of Arg-234, which is linked to Glu-50 in the R-state, and which is approximately 7.5 Å from its T-state position. The rearrangement of this loop is rather complex in that a number of specific interactions are broken and formed by the loop movement. Specifically, the intrachain interaction between Asp-271 and Tyr-240 and the intersubunit interactions between Glu-239 and both Lys-164 and Tyr-165 seem to restrain the position of the 240s loop in the T-state while the Arg-229—Glu-233 and Glu-50—Arg-234 salt links seem to stabilize the loop in the R-state. In order to acquire more details about the function of these interactions, we have used site-directed mutagenesis to perturb them. When phenylalanine was substituted for Tyr-240, a substantial reduction in cooperativity and a marked increase in the affinity for aspartate, but no alteration in the specific activity were observed (Figure 8) (Middleton & Kantrowitz, 1988). When Asp-271 was changed to Asn an enzyme was obtained having almost the identical properties as the enzyme in which Phe was substituted for Tyr-240. These results indicate that the interaction between Tyr-240 and Asp-271 is important but not critical for the stability of the low affinity– low activity T-state, and is unnecessary for the stability of the R-state.

A comparison of the T and R state structures reveals that the intersubunit interactions between Glu-239 of one catalytic subunit and both Lys-164 and Tyr-165 of the other catalytic subunit, which exist in the T-state, are completely absent in the R-state structure. Based on the structural data, these interactions seem to stabilize the more compact T-state. The replacement of Glu-239 by Gln causes a complete lost of cooperativity, no alteration in maximal velocity and an enhanced affinity for aspartate. Although kinetic experiments suggest that this enzyme is locked in the high affinity R-state, a preliminary X-ray diffraction study shows unit cell changes that suggest a form of the Gln-239 mutant which is between the T and R forms. These preliminary crystallographic studies have now been confirmed by low angle X-ray scattering experiments in the laboratory of G. Hervé by P. Vachette using synchrotron radiation at L.U.R.E. Detailed three-dimensional X-ray studies are now in progress to further characterize the Gln-239 mutant enzyme. The kinetic data nevertheless indicate that the simple replacement of Glu-239 by Gln is sufficient to destabilize the low affinity–low activity state making the high affinity–high activity state the only form in which this mutant enzyme can exist. Thus, the interactions between the catalytic chains of the upper and lower trimer are critical for the stability of the T-state. When these links are lost, the enzyme is incapable of remaining in the T-state.

The Allosteric Mechanism

Based on the structural data and the results of our site-directed mutagenesis experiments, the following somewhat speculative mechanism is proposed for the allosteric transition (Ladjimi & Kantrowitz, 1988). The binding of the first aspartate molecule (in the presence of carbamyl phosphate) causes two major structural changes in the enzyme. First, a closure of the two domains within a catalytic chain occurs with corresponding shifts in the 80s and 240s loops, thus establishing an active site with high affinity for the substrates and increased catalytic activity. Second, simultaneously with these tertiary changes there is a quaternary conformation change which results in the closure of the domains in all of the other catalytic chains and the conversion of all of the remaining active sites into the high affinity–high activity form. Perhaps the most important feature of this mechanism is an intrinsic difference in both substrate affinity and catalytic activity between the T and R states of the enzyme. The binding of substrates at one active site is sufficient to cause all of the remaining active sites to be converted, in a concerted manner, into a form which has both substantially enhanced affinity for the substrates and catalytic efficiency. Kinetic measurements of the reverse reaction support the requirement for a concerted transition (Foote & Lipscomb, 1981; Foote & Schachman, 1985).

What are the molecular level events that make this mechanism possible? For the wild type enzyme, there is ample evidence that the allosteric change is induced by the binding of substrates. Substrate binding not only causes tertiary conformational changes within a catalytic chain resulting in the closure of the two domains but also results in an expansion of the holoenzyme along its molecular 3-fold axis. The domain closure is a complex structural rearrangement which involves more than a simple hinge motion and which cannot take place without conformational changes at the quaternary level.

The key to the entire quaternary change is the rearrangement of the 240s loop of the aspartate domain. This loop not only undergoes a shift in position due to the movement of the aspartate domain but also undergoes a major reorientation, which is stabilized by new interdomain and intrachain interactions (see Figure 6). In the T-state, the 240s loop of C1 is maintained far apart from the carbamyl phosphate domain by both the Tyr-240—Asp-271 interaction and the Glu-239—Lys-164, Glu-239—Tyr-165 links between neighboring 240s loops. These interactions prevent the movement of the 240s loop toward the carbamyl phosphate domain, which would result in a closure of the domains and the formation of the high affinity–high activity state through the Glu-50—Arg-167, Glu-50–Arg-234 salt bridges. In order for domain closure to occur, upon aspartate binding, not only do the interactions which stabilize the 240s loop in the T–state have to be broken, but also steric constraints have to be overcome. This steric hindrance is relieved by a structural rearrangement of the loop which removes the Tyr-240—Asp-271 interaction and the two symmetrically related links between Glu-239 and Lys-164 and Tyr-165 (C1–C4) resulting in domain closure. However, this change cannot occur only in one C1–C4 pair, since the different catalytic chains of the upper and lower subunits are in fact held together partly by the favorable interactions between their respective 240s loops. Any change in the intersubunit interactions of one pair, say C1–C4, is transmitted to the C2–C5 and C3-C6 pairs as well. Thus the molecule is converted to the R conformation by the rotations and elongations previously described. The analysis of a double mutant, at both Glu-50 and Tyr-240 supports this model (Ladjimi & Kantrowitz, 1988).

We now consider the C:R interactions in the T and R structures (Honzatko et al., 1982; Kim et al., 1987; Krause et al., 1985, 1987; Voltz et al., 1986; Cherfils, 1987). A major change occurs in the C1-R4 (i.e. C6-R3) interface of the T form; this substantial interface disappears in the more expanded R form. Also, the C1-C4 interface decreases considerable from the T to R transition. Other interfaces such as C1-C2 (C5-C6) and C1-R1 (C6-R6) change by lesser amounts, while very little change occurs in the R1-R6 interface in the allosteric transition. These interfaces, and the interface between domains of the R chain have not yet been subjected to extensive single point mutations. The disappearance or decrease of these interfaces as the molecule changes from T to R is probably largely compensated by the binding interactions between enzyme and substrates, including the changes in tertiary structure which bring the two domains of the catalytic chain closer together. Clearly, the active site of the T form is more open than the active site of the R form of the enzyme.

So far we have discussed homotropic cooperativity. The heterotropic effects of inhibition by CTP and of activation by ATP involve the transmission of conformational information some 60 Å from the allosteric site to the nearest active sites (see Figure 3). Structural and mutagenic studies are currently in progress to identify possible distinct pathways for the heterotropic effects in the regulation of aspartate transcarbamylase.

ACKNOWLEDGEMENTS

This work was supported by grants from the National Institutes of Health (GM26237 and DK1429 ERK) and (GM 06920 WNL). E.R.K. would also like to thank the Camille and Henry Dreyfus Foundation for a fellowship. Reprinted with permission from *Science* 241:669 (1988) [copyright 1988 by the AAAS].

REFERENCES

Cherfils, J., 1987, Etude d'aspects structuraux de L-aspartate transcarbamylase de *E. coli*, Thése, Laboratoire de Biologie Physicochimique Bât. 433 Universite Paris-Sud, 91405, Orsay, France.

Collins, K.D. and Stark, G.R., 1969, Aspartate transcarbamylase: studies of the catalytic subunit by ultraviolet difference spectroscopy, *J. Biol. Chem.* 244:1869.

Collins K.D. and Stark, G.R., 1971, Aspartate transcarbamylase: interaction with the transition state analogue N-(phosphonacetyl)-L-aspartate, *J. Biol. Chem.* 246:6599.

Foote J. and Lipscomb, W.N. 1981, Kinetics of aspartate transcarbamylase from *Escherichia coli* for the reverse direction of reaction, *J. Biol. Chem.* 256:11428.

Foote, J. and Schachman, H.K., 1985, Homotropic effects in aspartate transcarbamoylase. What happens when the enzyme binds a single molecule of the bisubstrate analog N-phosphonacetyl-L-aspartate?, *J. Mol. Biol.* 186:175.

Gerhart, J.C. and Pardee, A.B., 1962, Enzymology of control by feedback inhibition, *J. Biol. Chem.* 237:891.

Gerhart, J.C. and Pardee, A.B., 1963, The effect of the feedback inhibitor CTP, on subunit interactions in aspartate transcarbamylase, *Cold Spring Harb. Symp. Quant. Biol.* 28:491.

Gerhart, J.C. and Schachman, H.K., 1965, Distinct subunits for the regulation and catalytic activity of aspartate transcarbamylase, *Biochemistry* 4:1054.

Gerhart, J.C. and Schachman, H.K., 1968, Allosteric interactions in aspartate transcarbamylase II. Evidence for different conformational states of the protein in the presence and absence of specific ligands, *Biochemistry* 7:538.

Gouaux, E. and Lipscomb, W.N., 1988, Three-dimensional structure of carbamyl phosphate and succinate bound to aspartate carbamyltransferase, *Proc. Natl. Acad. Sci. USA* 85:4205.

Greenwell, P., Jewett, S.L. and Stark, G. R., 1973, Aspartate transcarbamylase from *Escherichia coli*: The use of pyridoxal 5'-phosphate as a probe in the active site, *J. Biol. Chem.* 248:5994.

Griffin, J. H., Rosenbusch, J.P., Blout, E.R. and Weber, K.K., 1973, Conformational changes in aspartate transcarbamylase: II. Circular dichroism evidence for the involvement of metal ions in allosteric interactions, *J. Biol. Chem.* 248:5057.

Honzatko, R.B., Crawford, J.L., Monaco, H.L., Ladner, J.E., Edwards, B.F.P., Evans, D.R., Warren, S.G., Wiley, D.C., Ladner, R.C. and Lipscomb, W.N., 1982, Crystal and molecular structures of native and CTP-liganded aspartate carbamoyltransferase from *Escherichia coli, J. Mol. Biol.* 160:219.

Honzatko, R.B. and Lipscomb, W.N., 1982, Interactions of phosphate ligands with *Escherichia coli* aspartate cambamoyltransferase in the crystalline state, *J. Mol. Biol.* 160:265.

Howlett , G.J. and Schachman, H.K., 1977, Allosteric regulation of aspartate transcarbamoylase. Changes in the sedimentation coefficient promoted by the bisubstrate analogue N-(phosphonacetyl)-L-aspartate, *Biochemistry* 16:5077.

Hu, C.Y., Howlett, G.J. and Schachman, H.K., 1981, Spectral alterations associated with ligand-promoted gross conformational changes in aspartate transcarbamoylase, *J. Biol. Chem.* 256:4998.

Jones, M.E., Spector, L. and Lipmann, F., 1955, Carbamyl phosphate. The carbamyl donor in enzymatic citrulline synthesis, *J. Am. Chem. Soc.* 77:819.

Kantrowitz, E.R. and Lipscomb, W. N., 1976, An essential arginine residue at the active site of aspartate transcarbamylase, *J. Biol. Chem.* 251:2688.

Ke, H.-M., Honzatko, R.B. and Lipscomb, W.N., 1984, Complex of N-phosphonacetyl-L-aspartate with aspartate carbamoyltransferase: X-ray refinement, analysis of Conformational changes and catalytic and allosteric mechanisms, *Proc. Natl. Acad. Sci. USA* 81:4027.

Ke, H.-M., Lipscomb, W.N., Cho, Y. and Honzatko, R.B., 1988, Structure of unligated aspartate carbamoyltransferase of *Escherichia coli* at 2.6-Å resolution, *J. Mol. Biol.* 204:725.

Kempe, T.D. and Stark, G.R., 1975, Pyridoxal 5'-phosphate, a fluorescent probe in the active site of aspartate transcarbamoylase, *J. Biol. Chem.* 250:6861.

Kim, K.H., Pan, Z., Honzatko, R.B., Ke, H.-M. and Lipscomb, W.N., 1987, Structural asymmetry in the CTP-liganded form of aspartate carbamoyltransferase from *Escherichia coli*, *J. Mol. Biol.* 196:853.

Kirshner, M.W. and Schachman, H.K., 1971, Conformational changes in proteins as measured by difference sedimentation studies. II. Effect of stereospecific ligands on the catalytic subunit of aspartate transcarbamylase, *Biochemistry* 10:1919.

Kirshner, M.W. and Schachman, H.K., 1973, Local and gross conformational changes in aspartate transcarbamylase, *Biochemistry* 12:2997.

Koshland, D.E., Jr., Nemethy, G. and Filmer, D., 1966, Comparison of experimental binding data and theoretical models in proteins containing subunits, *Biochemistry* 5:365-385.

Krause, K.L., Voltz, K.W. and Lipscomb, W.N., 1985, Structure at 2.9-Å resolution of aspartate carbamoyltransferase complexed with the bisubstrate analogue N-(phosphonacetyl)-L-aspartate, *Proc. Natl. Acad. Sci. USA* 82:1643.

Krause, K.L., Voltz, K.W. and Lipscomb, W.N., 1987, 2.5 Å structure of aspartate carbamoyltransferase complexed with the bisubstrate analog N-(phosphonacetyl)-L-aspartate, *J. Mol. Biol.* 193:527.

Ladjimi, M.M. and Kantrowitz, E.R., 1988, A possible model for the concerted allosteric transition in *Escherichia coli* aspartate transcarbamylase as deduced from site-directed mutagenesis studies, *Biochemistry* 27:276.

Ladjimi, M.M., Middleton, S.A., Kelleher, K.S. and Kantrowitz, E.R., 1988, Relationship between domain closure and binding, catalysis, and regulation in *Escherichia coli* aspartate transcarbamylase, *Biochemistry* 27:268.

Lahue, R.S. and Schachman, H.K., 1984, The influence of quaternary structure of the active site of an oligomeric enzyme. Catalytic subunit of aspartate transcarbamoylase, *J. Biol. Chem.* 259:13906.

Lauritzen, A.M. and Lipscomb, W.N., 1982, Modification of three active site lysine residues in the catalytic subunit of aspartate transcarbamylase by D- and L-bromosuccinate, *J. Biol. Chem.* 257:1312.

Middleton, S.A. and Kantrowitz, E.R., 1986, Importance of the loop at residues 230-245 in the allosteric interactions of *Escherichia coli* aspartate transcarbamylase*Proc. Natl. Acad. Sci. USA* 83:5866.

Monod, J., Wyman, J., Changeux, J.-P., 1965, On the nature of allosteric transitions: A plausible model, *J. Mol. Biol.* 12:88.

Moody, M.F., Vachette, P. and Foote, A.M., 1979, Changes in the x-ray solution scattering of aspartate transcarbamylase following the allosteric transition, *J. Mol. Biol.* 133:517.

Reichard, P. and Hanshoff, G., 1956, Aspartate carbamyl transferase from *Escherichia coli*, *Acta Chem. Scand.* 10:548.

Robey, E.A. and Schachman, H.K., 1985, Regeneration of active enzyme by formation of hybrids from inactive derivates: Implications for active sites shared between polypeptide chains of aspartate transcarbamoylase, *Proc. Natl. Acad. Sci. USA* 82:361.

Robey, E.A., Wente, S.R., Markby, D.W., Flint, A, Yang, Y.R. and Schachman, H.K.,1986, Effect of amino acid substitutions on the catalytic and regulatory properties of aspartate transcarbamoylase, *Proc. Natl. Acad. Sci. USA* 83:5934.

Schachman, H.K., Pauza, C.D., Navre, M., Karels, M.J., Wu, L. and Yang, Y.R., 1984, Location of amino acid alterations in mutants of aspartate transcarbamoylase: Structural aspects of interallelic complementation, *Proc. Natl. Acad. Sci. USA* 81:11.

Voltz, K.W., Krause, K.L. and Lipscomb, W.N., 1986, The binding of N-(phosphoacetyl)-L-aspartate to aspartate carbamoyltransferase of *Escherichia coli, Biochem. Biophys. Res. Commun.* 136:822.

Wente, S.R. and Schachman, H.K., 1987, Shared active sites in oligomeric enzymes: Model studies with defective mutants of aspartate transcarbamoylase produced by site-directed mutagenesis, *Proc. Natl. Acad. Sci. USA* 84:31.

FUNDAMENTALS OF NEUTRON DIFFRACTION

V. Ramakrishnan

Biology Department, Brookhaven National Laboratory, Upton, NY 11973, USA

The purpose of this article is to describe the differences between neutron diffraction and X-ray diffraction, and to give specific examples of how these differences have been exploited in the study of biological structures. Those who wish a comprehensive treatment of this topic can turn to a number of reviews and symposium volumes that deal with the application of neutron scattering to biological problems (Schoenborn and Nunes, 1972; Engelman and Moore, 1975; Jacrot, 1976; Kneale et al., 1977; Moore, 1982; Schoenborn, 1984).

SCATTERING LENGTH AND SCATTERING LENGTH DENSITY

When radiation is incident on matter and is scattered elastically, the amplitude of the scattered radiation is given by

$$F(S) = \int \rho(r) e^{2\pi i r . S} d^3 r \qquad [1]$$

Here the scattering vector $S = s - s_0$ where s and s_0 are vectors of magnitude $1/\lambda$ in the directions of the scattered and incident radiation respectively, and λ is the wavelength of the incident radiation.

The term $\rho(r)$ is called the scattering length density. For the scattering of X-rays by ordinary matter, the scattering length density is proportional to e^2/mc^2 times the number density of the charged particles involved in the scattering. This dependence on the mass of the scatterer is the reason by nearly all the scattering of matter by X-rays comes from electrons; it is the reason we talk about "electron density" in X-ray crystallography.

If the object consists of discrete scattering centers, e.g. a molecule with atoms whose positions are given by r_j, then the amplitude for scattering from the molecule is given by

$$F(S) = \sum f_j e^{2\pi i r_j . S} \qquad [2]$$

The f_j in equation (2) are the atomic scattering factors given by

$$f_j(S) = \int_{\text{vol. of atom.}} \rho(r)e^{2\pi i r \cdot S} d^3 r \qquad [3]$$

where the integration is over the jth atom. The f_j's are also known as "scattering lengths." Equations 1-3 form the basis of all scattering and crystallography.

For X-rays, $\rho(r)$ is proportional to the electron density. In the case where S=0 (i.e. at zero scattering angle), the integral becomes equal to e^2/mc^2 times Z, the atomic number, and this is why heavy atoms scatter proportionately more than light ones. As a result, hydrogen scatters very little, and usually cannot be detected directly in an X-ray experiment. For non-zero $S, f_j(S)$ falls off with angle, because the dimensions of the atom are comparable to the wavelength of X-rays.

For neutrons, things are not as straightforward. Neutrons interact mainly with the nuclei of atoms, via a short range force, "the strong interaction." Thus the scattering lengths of various elements are not proportional to the atomic number, but rather depend on the properties of the particular nucleus. Further, since nuclei are several orders of magnitude smaller than typical neutron wavelengths, the scattering length density in equation 3 can be approximated by a delta function, with the result th at the atomic scattering lengths are isotropic: they do not show an angular dependence.

CONTRAST AND SMALL ANGLE SCATTERING

Equations 1-3 apply in a vacuum. When radiation strikes particles in solution, a large amount of scattered radiation is present at very low scattering angles. This phenomenon, known as small angle scattering, contains useful information about the global features of the particles. However, in this case, what matters is not the absolute density of the particle, but the contrast, which is the difference between the particle density and the density of the solvent. A dramatic visual demonstration of contrast is shown in Figure 1.

For particles in solution, therefore, equations 1 and 2 have to be amended. If ρ_s is the scattering length density of the solvent, and $\Delta\rho(r) = \rho(r) - \rho_s$ is the contrast of the particle then

$$F(S) = \sum f_j e^{2\pi i r_j \cdot S} = \int \Delta\rho(r)e^{2\pi i r \cdot S} d^3 r \qquad [4]$$

The intensity for particles in solution, which are randomly oriented, is

$$I(S) = <F^*(S) \cdot F(S)> \qquad [5]$$

where <...> denotes a spatial average over all possible orientations.

Debye (1915) has shown that equations 4 and 5 lead to

$$I(S) = \sum_{ij} f_i f_j \frac{\sin 2\pi r_{ij} S}{2\pi r_{ij} S} = \int \Delta\rho(r)\Delta\rho(r') \frac{\sin(2\pi S |r-r'|)}{2\pi S |r-r'|} d^3 r d^3 r' \qquad [6]$$

where r_{ij} is the distance between the ith and jth atoms, and the sum and integral are both over the entire macromolecule.

Figure 1. The importance of contrast. Both tubes contain two pyrex glass beads in a piece of glass wool, but the tube on the right contains 95% DMSO, which has the same refractive index as the glass wool. The glass wool is thus "contrast matched", allowing the pyrex beads to be visible. Photo courtesy of Professor D.M. Engelman of Yale University.

At low angles, a Taylor expansion of equation (6) gives (Guinier and Fournet, 1955),

$$I(S) = I(0) \ e^{-\frac{4\pi^2 R_g^2}{3}} \qquad\qquad [7]$$

where R_g is the radius of gyration. Both $I(0)$ and R_g are functions of contrast (Stuhrmann, 1974; Ibel and Stuhrmann, 1975):

$$I(0) = (<\Delta\rho>V)^2 \qquad\qquad [8]$$

$$R_g^2 = R_c^2 + \frac{\alpha}{<\Delta\rho>} + \frac{\beta}{<\Delta\rho>^2} \qquad\qquad [9]$$

where $<\Delta\rho>$ is the mean contrast between macromolecule and solvent and V is the particle volume.

The parameter α is a measure of how the density changes as a function of distance from the center of mass. For particles that are denser on the outside than in the core, $\alpha > 0$. The variation of the position of the apparent center of mass with contrast is measured by the parameter β. If a particle consists of two non-concentric components of different densities, then $\beta > 0$, and in fact the distance between the centers of mass of the two components can be estimated from β.

An alternative way to analyze small angle scattering data is by means of the length distribution function $P(r)$:

$$P(r) = \frac{8\pi r}{<\Delta\rho>^2} \int_0^\infty SI(S)\, \sin(2\pi rS)\, dS \qquad [10]$$

The original intensity can be calculated from $P(r)$ as follows:

$$I(S) = <\Delta\rho>^2 \int_0^\infty P(r) \frac{\sin 2\pi rS}{2\pi rS}\, dr \qquad [11]$$

$P(r)dr$ is the probability that any two points in the particle lie at a distance between r and $r+dr$ from each other. It is therefore the small angle scattering analogue of the Patterson function in crystallography. It contains all the information present in the intensity, and in principle, any parameter that can be calculated from $I(S)$ can also be calculated from $P(r)$. For example,

$$R_g^2 = \frac{1}{2} \int r^2 P(r)\, dr \qquad [12]$$

WHY NEUTRONS?

There are two interesting differences between X-ray and neutron scattering, which can be exploited for biological structure studies: (i) the scattering by an element can show a strong isotope effect; in particular, there is a large difference between the scattering lengths of hydrogen and deuterium; (ii) there is no systematic variation in the magnitude of the scattering length with elements of increasing atomic number, with the result that the scattering length of hydrogen is comparable to those of other elements. Table I lists the X-ray and neutron scattering lengths for various elements.

Table I. Neutron and X-ray scattering elements of various elements

Element	Neutron Scattering length (10^{-12} cms)	X-ray Scattering length (10^{-12} cms)
H	-0.38	0.28
D	0.65	0.28
C	0.66	1.7
N	0.94	2.0
O	0.58	2.3
P	0.51	4.2
S	0.28	4.5
Ni	-0.34	6.2
Ti	1.0	7.9

These two properties have been responsible for the vast majority of biological experiments using neutron scattering.

Crystallography

Approximately half the atoms in a protein are hydrogen atoms. These atoms cannot be directly visualized by X-ray crystallography; rather their positions are inferred from the location and orientation of the various chemical groups to which they are bonded. In many cases, however, the presence or absence of a hydrogen on a specific chemical group is important to understanding the functioning of an enzyme. Because the neutron scattering length of hydrogen is comparable to that of the other elements, it has been possible to locate specific hydrogens on proteins. If the hydrogen of interest can be replaced by deuterium, it is even easier to detect it, because of the large scattering length of deuterium.

There is a large difference between the scattering lengths of deuterium and hydrogen. It is possible to soak a protein crystal in D_2O, and determine by neutron diffraction which protons have been exchanged for deuterium. Data of this kind yield information about the dynamics of proteins: which parts of the protein are relatively stable, and which undergo conformational fluctuations that allow access to solvent?

The low resolution reflections in diffraction data on proteins contains information about the solvent density and structure. For example, it was noticed very early on that by changing the electron density of the mother liquor, it was possible to drastically change the intensity of the low angle reflections (Bragg and Perutz, 1952). In normal crystallographic analysis, these reflections are discarded. Recently, Cheng and Schoenborn (1989) have developed a method to include these reflections in their analysis and gain information about the solvent structure around proteins.

So far, phasing in neutron protein crystallography has not been done *ab initio*. Usually, a well refined X-ray structure is used to calculate neutron phases as a starting point for refinement. In the initial phase calculation, the contributions of the hydrogen atoms are often omitted, in order to avoid biasing the structure. The refined neutron structure can be used to further refine the original X-ray structure. Such a joint X-ray and neutron refinement has been done on ribonuclease A (Wlodawer and Sjolin, 1984).

Small Angle Scattering

The usefulness of neutrons in small angle scattering comes from the fact that it is possible to change contrast much more easily than when using X-rays. Contrast can be changed by two methods: one can change the density of the solvent or that of the macromolecule.

Figure 2 shows the neutron scattering length density of some common classes of biological macromolecules. By changing the H_2O/D_2O ratio of the buffer, the solvent scattering length density can be changed to span the entire range of densities of macromolecules, thus allowing contrast variation over a wide range. For example, nucleic acids are contrast matched in approximately 65% D_2O, while proteins are matched in approximately 42% D_2O. Thus, in a protein-nucleic acid complex, it is possible to enhance the contribution of one component or another by simply changing the D_2O content of the buffer

With neutrons, it is also possible to change the density of the macromolecule by deuteration. Biological macromolecules are normally deuterated biosynthetically, by the growth of algae or bacteria in deuterated media. The deuteration of eukaryotic macromolecules is more difficult, since eukaryotes do not grow well in D_2O. However, advances in genetic engineering have now made it possible to express many eukaryotic genes in *E. coli*, which grows well in D_2O.

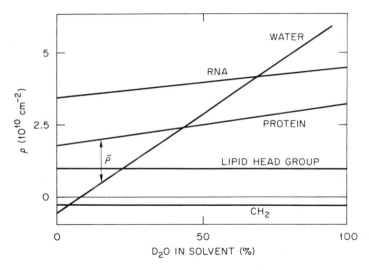

Figure 2. Scattering length densities of various types of macromolecules, and of water as a function of D_2O concentration in the solvent. For a given D_2O concentration, the difference between a point on the macromolecule line and the corresponding point on the water line is the contrast. The point where the water line intersects the line for a macromolecule shows the D_2O concentration in which the macromolecule is "contrast matched." Taken from Jacrot (1976).

If a multicomponent system can be reconstituted, and if individual components can be deuterated, it is then possible by neutron scattering to obtain information about the individual components, including the locations of their centers of mass in the complex. Thus specific deuteration greatly enhances the power of neutron scattering.

Inelastic Scattering

Thermal and cold neutrons have energies that are comparable to the vibrational energies of molecules. They therefore excite vibrational states in condensed matter, and are scattered inelastically. Measurement of the spectrum of inelastically scattered neutrons thus allows one to obtain information about the dynamics of the sample. Because the inelastic scattering cross-section of hydrogen is very large, with proteins the inelastic scattering arises mainly from the hydrogens in protein. So far, inelastic scattering experiments have been difficult to perform and interpret, and have not yet been used widely in biology. However, a recent experiment on dynamical transitions in myoglobin (Doster et al., 1989), which for the first time offered an experimental test of molecular dynamics simulations on the picosecond time scale, shows that the technique holds promise for the future.

INSTRUMENTATION

Most biological experiments to date have used reactors as a source of neutrons. In reactors, neutrons are produced by the fission of U^{235}. These neutrons are then cooled by scattering with the moderating fluid, which is often D_2O. Neutrons are then sampled through a port, and are monochromated and collimated as required.

Recently, neutrons have also been produced by colliding a proton beam with a stationary target of tungsten. These sources are called pulsed sources because they produce neutrons in bursts when protons in a storage ring are dumped onto the target. While the instantaneous flux is very high during the pulses, it remains to be seen whether the time-averaged flux will compare favorably with reactor fluxes. However, because the neutron pulses occur at precise times, monochromatization is not necessary and the entire spectrum of neutrons is used. This is because the time of arrival of a neutron at the detector allows its velocity, and hence its wavelength, to be calculated. By determining both the scattering angle and the time of arrival of a scattered neutron, the value of S for each neutron can be assigned, and the total scattered intensity can be measured as a function of S even with a polychromatic beam.

Even the most powerful reactors produce a flux of neutrons that is several orders of magnitude lower than a rotating anode X-ray generator. This means that sample sizes have to be much larger in order to obtain meaningful data. In order to maintain resolution, the instruments themselves have to be much larger. For small angle scattering, these instruments can be as long as 20-80 meters. Several neutron instruments are described by Bacon (1975).

Crystallography and small angle scattering impose different requirements for instrumentation. In crystallography, one needs a highly monochromatic beam in order to be able to collect high resolution data. This is done by allowing the white beam of neutrons from the reactor to be incident on a single crystal (e.g. of graphite), and choosing a strong Bragg reflection as a source of monochromatic neutrons. In small angle scattering, one can get away with as much as a 10% spread in the wavelength distribution of the neutron beam. It therefore pays to use devices which have a broader bandpass. One of these is the velocity selector, which consists of a series of blades with slits cut in them so that only neutrons within a certain range of velocities can pass through all the blades as the assembly is rotated. Wavelength selection is done by changing the rotation speed of the assembly. An alternative to the velocity selector is the multilayer monochromator, which consists of an optically flat glass substrate onto which are deposited alternate layers of Ni and Ti (Saxena and Majkrzak, 1984; Saxena, 1986). These two elements have very different scattering lengths, with the result that the multilayers act as one dimensional crystals and monchromate the incident beam to produce a wavelength spread of about ten percent.

In small angle scattering, it is also beneficial to have a higher wavelength of neutrons. The scattering pattern is spread out to more accessible angles, because increasing the wavelength means that the same value of S is measured at a higher scattering angle. Unlike the X-ray case, absorption does not become a serious problem as wavelength is increased. Most reactor neutron spectra peak at about 1-1.5 Å, which is ideal for crystallography but not for small angle scattering, where >5 Å neutrons are preferable. To overcome the scarcity of neutrons in the higher wavelength range, a cold moderator is often introduced into the reactor. This is a vessel containing liquid deuterium or hydrogen. Neutrons enter the vessel and lose energy by scattering with the cold fluid, and emerge at a lower energy and correspondingly higher wavelength. The introduction of cold moderators has made the available flux higher by almost an order of magnitude for small angle scattering experiments.

Area detectors, which have recently made such an impact on X-ray crystallography, have been in use in the neutron field for almost twenty years: because of the intrinsic weakness of neutron beams, most experiments would not be possible without them. Most neutron detectors are based on a multi-wire gas chamber, with He^3 as the detector gas. The advantage of He^3 is that it has a high cross-section combined with a high spatial resolution, so that detectors can be made that are suitable for both crystallography and small angle scattering.

EXAMPLES OF THE APPLICATIONS OF NEUTRON SCATTERING

For these examples, I have primarily drawn on work done using the neutron facilities at Brookhaven National Laboratories. A large number of equally important experiments have been done at other facilities, especially the Institut Laue-Langevin in Grenoble.

Crystallography

In the catalytic mechanism of trypsin, one question was which residue, His-57 or Asp-102, acted as the chemical base in the hydrolysis reaction. Kossiakoff and Spencer (1980, 1981) collected neutron data on a trypsin crystal that had been soaked in D_2O and by direct location of density corresponding to a deuterium atom were able to show that His-57 and not Asp-102 was the catalytic base.

Crystals grown in H_2O and subsequently soaked in D_2O have been studied by a large number of workers. Because hydrogens have a negative density and deuteriums have a large positive density, it is possible to determine which hydrogens have exchanged in the structure. In the case of trypsin (Kossiakoff, 1982) the pattern of hydrogen exchange showed that hydrogen bonding was the dominant factor in exchange. All those hydrogens which were bonded to water molecules were fully exchanged; however, those sites in β-sheet structures which are hydrogen bonded to main chain carbonyls and are flanked by similarly bound peptides, are to a large extent unexchanged.

NMR experiments indicate that the methyl groups of proteins spin rapidly about their axes. A 1.4 Å data set was collected for crambin by Teeter and Kossiakoff (1984). Analysis of the methyl groups from this data (Kossiakoff and Shteyn, 1984) showed that the methyl groups do not spin uniformly, but spend more time in certain preferred conformations, which contribute predominantly to a time averaged structure. Also, this preferred conformation in most cases turns out to be the energetically favored staggered conformation, showing that in general crystal packing effects do not impose additional torsional constraints.

In X-ray analysis, it is customary to ignore the contribution of bulk solvent by omitting the low-order reflections in refinement. Recently, Cheng and Schoenborn (1989) have re-analyzed data for carbonmonoxymyoglobin by taking into account the effect of the solvent. The contribution of the solvent to the low order reflections was evaluated by dividing the solvent volume into shells extending outwards from the surface of the protein. Each shell was assigned a scattering length density and a "liquidity" factor (which is the analogue of the temperature factor for atoms). Inclusion of these terms allowed all structure factors to be used in the refinement. By doing a least-squares refinement to the observed reflections, it was found that the B-factor for the shells had well defined minima at 2.4 and 4.0 Å, which correspond to the first and second hydration layers of water around the protein. This shows that proteins are surrounded by hydration layers that have lower mobility than bulk solvent. Also, many additional water molecules emerged in the course of the refinement. It should be pointed out that the emergence of additional water molecules is not a sign of refinement error: it is in fact a consequence of the inclusion of the additional 2000 low-order reflections, which amounts to inclusion of a significant amount of new information. In fact, all the water molecules were found to be bonded to polar groups. An internal check on the entire procedure was that when it was applied to data collected for crystals grown in 40%, 80% and 90% D_2O the densities of the solvent shells that emerged from the refinement agreed with the predicted values from the D_2O content.

Small Angle Scattering

One of the earliest successes of contrast variation using neutron scattering was in studies on the structure of the nucleosome (Pardon et al., 1975; Hjelm et al., 1977). The radius of gyration measured as a function of contrast gave a positive α (see equation [9]) implying that the density of the particle increased with the distance from the center of mass. The experiments showed conclusively that the nucleosomal DNA was wrapped around a histone core, and allowed a low resolution model of the nucleosome to be determined.

The small ribosomal subunit of *E. coli* consists of 21 proteins complexed with 16S RNA (Wittmann, 1982). Neutron scattering was used to determine the location of individual proteins, based on the idea (Engelman and Moore, 1972) that if pairs of proteins were labeled by deuteration in a reconstituted ribosome, the scattering from that ribosome contains a term that corresponds to interference between the two proteins. The experiment can also be done with a pair of unlabelled proteins reconstituted with all the other components deuterated. The interference term can be extracted by subtracting the scattering from the respective singly labeled proteins. By repeating the process with various pairs of proteins, and using triangulation, a three dimensional arrangement of the proteins could be determined. In practice this is done as follows:

If the interference term is denoted by $I(S)$, then the length distribution function $P(r)$ (see equation 10) now corresponds to the distribution of all possible lengths that go from one labeled protein to the other. The second moment of this distribution is given by (Moore and Weinstein, 1978):

$$M_{i\,j} = \int r^2 P(r)\,dr$$

$$= R_i^2 + R_j^2 + d_{i\,j}^2$$

$$= R_i^2 + R_j^2 + (x_i - x_j)^2 + (y_i - y_j)^2 + (z_i - z_j)^2 \qquad [13]$$

where x_i, y_i, z_i, are the coordinates of the center of mass of the ith protein, and R_i is its *in situ* radius of gyration. From equation [13], it is clear that with a sufficient number of measurements M_{ij}, it should be possible to determine the locations and radii of gyration of the individual proteins. As a result of this work, which took over a decade to carry out, we now know the locations of all 21 proteins of the small ribosomal subunit (Capel et al., 1987). Experiments to determine the spatial arrangement of the 50S ribosomal proteins by a similar procedure are in progress, with the locations of many proteins already determined (Novotny, et al., 1986).

PROBLEMS WITH NEUTRON SCATTERING

Unlike X-ray diffraction, which usually can be done in individual laboratories, neutron diffraction requires large and expensive facilities. Only a few such facilities exist in the world. The use of these facilities is justified only if neutron experiments provide data that cannot be obtained by other techniques, such as X-ray diffraction.

Neutron sources are intrinsically weak. For crystallography, this means that one requires very large crystals (preferably 1 mm or greater in each dimension) in order to be able to collect interpretable data. This requirement greatly limits the number of proteins that can be studied by neutron diffraction. On the other hand, because thermal neutrons have several orders of magnitude less energy than X-rays of the same wavelength, they do not produce free radicals, and there is negligible radiation damage to the crystal. Thus a complete data set can be collected on a single crystal, avoiding scaling problems.

For small angle scattering, the low flux of neutron beams means that larger amounts of samples are required. Selective deuteration, while extremely powerful, is expensive, and moreover is not always feasible, especially with proteins from eukaryotic sources. However, new methods for gene expression in prokaryotes should go a long way to alleviate this problem.

If the problems of obtaining suitable samples in sufficient quantity can be overcome, neutron diffraction reveals a wealth of detail that extends the information obtained using X-rays.

ACKNOWLEDGEMENTS

Work on the use of neutrons in biology at Brookhaven National Laboratory is supported by the Office of Health and Environmental Research of the United States Department of Energy.

REFERENCES

Bacon, G.E., 1975, Neutron Diffraction, Clarendon Press, Oxford.

Bragg, W.L. and Perutz, M.F., 1952, The external form of the haemoglobin molecule, *Acta Crystallogr.* 5:277.

Capel, M.S., Engelman, D.M., Freeborn, B.R., Kjeldgaard, M., Langer, J.A., Ramakrishnan, V., Schindler, D.G., Schneider, D.K., Schoenborn, B.P., Sillers, I.-Y., Yabuki, S. and Moore, P.B., 1987, A complete mapping of the positions of the proteins in the small ribosomal subunit of *E. coli*, *Science* 238:1327.

Debye, P., 1915, Zerstreuung von Rontgenstrahlen, *Ann. Phys.* 46:809.

Doster, W., Cusack, S. and Petry, W., 1989, Dynamical transition of myoglobin revealed by inelastic neutron scattering, *Nature* 337:754.

Cheng, X. and Schoenborn, B.P., 1989, Hydration in protein crystals, *Acta Crystallogr.*, in press.

Engelman, D.M. and Moore, P.B., 1972, A new method for the determination of biological quaternary structures by neutron scattering, *Proc. Natl. Acad. Sci. USA* 69:1997.

Engelman, D.M. and Moore, P.B., 1975, Determination of quaternary structure by small angle neutron scattering, *Ann. Rev. Biophys. Bioeng.* 4:219.

Guinier, A. and Fournet, G., 1955, "Small angle scattering of X-rays," Wiley, New York.

Hjelm, R.P., Kneale, G.G., Suau, P., Baldwin, J.P., Bradbury, E.M. and Ibel, K., 1977, Small angle neutron scattering studies of chromatin subunits in solution, *Cell* 10:139.

Ibel, K. and Stuhrmann, H.B., 1975, Comparison of neutron and X-ray scattering of dilute myoglobin solutions, *J. Mol. Biol.* 93:255.

Jacrot, B., 1976, The study of biological structures by neutron scattering from solution, *Rep. Prog. Phys.* 39:911.

Kneale, G.G., Baldwin, J.P. and Bradbury, E.M., 1977, Neutron scattering studies of biological macromolecules in solution, *Q. Rev. Biophys.* 10:485.

Kossiakoff, A.A., 1982, Protein dynamics investigated by the neutron diffraction-hydrogen exchange technique, *Nature* 296:713.

Kossiakoff, A.A. and Shteyn, S., 1984, Effect of packing structure on side-chain methyl rotor conformations, *Nature* 311:582.

Kossiakoff, A.A. and Spencer, S.A., 1980, Neutron diffraction identifies His-57 as the catalytic base in trypsin, *Nature* 288:414.

Kossiakoff, A.A. and Spencer, S.A., 1981, Direct determination of the protonation states of aspartic acid-102 and histidine-57 in the tetrahedral intermediate of serine proteases: neutron structure of trypsin, *Biochemistry* 20:6462.

Moore, P.B., 1982, Small angle scattering techniques for the study of biological macromolecules and macromolecular aggregates, *Meth. Exp. Phys.* 20:3370.

Moore, P.B. and Weinstein, E., 1979, On the estimation of the locations of subunits within macromolecular aggregates from neutron interference data, *J. Appl. Cryst.* 12:321.

Novotny, V., May, R.P. and Nierhaus, K.H., 1986, *in:* "Structure, Function and Genetics of Ribosomes," Hardesty, B. and Kramer, G., eds., Springer-Verlag, Berlin.

Pardon, J.F., Worcester, D.L., Wooley, J.C., Tatchell, K., Van Holde, K.E. and Richards, B.M., 1975, Low angle neutron scattering from chromatin subunit particles, *Nucl. Acids Res.* 2:2163.

Schoenborn, B.P. and Nunes, A.C., 1972, Neutron Diffraction, *Ann. Rev. Biophys. Bioeng.* 1:529.

Schoenborn, B.P., ed., 1975, "Neutron scattering for the analysis of biological structures," Brookhaven Symposia in Biology, volume 27, U.S. Government Printing Office.

Schoenborn, B.P., ed., 1984, "Neutrons in Biology," Plenum Press, New York.

Stuhrmann, H.B., 1974, Neutron small angle scattering of biological macromolecules in solution, *J. Appl. Cryst.* 7:173.

Teeter, M. and Kossiakoff, A.A., 1984, The neutron structure of the hydrophobic plant protein crambin, *in:* "Neutrons in Biology," Schoenborn, B.P., ed., Plenum Press, New York.

Wittmann, H.G., 1982, Components of Bacterial Ribosomes, *Ann. Rev. Biochem.* 51:155.

Wlodawer, A. and Sjolin, L., 1984, Application of joint neutron and X-ray refinement to t he investigation of the structure of ribonuclease A at 2.0 Å resolution, *in:* "Neutrons in Biology," Schoenborn, B.P. ed., Plenum Press, New York.

PURIFICATION OF COMPLEXES BETWEEN PEPTIDE ANTIGENS AND CLASS II MAJOR HISTOCOMPATIBILITY COMPLEX ANTIGENS USING BIOTINYLATED PEPTIDES

Theodore Jardetzky[1,2], Joan Gorga[1], Robert Busch[3], Jonathan Rothbard[1], Jack L. Strominger[1,2], and Don Wiley[1,2]

[1]Department of Biochemistry and Molecular Biology, Harvard University Cambridge, USA, [2]Howard Hughes Medical Institute and [3]Laboratory of Molecular Immunology, Imperial Cancer Research Institute, London, UK

INTRODUCTION

The Major Histocompatability Antigens (MHC) are cell surface glycoproteins which are involved in the regulation of the immune response to foreign antigens. The MHC proteins bind peptides generated by cellular processing of native proteins (Babbitt et al., 1985; Buus et al., 1986; Watts & McConnell, 1986; Chen & Parham, 1989), and the peptide-MHC complexes are presented on cell surfaces to T lymphocytes (Schwartz, 1985). Class I MHC proteins maintain surveillance over proteins which are found in the cell cytoplasm, such as newly synthesized viral proteins in an infected cell (Morrison et al., 1987; Moore et al., 1988). The foreign peptide-MHC complexes are recognized by cytotoxic T lymphocytes, which respond by lysing the infected cell.

Class II MHC proteins provide a different role. They are designed to bind fragments of proteins which are exogenous to the cell and regulate the proliferation of specific antibody producing cells. One possible route for this process involves the binding of native antigen by cell surface antibodies, internalization of antigen and proteolytic breakdown, followed by the formation of foreign peptide class II MHC complexes (Ziegler and Unanue, 1982; Shimonekevitz et al., 1984; Lanzavecchia et al., 1988). The presentation of these complexes on the B-cell surface stimulates T helper lymphocytes to release lymphokines which affect the proliferation and maturation of the antibody cells. The division of the immune system into cellular (cytotoxic) and humoral (antibody) responses is also evident in the two classes of MHC molecules and the functional surveillance which they carry out.

Any one individual has a fixed set of Class I and Class II molecules (six each), and the set of MHC molecules is unique to that individual due to the high polymorphism of the MHC genes. These differences allow for the discrimination between self and non-self and are responsible for graft rejections in transplantation. Because of the limited number of MHC molecules per individual, each one must be capable of binding many different peptides, in order to ensure the proper immune response to an immense number of foreign antigens. Binding studies have shown that the peptide-class II MHC complexes have extraordinary stability (Buus et al., 1986; Sadegh-Nasseri & McConnell, 1989; Jardetzky et al., 1989), with half lives of dissociation on the order of days. It has also been observed in the peptide-MHC interactions studied to date that the binding affinity is determined by only a few amino acids in the peptides (Allen et al., 1987; Sette et al., 1987), with perhaps as few as one amino acid side chain embedded in polyalanine providing for the full affinity (Jardetzky et al., 1989). The remaining amino acid positions of the peptides can be substituted without affecting MHC binding. Although such studies suggest that the recognition of certain sequence motifs or particular amino acids may provide a means for the recognition of many sequence disparate peptides, many questions remain to be answered about the recognition events which create such stable complexes. The binding must have high energy barriers for dissociation, yet be able to accomodate variable sequences. Whether this involves the interaction of peptide main chain with the MHC site, or MHC conformational adjustments to complement each new peptide sequence, is not clear.

The solution of the HLA-A2 crystal structure (Bjorkman et al., 1987a,b) has provided a view of the class I MHC antigen binding site, as well as a model for the class II structure (Brown et al., 1988). The binding site is a long groove formed by a beta sheet as the base and two alpha helices as the sides. From this structure it is not possible to conclude in detail how peptides are recognized. Unfortunately, the ability to further explore the mechanism of peptide binding to HLA-A2 crystallographically has been hampered by the difficulty in isolating defined peptide-MHC complexes. Recent work (Chen & Parham, 1989; Bjorkmann et al., 1987a) suggests that peptide binding to class I molecules may occur at a barely detectable level due to the presence of tightly bound antigens which copurify with the molecule. It has proven easier to detect peptide binding to class II MHC molecules *in vitro* in a number of systems (Babbitt et al., 1985; Buus et al., 1987; Sadegh-Nasseri & McConnell, 1989), where between 10% to 70% of the protein has proven to be active in peptide binding.

Here we describe an approach to isolating pure, defined peptide-MHC complexes using a class II MHC molecule (HLA-DR1) and biotinylated peptide. By taking advantage of the essentially irreversible interaction of biotin with the proteins avidin and streptavidin (K_d = 10^{-15} M, Hoffman et al., 1980), and the stability of the peptide-MHC interactions, a properly biotinylated peptide should allow one to separate bound MHC complexes, from inactive or blocked MHC molecules. We have biotinylated an antigenic peptide derived from the influenza virus hemagglutinnin, which we have previously shown binds to HLA-DR1 (Jardetzky et al., 1989). The biotinylated peptide binds HLA-DR1 with an affinity similar to that of the parent peptide, and allows the separation of bound HLA-DR1 from unbound HLA-DR1 using avidin or streptavidin. This demonstrates that the isolation of peptide-HLA-DR1 complexes using biotinylated peptides is feasible, and should allow one to construct defined complexes for crystallographic and biochemical analysis. The method also provides a simple assay to screen for conditions which may increase the percentage of active HLA-DR1 molecules.

RESULTS

Biotinylated Peptide Binds to HLA-DR1

The influenza hemagglutinnin derived peptide HA 306-318 shown in Figure 1 stimulates an HLA-DR1 restricted T cell response in humans (Lamb et al., 1982a,b; Rothbard et al., 1988) and it has also been possible to demonstrate a direct binding interaction with purified HLA-DR1 *in vitro* (Jardetzky et al., 1989). An analog of this peptide was synthesized in which the three lysines were substituted by arginines and the N-terminal proline was substituted by lysine. Based on a study of 39 single amino acid substitutions of this peptide, these substitutions would not be expected to disrupt the binding to HLA-DR1 (Jardetzky et al., 1989). The peptide was acetylated at the N-terminus, followed by biotinylation, using the N-hydroxy-succinimide-long chain biotin reagent, to give a peptide uniquely biotinylated at lysine 306, as shown in Figure 1.

The biotinylated peptide (HA-Bio306) was tested for binding to HLA-DR1 in an inhibition assay. This assay tests for the ability of peptides to inhibit the binding of a radioactively

306   318

NH$_2$-P K Y V K Q N T L K L A T-COOH

Ac-(N-ε-Biotinyl-K) R Y V R Q N T L R L A T-COOH

Figure 1.   Sequences of the influenza hemagglutinnin peptides. The sequence of residues 306-318 and the biotinylated analog HA-Bio306 are shown above. Peptides were synthesized and purified by HPLC as described previously (Rothbard et al., 1988).

labeled peptide (influenza matrix protein 17-31) to HLA-DR1, using native polyacrylamide gel electrophoresis to separate bound from unbound peptide (Jardetzky et al., 1989). The results of a titration of HA-Bio306 are shown in Figure 2a. The bands containing DR1 and peptide were cut from the gel and the level of inhibition quantitated. Figure 2b shows a comparison titration between HA 306-318 and the biotinylated derivative HA-Bio306. This demonstrates that the multiple substitutions of lysine to arginine and proline to biotinylated lysine do not significantly alter the peptide binding to HLA-DR1.

Biotinylated Peptide-HLA-DR1 Complexes Bind Avidin

In order for HA-Bio306 to be a useful reagent for the isolation of peptide-MHC complexes, the biotin must be accessible enough to be bound by avidin. To test this, HLA-DR1 was radioactively labeled and incubated with an excess of HA-Bio306 for two days. At the end of the incubation period, avidin was added to a five-fold excess of sites over HA-Bio306. The samples were subsequently analyzed by native 10% polyacrylamide gel electrophoresis.

In Figure 3, lane 5, a new, slower moving band appears under these conditions. This band is not evident with (a) HLA-DR1 alone (Figure 3, lane 1), (b) HLA-DR1 with HA-Bio306 without avidin (lane 2), or (c) HLA-DR1 alone with avidin (lane 3). The extra band can also be inhibited by including an excess of the peptide HA 306-318 at the time of incubation (lane 4). These data suggest that the biotinylated peptide is capable of interacting with avidin while bound to the HLA-DR1 molecule. The low percentage of HLA-DR1 which is shifted under these conditions is also consistent with the expectation that the majority of the protein is blocked or inactive in peptide binding (Buus et al., 1988), although it is also possible that the labeling procedure contributes to the inactivity of the protein. However, this gel based assay should allow one to screen many different incubation conditions, in the expectation that more protein could be induced to bind peptides.

Precipitation of HLA-DR1 by Streptavidin-Agarose

In order to investigate the feasibility of the purification of peptide-HLA-DR1 complexes, the precipitation of the preformed HA-Bio306-HLA-DR1 complexes by streptavidin linked to agarose was tested. In these experiments, radioactive HLA-DR1 was incubated in the presence and absence of HA-Bio306, as in the experiments shown in Figure 3. The samples were added to 25 μl streptavidin-agarose, and incubated for a further 3 hours. After washing, the pellets were counted directly, giving the results shown in Figure 4a. HLA-DR1 could be specifically bound to the streptavidin agarose column, by preincubation with the biotinylated peptide HA-Bio306. The precipitation of HLA-DR1 could be inhibited by excess HA 306-318 and excess free biotin. The pellets were tested further for the precipitation of both HLA-DR1 heavy and light chains, by boiling in SDS-PAGE sample buffer followed by analysis of the supernatants by gel electrophoresis. The results shown in Figure 4b demonstrate that both chains of the HLA-DR1 molecule are precipitated in the experiment.

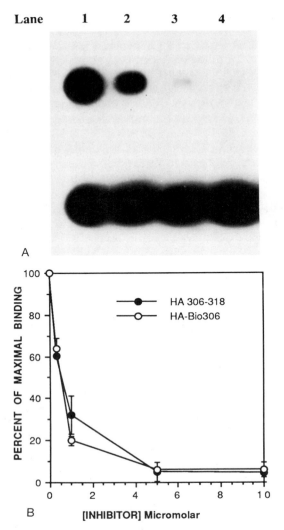

Figure 2. Binding of HA-Bio306 to HLA-DR1.

(a) Inhibition of iodinated matrix peptide 17-31 binding to HLA-DR1; lane 1: 0.3 µM HA-Bio306; lane 2: 1.0 µM HA-Bio306; lane 3: 5.0 µM HA-Bio306; lane 4: 10.0 µM HA-Bio306. Immunoaffinity purified and papain cleaved HLA-DR1 (Gorga et al., 1987) was incubated at a concentration of 2.0 µM with approximated 0.2 µM labeled peptide (Bolton, 1986) at 37°C for 48 hours, in phosphate buffered saline, 1mM EDTA, 1mM phenylmethylsulfonyl fluoride, 1mM iodoacetamide, pH 7.0. Samples were electrophoresed through 12% acrylamide Laemmli gels without SDS, fixed 30 minutes in 10% acetic acid, and exposed to Kodak X⁻OMAT AR film at -70°C.

(b) Relative binding affinities of peptides HA 306-318 and HA-Bio306. After autoradiography, the radioactive peptide HLA-DR1 complex bands were excised from the gels and the number of associated counts determined. Values were normalized to triplicate determinations of the maximal amounts found in the absence of inhibitory peptides and corrected for the background found with radioactive peptide in the absence of HLA-DR1 (generally <5% of total signal).

Lane   1   2   3   4   5

Figure 3.   Gel shift assay for avidin binding to HLA-DR1-peptide complexes.
Lane 1 HLA-DR1; lane 2: HLA-DR1 + 10 μM HA-Bio306; lane 3: HLA-DR1 +
00 μM avidin; lane 4: HLA-DR1 + 10 μM HA-Bio306 + 600 μM HA 306-318 +
100 μM avidin; lane 5: HLA-DR1 + 10μM HA-Bio306 + 100 μM avidin.
Approximately 100 nM HLA-DR1, iodinated with Bolton-Hunter reagent
(Bolton,1986), was incubated with 10 μM HA-Bio306 for 48 hours at 37°C. Avidin
was added to the indicated samples to a final concentration of 100 μM (monomer)
and incubated an additional 20 minutes at 37°C. Samples were run on a native 10%
polyacrylamide gel until the dye front reached the bottom. The gel was fixed, dried
and exposed as described in Figure 2.

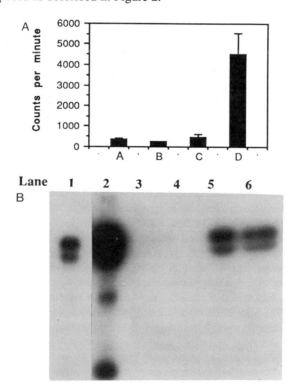

Figure 4.   Streptavidin agarose precipitation of HLA-DR1.
(a) Counts per minute recovered in agarose pellets. (A) : HLA-DR1; (B) : HLA-DR1
+ 10 μM HA-Bio306 + 380 μM free biotin; (C) : HLA-DR1 + 10 μM HA-Bio306 +

600 µM HA 306-318; (D) : HLA-DR1 + 10 µM HA-Bio306. Samples were incubated as described in Figure 3, added to 50 µL of a 1:1 slurry of streptavidin-agarose (Bethesda Research Labs) in TTBS (0.02 M Tris, pH 7.5, 0.5 M NaCl, 0.2% Tween-20), and further incubated at room temperature for 3 hours. The samples were washed four times with 1 ml of TTBS and the pellets counted in a gamma counter.

(b) SDS-PAGE of precipitated HLA-DR1. Lanes 1-2: HLA-DR1 control loaded directly on gel; lane 1 exposed 12 hrs, lane 2 exposed 70 hours. Lane 3: precipitation A; lane 4: precipitation B; lanes 5 and 6: precipitation D. Pellets from samples A, B, and D were boiled in SDS-PAGE sample buffer and the supernatants were electrophoresed. The gels were fixed, dried and exposed as in Figure 2 for 70 hours.

DISCUSSION

The work presented here indicates that an antigenic peptide can be biotinylated and retain both the affinity for the MHC molecule to which it is restricted and to the proteins avidin and streptavidin. The assays described should allow an exploration of conditions which may lead to an increase in the peptide binding activity of the HLA-DR1 preparations. In addition, these experiments open up the possibility of further exploring the very slow kinetics of the MHC peptide interaction, starting from a well defined set of molecules. It is not yet clear if the observed kinetics are due to slow exchange reactions of peptides already bound to the MHC, or if the slow rates reflect the complexity of the intrinsic peptide binding mechanism (Buus et al., 1986, Sadegh-Nasseri & McConnell, 1989). At the present time it is possible to demonstrate that between 10 and 20% of the HLA-DR1 can be complexed with peptides under the incubation conditions used here. However, low pH treatment of mouse class II MHC molecules has been reported to increase the percentage of active protein molecules (Buus et al., 1988). The increases in activity were irreproducible, most likely due to the denaturation or aggregation of protein. It will be important therefore to find some appropriate compromise between the dissociation of endogenously bound peptides and the irreversible denaturation of the HLA-DR1 molecule. Such conditions may allow one to isolate HLA completely free of bound peptides.

The biotin-avidin interaction has been used previously for the isolation of receptors, in immunoassays, and as an alternative to radioactive labeling in many molecular biological techniques. Although not presented here, two methods have been developed for the release of affinity purified molecules from streptavidin- or avidin-agarose, without denaturation of the protein of interest. The guanido analog of biotin, 2-imino-biotin (Hoffman et al., 1980; Orr, 1981), is bound to avidin at higher pH values (>9.5), and can be eluted at pH values of 4 or lower. The alternative is to use a biotin analog with a cleavable linker arm containing a disulfide bond. Proteins can then be released under mildly reducing conditions. The ability to purify MHC-peptide complexes provides a well defined starting point for crystallization and biochemical experiments on the mechanism of peptide binding.

REFERENCES

Allen, P., Matsueda, GH., Evans, R., Dunbar, J. Marshall, G. and Unanue, E., 1987, Identification of the T-cell and Ia contact residues of a T-cell antigenic epitope, *Nature* 327:713.

Babbitt, B.P., Allen, P.M., Matsueda, G., Haber, E. and Unanue, E.,1985, Binding of immunogenic peptides to Ia histocompatibility molecules, *Nature* 317:35Bjorkman, P.J.,

Saper, M.A., Samraoui, B., Bennett, W.S., Strominger, J.L. and Wiley, D.C., 1987a, Structure of the human class I histocompatibility antigen HLA-A2, *Nature* 329:506.

Bjorkman, P.J., Saper, M.A., Samraoui, B., Bennett, W.S., Strominger, J.L. and Wiley, D.C., 1987b, The foreign antigen binding site and T cell recognition regions of class I histocompatibility antigens, *Nature* 329:512.

Bolton, A.E., 1986, Comparative methods for the radiolabeling of peptides, *Methods in Enzymology* 124:18.

Brown, J.H., Jardetzky, T., Saper, M.A., Samraoui, B., Bjorkman, P.J. and Wiley, D.C., 1988, A hypothetical model of the foreign antigen binding site of class II histocompatibility molecules, *Nature* 332:845.

Buus, S., Sette, A., Colon, S.M., Jenis, D.M. and Grey, H.M., 1986, Isolation and characterization of antigen-Ia complexes involved in T cell recognition, *Cell* 47:1071.

Buus, S., Sette, A., Colon, S.M. and Grey, H.M., 1988, Autologous peptides constitutively occupy the antigen binding site on Ia, *Science* 242:1045.

Chen, B. and Parham P., 1989, Direct binding of influenza peptides to class I HLA molecules, *Nature* 337:743.

Gorga, J.C., Horejsi, V., Johnson, D.R., Raghupathy, R. and Strominger, J.L., 1987, Purification and characterization of class II histocompatibility antigens from a homozygous human B cell line, *J. Biol. Chem.* 262:16087.

Hoffman, K., Wood, S.W., Brinion, C.C., Moutibeller, J.A. and Finn, F.M., 1980. Iminobiotin affinity columns and their application to retrieval of strptavidin, *Proc. Natl. Acad. Sci. USA* 77:4666.

Jardetzky, T., Gorga, J., Busch, R., Rothbard, J., Strominger, J. and Wiley, D., 1989, manuscript in preparation.

Lamb, J., Eckels, D., Lake, P., Woody, J. and Green, M., 1982a, Human T cell clones recognize chemically synthesized peptide of influenza hemagglutinin, *Nature* 300:66.

Lamb, J., Eckels, D., Phelan, M., Lake, P. andWoody, J., 1982b, Antigen specific T cell clones: viral antigen specificity of influenza virus immune clones, *J. Immunol.* 128:1428.

Lanzevecchia, A., 1988, Clonal sketches of the immune response, *EMBO J.* 7:2945.

Moore, M.W., Carbone, F.R. and Bevan, M.J., 1988, Introduction of soluble protein into the class I pathway of antigen processing and presentation, *Cell* 54:777.

Morrison, L.A., Lukacher, A.E., Braciale, V.L., Fan , D.D. and Braciale, T.J., 1986, Differences in antigen presentation to MHC class I and class II restricted influenza virus specific cytolytic T lymphocyte clones, *J. Exp. Med.* 163:903.

Orr, G., 1981, The use of the 2-iminobiotin-avidin interaction for the selective retrieval of labeled plasma membrane components, *J. Biol. Chem.* 256:761.

Rothbard, J.B., Lechler, R.I., Howland, K., Bal, V., Eckels, D.D., Sekaly, R., Long, E., Taylor, W.R. and Lamb, J., 1988, Structural model of HLA-DR1 restricted T cell antigen recognition, *Cell* 52:515.

Sadegh-Nasseri, S. and McConnell, H.M., 1989, A kinetic intermediate in the reaction of an antigenic peptide and I-E^k, *Nature* 337:274.

Schwartz, R.H., 1985, T-lymphocyte recognition of antigen in association with gene products of the major histocompatibility complex, *Ann. Rev. Immun.* 3:237.

Sette, A., Buus, S., Colon, S., Smith, J., Miles, C. and Grey, H.M., 1987, Structural characteristics of an antigen required for its interaction with Ia and recognition by T cells, *Nature* 328:395.

Shimonkevitz, R., Colon, S.,Kappler, J., Marrack, P. and Grey, H., 1984, Antigen recognition by H-2 restricted T cells. II. A tryptic ovalbumin peptide that substitutes for processed antigen, *J. Immunol.* 133:2067.

Watts, T.H. and McConnell, H.M., 1986, High affinity fluorescent peptide binding to I-A(d) in lipid membranes, *Proc. Natl. Acad. Sci. USA* 83:9660.

67

Ziegler, H.K. and Unanue, E.R., 1982, Decrease in macrophage antigen catabolism caused by ammonia and chloroquine is associated with inhibition of antigen presentation to T cells, *Proc. Natl. Acad. Sci. USA* 79:175.

NMR METHOD FOR PROTEIN STRUCTURE DETERMINATION IN SOLUTION

Kurt Wüthrich

Institut für Molekularbiologie und Biophysik, Eidgenössische Technische Hochschule-Hönggerberg, CH-8093 Zürich, Switzerland

INTRODUCTION

The introduction of Nuclear Magnetic Resonance as a second method for protein structure determination, besides X-ray diffraction in single crystals, has already helped to significantly increase the number of known protein structures. However, it is also of more fundamental interest, since it provides data that are in many ways complementary to those obtained from X-ray crystallography. Its use thus promises to widen our view of protein molecules with regard to a better grasp of the relations between structure and function. The complementarity of the two methods results from the facts that the time scales of the two types of measurements are widely different, and that, in contrast to the need of single crystals for diffraction studies, the NMR measurements use proteins in solution or other noncrystalline states.

The NMR method for protein structure determination (Wüthrich, 1986; 1989b) is of recent origin, and on a general level its development was closely linked with that of NMR instrumentation and NMR methodology (Wüthrich, 1989a). More specifically, the method was originally based on three elements: (i) Nuclear Overhauser enhancement (NOE) experiments enabling measurements of ^{1}H-^{1}H distances in macromolecules (Gordon and Wüthrich, 1978; Wagner and Wüthrich, 1979). (ii) An efficient technique for obtaining sequence-specific assignments of the many hundred to several thousand NMR lines in a protein (Dubs et al., 1979; Billeter et al., 1982; Wagner and Wüthrich, 1982; Wider et al., 1982). (iii) A distance geometry algorithm (Blumenthal, 1970; Crippen and Havel, 1988) specifically implemented for computing three-dimensional polypeptide structures from NMR data (Braun et al., 1981; Havel and Wüthrich, 1984). Of these three elements, ^{1}H-^{1}H distance measurements had become feasible before resonance assignments were readily available. The ensuing situation was similar to that of protein crystallography at the time before the heavy-atom derivative technique presented an avenue for solving the phase problem (Blundell and Johnson, 1976; Wüthrich et al., 1982). Both the sequential resonance assignment technique (Dubs et al., 1979) and three-dimensional polypeptide structure determination by NMR and distance geometry (Braun et al., 1981) were developed and applied before two-dimensional (2D) NMR experiments were ready for the relevant measurements with proteins. However, full realization of the potentialities of the method was achieved only with the greatly improved resolution and efficiency afforded by 2D NMR spectroscopy (Ernst et al., 1987; Wüthrich, 1986). It was therefore of great importance for the subsequent rapid progress that adaptation of existing 2D NMR experiments and the development of special new techniques for use with biomacromolecules was started immediately after the fundamental work on 2D NMR by Ernst and coworkers (Aue et al., 1976). The first 2D NMR spectra of a protein were recorded in 1977 (Nagayama et al., 1977), by 1980 2D NMR experiments could provide the information needed for protein structure determination (Anil-Kumar et al., 1980), and in 1984 the first determination of the complete structure of a globular protein in solution was completed (Williamson et al., 1985).

SURVEY OF THE NMR METHOD FOR PROTEIN STRUCTURE DETERMINATION

In present practice the data for protein structure determination are collected using 2D NMR experiments (Figure 1). 2D NMR spectra (Ernst et al., 1987; Wüthrich, 1986) are presented in a two-dimensional frequency plane, with the two frequency axes ω_1 and ω_2. The spectra of prime

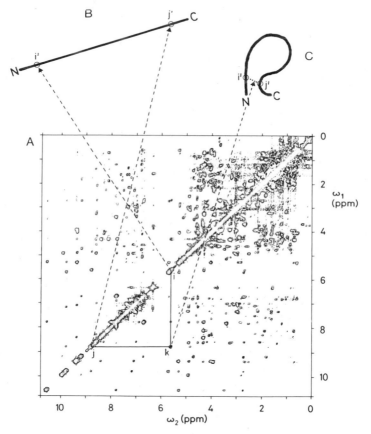

Figure 1.   Survey of the NMR method for protein structure determination in solution. (A) Contour plot of a 500 MHz ^{1}H NOESY spectrum of the protein basic pancreatic trypsin inhibitor (BPTI), with the two frequency axes ω_1 and ω_2. A cross peak k is connected by a horizontal and a vertical line with the diagonal peaks i and j. (B) Straight line representing a polypeptide chain. Two protons are identified by circles and the letters i' and j'. Two broken arrows connect the corresponding diagonal peaks i and j in the NOESY spectrum with these protons. (C) Schematic presentation of the loop structure imposed on the polypeptide chain by the fact that the protons i' and j' are separated by a distance of less than 5.0 Å, as is evidenced by the appearance of the NOESY cross peak k.

importance for work with proteins contain an array of diagonal peaks with $\omega_1 = \omega_2$, which correspond largely to the conventional one-dimensional (1D) spectrum and display the chemical shift positions of the resonance lines. In addition, there are a large number of cross peaks with $\omega_1 \neq \omega_2$. Through simple geometric patterns, as indicated in Figure 1A for the peaks labeled i,j and k, each cross peak establishes a correlation between two diagonal peaks. In 2D nuclear Overhauser enhancement spectroscopy (NOESY), which is the pivotal experiment in studies of protein structures, the cross peaks represent NOE's and indicate that the protons corresponding to the two correlated diagonal peaks are separated only by a short distance, say less than 5.0 Å. This information is used in the following way: First, sequence-specific resonance assignments must be obtained, i.e. for all the protons in the polypeptide chain the corresponding diagonal peaks must be identified. This is indicated in Figure 1A and B, by the two dashed arrows linking the protons i' and j' with their diagonal positions i and j in the NOESY spectrum. Each NOESY cross peak then tells us that two protons in known locations along the polypeptide chain are separated by a distance of less than approximately 5.0 Å in the three-dimensional protein structure. Since the overall length of an extended polypeptide chain with n residues is n*3.5 Å, the NOE's may thus impose stringent constraints on the polypeptide conformation. In Figure 1C this is schematically indicated by the formation of a loop through the near approach of the two protons i' and j', which are correlated via the NOESY cross peak k. Typically, the NOESY spectra of proteins contain many hundred cross peaks (Figure 1A), which shows that the polypeptide chain in the three-dimensional structure contains a large number of loops of the type shown in Figure 1C. For the structural interpretation of these data mathematical techniques, in particular distance geometry, are available to identify those three-dimensional arrangements of the linear polypeptide chain which satisfy all the experimental constraints (Havel and Wüthrich, 1984; Braun and Go, 1985; Wüthrich, 1986; Braun, 1987). Figures 2 and 3 show two structures determined using this approach.

Figure 2 shows a superposition for minimal root mean square deviation (RMSD) of a group of structures for an undecapeptide segment of the polypeptide hormone glucagon. Each of these structures is the result of a distance geometry calculation using the experimental NMR constraints (Braun et al., 1983). The different structures were obtained by repeating the same calculation with different, randomly generated starting conditions. The experimental constraints used represented only an incomplete set of intramolecular interatomic connections, and each individual constraint was defined such that it described the allowed distance range from the van der Waals contact distance of 2.0 Å to an upper limit derived from the NOE's, rather than a

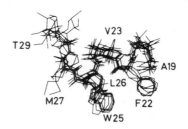

Figure 2. Three-dimensional structure of the segment consisting of the residues 19-29 in the polypeptide hormone glucagon bound to the lipid-water interphase on the surface of dodecylphosphocholine micelles. 6 conformers have been superimposed for minimal RMSD, each of which was computed with a metric matrix distance geometry algorithm from the same NMR data but with different starting conditions. The side chains of the residues Gln 20, Asp 21, Gln 24 and Asn 28 are not shown, since they were not constrained by the experimental data. (Reproduced from Braun et al., 1983).

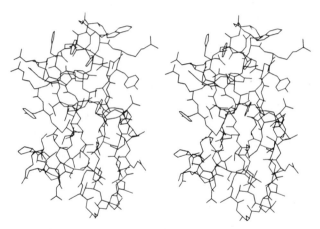

Figure 3. Stereo view of the three-dimensional structure of the α-amylase inhibitor Tendamistat determined from NMR measurements in aqueous solution. All bonds connecting heavy atoms are shown for the residues 5-73 (Kline et al., 1988).

precise value for the distance. Therefore, the individual structures are similar but not identical. The result of a structure determination by NMR and distance geometry is commonly represented by a group of conformers, each of which represents a solution to the geometric problem of fitting the polypeptide chain to the ensemble of all experimental constraints. The spread among the different structures, usually expressed by the average of the pairwise RMSD's (Braun et al., 1983; Havel and Wüthrich, 1984; Wüthrich, 1986; Braun, 1987), gives an indication of the precision of the structure determination. Figure 3 shows an individual conformer of Tendamistat taken from a group of structures such as that shown in Figure 2 (Kline et al., 1988).

Similar to the practice with protein crystal structures (e.g., Hendrickson and Konnert, 1981), the solution conformations resulting from distance geometry calculations are now routinely subjected to energy minimization, either by molecular dynamics techniques (Brünger et al., 1986; Kaptein et al., 1988) or by molecular mechanics calculations (Williamson et al., 1985; Billeter et al., 1989a). Although these restrained refinement procedures do not usually lead to significant changes in the molecular geometry obtained by distance geometry, they ensure proper covalent bond length and bond angles, remove bad nonbonding interatomic contacts, and overall result in a "reasonable" conformational energy. The mathematical principles and computational techniques for the structural analysis of NMR data with proteins are at present the subject of intensive research efforts, and procedures that are fundamentally different from both distance geometry and molecular dynamics have also been proposed (Duncan et al., 1987; Billeter et al., 1987). For completeness' sake it may also be added that empirical pattern recognition methods (Billeter et al., 1982; Wüthrich et al., 1984; Wüthrich, 1986) prove to be highly successful as an efficient means for the identification of regular polypeptide secondary structures in proteins.

NMR ASSIGNMENTS FOR BIOPOLYMERS BY HOMONUCLEAR AND HETERONUCLEAR EXPERIMENTS

The decisive role of sequence-specific resonance assignments is readily appreciated from inspection of Figure 1: Without the resonance assignments the size and the location in the amino acid sequence of the loop indicated by the cross peak k would not be defined, and hence there would be no way to determine a molecular structure from the information that can be derived from NOE's (Wüthrich et al., 1982; Wüthrich, 1986).

Besides NOESY, obtaining resonance assignments in a protein makes use of NMR experiments which delineate scalar, through-bond relations between nuclear spins, for example, 2D correlated spectroscopy (COSY), relayed coherence transfer spectroscopy (RELAYED-COSY), total correlation spectroscopy (TOCSY), multiple quantum spectroscopy, and a catalogue of far over 100 variants of these basic experiments (Ernst et al., 1987). In proteins these techniques can delineate networks of scalar spin-spin couplings between hydrogen atoms that are separated from each other by three or less covalent bonds. In small chemical compounds, complete NMR assignments can usually be obtained from such networks of scalar spin-spin couplings. This applies also for the individual amino acids. In contrast, in polypeptide chains the elucidation of sequence-specific resonance assignments is not straightforward because typical proteins contain multiple copies of individual ones of the 20 common amino acids. The identification of the groups of scalar-coupled spins belonging to individual amino acid residues is therefore in general not sufficient to define a unique sequence location. The problem was solved using the sequential assignment strategy (Billeter et al., 1982; Wagner and Wüthrich, 1982; Wider et al., 1982) illustrated in Figure 4. The dotted lines indicate the ^{1}H-^{1}H connectivities within the amino acid residues, which can be established via scalar spin-spin couplings. In the example of Figure 4, these spin systems unambiguously identify the amino acid residues alanine and valine (Wüthrich, 1986). Relations between protons in sequentially neighboring amino acid residues are established by "sequential" NOE's manifesting close approach between αCH(i) and NH(i+1), or NH(i) and NH(i+1), or both. In Figure 4, we obtain the result that the protein studied contains a dipeptide segment Ala-Val. This dipeptide is then matched against the independently known amino acid sequence. If the latter contains Ala-Val only once, the assignment problem is solved. Otherwise, to distinguish between the different Ala-Val sites, tri- or tetrapeptide segments including Ala-Val must be identified by NMR and matched against the amino acid sequence. In globular proteins there is only a small probability that identical tri- or tetrapeptide segments occur repeatedly (Wüthrich, 1986), so that this assignment strategy is quite generally applicable.

The sequential assignment approach has so far been successfully applied with more than 100 proteins in the size range up to a molecular weight of approximately 15,000. In favorable cases, somewhat bigger proteins may also be assigned, but usually the technical difficulties increase significantly with increasing size beyond approximately 12,000. Thereby it is quite typical that spectral overlap and line broadening impose limitations on the extent to which the scalar coupling networks can be unravelled, while the sequential NOE connectivities are still

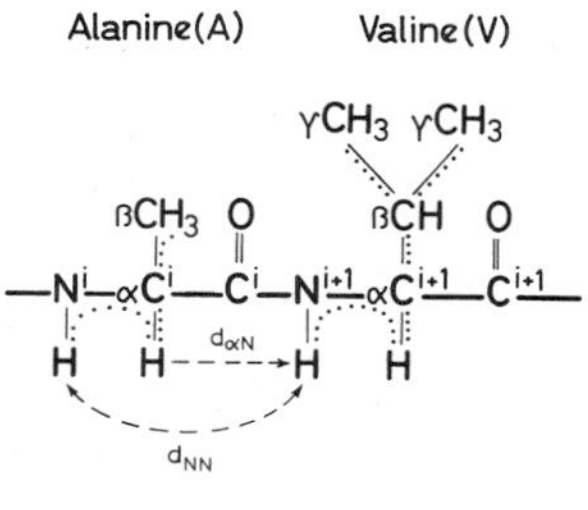

Figure 4. Illustration to the description of sequential resonance assignments. In the dipeptide segment -Ala-Val- the dotted lines indicate ^{1}H-^{1}H relations which can be established by the scalar spin-spin couplings observed, for example, in COSY. The broken arrows indicate relations between protons in sequentially neighboring residues, which can be established by NOESY cross peaks manifesting short sequential distances $d_{\alpha N}$ (between C$^\alpha$H and the amide proton of the following residue) and d_{NN} (between the amide protons of neighboring residues).

amenable for complete analysis. For such cases a "main-chain-directed strategy for the assignment of ^{1}H NMR spectra of proteins" was proposed (Englander and Wand, 1987), which exploits knowledge on typical patterns of NOE connectivities in regular polypeptide secondary structures (Billeter et al., 1982; Wüthrich et al., 1984) and would not depend on extensive identifications of the amino acid side chain spin systems. This approach has attractive traits with regard to computer-aided, automated assignment procedures (e.g., Billeter et al., 1988). It remains to be seen if it is really applicable for bigger proteins than the conventional sequential assignment strategy, which has also been adapted for situations where only incomplete spin system identifications are available (Wüthrich 1983; 1986).

Isotope labelling with ^{2}H, ^{13}C or ^{15}N has long been recognized as a quite obvious means of either simplifying complex NMR spectra of proteins (e.g., Markley et al., 1986), or identifying distinct amino acid residues or groups of amino acid residues (e.g., LeMaster and Richards, 1985; Griffey et al., 1986). Following recent advances in both isotope labelling techniques and NMR experiments, there are now strong indications that the combined use of isotope labelling and the sequential assignment strategy will enable detailed structural studies of larger proteins, extending the upper size limit for structure determinations in solution to 30,000, or perhaps even beyond. Three approaches appear particularly promising for applications in different situations: (i) Residue-selective labelling with ^{15}N and/or ^{13}C combined with homonuclear ^{1}H NMR using heteronuclear filtering techniques (Griffey and Redfield, 1987; Senn et al., 1987; Otting et al., 1986; Wörgötter et al., 1986, 1988). (ii) Nonselective labelling with ^{15}N combined with three-dimensional heteronuclear NMR experiments (Fesik et al., 1989; Marion et al., 1989; Weber et al., 1989). (iii) Random fractional deuteration combined with homonuclear ^{1}H NMR (LeMaster and Richards, 1988).

Sequence-specific resonance assignments obtained with the aforementioned techniques do not usually include stereospecific assignments of prochiral groups of protons, so that the input for the structure calculations has to use pseudoatoms (Wüthrich et al., 1983), with concomitant loss of precision of the input data (Wüthrich, 1986). In the past, stereospecific assignments were obtained in some special situations (Senn et al., 1984; Zuiderweg et al., 1985), and more recently more systematic approaches were introduced (e.g., Hyberts et al., 1987; Wagner et al., 1987; Arseniev et al., 1988) and also largely automated (Weber et al., 1988; Guntert et al., 1989). As it has been clearly demonstrated that stereospecific assignments can significantly improve the precision of a protein structure determination (e.g., Kline et al., 1986, 1988; Guntert et al., 1989; Driscoll et al., 1989), they are commonly included in current structure determinations by NMR. Support for this part of the spectral analysis may come from improved NMR experiments as well as from the use of nonrandom isotope labelling techniques (Senn et al., 1989).

COMPLEMENTARY INFORMATION FROM NMR IN SOLUTION AND X-RAY DIFFRACTION IN SINGLE CRYSTALS

Since protein structure determinations by NMR or by X-ray diffraction can be performed completely independently, one now has, for the first time, a basis for meaningful comparisons of corresponding structures in single crystals and in noncrystalline states. This is highly relevant, since the solution conditions for NMR studies can often be chosen so as to coincide closely with the natural, physiological environment of the protein (Wüthrich, 1986). There are presently two well-documented examples where different molecular architectures were reported for the same protein in single crystals and under nondenaturing solution conditions. One of these is the polypeptide hormone glucagon (Figure 2), which appears to have a pronounced tendency for adopting different conformations in different environments (Sasaki et al., 1975;

Braun et al., 1983). Metallothionein-2 from rat liver is a protein showing substantial, fundamental differences between the global molecular architectures seen in aqueous solution (Schultze et al., 1988) and in a recently published crystal structure (Furey et al., 1987). In contrast to the forementioned examples, the α-amylase inhibitor Tendamistat (Figure 3) is a protein for which the global molecular architecture is the same in crystals and in solution (Kline et al., 1986; Pflugrath et al., 1986), and similar behavior was reported for other globular proteins (e.g., Clore et al., 1987; Wagner et al., 1987). Nonetheless, even for these globular proteins, detailed comparisons using refined structures in crystals and in solution (Kline et al., 1988; Pflugrath et al., 1986; Billeter et al., 1989b) reveal a wealth of more subtle differences. This may involve individual amino acid residues or groups of spatially adjacent residues, and are mostly on or near the protein surface. This is again of special interest, since the molecular surface is generally most directly related with the functional properties of the molecules, and often includes the active site.

ACKNOWLEDGMENTS

The author's laboratory is supported by the Schweizerischer Nationalfonds (project 31.25174.88) and special grants of the ETH Zürich. I thank Mr. R. Marani for the careful processing of the typescript.

REFERENCES

Anil-Kumar, Wagner, G., Ernst, R.R. and Wüthrich, K., 1980, Studies of J-connectivities and selective ^{1}H-^{1}H Overhauser effects in H_2O solutions of biological macromolecules by two-dimensional NMR experiments, *Biochem. Biophys. Res. Comm.* 96:1156.

Arseniev, A., Schultze, P., Wörgötter, E., Braun, W., Wagner, G., Vasák, M., Kägi, J.H.R. and Wüthrich, K., 1988, Three-dimensional structure of rabbit liver [Cd^7]-metallothionein-2a in aqueous solution determined by nuclear magnetic resonance, *J. Mol. Biol.* 201:637.

Aue, W.P., Bartholdi, E. and Ernst, R.R., 1976, Two-dimensional spectroscopy: application to NMR, *J. Chem. Phys.* 64:2229.

Billeter, M., Braun, W. and Wüthrich, K., 1982, Sequential resonance assignments in protein ^{1}H nuclear magnetic resonance spectra, computation of sterically allowed proton-proton distances and statistical analysis of proton-proton distances in single crystal protein conformations, *J. Mol. Biol.* 155:321.

Billeter, M., Basus, V.J. and Kuntz, I.D., 1988, A program for semi-automatic sequential resonance assignments in protein ^{1}H nuclear magnetic resonance spectra, *J. Magn. Res.* 76:400.

Billeter, M., Havel, T.F. and Wüthrich, K., 1987, The ellipsoid algorithm as a method for the determination of polypeptide conformations from experimental distance constraints and energy minimization, *J. Comp. Chem.* 8:132.

Billeter, M., Schaumann, Th., Braun, W. and Wüthrich, K., 1989a, Restrained energy refinement with two different algorithms and force fields of the structure of the α-amylase inhibitor Tendamistat determined by NMR in solution, *Biopolymers*, in press.

Billeter, M., Kline, A.D., Braun, W., Huber, R. and Wüthrich, K., 1989b, Comparison of the high-resolution structures of the α-amylase inhibitor Tendamistat determined by NMR in solution and by X-ray diffraction in single crystals, *J. Mol. Biol.* 206:677.

Blumenthal, L.M., 1970, "Theory and application of distance geometry," Chelsea, New York.

Blundell, T.L. and Johnson, L.N., 1976, "Protein crystallography," Academic Press, New York.

Braun, W., 1987, Distance geometry and related methods for protein structure determination from NMR data, *Quart. Rev. Biophys.* 19:115.

Braun, W. and Go, N., 1985, Calculation of protein conformations by proton-proton distance constraints, a new efficient algorithm, *J. Mol. Biol.* 186:611.

Braun, W., Bösch, C., Brown, L.R., Go, N. and Wüthrich, K., 1981, Combined use of proton-proton Overhauser enhancements and a distance geometry algorithm for determination of polypeptide conformations. Application to micelle-bound glucagon, *Biochim. Biophys. Acta* 667:377.

Braun, W., Wider, G., Lee, K.H. and Wüthrich, K., 1983, Conformation of glucagon in a lipid-water interphase by ^{1}H nuclear magnetic resonance, *J. Mol. Biol.* 169:921.

Brünger, A.T., Clore, G.M., Gronenborn, A.M. and Karplus, M., 1986, Three-dimensional structure of proteins determined by molecular dynamics with interproton distance restraints: Application to crambin, *Proc. Natl. Acad. Sci. USA* 83:3801.

Clore, G.M., Gronenborn, A.M., James, M.N.G., Kjaer, M., McPhalen, C.A. and Poulsen, F.M., 1987, Comparison of the solution and X-ray structures of barley serine proteinase inhibitor-2, *Protein Engineering* 1:313.

Crippen, G.M. and Havel, T., 1988, "Distance geometry and molecular conformation," Wiley, New York.

Driscoll, P.C., Gronenborn, A.M. and Clore, G.M., 1989, The influence of stereospecific assignments on the determination of three-dimensional structures of proteins by NMR spectroscopy. Application to the sea anemone protein BDS-I, *FEBS Lett.* 243:223.

Dubs, A., Wagner, G. and Wüthrich, K., 1979, Individual assignments of amide proton resonances in the proton NMR spectrum of the basic pancreatic trypsin inhibitor, *Biochim. Biophys. Acta* 577:177.

Duncan, B., Buchanan, B.G., Hayes-Roth, B., Lichtarge, O., Altman, R., Brinkley, J., Hewett, M., Cornelius, C. and Jardetzky, O., 1987, PROTEAN: A new method for deriving solution structures of proteins, *Bull. Magn. Reson.* 8:111.

Englander, S.W. and Wand, A.J., 1987, Main-chain-directed strategy for the assignment of ^{1}H NMR spectra of proteins, *Biochemistry* 26:5953.

Ernst, R.R., Bodenhausen, G., Wokaun, A., 1987, "Principles of nuclear magnetic resonance in one and two dimensions," Clarendon Press, Oxford.

Fesik, S.W., Gampe, R.T., Jr., Zuiderweg, E.R.P., Kohlbrenner, W.E. and Weigl, D., 1989, Heteronuclear three-dimensional NMR spectroscopy applied to CMP-KDO synthetase (27.5 kD), *Biochem. Biophys. Res. Comm.* 159:842.

Furey, W.F., Robbins, A.H., Clancy, L.L., Winge, D.R., Wang, B.C. and Stout, C.D., 1986, Crystal structure of Cd,Zn metallothionein, *Science* 231:704.

Gordon, S.L. and Wüthrich, K., 1978, Transient proton-proton Overhauser effects in horse ferrocytochrome c, *J. Amer. Chem. Soc.* 100:7094.

Griffey, R.H. and Redfield, A.G., 1987, Proton-detected heteronuclear edited and correlated nuclear magnetic resonance and nuclear Overhauser effect in solution, *Quart. Rev. Biophys.* 19:51.

Griffey, R.H., Redfield, A.G., McIntosh, L.P., Oas, T.G. and Dahlquist, F.W., 1986, Assignment of proton amide resonances of T4 lysozyme by ^{13}C and ^{15}N multiple isotopic labeling, *J. Amer. Chem. Soc.* 108:6816.

Güntert, P., Braun, W., Billeter, M. and Wüthrich, K., 1989, Automated stereospecific ^{1}H-NMR assignments and their impact on the precision of protein structure determinations in solution, *J. Amer. Chem. Soc.* 111:3997.

Havel, T.F. and Wüthrich, K., 1984, A distance geometry program for determining the structures of small proteins and other macromolecules from nuclear magnetic resonance measurements of intramolecular ^{1}H-^{1}H proximities in solution, *Bull. Math. Biol.* 46:673.

Hendrickson, W.A. and Konnert, J.H., 1981, Stereochemically restrained crystallographic least-squares refinement of macromolecular structures, *in:* "Biomolecular Structure,

Conformation, Function and Evolution," Vol.1, R. Srinivasan, N. Yathindra and E. Subramanian, eds., Pergamon Press, New York.

Hyberts, S.G., Märki, W. and Wagner, G., 1987, Stereospecific assignments of side-chain protons and characterization of torsion angles in eglin c, *Eur. J. Biochem.* 164:625.

Kaptein, R., Boelens, R., Scheek, R.M. and van Gunsteren, W.F., 1988, Protein Structures from NMR, *Biochemistry* 27:5389.

Kline, A.D., Braun, W. and Wüthrich, K., 1986, Studies by ^{1}H nuclear magnetic resonance and distance geometry of the solution conformation of the α-amylase inhibitor Tendamistat, *J. Mol. Biol.* 189:377.

Kline, A.D., Braun, W. and Wüthrich, K., 1988, Determination of the complete three-dimensional structure of the α-amylase inhibitor Tendamistat in aqueous solution by nuclear magnetic resonance and distance geometry, *J. Mol. Biol.* 204:675.

LeMaster, D.M. and Richards, F.M., 1985, ^{1}H-^{15}N heteronuclear NMR studies of *Escherichia coli* thioredoxin in samples isotopically labeled by residue type, *Biochemistry* 24:7263.

LeMaster, D.M. and Richards, F.M., 1988, NMR sequential assignment of *Escherichia coli* thioredoxin utilizing random fractional deuteration, *Biochemistry* 27:142.

Marion, D., Kay, L.E., Sparks, S.W., Torchia, D.A. and Bax, A., 1989, Three-dimensional heteronuclear NMR of ^{15}N-labeled proteins, *J. Amer. Chem. Soc.* 111:1515.

Markley, J.L., Putter, I. and Jardetzky, O., 1968, High-resolution nuclear magnetic resonance spectra of selectively deuterated staphylococcal nuclease, *Science* 161:1249.

Nagayama, K., Wüthrich, K., Bachmann, P. and Ernst, R.R., 1977, Two-dimensional J-resolved ^{1}H NMR spectroscopy of biological macromolecules, *Biochem. Biophys. Res. Comm.* 78:99.

Otting, G., Senn, H., Wagner, G. and Wüthrich, K., 1986, Editing of 2D ^{1}H NMR spectra using X half-filters. Combined use with residue-selective ^{15}N labeling of proteins, *J. Magn. Res.* 70:500.

Pflugrath, J.W., Wiegand, G., Huber, R. and Vértesy, L., 1986, Crystal structure determination, refinement and the molecular model of the α-amylase inhibitor HOE 467A, *J. Mol. Biol.* 189:383.

Sasaki, K., Dockevill, S., Ackmiak, D.A., Tickle, I.J. and Blundell, T.L., 1975, X-ray analysis of glucagon and its relationship to receptor binding, *Nature* 257:751.

Schultze, P., Wörgötter, E., Braun, W., Wagner, G., Vasák, M., Kägi, J.H.R. and Wüthrich, K., 1988, The conformation of [Cd$_7$]-metallothionein-2 from rat liver in aqueous solution determined by nuclear magnetic resonance, *J. Mol. Biol.* 203:251.

Senn, H., Billeter, M. and Wüthrich, K., 1984, The spatial structure of the axially bound methionine in solution conformations of horse ferrocytochrome c and pseudomonas aeruginosa ferrocytochrome c-551 by ^{1}H NMR, *Eur. Biophys. J.* 11:3.

Senn, H., Otting, G. and Wüthrich, K., 1987, Protein structure and interactions by combined use of sequential NMR assignments and isotope labeling, *J. Amer. Chem. Soc.* 109:1090.

Senn, H., Werner, B., Messerle, B.A., Weber, C., Traber, R. and Wüthrich, K., 1989, Stereospecific assignment of the methyl ^{1}H NMR lines of valine and leucine in polypeptides by nonrandom ^{13}C labelling, *FEBS Lett.* 249:113.

Wagner, G. and Wüthrich, K., 1979, Truncated driven nuclear Overhauser effect (TOE), a new technique for studies of selective [1]H-[1]H Overhauser effects in the presence of spin diffusion, *J. Magn. Reson.* 33:675.

Wagner, G. and Wüthrich, K., 1982, Sequential resonance assignments in protein ^{1}H nuclear magnetic resonance spectra. Basic pancreatic trypsin inhibitor, *J. Mol. Biol.* 155:347.

Wagner, G., Braun, W., Havel, T.F., Schaumann, Th., Go, N. and Wüthrich, K., 1987, Protein structures in solution by nuclear magnetic resonance and distance geometry. The polypeptide fold of the basic pancreatic trypsin inhibitor determined using two different algorithms, DISGEO and DISMAN, *J. Mol. Biol.* 196:611.

Weber, P.L., Morrison, R. and Hare, D., 1988, Determining stereo-specific ^{1}H NMR assignments from distance geometry calculations, *J. Mol. Biol.* 204:483.

Weber, P.L. and Mueller, L., 1989, Use of ^{15}N labeling for automated three-dimensional sorting of cross peaks in protein 2D NMR spectra, *J. Magn. Res.* 81:430.

Wider, G., Lee, K.H. and Wüthrich, K., 1982, Sequential resonance assignments in protein ^{1}H nuclear magnetic resonance spectra. Glucagon bound to perdeuterated dodecylphosphocholine micelles, *J. Mol. Biol.* 155:367.

Williamson, M.P., Havel, T.F. and Wüthrich, K., 1985, Solution conformation of the proteinase inhibitor IIA from bull seminal plasma by ^{1}H nuclear magnetic resonance and distance geometry, *J. Mol. Biol.* 182:295.

Wörgötter, E., Wagner, G. and Wüthrich, K., 1986, Simplification of two-dimensional ^{1}H NMR spectra using an X-filter, *J. Amer. Chem. Soc.* 108:6162.

Wörgötter, E., Wagner, G., Vasák, M., Kägi, J.H.R. and Wüthrich, K., 1988, Heteronuclear filters for two-dimensional ^{1}H NMR. Identification of the metal-bound amino acids in metallothionein and observation of small heteronuclear long-range couplings, *J. Amer. Chem. Soc.* 110:2388.

Wüthrich, K., 1983, Sequential individual resonance assignments in the ^{1}H NMR spectra of polypeptides and proteins, *Biopolymers* 22:131.

Wüthrich, K., 1986, "NMR of proteins and nucleic acids," Wiley, New York.

Wüthrich, K., 1989a, The development of nuclear magnetic resonance spectroscopy as a technique for protein structure determination, *Accts. Chem. Res.* 22:36.

Wüthrich, K., 1989b, Protein structure determination in solution by NMR spectroscopy, *Science* 243:45.

Wüthrich, K., Wider, G., Wagner, G. and Braun, W., 1982, Sequential resonance assignments as a basis for determination of spatial protein structures by high resolution proton nuclear magnetic resonance, *J. Mol. Biol.* 155:311.

Wüthrich, K., Billeter, M. and Braun, W., 1983, Pseudo-structures for the 20 common amino acids for use in studies of protein conformations by measurements of intramolecular proton-proton distance constraints with nuclear magnetic resonance, *J. Mol. Biol.* 169:949.

Wüthrich, K., Billeter, M. and Braun, W., 1984, Polypeptide secondary structure determination by nuclear magnetic resonance observation of short proton-proton distances, *J. Mol. Biol.* 180:715.

Zuiderweg, E.R.P., Boelens, R. and Kaptein, R., 1985, Stereospecific assignments of ^{1}H-NMR methyl lines and conformation of valyl residues in the *lac* repressor headpiece, *Biopolymers* 24:601.

DETERMINATION OF STRUCTURAL UNCERTAINTY FROM NMR AND OTHER DATA:

The *Lac* Repressor Headpiece

Russ B. Altman, Ruth Pachter and Oleg Jardetzky

Stanford Magnetic Resonance Laboratory, Stanford University Medical Center
Stanford, CA 94305-5055, USA

INTRODUCTION

The static high resolution structures determined by x-ray crystallography have provided most of our structural information about protein structure (Blundell and Johnson, 1972). While these structures have provided a wealth of data about common motifs and folding patterns within proteins, there remains an important unresolved issue: is the set of crystallizable proteins representative of all proteins, or are we necessarily seeing a biased sample whose motifs and folding patterns are themselves part of the necessary determinants for crystallization? It is possible, perhaps likely, that there are important protein structural motifs that have not been observed in crystalline structures precisely because the properties of the motifs (for example, in the *lac*-repressor the high degree of hinge motion (Pilz et al., 1980) makes crystallization difficult or impossible. These considerations have lead many investigators to techniques such as high resolution NMR spectroscopy to study the structures of proteins in solution (for summaries, see Jardetzky and Roberts, 1981; Wüthrich, 1986). It is hoped that these studies will furnish complementary data about the structure and dynamics of proteins.

The results of NMR and other solution studies of protein structure are important not only to understand the solution activity of protein structures, but are also critical to tasks of protein structure design. Most protein design efforts to date have been based on the perturbation of known structural motifs in order to produce related structures with slightly different structural or functional characteristics (Sauer et al., 1986; Lim and Sauer, 1989) Static crystallographic structural data have been used as starting points to predict the resulting structure after substitution, deletion or augmentation of the primary sequence of amino acids. These methods are likely to be sensitive to the dynamics of the starting structures: there may be important differences between making sequence changes in a relatively fixed portion of the protein (whose structures is well-determined by the available data) versus making changes in a portion of the protein which is more uncertain (because the structure is mobile or because there is simply not enough data to uniquely determine the configuration of atoms). It is therefore important that studies of protein structure in solution provide information about the relative certainty in the positions of the atoms within a protein molecule.

Over the last decade, a number of methods for the determination of protein structure from non-crystallographic experimental data have been reported (Braun and Go, 1985; Havel and Wüthrich et al., 1984; Williamson et al., 1985; Gunsteren and Karplus, 1982, Duncan et al.,

1986). Each of these methods makes use of the distance information inherent in NMR nuclear Overhauser enhancement (NOE) measurements to deduce the structure of proteins. For understandable reasons, early methods were modelled on crystallographic refinement techniques in which a target function optimization is performed. Unfortunately, these techniques often require data of high precision in order to consistently converge; optimizations based on highly uncertain data are subject to problems of multiple minima or gradient search difficulties. Crystallographic data is generally of high abundance and precision, but many NMR data sets simply do not contain information of sufficient content or precision to allow straightforward application of optimization techniques. In general, information must be added to these data sets in order to make them suitable for these techniques. One source of information is the addition of energy equations to describe the interactions between bonded and non-bonded atoms, the NOE constraints then become pseudo-energy terms which have very strong guiding effects over the minimization or simulation (Clore et al., 1986). A second source of information is provided by very precise interpretation of the NOE measurement (Williamson et al., 1985). If an NOE can be taken as a well-calibrated distance with little uncertainty, the constraining power on a protein structure becomes much greater. Therefore, there have been many efforts to more precisely interpret the distance between protons implied by the observation of an NOE between them. Each of these information sources have indeed been added to NMR data sets and the result is uniform and quite striking: the structures determined by NMR *using additional information or not* are in excellent agreement about the backbone fold of most proteins. However, the *detailed packing* of sidechains is more sensitive to the sources of information used. In this paper, we review the work on the *lac* repressor headpiece and report a comparison of the structure of this protein determined by three different methods, including a method that allows simultaneous determination of structure as well as the uncertainty in the structure. We show that the NMR data reveals areas of structural uncertainty that may have important implications for understanding the function of this protein, as well as possible strategies for modification or design of similar proteins.

THE PROTEIN

The *lac* repressor headpiece is a fragment of the 51 N-terminal amino acids of the repressor molecule. The *lac* repressor is a DNA binding protein and has been the subject of study by NMR methods for about a decade (Pilz et al., 1980). Assignments of the 2DFT NMR spectra peaks to individual protons have been reported (Ribeiro et al., 1981; Zuiderweg et al., 1983; Zuiderweg et al, 1985) . The earliest papers addressing the structure of the protein focused on location of the secondary structures which have been assigned (usually within one or two amino acids at each end) to amino acids 6-14, 16-25 and 34-45 (Zuiderweg et al., 1983 and 1985). Early structures based on the location of the helices, short covalent tether between the first two helices and a few important long range NOEs provided stylized views of a three-helix triangle similar to the helix-turn-helix motif seen in crystallized repressor molecules (Jardetzky, 1984; Zuiderweg et al., 1984; Ohlendorf and Matthews, 1983). It was observed at this time that the NMR data may have been consistent with more than one conformation and that there were probably a set of contiguous locations for each of the helices that remained consistent with the data. This observation provided the impetus for the development of the PROTEAN program described in detail below (Altman and Jardetzky, 1986; Duncan et al., 1986). Early versions of this program produced topologies of the headpiece at a gross level (equivalent to a crystallographic map of 5-9 Å resolution) that more formally circumscribed the boundaries of freedom for each of the secondary structural elements within the headpiece.

Distance geometry methods, which could produce a detailed full atomic representation of the structure based on data were first reported in 1985 (Kaptein et al., 1985; Frayman, 1985).

These structures elaborated the topology in a more detailed way, but did not always precisely agree on the detailed arrangement of the atoms. These results were troubling because the relatively sparse data set might have been consistent with other structures and there was little information on the convergence properties of these methods and whether or not a complete sample of the conformational space had been obtained.

The availability of improved NMR data sets has enabled these structural pictures to be refined. A restrained molecular dynamics structure has been obtained using both energy terms as well as pseudo-energies representing the NOE distance as a energy potential (Scheek, R. personal communication). A new structure of the *lac* repressor has been determined in full atomic detail using a method which produces an upperbound on the set of atomic positions that are compatible with the data. This method, described in detail in (Altman and Jardetzky, 1989; Altman, 1989) determines which regions of the protein are relatively well defined and which regions remain considerably uncertain. This information can be used in a number of ways: 1) to identify regions of the protein that must be studied more intensively in order to define the source of uncertainty (inadequate data vs. mobile segments of peptide) and 2) to evaluate other proposed structures to ensure that they fall within the expected range of atomic positions and 3) to identify regions that are sufficiently well suited to sequence modification.

METHODS AND RESULTS

Secondary Structure Determination

The secondary structure of the *lac*-repressor headpiece has been the subject of a number of reports (Ribeiro et al., 1981; Zuiderweg et al., 1983 and 1985). We have constructed a program, ABC, which identifies the location of α-helices and β-strands based on repeating patterns of NOEs within the NOESY spectra (Brugge et al., 1987). This program does not perform any numerical calculations, but instead analyzes the spectra for characteristic patterns of NOEs between the backbone protons within a polypeptide. For example, the protons on α-carbons of amino acid i will tend to have NOEs with the protons of the amide nitrogens of amino acid $i+3$. Similar repeat patterns can be found in β-strands. By identifying the location of these patterns, the location of secondary structures within the protein can be determined from the NMR spectra. Results obtained with the ABC program demonstrate that the precise boundaries of secondary structure location are very sensitive to the rules of interpretation that are provided to the program. It is important, therefore, to have an automated method for evaluating the spectral data in an systematic and objective way in order to evaluate the effects of modifying the rules.

In the case of the *lac*-repressor, our program identifies three α helices with boundaries at amino acids 6-13, 15-25 and 35-45 as shown in Figure 1. This is in good general agreement with the manual determination of secondary structure reported previously. The differences in the location of the beginning or end of these helices are less than 2 amino acids in all cases. Identification of the location and type of the secondary structure within a molecule allows us to use this information to simplify the calculation of initial topology. We have shown that the variation of backbone atom location in secondary structures (taken from the protein data bank) is relatively low and that an idealized secondary structure can be used for the purposes of initial placement of secondary structures (Altman and Jardetzky, 1986). Different secondary structures show a different range of variability. The backbone atoms in an α-helix deviate from ideal by an average of 0.5Å, whereas the backbone atoms of antiparallel β-strands deviate by an average of 1.5Å. This information is used to keep track of the error in our approximations.

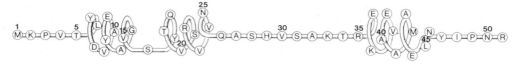

Figure 1. Secondary structure of the *lac* repressor headpiece defined by NMR. The primary structure of the *lac* repressor headpiece is shown schematically along with the location of the three α helices within the sequence.

Determination of General Topology

In order to determine the relative location of the secondary structures within the *lac* repressor, we used the solid level functionality of the PROTEAN system described in detail in (Altman et al., 1987; Brinkley et al., 1988). This set of functions allows us to model secondary structures as solid cylinders and spheres. The system establishes a global coordinate system around which one secondary structural element is fixed. For each remaining secondary structure, the system then discretely samples conformational space (usually at a resolution of 1 or 2 Å in position and 10° in orientation) and tests the compatibility of different locations with the experimental constraints. Since abstract objects such as cylinders and spheres are used to represent the elements of the protein, the experimental constraints (which are between protons at the end of amino acid sidechains) are not used directly, instead equivalent distance ranges on the backbone atoms to which they are connected are used. This results in substantial loss of information, but not so much that the value of the experimental constraints is lost. In fact, we have found that the topology of the protein molecule is clearly determined even using this abstraction of the detailed NOE information.

In the case of the *lac*-repressor, we model the three previously determined α-helices as cylinders with radius 3.5 Å and length of 1.5 Å times the number of amino acids within the helices. This represents only a core segment of the helices. The location of backbone atoms on these helices can easily be calculated. Given the positions and orientations of two helices, we can check the distance between the backbone atoms, correct for the uncertainty in the idealized approximation and test for compatibility with the abstracted distance constraint. We also check that the cylinders do not overlap in space, a rough check for van der Waals overlap. The result of these operations is a list of locations for each cylinder which describes the space of positions which it can occupy. If one location is selected from each list for each secondary structure, then one candidate conformation is generated. Using constraint satisfaction techniques described in detail in (Brinkley et al., 1988), we can reduce the size of these lists of positions (or accessible volumes) so that they are a tight upperbound on the set of true relative positions occupied by the secondary structures.

The results for the *lac* repressor are shown in Figure 2. The detailed structure of the protein is not clear, but we have generated an upperbound on the location of each secondary structure which is a useful starting point for further refinement. In order to consider the detailed locations of each atom, we move away from a discretely sampled approach. The number of possible locations for each atom is simply too large to track individually. Instead, we use a parametric representation in which we compute the mean and variance of each atom. Using the constraints (now in a form in which most or all of the experimental information is preserved) we can refine the estimates of mean and variance for each atom using probabilistically sound updating mechanisms. Table 1 illustrates how this conversion is performed.

Figure 2. The sampled structure of the *lac* repressor headpiece. Helix 1 is fixed in the coordinates system with its long axis oriented along the z-axis. Sampled accessible volumes for helices 2 and 3 are shown (top, bottom). The structure is derived using ideal secondary structure approximations and relaxed distance constraints. It represents an upperbound on the possible locations for the secondary structures. This result is used as a starting point for further refinement steps.

Table 1. Converting Sampled Locations for Helices into atomic positions for backbone atoms. The columns xp, yp, zp, ϕ, ψ, w show six coordinates used to place an idealized secondary structure in space. These have been chosen arbitrarily from a list of 348 locations found in positioning helix-3 relative to the fixed helix-1. xp, yp, zp are cartesian coordinates used to place the center of the secondary structure. ϕ, ψ and w are the Euler angles used to describe the orientation of the secondary structure. These six numbers are sampled systematically and tested by PROTEAN. They imply a unique transformation matrix which allows us to position atoms within the secondary structure in space around the fixed global coordinate system. The columns x-39, y-39 and z-39 show the coordinates of α carbon 39 for each location. α carbon 39 is at position (2.3 0 0) in the local coordinate system defined around helix-3 and this position has been transformed by each of the locations below.

xp	yp	zp	ϕ	ψ	w	x-39	y-39	z-39
4.0	-6.0	-2.0	210	90	120	2.9959	-8.5750	5.9918
6.0	-6.0	2.0	180	150	60	4.0000	-10.3000	-0.0000
4.0	-8.0	0.0	150	150	90	-1.7125	-9.4938	3.7250
0.0	-6.0	6.0	300	150	240	-1.8584	-9.2270	4.5750
6.0	-4.0	4.0	0	180	270	4.5750	-6.9959	5.9918
4.0	-6.0	4.0	210	90	150	6.0000	-6.3000	4.0000
0.0	-8.0	4.0	150	150	120	-1.2270	-7.8584	6.5750
-2.0	-8.0	2.0	150	60	150	2.8500	-9.9918	-0.0000
4.0	-8.0	0.0	180	120	90	6.9959	-7.9918	1.4250
2.0	-8.0	4.0	150	60	180	4.9959	-7.7250	-0.8500

Probabilistic Refinement

The list of locations for each secondary structure implies a set of locations for each of the constituent backbone atoms within the secondary structure. We generated a list of positions for each α carbon within a helix based on the accessible volumes for the helices. We generated the mean position and variance for each α carbon as a starting point for refinement (Table 2). α carbons that were not part of a helix were placed at the origin of the coordinate system with a large variance to account for the uncertainty in their position.

Table 2. Statistical Summary of Sampled Positions for α carbons of Helix-3. For each of the sampled cartesian locations of each α carbon in helix-3 (as illustrated in Table 1), we have calculated the mean value and covariance matrix for the coordinates. These values are a parametric summary of each atom location which is used as a starting point for refinement by the double iterated Kalman filter. The columns show (from left to right) amino acid number, x, y, z, variance of x, variance of y, variance of z, covariance of xy, covariance of xz and covariance of yz.

AA	X	Y	Z	V(x)	V(y)	V(z)	V(xy)	V(xz)	V(yz)
34	6.6308	−5.1527	8.1229	13.82	14.39	19.71	−0.11	−11.41	3.06
35	7.0535	−6.6158	5.9839	10.32	14.62	21.65	3.41	−6.57	3.24
36	5.3805	−8.5642	6.7003	15.12	9.09	15.82	2.79	−8.56	3.46
37	3.5046	−6.4646	6.7131	10.34	6.18	7.07	−0.27	−5.81	2.51
38	3.7949	−5.2854	4.1147	4.96	4.43	6.65	1.02	−2.21	1.12
39	3.5357	−7.8347	3.1269	11.08	4.77	8.83	3.92	−2.60	0.62
40	1.3013	−8.1686	4.1909	11.66	2.72	6.52	−0.44	−4.44	2.23
41	0.3022	−5.5435	2.9316	4.68	1.36	2.11	−0.37	−1.45	0.69
42	0.8494	−6.1614	0.4276	8.90	2.25	3.33	2.01	−1.20	−0.42
43	−0.3758	−8.6121	0.6789	17.77	3.08	9.36	1.03	−5.78	−0.14
44	−2.5317	−7.1832	1.2183	11.27	5.23	10.14	−2.64	−4.96	1.92
45	−2.5920	−5.2689	−1.0979	9.80	3.96	7.04	−0.39	−4.64	0.08

We then used a technique called the double iterated Kalman filter (DIKF), described in detail in (Altman and Jardetzky, 1989), to refine the estimates of mean and variance for each α carbon. The DIKF is a conditional probability based update mechanism which compares the value of experimental measurements with the predicted value of the measurements based on a mathematical model of the experiment and its variance (or noise) as well as the current structural model of the protein and its variance. If the predicted value and the experimental value are not the same, then the structural model (estimate of mean positions for all atoms and their variance) must be updated. The magnitude of the adjustment depends on the relative uncertainty of the model an the experimental data:

1. If the variance of the experimental measurement is small relative to the variance of the predicted measurement as calculated from the current structural model, then the structure model is updated to bring it into agreement with the experimental data.

2. If the variance of the experimental measurement is large relative to the variance of the predicted measurement, then the model is updated only slightly to reflect greater confidence in the model than on the single piece of uncertain data.

3. In general, the ratio of the variances is used to determine the magnitude of the adjustment to the structural model. The equation for updating the state estimate and covariance given the model of the data is given by the linear Kalman filter (Gelb, 1984).

$$\vec{x}_{new} = \vec{x}_{old} \, K \left[z - h(\vec{x}_{old}) \right]$$

Where x is a vector of the cartesian coordinates of all atoms, z is an observed distance between atoms and h(x) is the calculated distance between the atoms. The new state vector is the sum of the old state vector plus a weighted difference (with gain matrix, K, described below) between the measured data, z, and the predicted data h(x). C(x) is the variance-covariance matrix for the elements in vector x. The new covariance matrix is the difference between the old covariance matrix and a weighted product of the gain matrix, K, derivative of the data model, H and the old covariance matrix.

$$C(\vec{x}_{new}) = C(\vec{x}_{old}) - K \, H \, C(\vec{x}_{old})$$

where

$$K = C(x_{old}) \, H^T \left[HC(x_{old})H^T + C(v) \right]^{-1}$$

K is the "Kalman gain matrix" and is simply the ratio between the certainty in the estimate of the data by the old model and the certainty in the measurement itself. H^T is the transpose of matrix H. Note that the term within the inverse is the first order Taylor approximation to the variance of the measurement, h(x):

$$C(h(x)) = HC(x)H^T + \dots$$

where H is given by

$$H = \frac{\partial h(\vec{x})}{\partial \vec{x}} \Big|_{\vec{x}_{old}}$$

and therefore

$$C(z) = C(h(x)) + C(v)$$

or, inserting the above

$$C(z) = H \, C(x)H^T + C(v)$$

The derivative of the data model, or h(x), is used to calculate the direction of updating when moving the state vector values. It shows how the changes in the values of state variables effect the state estimate of the data. If h(x) is not linear in the state vector, then an iterated Kalman filter can be used to minimize errors in the update:

$$x_{new_i} = x_{old} + K \left[z - (h(x_{new_{i-1}}) + H(x_{old} - x_{new_{i-1}})) \right]$$

and

$$C(x_{new}) = C(x_{old}) - K_i HC(x_{old})$$

where

$$K_i = C(x_{old})H^T \left[HC(x_{old})H^T + C(v) \right]^{-1}$$

As more data are added to the system, the estimate of the mean value, x, is refined and the uncertainty in this estimate, $C(x)$, decreases as a function of the constraint uncertainties. Even with the iterated Kalman filter, we have found that the filter can have residual inaccuracies due to the nonlinearities in the system. After sequential introduction of all new constraints, we have an estimate of the mean values for x which is better than the starting means, but which may not yet be the best estimate. The mean positions can, however, be used as starting locations for another round of updating. In order to allow the atoms freedom to move in response to the constraints, all the covariances are reset to their initial (large) values and all constraints are re-introduced into the system. By repeating this operation, we perform an operation that has some similarity to simulated annealing (van Laarhoven, 1987) in the case of first order optimization methods. Resetting the covariance after each iteration to its initially large size is equivalent to "heating up" the system. As more constraints are introduced the mean position is refined and the covariance matrix settles down to a "cool" low value. If the mean position does not satisfy the constraints to within tolerable number of standard deviations, then the cycle is repeated until all constraints are satisfied to within some threshold. This is typically set to be one standard deviation. We refer to this method as the *double iterated Kalman filter* (DIKF). It is necessary because the essential non-linear nature of the protein structure problem requires that a system based on linear estimates be iterated in order to reduce residual errors.

For the *lac* repressor, each distance constraint was abstracted into a distance distribution function between the relevant α carbons. For example, in the case of NOE measurements, the mean distance was a function of the NOE distance (2-4 Å) and the average distance between the α carbon and the involved proton (1-8Å, depending on the amino acid sidechain). The variance is simply the sum of the variances of these parameters. After the DIKF is used to refine the initial estimate derived from the solid level, the mean positions of the α carbons are determined with much smaller variance and the topology of the molecule becomes clear, as shown in Figure 3.

We have compared the locations of the α carbons as calculated with the DIKF to the locations of α carbons calculated by methods using a full atomic representation to calculate the structure of the *lac* repressor headpiece. We extracted only the α carbons from each of these structures and used the standard unweighted root mean square (RMS) algorithm to calculate the average distance of the corresponding atoms within two structures (Arun et al., 1987). Only α carbons 4 through 49 were compared because all three methods found amino acids 1,2,3,50 and 51 virtually unconstrained and with very high variance. The α carbons calculated by the DIKF are 2.2 Å RMS from the α carbons provided to us by Dr. Ruud Scheek of Groningen and derived from restrained molecular dynamics using the same data set. Our structure is 2.6 Å RMS from the structure provided by Dr. Felix Frayman using a distance geometry-like technique and a slightly different data set. The Frayman and Groningen structures are 2.4 Å apart. As expected, the structures determined by Frayman and the Groningen group fall within a small number of standard deviations from our mean positions. Figure 4 provides a visual comparison of the structures.

Full Atomic DIKF

Having determined the topology of the molecule using an α carbon representation, it becomes possible to perform the final refinement to a full atomic representation using the DIKF. Using the mean positions and variances calculated in the previous step, we interpolated between α carbons to generate a starting position for each backbone atom (N, C and O). Variances were

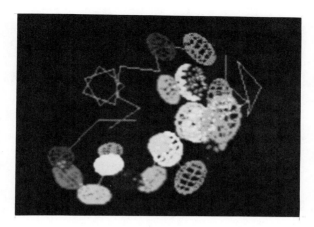

Figure 3. The structure of the *lac* repressor headpiece as calculated by the DIKF using only α carbons. The mean positions for the carbons are shown along with ellipsoids of uncertainty for some atoms drawn from the resulting variances at two standard deviations.

set to twice the variance of the α carbons. In addition, a pseudoatom representation (Wüthrich et al., 1983) was used for the remaining atoms within each amino acid sidechain and positioned arbitrarily within 5 Å of the associated α carbon with a variance of 49 Å^2 to account for uncertainty in position. The total number of atoms and pseudoatoms for the calculation of the *lac*-repressor headpiece is 399. We introduced 858 explicit distance constraints; including 404 covalent constraints, 16 hydrogen bonds, 152 nonbonded backbone distances, 150 short range NOEs, 21 medium range NOEs and 113 long range NOEs. The NOE distance constraints are abstracted more precisely than previously so that only the errors inherent in the pseudoatom approximation, in addition to the native uncertainty of the NOE measurement, are present in the estimate of mean and variance between the two atoms. The program also introduces a dynamic van der Waals constraint by checking interatomic distances for violations of hard shell radii and adding a constraint only for those pairs of atoms violating these radii.

Table 3. The intermediate mean positions and variances for the α carbons of helix-3 after the applying the DIKF to the full set of α carbons of the *lac* repressor with abstracted constraints. It can be seen clearly in comparison with Table 2 that there has been a refinement in both the positions of the atoms (x, y, z) and the uncertainty in those positions V(x), V(y), V(z), V(xy), V(xz) and V(yz).

AA	X	Y	Z	V(x)	V(y)	V(z)	V(xy)	V(xz)	V(yz)
34	11.2926	-4.5388	4.7206	0.606	1.064	1.272	-0.123	-0.4569	-0.170
35	11.0946	-7.7086	2.5390	0.415	0.930	1.094	0.163	-0.2064	0.044
36	8.8872	-9.4609	5.0295	0.719	0.774	1.086	0.153	-0.5052	0.230
37	6.4654	-6.5194	5.2132	0.580	0.725	0.451	0.001	-0.2283	0.067
38	6.0862	-6.3402	1.3198	0.338	0.298	0.434	0.061	-0.0620	0.005
39	5.2573	-10.1670	1.2384	0.533	0.401	0.669	0.206	-0.1377	0.118
40	2.5781	-9.9554	4.0291	0.663	0.336	0.545	-0.014	-0.1879	0.200
41	0.8126	-7.1224	2.2577	0.408	0.171	0.318	0.021	-0.0838	0.011
42	0.6679	-8.9782	-1.0843	0.536	0.257	0.361	0.030	-0.1294	0.048
43	-0.9058	-12.0280	0.6380	0.729	0.209	0.750	-0.065	-0.1827	0.036
44	-3.6426	-9.7508	2.1665	0.531	0.513	0.726	-0.184	-0.0500	0.019
45	-4.5627	-8.2619	-1.3299	0.515	0.386	0.613	-0.070	-0.2543	0.052

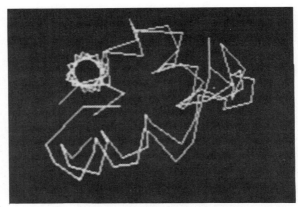

Figure 4. The mean positions of α carbons as calculated by the DIKF superimposed upon the mean positions of 10 structures provided by the Groningen group. The backbones are 2.4 Å apart.

The DIKF was again used, as described above, to update the estimates for mean position and variance for each atom. The starting error for the constraints was an average of seven standard deviations (SD) (that is, the distances predicted from the starting positions were an average of seven SD from the measured experimental value--using the known variance of the measurement) with a maximum error of 60 SD. After 23 cycles of refinement the average error for all constraints was 0.10 SD with a maximum of 3 SD. The result is shown in Figure 5. The helical backbone atoms (as illustrated in Table 4) have the lowest variance of approximately 0.2 Å^2. Atoms in the coil regions between helices have an average variance of 3 Å^2, while atoms in the unconstrained C- and N-termini have an average variance of 10 Å^2. We compared the mean positions calculated here with the structures provided to us by the Groningen group and Frayman. Once again, the unconstrained C- and N-termini were not included in the comparison

Table 4. The final mean positions and variances of the α carbons of helix-3 in the *lac* repressor headpiece. These are the positions after the full atomic representation was used in the DIKF to refine the estimates produced by the α carbon representation and summarized in Table 3. It is clear that, relative to Table 3, the mean positions and variances have changed little compared to the change observed between the start of the calculation, Table 2 and the intermediate result of Table 3. The columns are as labelled previously.

AA	X	Y	Z	V(x)	V(y)	V(z)	V(xy)	V(xz)	V(yz)
34	10.649	-3.282	6.174	0.531	0.750	0.814	-0.005	0.028	-0.208
35	10.906	-5.563	3.221	0.335	0.565	0.553	-0.021	-0.017	0.075
36	9.345	-8.533	4.978	0.448	0.468	0.757	0.014	0.075	0.018
37	6.512	-6.280	6.155	0.429	0.574	0.348	0.030	-0.017	0.024
38	5.908	-5.299	2.749	0.238	0.210	0.326	-0.012	-0.006	-0.002
39	5.885	-8.461	1.451	0.423	0.283	0.472	-0.049	0.022	0.000
40	3.355	-9.620	4.057	0.398	0.252	0.474	0.018	0.046	0.045
41	1.124	-6.861	3.128	0.328	0.153	0.224	0.009	-0.005	0.016
42	1.061	-7.885	-0.440	0.362	0.207	0.241	-0.049	-0.033	0.006
43	0.347	-11.286	0.472	0.473	0.174	0.603	-0.032	0.135	-0.017
44	-2.572	-9.921	2.405	0.312	0.404	0.476	0.032	0.041	0.059
45	-3.682	-7.916	-0.468	0.420	0.333	0.467	-0.006	-0.054	0.005

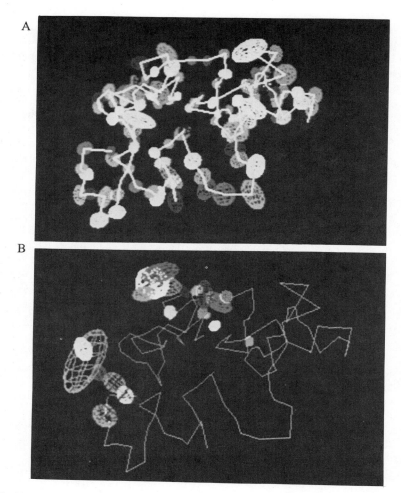

Figure 5. (A) The full peptide backbone of the *lac* repressor headpiece as calculated by the DIKF along with ellipsoids showing two standard deviations of uncertainty. (B) Mean positions of the backbone of the *lac* repressor headpiece are shown along with the mean positions and ellipsoids of uncertainty (at 2 SD) for two sidechains (7 and 25).

because of the great variance in position calculated by all three methods. The mean positions of the backbone atoms in our calculated structure are 2.2 Å from the backbone atoms of the Groningen structures. The mean positions of backbones and sidechain atoms are 2.8 Å from the corresponding atoms in the Groningen structures. The backbone of the Frayman structure is 3.6 Å from the calculated mean positions and is 2.4 Å from the Groningen backbones.

Comparison of Predicted NMR Spectra

In order to compare the ability of the Groningen structures and our structures to predict the observed NMR spectra, theoretical NOESY spectra were generated by solving the generalized Bloch equations (Macura and Ernst, 1980). Proton coordinates were generated by adding protons with standard bond lengths and angles to the mean positions calculated by the DIKF or to a selected Groningen structure. The program BLOCH written by Madrid was used to solve the Bloch equations (Madrid and Jardetzky, 1988), neglecting spins further than 6.0 Å. The minimum allowed distance between protons was set to 1.8 Å (artifactual distances closer than this were rounded up to 1.8 Å). Methyl groups are considered to be single nuclei with spin, I, of 3/2. A uniform correlation time was estimated from the molecular weight of the protein (5,500 daltons).

All other experimental parameters were as reported in (Zuiderweg et al., 1985). For these calculations, we did not use the uncertainty in atomic position as calculated by the DIKF, but relied solely on the mean positions.

Spectra calculated with the two structures were compared to the NOEs identified experimentally. The results are summarized in Table 5.

Table 5. Comparison of the number of experimental NOEs and the number calculated from two proposed structures (A = DIKF mean position, B = arbitrarily selected Groningen structure) of the *lac* repressor headpiece. For backbone and medium range NOEs, numbers in parentheses () denote NOEs calculated, but not reported experimentally. For long range NOEs, numbers in brackets [] denote NOEs calculated between protons in correct residue pairs, but between incorrect protons. These represent approximately correct NOEs in the context of appropriate, but slightly inaccurate, sidechain contacts. Tm = mixing time for experiment, or for calculation.

	Number of NOEs		
	Experimental	Calculated	
		A	B
Backbone NOEs (Tm = 100 ms)			
α-amide (i,i+1)	35	13(3)	15(3)
amide-amide (i,i+/-1)	34	23(6)	30(5)
β-amide (i,i+/-1)	33	24(9)	27(8)
Medium Range NOE Connectivities (Tm = 100 ms)			
α-amide (i,i+3/4)	15	10(6)	4(3)
α-β (i,i+3)	6	4(4)	8(5)
Long Range NOE Connectivities			
Tm = 50 ms	124	16[100]	17[57]
Tm = 100 ms		26[142]	38[69]
Tm = 200 ms		28[157]	53[198]

DISCUSSION

The sequence of steps outlined here: secondary structure → rough sampled topology → α-carbon topology → full atomic topology is not fixed. It is possible to go directly to the α carbon representation from the start or even to the full atomic representation. We choose to use these steps in the solution of the structure because each phase gives the next a rigorous upper bound with which to start the next stage of refinement. This markedly increases the efficiency of the solution, as detailed in (Altman and Jardetzky, 1989).

The secondary structure determination, when possible, is very useful for allowing us to estimate the rough topology using the solid level. We have shown that this method works well for β-strands as well as α-helices in Altman and Jardetzky, 1986. If we are unable to identify clear areas of secondary structure, the α carbon representation (Table 1) within the DIKF simply starts with larger estimates of the variance of each atom. We have found that when a good set of secondary structures is obtained from ABC, the topology of many molecules becomes immediately clear. This has been demonstrated with myoglobin (Lichtarge et al., 1987), cytochrome-b562 (Brinkley et al., 1988) and the first domain of T4 Lysozyme (Altman and Jardetzky, 1986). In cases where the data sets is weak, the solid level usually produces a more amorphous appearing structure with large uncertainties which are nevertheless rigorous upperbounds on the structure and can be refined using the DIKF.

It is somewhat surprising that the α carbon representation used in the first refinement with the DIKF produces a structure in good agreement with other structures produced using a full atomic representation. This result supports the notion that the *determination* of backbone topologies is not exquisitely sensitive to the detailed packing of amino acid sidechains. This does not imply that the *folding* of backbone topologies is not sensitive to the detailed amino acid sidechains. There is some evidence, however, that some abstractions of the primary sequence may contain sufficient information to determine backbone folds with reasonable precision (Skolnick et al., 1989).

The correct way to compare structures produced by this method with others is not absolutely clear. We have presented here RMS distances between mean positions for atoms as determined by us and point locations as determined by other methods. In general, a mean position need not satisfy detailed chemical and geometrical constraints since more than one conformation may be contributing. In cases where the variance of an atom is quite low (such as within the helix backbone segments), the mean positions are very close to point locations and satisfy many of the detailed instantaneous constraints on bond length and bond angle. However, in cases when the variance of atom is larger, the significance of comparing the mean of a distribution with structures which are essentially samples from that distribution is difficult to assess. In these cases it would be preferable to develop comparison algorithms which use the estimates of uncertainty in position to weight a least squares determination of mean error. Such algorithms could also be used to compare the results of multiple runs of one method with multiple runs of another. We believe that the development of a method which simultaneously calculates both position and variance in position is a valuable first step towards the development of such analysis techniques.

It is encouraging that a mean/variance representation of the molecule is sufficient to provide enough detail to partially explain the observed NMR spectra. As shown in Table 5, the number of NOEs that are predicted exactly is not large for any of the structures analyzed in this paper (30% for our structure, 31% for the Groningen structures). However, the number of closely related NOEs observed in our structures is quite high. The failure of the sample structures defined by approximate methods (either DIKF or by restrained molecular dynamics) to completely reproduce the original data set used in the structure determination requires further investigation.

There are at least three sources of error that could account for this failure:

1. In the calculation of the NOE using the Bloch equations, it is assumed that the protein is rigid and has a single correlation time, $\tau_c = 2.8$ ns. The NOE intensities are sensitive to the correlation time (Madrid et al., 1989) - and to internal motion (Lane, 1988).

2. The mean structure calculated by our method or the sample structure calculated by molecular dynamics will most likely closely resemble, but not correspond exactly to the "real" structure as

it exists in solution. The NOE intensities calculated using the Bloch equations are very sensitive to small differences in the coordinates (Madrid and Jardetzky, 1988).

3. There may be a population of structures contributing to the spectra which are not well modelled by a single (mean or otherwise) structure. This phenomenon has been demonstrated in the case of the protein BPTI (Madrid and Jardetzky, 1988).

Several groups have introduced the iterative solution of the Bloch equations as a refinement procedure (Keepers and James, 1984; Jardetzky, 1985, Lefèvre et al., 1987, Boelens et al., 1989). The potential inaccuracies inherent in this approach have recently been discussed by Clore and Gronenborn (1989) and Koehl et al. (this volume).

The agreement of at least three independent structure determination methods on the general topology of the *lac* repressor headpiece allows us to draw this conclusion: NMR data are sufficient to determine general backbone topology. It appears as if any reasonable method of data analysis will converge to a small set of topologies, given a reasonable data set. However, the detailed packing is difficult to determine unambiguously with data sets typical of current NMR experiments. It is this observation which has guided our work towards methods, such as DIKF which simultaneously calculate the "average structure" as well as the uncertainty in the positions of each atom. There are certainly many refinement programs that can start with a detailed topology such as that produced by the DIKF and compute a refined single conformation--subject to the assumptions contained within these refinement programs. We are more interested in describing the bounds on the structures that these methods will produce than in converging on a single structure of undefinable accuracy.

In the case of the *lac* repressor headpiece, the data argue unequivocally for relatively well defined helical elements in a closely packed triangle. The turn between helix 1 and helix 2 is also clearly defined with small variances. However, there is striking uncertainty (both by our calculation and by the repeated restrained molecular dynamics calculations) in the position of the N-terminus, the C-terminus and in the long coil region between helices 2 and 3. The C-terminal uncertainty is probably a by product of the proteolytic cleavage which has left it without its normal polypeptide neighbors (this section of the protein may also act as a hinge region to the core region of the repressor molecule). The N-terminal uncertainty is understandable as a possible freely mobile hydrophilic segment at the end of the polypeptide that associates with the solution in relatively unstructured way. The same might be argued about the coil region between the second and third helices. It is interesting that this region is relatively underconstrained by the available data. The position of helix 3 relative to the other two helices is well-defined despite the apparent disorder of the intervening coil. This disorder may reflect a functionally important mobility that is used in the process of binding nucleic acids. Along with the other hinge region between the first and second helices, it may be a primary reason for the difficulty in crystallizing the *lac*-repressor headpiece. These results lead us to predict that a crystal structure of the *lac*-repressor headpiece would likely reflect a conformation of the loop between helices 2 and 3 that was *crystallographically* optimal, but perhaps not functionally significant. We would expect, however, that the detailed contacts between helices in such a crystal structure would be relatively constant (since they are well defined even under NMR experimental conditions).

SUMMARY

We have used the *lac* repressor as a case study of the progress of protein structure determination from experimental, principally NMR, data obtained in solution. Early efforts on this molecule showed that the secondary structure could be reasonably well defined, but that there may be a range of tertiary structures consistent with the data. Our results and the recent results of others, have verified this prediction and have shown specifically that the relative positions of the secondary structures and the general topology of the molecule are well established by the NMR data, but that the C-termini, N-termini and coil region between the second and third helices are less well defined. We consider these results to be important grounds for suggesting that structure determination methods using solution data should not be modelled on crystallographic refinements, whose primary goal is a single "optimal" structure. Whereas all methods have shown good agreement in the general topology of the *lac* repressor, specific atomic configurations are more difficult to define uniquely from experimental data alone (and even with supplemental energy terms). In light of these observations, structure determination programs which include explicit estimates of uncertainty provide important information.

ACKNOWLEDGEMENTS

This research was supported by NIH grant RR02300.

REFERENCES

Altman, R.B., 1989, "Exclusion methods for the determination of protein structure from experimental data," Ph.D. Thesis, Medical Information Sciences, Stanford University, Stanford, CA.

Altman, R., Duncan, B., Brinkley, J.F., Buchanan, B. and Jardetzky, O., 1987, Determination of the spatial distribution of protein structure using solution data, *in:* "NMR Spectroscopy in Drug Research," Monksgaard, Copenhagen, pages 214.

Altman, R. and Jardetzky, O., 1986, New strategies for the determination of macromolecular structure in solution, *J. Biochemistry* 100(6):1403.

Altman, R. and Jardetzky, O., 1989, The heuristic refinement method for the determination of the solution structure of proteins from NMR data, *Methods in Enzymology* 177:218.

Arun, K.S., Huang, T.S. and Blostein, S.D., 1987, Least-squares fitting of two 3-D point sets, *IEEE Transactions on Pattern Analysis and Machine Intelligence* 9(5):698.

Blundell, T.L. and Johnson, L.N., 1972, "Protein Crystallography," Academic Press, New York.

Boelens, R., Koning, T.M.G., van der Marel, G.A., van Boom, J.H. and Kaptein, R., 1989, Iterative procedure for structure determination from proton-proton NOEs using a full relaxation matrix approach. Application to a DNA octamer, *J. Magn. Res.* 82:290.

Braun, W. and Go. N., 1985, Calculation of protein conformations by proton-proton distance constraints: a new efficient algorithm, *J. Mol. Biol.* 186:611.

Brinkley, J.F., Altman, R., Duncan, B.S., Buchanan, B.S. and Jardetzky, O., 1988, The heuristic refinement method for the derivation of protein solution structures: validation on cytochrome-b562, *J. Chem. Inf. Comp. Sci.* 28(4):194.

Brugge, J.A., Buchanan, B.G. and Jardetzky O., 1987, Toward automating the process of determining polypeptide secondary structure from ^{1}H NMR data, *J. Comp. Chem.* 9(6):662.

Clore, G., Brünger, A., Karplus, M. and Gronenborn, A., 1986, Application of molecular dynamics with interproton distance restraints to three-dimensional protein structure determination: a model study of crambin, *J. Mol. Biol.* 191:523.

Clore, G. and Gronenborn, A.M., 1989, How accurately can interproton distances in macromolecules really be determined by full relaxation matrix analyses of nuclear Overhauser enhancement data? *J. Magn. Res.* 84:398.

Duncan, B., Buchanan, B., Lichtarge, O., Altman, R., Brinkley, J., Hewett, M., Cornelius, C. and Jardetzky, O., 1986, PROTEAN: a new method of deriving solution structures of proteins, *Bull. Magn. Res.* 8:111.

Frayman, F., 1985, "PROTO: An Approach for Determining Protein Structures," Ph.D. Thesis, Computer Science Department, Northwestern University, Evanston, IL.

Gelb, A., 1984, "Applied Optimal Estimation," Massachusetts Institute of Technology Press, Cambridge, MA.

Gunsteren, W. and Karplus, M., 1982, Protein dynamics in solution and in crystalline environment: a molecular dynamics study, *Biochemistry* 21(10):2259.

Havel, T. and Wüthrich, K., 1984, A distance geometry program for determining the structures of small proteins and other macromolecules from nuclear magnetic resonance measurements of intramolecular H-H proximities in solution, *Bull. of Math. Biol.* 46(4):673.

Jardetzky, O., 1984, A method for the definition of the solution structure of proteins from NMR and other physical measurements: the *lac*-repressor headpiece, *in:* "Progress in Bioorganic Chemistry and Molecular Biology," Yu. A. Ovchinnikov, editor, Elsevier Science Publishers B.V., Amsterdam.

Jardetzky, O., 1985, New strategies for the determination of protein and nucleic acid structures in solution by nuclear magnetic resonance, *in:* "Seikagaku" (Proceedings of the Japanese Biochemical Society) 57(8), p. Roman 4.

Jardetzky, O. and Roberts, G.C.K., 1981, "NMR in Molecular Biology," Academic Press, New York.

Kaptein, R., Zuiderweg, E.R.P., Scheek, R.M. and Boelens, R., 1985, A protein structure from nuclear magnetic resonance data: *lac* repressor headpiece, *J. Mol. Biol.* 182:179.

Keepers, J.W. and James, T.L., 1984, A theoretical study of distance determination from NMR two-dimensional nuclear Overhauser effect spectra, *J. Magn. Res.* 57:404.

Koehl, P., Kieffer, B. and Lefèvre, J-F., 1990, The dynamics of oligonucleotides and peptides determined by proton NMR, *in:* "Protein Structure and Engineering," (O. Jardetzky , ed.), Plenum Press, NY.

Lane, A.N., 1988, The influence of spin diffusion and internal motions on NOE intensities in proteins, *J. Magn. Res.* 78:425.

Lefèvre, J-F., Lane, A.N. and Jardetzky, O., 1987. Solution structure of the *trp* operator of *Escherichia coli* determined by NMR, *Biochemistry* 26:5076.

Lichtarge, O., Cornelius, C.W., Buchanan, B. and Jardetzky, O., 1987, Validation of the first step of the heuristic refinement method for the derivation of solution structures of proteins from NMR data," *Proteins* 2:340.

Lim, W.A. and Sauer, R.T., 1989, Alternative packing arrangements in the hydrophobic core of λ repressor, *Nature* 339:31-36.

Macura, S. and Ernst, R.R., 1980, Elucidation of cross relaxation in liquids by two-dimensional NMR spectroscopy, *Mol. Phys.* 41:95.

Madrid, M. and Jardetzky, O., 1988, Comparison of experimentally determined protein structures by solution of Bloch equations, *Biochim. Biophys. Acta* 953:61.

Madrid, M., Mace, J.E. and Jardetzky, O., 1989, Consequences of magnetization transfer on the determination of solution structures of proteins, *J. Mag. Res.* 83:267.

Ohlendorf, D.H. and Matthews, B.W., 1983, Structural studies of protein-nucleic acid interactions, *Ann. Rev. Biophys. Bioeng.* 12:259.

Pilz, I., Goral, K., Kratky, O., Bray, R.P., Wade-Jardetzky, N. and Jardetzky O., 1980, Small-angle X-ray studies of the quaternary structure of the *lac* repressor from *Escherichia coli*, *Biochemistry* 19:4087.

Ribeiro, A.A., Wemmer, D., Bray, R.P. and Jardetzky, O., 1981, A folded structure for the *lac* repressor headpiece, *Biochem. Biophys. Res. Comm.* 99(2):668.

Sauer, R.T., Hehir, K., Stearman, R.S., Weiss, M.A., Jeitler-Nilsson, A., Suchanek, E.G. and Pabo, C.O., 1986, An engineered intersubunit disulfide enhances the stability and DNA biding of the N-terminal domain of λ repressor, *Biochemistry* 25(20):5992.

Skolnick, J., Kolinski, A. and Yaris, R., 1989, Dynamic Monte Carlo study of the folding of a six-stranded Greek key globular protein, *Proc. Natl. Acad. Sci. USA* 86:1229.

van Laarhoven, P.J.M., 1987, "Theoretical and Computational Aspects of Simulated Annealing," Ph.D. Thesis, Department of Computer Science, Erasmus University, Rotterdam, The Netherlands.

Williamson, M.P., Havel, T.F. and Wüthrich, K., 1985, The solution conformation of proteinase inhibitor IIa from bull seminal plasma, *J. Mol. Biol.* 182:295.

Wüthrich, K., 1986, "NMR of proteins and nucleic acids," Wiley, New York.

Wüthrich, K., Billeter, M. and Braun, W., 1983, Psuedo-structures for the 20 common amino acids for use in studies of protein conformations by measurements of intramolecular proton-proton distance constraints with nuclear magnetic resonance, *J. Mol. Biol.* 169:949.

Wüthrich, K., Billeter, M. and Braun, W., 1984, Polypeptide secondary structure determination by nuclear magnetic resonance observation of short proton-proton distances, *J. Mol. Biol.* 180:715.

Zuiderweg, E., Billeter, M., Boelens, R., Scheek, R., Wüthrich, K. and Kaptein, R., 1984, Spatial arrangement of the three alpha-helices in the solution conformation of *lac* repressor headpiece, *FEBS 1746* 174(2):243.

Zuiderweg, E., Kaptein, R. and Wüthrich, K., 1983, Sequence-specific resonance assignments in the H-NMR Spectrum of the *lac* repressor DNA binding domain 1-51 from *E. coli* by two dimensional spectroscopy. *Eur. J. Biochem.* 137:279.

Zuiderweg, E., Scheek, R.M. and Kaptein, R., 1985, Two-dimensional H-NMR studies on the *lac* repressor DNA binding domain: further resonance assignments and identification of nuclear Overhauser enhancements, *Biopolymers* 24(12):2257.

THE GENERATION OF THREE-DIMENSIONAL STRUCTURES FROM NMR-DERIVED CONSTRAINTS

Denise D. Beusen and Garland R. Marshall

Center for Molecular Design, Washington University, St. Louis, Missouri 63130, USA

INTRODUCTION

NMR is capable of providing many types of information about ligands, macromolecules, and their complexes. In recent years, the generation of solution structures based on NMR observations has become widespread. While these methods hold the promise of providing nearly as precise information for molecules in solution as X-ray methods do for crystals, they differ from X-ray in that the experimental observations do not reveal the ensemble of positions of all the heavy atoms, but rather the distances between certain pairs of atoms. Consequently, one needs to find tools that enable transformation from a set of pairwise distances [distance space] to Cartesian coordinate space in order to build usable structural models. These tools rely on an existing body of knowledge that describes reasonable geometries for covalently bonded atoms in amino acids and nucleotides. Most methods currently in use have their roots in molecular modeling, where computational methods have been developed to study the conformation of molecules and the relationship of conformation to biological activity.

INFORMATION USED AS INPUT IN STRUCTURE GENERATION METHODS

Distance Constraints Arising from the Nuclear Overhauser Effect (Noggle and Schirmer, 1971)

Since NMR is an absorption phenomenon, the intensity of the observed signal depends on the population difference between the ground and excited states. The rate of transition between states depends on the spin-lattice relaxation time T_1, a measure of the time required for nuclear spins to dissipate their energy into their surroundings, or lattice, which is normally dominated by the translational and rotational degrees of freedom of the molecules in which the nuclei are located. For two protons i and j, the possible states and their relative energies are given below:

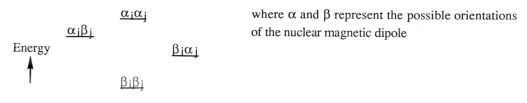

where α and β represent the possible orientations of the nuclear magnetic dipole

At equilibrium, the population of these states is determined by the Boltzmann distribution and a characteristic signal intensity is observed. By irradiating proton i, the population of states involving transitions of i are equalized. The system is no longer at equilibrium, and transitions between states to restore equilibrium are stimulated. The changes in population lead to a decrease or increase in signal intensity known as the nuclear Overhauser effect (NOE). For macromolecules, the new population distribution is established by mutual spin flips of the type $\alpha_i\beta_j <\!-\!-\!> \beta_i\alpha_j$. This process is known as cross-relaxation and is induced by molecular motion. The rate constant for the buildup in signal intensity perturbation is given by (Wüthrich, 1986):

$$\sigma_{ij} = \frac{\gamma^4\hbar^2}{10<r_{ij}^6>}\left(\frac{6\tau_{eff}}{1+4\omega^2\tau_{eff}^2} - \tau_{eff}\right)$$

where τ_{eff} = effective correlation time for the i-->j interproton vector
(approximate time required for reorientation of this vector by one radian)

r_{ij} = distance between two protons i and j

γ = gyromagnetic ratio of proton

$\hbar$ = h/2π

The first half of the expression on the right reveals the r^{-6} dependence of the NOE, while the second half contains the motional dependence. In the isolated spin pair approximation, the ratio of two interatomic distances (i-->j and k-->l) can be obtained from:

$$\frac{r_{ij}}{r_{kl}} = \left(\frac{\sigma_{kl}}{\sigma_{ij}}\right)^{\frac{1}{6}}$$

In practice, one measures the NOE between two protons whose internuclear distance is fixed (e.g., the geminal protons of a methylene group or adjacent protons on an aromatic ring). Knowing σ_{kl} and r_{kl}, one can measure σ_{ij} and from it calculate r_{ij}. The r^{-6} dependence of the NOE means that it falls off rapidly with distances, so under the best of conditions, the largest distance that can be measured is 5 Å.

There are several simplifying assumptions made in this approach, the first being that the correlation time for all interproton vectors is the same. In other words, there is no independent movement of atoms or groups within the molecule. Second, the possibility that a third (or more) nucleus is undergoing dipolar-relaxation with the proton pair of interest is ignored. Third, the assumption is made that dipolar-relaxation is the primary means of relaxation for the nuclei of concern. Early studies attempted to get around these shortcomings by simply grouping the calculated distances into short, medium, and long categories [corresponding roughly to 2.5, 3.0 or 3.5, and 4.0 or 5.0 Å] with a large error range. In fact, some investigators set all observable distances to a single range, 2.0 --> 4.0 Å (Wüthrich, 1989a,b). The errors introduced by the simplifying assumptions and efforts to more rigorously interpret the data will be discussed in a later section.

Distance Constraints Arising from Paramagnetic Enhancement of Relaxation

The unpaired electrons in spin labels or paramagnetic ions, because of their large magnetic moment, facilitate both the longitudinal [T_1] and transverse [T_2] relaxation of nearby nuclei.

T_{1p}, the perturbation in T_1 due to the paramagnetic species, is determined by subtracting T_1 measured in the presence of a diamagnetic species from that observed in the analogous paramagnetic system. The distance between the observed nucleus and the paramagnetic species is related to T_{1p} by

$$\frac{1}{T_{1p}} = \frac{K}{r^6}\left(\frac{3\tau_c}{1+\omega_I^2\tau_c^2} + \frac{7\tau_c}{1+\omega_S^2\tau_c^2}\right)$$

where

K =   term incorporating nuclear and electronic moments, stoichiometry and lifetime of the complex, fraction of ligand bound

ω_I, ω_S =   precession frequencies of the nucleus and electron, respectively

τ_c, the correlation time for the nuclear-electron dipolar interaction, must be determined independently. Several simplifying assumptions enter into this expression (Mildvan and Gupta, 1978), including fast exchange of the paramagnetic ligand. Determination of the bound conformation of enzyme substrates and inhibitors using this approach has recently been reviewed by Mildvan (1989).

Experimental Distances Derived from Scalar Couplings in Solids

A new spectroscopic technique, rotational-echo double-resonance (REDOR) NMR, has recently been described by Gullion and Schaefer (1989). This technique for solids utilizes magic-angle spinning and directly measures the dipolar coupling of labeled nuclei, allowing interatomic distances to be measured without the need to invoke simplifying assumptions. In a solid with ^{13}C-^{15}N dipolar coupling, the ^{13}C rotational spin echoes which form each rotor period following a ^1H-^{13}C cross-polarization transfer can be prevented from reaching full intensity by inserting two ^{15}N π pulses per period. One of the π pulses is synchronized with the completion of the rotor period, and the other with a time less than or equal to half the rotor period. The difference between a ^{13}C NMR spectrum obtained under these conditions, and one obtained with no ^{15}N π pulses, measures the ^{13}C-^{15}N coupling. The method offers high precision due to its simplicity of mechanism: the dephasing of magnetization of one spin in the presence of the local dipolar field of the other spin. This technique has recently been used to measure an intra-atomic distance in a helical peptide in the crystal (Marshall et al., in press), and would seem applicable to aggregated states such as found with integral membrane proteins. The major drawback is the requirement for incorporation of specific labels and the need for large amounts of samples due to the lower sensitivity of solid state NMR.

Other Miscellaneous Constraints

Virtually any information that can be translated into distance or torsional angle constraints can be incorporated into the methods described in the next section. Studies that measure the exchange rate of labile protons combined with qualitative evidence for secondary structures derived from NOESY spectra can yield distance constraints based on hydrogen bonds. Salt bridges and disulfide linkages can be similarly incorporated.

Coupling constants between protons separated by three bonds can be used to define the H-C-C-H dihedral angle (θ) based on the Karplus equation (Karplus, 1959, 1963):

$$^3J_{HH} = A\cos^2\theta + B\cos\theta + C$$

Pardi et al. (1984) calibrated the $^3J_{NH-\alpha H}$ observed in bovine pancreatic trypsin inhibitor (BPTI) versus angles in the crystal structure with the best fit corresponding to A = 6.4, B = -1.4, and C = 1.9. A similar calibration (Kopple et al., 1973; DeMarco et al., 1978; Cung and Marraud, 1982) has been done for the $^3J_{H\alpha H\beta}$ coupling constants of proteins. Proton coupling constants do not give information on the backbone ψ torsion angle. Consequently, the measurement of heteronuclear coupling constants such as that between $H^{\alpha i}$ and N^{i+1} is receiving increased attention (Montelione et al., 1989). N-H^β and H^β-C' coupling constants, in combination with $^3J_{H\alpha H\beta}$, should allow for stereospecific assignment of β hydrogens and unambiguous determination of χ_1 (Kessler et al., 1987).

METHODS FOR GENERATING STRUCTURES FROM NMR DATA

In reality, for any given set of distance data, there is probably no single structure that uniquely fits the distance constraints. Therefore, the goal of any structure generation method is to sample all of conformational space in order to discover all possible structures that are consistent with the constraints.

Distance Geometry (Havel et al., 1983)

Distance geometry methods are possibly the most widely used for solving structures, and the best known distance geometry program is DISGEO (Braun et al., 1981; Havel and Wuthrich, 1984). This method attempts to randomly sample all conformations which satisfy the inequality

$$L_{ij} \leq D_{ij} \leq U_{ij}$$

where D_{ij} is the distance between all pairs of points i and j, and U and L are the upper and lower limits, respectively, on all possible pairwise distances between the N atoms of the molecule. Once the sequence of the molecule is provided as input, the program selects L_{ij} based on standard bond lengths, bond angles, and van der Waals radii given in a library of standard amino acids or nucleotides. The NOE distances plus some estimate of error are used as U_{ij}. If no U_{ij} or L_{ij} is available from experimental data or the covalent structure, U_{ij} is set to a value large enough that it effectively imposes no constraint and L_{ij} is set equal to the sum of van der Waals radii for atoms i and j. The limits are initially checked to insure that they satisfy the following inequalities, and are adjusted until they do:

$$U_{ij} \leq U_{ik} + U_{jk}$$

$$L_{ij} \geq L_{ik} - U_{jk} \qquad \text{for all } i,j,k$$

The internuclear distances D_{ij} are randomly chosen within the refined limits and adjusted until they also satisfy the inequalities above. Once the matrix of internuclear distances **D** is determined, the distance of each point to the molecule's center of mass (e.g., d_{io}, d_{jo}) can be calculated from the formula below.

$$d_{io}^2 = n^{-1} \sum_{j=1}^{n} d_{ij}^2 - n^{-2} \sum_{j=2}^{n} \sum_{k=1}^{j-1} d_{jk}^2$$

Knowing the distance of each atom from the center of mass, the metric matrix can be calculated, each entry of which is defined as follows:

$$G = \frac{1}{2}\left(d_{io}^2 + d_{jo}^2 - D_{ij}^2\right)$$

From the eigenvalues of this matrix, the Cartesian coordinates of each point can be calculated. The process just described is known as EMBEDing and is often done in two stages. In the first stage, only a subset of atoms is embedded (e.g., backbone atoms in the protein). The interatomic distances in the resulting substructure serve as input in the initial setting of bounds for EMBEDing all atoms in the second stage. The structure is optimized by minimizing (with respect to the x,y,z coordinates) an error function of the type:

$$\sum_{i<j} \max\left[0, \frac{D_{ij}^2}{U_{ij}^2} - 1\right] + \max\left[0, \frac{L_{ij}^2}{D_{ij}^2} - 1\right]$$

Often, the energy of structures generated by distance geometry is very poor. This is a consequence of the fact that all distance violations, such as bond lengths and NMR constraints, have the same weight. Thus, a structure that appears to be a reasonable solution (based on the residual of the error function) may have a very high energy because two nonbonded atoms are partially overlapping. Havel and Wuthrich (1985) simulated several NMR data sets from the crystal structure of BPTI and used them as input to DISGEO. These data sets varied in the precision of the distances and whether or not short-range (with respect to sequence) distances were included. Upon comparison of the structures with the crystal structure, they concluded the following:

- Distance geometry structures tend to be expanded relative to the crystal, especially with less constrained data sets.

- All data sets produced the same elements of secondary structure; therefore, short-range constraints contribute relatively little to the determination of global conformation.

- The local conformation depends on having precisely defined short- range constraints. This is a problem, since sequential alpha to amide hydrogens range from 2.2 to 3.6 Å apart and distances derived using the simplified approach given earlier are generally regarded as having errors of ±0.5 Å or more.

- Short interproton distances, significantly smaller than the protein dimensions, can define the global fold of a protein.

One known shortcoming of distance geometry is its proclivity to generate structures having a local conformation that is the mirror image of that seen in the corresponding crystal structure. Ideally, if these aberrant structures appear, they will be in the minority and can be discarded after refinement by other methods (molecular dynamics, energy minimization) that indicate their energy to be significantly higher than correctly folded forms. Recently, some investigators have suggested that in the absence of sufficient constraints, distance geometry produces fully extended structures that are nearly identical (A. Pardi, personal communication). This suggests that conformational sampling by distance geometry is not random; therefore, care must be exercised in overinterpreting the results of computations for structures or regions of structures that are underdetermined. Nevertheless, distance geometry has been used extensively in solving the structures of both DNA and proteins in solution (Braun, 1987; Clore and Gronenborn, 1987; Patel et al., 1987; Wüthrich, 1989a,b).

DISMAN (Braun and Go, 1985)

In this method, only dihedral angles (and not coordinates) are treated as independent variables. As in distance geometry, a library of amino acids having standard bond angles, bond lengths, and van der Waals radii is used. Once the primary sequence is provided, a random starting conformation for the molecule is generated. The target function to be minimized in this case is the sum of many terms, similar in form to

$$\frac{(d_{\alpha\beta} - X_{\alpha\beta})(d_{\alpha\beta}2 - X_{\alpha\beta})^2}{X_{\alpha\beta}^2}$$

where $d_{\alpha\beta}$ is the distance between α and β in the model, and $X_{\alpha\beta}$ is the experimentally derived upper distance limit between α and β. The goal is to find the global minimum of the target function; that is, to perturb the dihedral angles of the starting conformation until $d_{\alpha\beta}$ are within the limits $X_{\alpha\beta}$ determined by NMR. As in any minimization process, there is a danger of becoming trapped in a local minimum--a conformation in which some constraints are met, but from which one cannot move to a conformation in which all constraints are met. To get around this problem, an index k is defined as the difference in sequence number between the residues to which α and β belong. At first, only distances for which k is small are considered and gradually k is increased. The minimized conformation at one value of k is used as a starting conformation for minimization at the next value of k until eventually all distance constraints are included in the minimization. In this way, locally correct structures are first made and then combined to give correct structures. As was done for DISGEO, Braun and Go (1985) prepared simulated NMR data sets from the crystal structure of BPTI and used DISMAN to generate structures. Generally, their findings were similar to those of Havel and Wuthrich (1985).

Restrained Molecular Dynamics

This approach simulates the motion of molecules by first assigning velocities to a starting array of atoms [built from standard amino acid or nucleotide fragments] and subsequently integrating the classical Newtonian equations of motion to determine the position of atoms at later time points (McCammon and Harvey, 1987). The energy of the molecule is the sum of its kinetic and potential energies, with the latter evaluated from a force field that describes equilibrium bond lengths, bond angles, and torsional angles. By increasing the kinetic energy [temperature] of the system, one can in principle overcome conformational barriers and the problem of local minima. In order for the simulation to converge on a folded structure consistent with the NMR data, additional terms are added to the force field:

$$E_{NOE} = \frac{kTS(r_{ij} - r_{ij}^o)^2}{2\Delta^2}$$

where
r_{ij} = interproton distance in model
r_{ij}^o = target interproton distance from NOE data
T = temperature of simulation
S = arbitrary scaling factor
Δ = error estimate on interproton distance
k = Boltzmann constant

During the course of the simulation, these terms act as an attractive potential to pull the proton pairs together. The scaling factor S starts at very low values and gradually increases as the simulation proceeds. In one study, a simulated NOE data set derived from a crystal structure was used in a restrained molecular dynamics simulation to fold the protein crambin (Clore et al., 1986). The method proved capable of reproducing the overall fold observed in the crystal structure. However, the final structure depended on the path taken during the simulation. If all the constraints were implemented at once, an incorrectly folded structure resulted. In contrast, if only the short-range distances were implemented initially and then the long-range constraints were invoked, the correct structure could be obtained. Thus, it seems that secondary structure must be defined prior to enforcing the constraints that have the greatest impact in determining the overall protein fold. Aside from the problem of path dependency, another serious drawback to this method is the need for massive amounts of CPU time compared to distance geometry, especially for large molecules. Although restrained molecular dynamics has been used by itself in generating solution structures for both DNA and proteins, its primary use is in the refinement of structures generated by other means, such as distance geometry or DISMAN (Clore and Gronenborn, 1987). The combination of approaches ameliorates the limitations of both methods. By using distance geometry, the path-dependent folding of molecular dynamics is eliminated; likewise, restrained molecular dynamics regularizes the poor geometries of DISGEO structures.

The complexity of molecular dynamics calculations means that they are quite time consuming, and consequently it is not possible to run simulations long enough to insure that all of the conformational space is sampled. In addition, the relative weighting of the NOE constraints and the normal force field terms begs the question of what is dominating the computation, and is the resulting hybrid likely to produce a meaningful result. Too strong a scaling factor for the NOE term will certainly perturb the calibrated force field, while too much weighting of the force field could bias the conformational search and yield results that are not meaningful. Certainly, this latter concern is true for electrostatic terms that are effective over much longer distances than most of the other terms. These concerns have resulted in another hybrid approach known as dynamical simulated annealing (Nilges et al., 1988). Only a subset of atoms (primarily backbone plus $C\beta$) is embedded; the remaining sidechain atoms are added in fully extended form and the ensemble is energy minimized without restraints. The resulting structures are subjected to restrained molecular dynamics at a very high temperature (1000 K) using a force field where all nonbonded interactions are rolled into a single term (dihedral, van der Waals, electrostatic, and hydrogen bonding); restrained dynamics at a reduced temperature slowly cools the system. The process is designed to speed the search of conformational space by efficiently overcoming energy barriers on the path to the global minimum.

Systematic Conformational Search

As in DISMAN, a rigid model of the molecule is built from fragments having reasonable geometries, but all torsional angles are systematically varied. Conformations having van der Waals contacts are eliminated, as are those which violate the distance or coupling constant constraints. In a tripeptide for which ϕ and ψ are allowed to rotate in $10°$ increments, there are $6^{36}=10^{28}$ conformations to evaluate. Therefore, any application of this method to proteins will require many precisely known distance constraints as well as a knowledge of the conformational preferences of amino acids in order to reduce the computation to a manageable size. This approach was first used to determine the conformation of peptides complexed with lanthanides (Barry et al., 1971). Since the method uses the rigid geometry approximation in order to minimize the number of variables to be scanned, the van der Waals radii must be appropriately calibrated to reproduce the experimental crystal data (Iijima et al., 1987). This method has been used to solve solution structures of the cyclic peptide, cyclosporin A (Beusen et al., in press). As the number of constraints which can be determined increases with the application of new NMR experiments, the problem may change from an underdetermined one with multiple solutions to one in which some subdomains of the protein may, in fact, be overdetermined. This would reduce the inherent combinatorials of systematic search and make it feasible to apply to larger systems.

Other Methods

Other *optimization methods* have been used in some studies to satisfy NOE constraints. The force field is modified by a term similar to that in restrained molecular dynamics. Unfortunately, the problem of becoming trapped in a local minimum means the result is highly dependent on the starting structure. One study (Holak et al., 1987) used a complex force field term for the NOE constraint and demonstrated convergence to a single structure from several very different starting geometries for a small peptide. The ellipsoid algorithm (Billeter et al., 1987), using large and discontinuous steps, may evade difficulties with local minima.

Monte Carlo simulations are a means of conformational sampling used in molecular modeling which have just recently been applied to NMR data (Bassolino et al., 1988). Like DISMAN, this technique uses a rigid geometry and operates in torsional space. Dihedral angles are randomly varied to generate a new conformation which is energetically evaluated using terms representing nonbonded interactions and NMR distance constraints. If an acceptance criterion based on the energy of the new conformation is met, the new conformation serves as a starting point for the next iteration. Not unexpectedly, convergence to a structure that satisfied the distance constraints required fine tuning of the relative weighting between the distance and nonbonded interaction terms, and the initial incorporation of short-range constraints followed by the long-range constraints.

The *Build-up procedure* has been used to predict structures of protein and peptide sequences. In this approach, the lowest energy conformations of an amino acid are combined with those of its neighboring amino acid, and the energy of the pair is evaluated. Only the lowest energy dipeptides are retained for the next step of the calculation. This process is repeated through the sequence, generating a family of low-energy structures. By incorporating distance constraints into the calculation, the structure of BPTI has been constructed from a data set simulated from the crystal structure (Vasquez and Scheraga, 1988).

Heuristic refinement (Brinkley, 1988) is a method in which secondary structures are represented as solid objects (e.g., a helix as a cylinder) and all possible orientations of objects with respect to one another are evaluated in order to find those consistent with the NMR data. The sequences connecting secondary structure elements are built by optimizing a localized target function.

Manual model building with a sufficiently sophisticated computer graphics system is certainly possible. Qualitative assessments of secondary structure derived from NMR spectra can be imposed on the primary sequence of a protein. With a small molecule, torsional angles can be individually adjusted. However, one cannot possibly hope to satisfy all of the constraints generated by NMR in any reasonable amount of time, and this is certainly not an effective way to randomly sample the conformational space defined by the constraints. Computer graphics systems have been used to generate starting conformations for restrained molecular dynamics into which the qualitatively assigned elements of secondary structure have been incorporated.

Comparison of Methods

No single NMR data set (either simulated or real) has been tested by all possible methods. However, some head-to-head comparisons have been made. Both DISMAN and DISGEO have been used to solve the solution structure of BPTI (Wagner et al., 1987). As expected, the DISGEO structures were expanded relative to the DISMAN structures, although both sets satisfied the distance constraints very well. Constraints consistently violated by one method also tended to be violated in the other, suggesting error in the constraints. The DISMAN structures seemed to center better around the BPTI crystal structure, while the DISGEO structures fit the crystal coordinates slightly better. In all cases, areas of greatest variability corresponded to the lowest density of constraints. DISGEO and restrained molecular dynamics were used to generate crambin structures from a simulated NMR data set (based on the crystal structure) (Clore et al., 1987a). Both methods reproduced the crystal structure well. DISGEO structures were expanded relative to the crystal, while the dynamics structures were contracted (due to attractive van der Waals interactions). Local structures were better reproduced by dynamics, probably because the effect of energetics is included.

One small cyclic peptide, cyclosporin, has been examined by constrained molecular dynamics (Lautz et al., 1987) and systematic search (Beusen et al., in press). The NMR data from the laboratory of Horst Kessler on cyclosporin, a cyclic 11-residue peptide, was used because it represented the state of the art when it was published in 1985 (Kessler et al., 1985; Loosli et al., 1985). The structure calculated by restrained molecular dynamics was nearly identical to that determined by crystallography (Petcher et al., 1976). When the 22 Φ,Ψ torsional variables in the backbone were systematically evaluated at a $10°$ increment, over 2,750,000 conformations consistent with the NMR data were found. While this number may seem startling large, if one assumes only two values for each torsional variable, over 4 million combinations (2^{22}) are possible. This implies that, in fact, the individual variables were tightly constrained. Most of the conformers were similar in structure to that seen in the crystal, but approximately 5% showed a different β-turn (type I versus type II' at the Sar residue). This indicates that restrained molecular dynamics did not adequately sample all of the conformational space in the search for structures consistent with the experimental data.

EVALUATION AND REFINEMENT OF STRUCTURES GENERATED FROM NMR DATA

Given the approximations used in measuring interproton distances and the problems of local minima and path dependency described in the computational methods, a legitimate concern is whether or not the structure generated is a correct one. Thus far, this question has been addressed by relying on a comparison of the calculated solution structure with a solved crystal structure. Most comparisons to date have involved globular proteins and seem to suggest that the NMR-derived structures are an accurate, if low-resolution, representation of the crystal structure. However, conformation is dependent on environment and the difference between solution and solid states argues that as the number of solution structures determined increases, cases similar to that of metallothionein (Braun, 1985; Furey, 1986), where the two structures differ, will increase in frequency.

In the absence of a corresponding crystal structure, investigators have addressed this issue by generating many structures from independent DISGEO or DISMAN calculations. The convergence of these calculations is monitored by the residual of the target or error function, the number and amount of distance constraint violations, and, finally, by a root-mean square deviation (RMSD) comparison of the structures. With a good set of distance constraints, a family of structures can be generated for which the RMSD of backbone atoms is 1-2 Å. When sidechain atoms are considered, the RMSD increases, possibly reflecting the greater mobility of these groups, resulting in error in the interproton distance estimate. A visual analysis and RMSD comparisons of atom subsets are essential in assessing the quality of structures. An interesting problem occurred in solving the solution structure of hirudin (Clore et al., 1987b). This small protein consists of three domains. While the structure of each domain was well defined, the orientation of the domains with respect to each other could not be determined because of a lack of interdomain NOE constraints. Consequently, an RMSD comparison of the entire molecule shows regions that fit poorly; however, the fit between the individual domain structures was very good. The ability of an analytical paradigm to distinguish between well-defined subdomains and areas undergoing conformational averaging may become crucial as new NMR techniques increase the number of experimental constraints which can be determined.

The question of how correctly an NMR-derived structure represents a molecule in solution really subdivides into two separate questions. First, do the algorithms accurately reproduce structures, and, second, how accurate are the calculated interproton distances and J values. The studies cited earlier demonstrated the ability of algorithms to reproduce global folding from simulated data sets. A recent study (Pardi et al., 1988) suggests that distance geometry will never be able to reproduce structures exactly based on the type of information presently available by NMR and points to the need for more information, either from NMR or from independent methods. A reasonable consideration that has not been adequately addressed to this point is how to code the uncertainty in atom position (analogous to temperature factor in X-ray). Some have advocated the use of thermal ellipsoids (Brinkley et al., 1988), but no suggestion has been made as to distinguishing areas of large RMSD arising from a lack of NOE constraints versus those due to areas possessing NOE constraints with a large error limit. It is worth noting that disordered areas are usually not visible by X-ray; such areas will always have short-range constraints observable by NMR but may not have NOEs that position them with respect to the remainder of the molecule.

Refinement of distance and J-value constraints is an active area of research. By measuring ^{13}C relaxation rates, one can determine τ_{eff} for local segments of the molecule and consequently refine the distance calculation. Spectral simulations are being used to define J values more accurately. Matrix methods, which completely define the relaxational properties of a molecule by determining σ for every proton combination, result in more precise distance estimates (Keepers and James, 1984; Borgias and James, 1988; Boelens et al., 1989). For areas of a molecule lacking NOE constraints, databases describing conformational preferences of short sequences are probably better at defining local conformation than the algorithms in use. Presently, the field is moving to a computational paradigm that is similar to that employed by crystallographers. Trial structures are initially generated and subsequently refined by fitting the structure's simulated NOESY spectrum to the observed by modifying distances in the trial structure (Hendrickson, 1989; Summer et al., 1989).

Realistically, the process of refinement may reach a limit imposed by the NMR methodology itself. Many motions are faster than can be observed by NMR, and some uncertainty in quantitating interatomic distances will persist simply because one can only hope to observe an ensemble of structures (Jardetzky, 1980). In these cases, the observed structure can only be interpreted as a weighted average of structures (Nikiforovich et al., 1988) and the challenge becomes one of identifying those structures. The databases mentioned previously may aid in identification; alternatively, one is forced to consider all possible candidates--a problem that in most cases could prove intractable.

REFERENCES

Barry, C.D., North, A.C.T., Glasel, J.A., Williams, R.J.P. and Xavier, A.V., 1971, *Nature* 232:236.
Bassolino, D.A., Hirata, F., Kitchen, D.B., Kominos, D., Pardi, A. and Levy, R.M., 1988, Determination of protein structures in solution using NMR data and impact, *Int. J. Supercomputer Applications* 2:41.

Beusen, D.D., Iijima, H. and Marshall, G.R., Structures from NMR distance constraints, *Biochem. Pharm.*, in press.

Billeter, M., Havel, T.F. and Wüthrich, K., 1987, The ellipsoid algorithm as a method for the determination of polypeptide conformations from experimental distance constraints and energy minimization, *J. Comp. Chem.* 8:132.

Boelens, R., Koning, T.M.G., van der Marel, G.A., van Boom, J.H. and Kaptein, R., 1989, Iterative procedure for structure determination from proton-proton NOEs using a full relaxation matrix approach. Application to a DNA octamer, *J. Magn. Reson.* 82:290.

Borgias, B.A. and James, T.L., 1988, COMATOSE, a method for constrained refinement of macromolecular structure based on two-dimensional nuclear Overhauser effect spectra, *J. Magn. Reson.* 79:493.

Braun, W., 1987, Distance geometry and related methods for protein structure determination from NMR data, *Quarterly Reviews of Biophysics* 19:115.

Braun, W. and Go, N., 1985, Calculation of protein conformations by proton-proton distance constraints, *J. Mol. Biol.* 186:611.

Braun, W., Bosch, C., Brown, L.R., Go, N. and Wüthrich, K., 1981, Combined use of proton-proton Overhauser enhancements and a distance geometry algorithm for determination of polypeptide conformations, *Biochim. Biophys. Acta* 667:377.

Braun, W., Wagner, G., Wörgötter, E., Vasak, M., Kagi, J. and Wüthrich, K., 1985, *J. Mol. Biol.* 186:611.

Brinkley, J.F., Altman, R.B., Duncan, B.S., Buchanan, B.G. and Jardetzky, O., 1988, Heuristic refinement method for the derivation of protein solution structures: Validation on cytochrome b562, *J. Chem. Inf. Comput. Sci.* 28:194.

Clore, G.M. and Gronenborn, A.M., 1987, Determination of three-dimensional structures of proteins in solution by nuclear magnetic resonance spectroscopy, *Prot. Eng.* 1:275.

Clore, G.M., Brünger, A.T., Karplus, M. and Gronenborn, A.M., 1986, Application of molecular dynamics with interproton distance restraints to three-dimensional protein structure determination, *J. Mol. Biol.* 191:523.

Clore, G.M., Nilges, M., Brünger, A.T., Karplus, M. and Gronenborn, A.M., 1987a, A comparison of the restrained molecular dynamics and distance geometry methods for determining three-dimensional structures of proteins on the basis of interproton distances, *FEBS Lett.* 213:269.

Clore, G.M., Sukumaran, D.K., Nilges, M., Zarbock, J. and Gronenborn, A.M., 1987b, The conformations of hirudin in solution: A study using nuclear magnetic resonance, distance geometry and restrained molecular dynamics, *EMBO J.* 6:529.

Cung, M.T. and Marraud, M., 1982, Conformational dependence of the vicinal proton coupling constant for the $C\alpha$-$C\beta$ bond in peptides, *Biopolymers* 21:953.

DeMarco, A., Llinas, M. and Wüthrich, K., 1978, Analysis of the H - NMR spectra of ferrichrome peptides. I. The non-amide protons, *Biopolymers* 17:617.

Furey, W.F., Robbins, A.H., Clancy, L.L., Winge, D.R., Wand, B.C. and Stout, C.D., 1986, *Science* 231:704.

Gullion, T. and Schaefer, J., 1989, Rotational-echo double-resonance NMR, *J. Magn. Reson.* 81:196.

Havel, T. and Wüthrich, K., 1984, A distance geometry program for determining the structures of small proteins and other macromolecules from nuclear magnetic resonance measurements of intramolecular H-H proximities in solution, *Bull. Math. Biol.* 46:673.

Havel, T. and Wüthrich, K., 1985, An evaluation of the combined use of nuclear magnetic resonance and distance geometry for the determination of protein conformation in solution, *J. Mol. Biol.* 182:281.

Havel, T.F., Kuntz, I.D. and Crippen, G.M., 1983, The theory and practice of distance geometry, *Bull. Math. Biol.* 45:665.

Hendrickson, W.A., 1989, NMR structural analysis from the perspective of a protein crystallographer, *J. Cell. Biochem.* 13A:12.

Holak, T.A., Prestegard, J.H. and Forman, J.D., 1987, NMR-Pseudoenergy approach to the solution structure of acyl carrier protein, *Biochemistry* 26:4652.

Iijima, H., Dunbar, J.B., Jr. and Marshall, G.R., 1987, The calibration of effective van der Waals atomic contact radii for proteins and peptides, *Proteins: Struct. Funct. Genet.* 2:330.

Jardetzky, O., 1980, On the nature of molecular conformations inferred from high-resolution NMR, *Biochim. Biophys. Acta* 621:227.

Karplus, M., 1959, Contact electron-spin coupling of nuclear magnetic moments, *J. Chem. Phys.* 30:11.

Karplus, M., 1963, Vicinal proton coupling in nuclear magnetic resonance, *J. Am. Chem. Soc.* 85:2870.

Keepers, J.W. and James, T.L., 1984, A theoretical study of distance determinations from NMR two-dimensional nuclear Overhauser effect spectra, *J. Magn. Reson.* 57:404.

Kessler, H., Loosli, H.R., Oschkinat, H. and Widmer, A., 1985, Assignment of the ^{1}H-, ^{13}C- and ^{15}N-NMR spectra of cyclosporin A in $CDCl_3$ and C_6D_6 by a combination of homo- and heteronuclear two-dimensional techniques, *Helv. Chim. Acta* 68:661.

Kessler, H., Griesinger, C. and Wagner, K., 1987, Peptide conformations. 42. Conformation of side chains in peptides using heteronuclear coupling constants obtained by two-dimensional NMR spectroscopy, *J. Am. Chem. Soc.* 109:6927.

Kopple, K.D., Wiley, G.R. and Tauke, R., 1973, A dihedral angle-vicinal proton coupling constant correlation for the α–β bond of amino acid residues, *Biopolymers* 12:627.

Lautz, J., Kessler, H., Kaptein, R. and van Gunsteren, W.F., 1987, Molecular dynamics simulation of cyclosporin A: The crystal structure and dynamic modelling of a structure in apolar solution based on NMR data, *J. Comput.-Aided Mol. Design* 1:219.

Loosli, H.R., Kessler, H., Oschkinat, H., Weber, H.-P., Petcher, T.J. and Widmer, A., 1985, The conformation of cyclosporin A in the crystal and in solution, *Helv. Chim. Acta* 68:682.

Marshall, G.R., Beusen, D.D., Kociolek, K., Redlinski, A.S., Leplawy, M.T., Pan, Y. and Schaefer, J., Determination of a precise interatomic distance in a helical peptide by REDOR NMR, *J. Am. Chem. Soc.*, in press.

McCammon, J.A. and Harvey, S.C., 1987, "Dynamics of Proteins and Nucleic Acids," Cambridge University Press, Cambridge, UK.

Mildvan, A.S., 1989, NMR studies of the interaction of substrates with enzymes and their peptide fragments, *FASEB J.* 3:1705.

Mildvan, A.S. and Gupta, R.K., 1978, Nuclear relaxation measurements of the geometry of enzyme-bound substrates and analogues, *Methods Enzymol.* 49G:322.

Montelione, G.T., Winkler, M.E., Rauenbuehler, P. and Wagner, G., 1989, Accurate measurements of long-range heteronuclear coupling constants from homonuclear 2D NMR spectra of isotope-enriched proteins, *J. Magn. Reson.* 82:198.

Nikiforovich, G.V., Vesterman, B.G. and Betins, J., 1988, Combined use of spectroscopic and energy calculation methods for the determination of peptide conformation in solution, *Biophys. Chem.* 31:101.

Nilges, M., Clore, G.M. and Gronenborn, A.M., 1988, Determination of three-dimensional structures of proteins from interproton distance data by hybrid distance geometry-dynamical simulated annealing calculations, *FEBS Lett.* 229:317.

Noggle, J.H. and Schirmer, R.E., 1971, "The Nuclear Overhauser Effect," Academic Press, New York.

Pardi, A., Billeter, M. and Wüthrich, K., 1984, Calibration of the angular dependence of the amide proton-Ca proton coupling constants in a globular protein, *J. Mol. Biol.* 180:741.

Pardi, A., Hare, D.R. and Wang, C., 1988, Determination of DNA structures by NMR and distance geometry techniques: A computer simulation, *Proc. Natl. Acad. Sci. USA* 85:8785.

Patel, D.J., Shapiro, L. and Hare, D., 1987, Nuclear magnetic resonance and distance geometry studies of DNA structures in solution, *Ann. Rev. Biophys. Biophys. Chem.* 16:423.

Petcher, T.J., Weber, H.-P. and Ruegger, A., 1976, Crystal and molecular structure of an iodo-derivative of the cyclic undecapeptide cyclosporin A, *Helv. Chim. Acta* 59:1480.

Summer, M.F., Hare, D., South, T.L. and Kim, B., 1989, Structure of a retroviral zinc finger: 2D NMR spectroscopy and distance geometry calculations on a synthetic finger from HIV-1 nucleic acid binding protein, p7, *J. Cell. Biochem.* 13A:17.

Vasquez, M. and Scheraga, H.A., 1988, Calculation of protein conformation by the build-up procedure. Application to bovine pancreatic trypsin inhibitor using limited simulated nuclear magnetic resonance data, *J. Biomol. Struct. Dynamics* 5:705.

Wagner, G., Braun, W., Havel, T.F., Schaumann, T., Go, N. and Wüthrich, K., 1987, Protein structures in solution by nuclear magnetic resonance and distance geometry - The polypeptide fold of the basic pancreatic trypsin inhibitor determined using two different algorithms, DISGEO and DISMAN, *J. Mol. Biol.* 196:611.

Wüthrich, K., 1986, "NMR of Proteins and Nucleic Acids," John Wiley and Sons, New York.

Wüthrich, K., 1989a, Protein structure determination in solution by nuclear magnetic resonance spectroscopy, *Science* 243:45.

Wüthrich, K., 1989b, The development of nuclear magnetic resonance spectroscopy as a technique for protein structure determination, *Acc. Chem. Res.* 22:36.

2D-NMR FOR 3D-STRUCTURE OF MEMBRANE SPANNING POLYPEPTIDES:

Gramacidin A and Fragments of Bacteriorhodopsin

V.F. Bystrov, A.S. Arseniev, I.L. Barsukov, A.L. Lomize, G.V. Abdulaeva, A.G. Sobol, I.V. Maslennikov, and A.P. Golovanov

Shemyakin Institute of Bioorganic Chemistry, USSR Academy of Sciences, Ul. Miklukho-Maklaya 16/10, Moscow V-437, USSR

ABSTRACT

Spatial structures of gramicidin A (GA) and the bacteriorhodopsin (BR) fragments in membrane mimetic milieu have been determined by NMR spectroscopy. GA incorporated into SDS micelles adopts the conformation of the monovalent cation selective transmembrane channel, i.e. head-to-head dimer formed by two right-handed $\pi_{L,D}^{6.3}$-helices. Interaction of impermeable Mn(II) ions with the channel was studied by paramagnetic induced proton relaxation and their preferred binding sites were evaluated. Additional results on synthetic GA analogs reveal structure-functional relationships and mechanism of channel unfolding. In contrast to micelles, organic solvents induce formation of four distinct GA double helices (parallel and antiparallel) with 5.6 residues per turn. In organic solvent in the presence of caesium salt the right-handed antiparallel double helix with 7.2 residues per turn becomes a dominant species. Discrepancies and similarities with the results obtained by other techniques are discussed.

BR and its proteolytic fragments when extracted by gel chromatography in methanol-chloroform mixture retain unique native conformations as detected by NMR and CD spectra. The tertiary structure of BR was probed by ^{19}F NMR of [5-F]Trp BR. 2D NMR unequivocally defines conformations of the BR membrane spanning proteolytic and synthetic fragments as right-handed α-helices of about 20 residues each.

NMR strategy for spatial structure analysis of transmembrane polypeptides is discussed. The GA and BR study illustrates two feasible NMR approaches to the conformational analysis of intrinsic membrane polypeptides. The first utilizes the artificial quasimembrane system in which the molecule adopts its native structure. In the second use is made of the properly chosen solvent which preserves the conformation formed by the molecule in the natural membrane environment.

INTRODUCTION

Slightly over 30 years have passed since the first NMR spectra of proteins in solution were recorded (McDonald and Phillips, 1967). Since that time this physical technique passed through many exciting steps and produced an important impact on the conformational study of peptides and proteins in solution, competing even with X-ray crystallography in thoroughness and accuracy (although at present the NMR techniques are limited to the relatively small size of the proteins).

It is well recognized that biological activity of peptides and proteins is governed by their chemical and spatial structure. Determination of the solution three-dimensional conformation of polypeptides, the key to their mode of action and the basis for predetermined synthesis of new physiologically active molecules is one of the most important objectives bioorganic chemistry and molecular biology have set before physico-chemical methodology. Of equal importance are problems concerning the dynamic aspects - degree of flexibility of molecular spatial structure under different physiological conditions of environment. Even minor conformational transitions could play a significant role in the polypeptide biological mechanism of action. Thus it is vitally significant to evaluate not only the structure, for instance, of the binding site in the ligand-receptor interaction, but also to present data on conformational changes, if any, that could occur in the course of interaction.

Of the various spectroscopic techniques available, NMR spectroscopy is the choice in solution for:
- disclosing of the conformational perturbations and equilibria under various conditions of medium, including those closely resembling fluid physiological environment;
- delineation of the predominant species, the structure and conformational transitions of which are to be established, taking into account that a possible difference between the structure predominant in solution and that in crystalline state;
- shedding light on the specific intramolecular interactions leading to the given structural features, both conformational and dynamic;
- identification of the particular chemical groups of the molecule participating in intermolecular interaction with biological target, for instance, with receptors, ligands, metal cations;
- revealing the role of the molecular spatial structure and its dynamic properties in particular biological interaction and process, performed by the molecule, that is, finally, to answer the question of just why Nature has constructed a particular molecule for a given task and not another.

It should be stressed, however, that the conformational analysis of even simple oligopeptides can best be accomplished in the most effective and economical mode by combining NMR spectroscopy with other techniques in solution, namely, circular dichroism, infrared and Raman spectroscopy, selective spin and fluorescence labeling, theoretical conformational analysis, etc. (Ovchinnikov and Ivanov, 1975). In the composite approach the results obtained by each technique are used to provide the information they are best fitted to deal with, and thus permit the overall problem to be solved in the most efficient manner.

For the last 5-6 years high resolution NMR spectroscopy, thanks to advent of very high field superconducting solenoids and the two-dimensional, on-line computer based methods, is becoming as an alternative technique to the X-ray crystallography for evaluating molecular spatial structure of large peptides and reasonably small (up to 10-20 kD) proteins. A fair number of structures of globular proteins in solution have been already determined (Gronenborn and Clore, 1989).

As to the non-globular polypeptides, remarkable progress is achieved for molecules tended to be adsorbed at the lipid-water interface, such as glucagon (Braun et al., 1981; 1983), mellitin (Brown et al., 1982; Higashijima et al., 1988), mastoparans (Wakamatsu et al., 1984), α-mating factor (Wakamatsu et al., 1987), δ-hemolysin (Lee et al., 1987) and enkephalin (Higashijima et al., 1988) by using lipid micelles and vesicles to mimic the membrane surface.

Nevertheless, NMR experiences fundamental and practical problems with integral membrane polypeptides and proteins, because in the natural environment the resonances are too broad that prevents the use of high resolution NMR conformational analysis.

In this contribution we would like to present two palliative approaches (Figure 1) to cure this drawback by careful selection of artificial, membrane mimicing environment suitable for high resolution NMR study, with proper control of preservation of the native membrane conformation.

In the case of gramicidin A, a transmembrane ion channel, we incorporated the peptide into the membrane mimicing environment - sodium dodecyl sulphate micelles, which forces the molecule to adopt the intrinsic membrane conformation feasible for NMR study. In another case - bacteriorhodopsin, a bacterial membrane light-driven proton pump, the purple membrane was solubilized at the mild conditions which prevent unfolding the molecule and then delipidated and purified by gel filtration in the same medium to obtain the NMR sample with retained native-like spatial structure.

GRAMICIDIN A

Gramicidin A (GA) produced by Bacillus brevis (Dubos, 1939) is a 15-membered linear polypeptide formed by alternating L- and D-amino acid residues with blocked end groups (Sarges and Witkop, 1965):

$$HCO - L\text{-}Val - Gly - L\text{-}Ala - D\text{-}Leu - L\text{-}Ala - D\text{-}Val -$$
$$L\text{-}Val - D\text{-}Val - L\text{-}Trp - D\text{-}leu - L\text{-}Trp - D\text{-}Leu - L\text{-}Trp -$$
$$D\text{-}Leu - L\text{-}Trp - NH(CH_2)_2OH$$

The unique property of GA is the ability to induce selective transport of univalent cations across biological and model membranes (for review, see Andersen, 1984). In essence the study of the GA transmembrane channel organization and of the mechanism that controls its cation conductance provides a basis to gain insight into protein directed ion translocation in general.

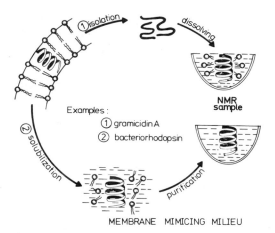

Figure 1. Two currently used approaches for incorporation of membrane spanning polypeptides into membrane mimicing environment.

The problem of the channel spatial structure, that is of vital importance for analysis of the translocation mechanism, is dramatically complicated because of diversity of allowed conformations of gramicidin A promoted by alternation of L- and D-amino acid residues (Ramachandran and Chandrasekharan, 1972). Although the dimeric structure is well established (Veatch and Stryer, 1977) still many forms should be considered, including single stranded and double stranded helices, with different relative orientation of N- and C-terminals, different handedness, and different number of residues per turn. Altogether there are more than 40 energetically allowed symmetrical backbone structures, plus a number of unsymmetrical double helices with shifted backbone chains.

Now it is well recognized that the dimers adopt principally different conformations in heterogeneous environment (membranes and micelles) and in homogeneous media (organic solvents).

Gramicidin A Conformational Diversity in Organic Solvents

In organic solvents (ethanol, dioxan, methanol, methanol-chloroform) gramicidin A forms an equilibrated set of at least four organized conformations (species) (Veatch et al., 1974) with so slow rate of interconversion that all of them are clearly observed in the proton 2D NMR spectrum, as shown in Figure 2 for ethanol solution, for instance. The exhaustive analysis of phase sensitive DQF-COSY, RELAY, HOHAHA and NOESY spectra in conjunction with a priori consideration of NOE for different basic conformations (Barsukov et al., 1987a; Bystrov and Arseniev, 1988) leads to the conclusion that gramicidin A in ethanol is represented by four double stranded intertwined helical dimers with 5.6 residues per turn, but with different handedness and relative orientation of N- and C-terminal (Figure 3). In fact, these

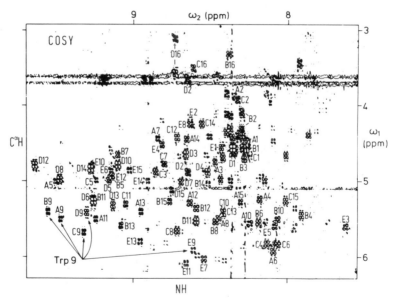

Figure 2. Fingerprint region of a 500 MHz phase sensitive absorption mode double quantum filtered COSY spectrum of a 30 mM gramicidin A in CD_3CD_2OH at 30°C. Cross peaks of NH/C$^\alpha$H protons are identified by letters A (species 1), B and C (two unsymmetrical polypeptide chains of species 2), D (species 3), and E (species 4), and by numbers corresponding to amino acid residue positions in the gramicidin A primary structure.

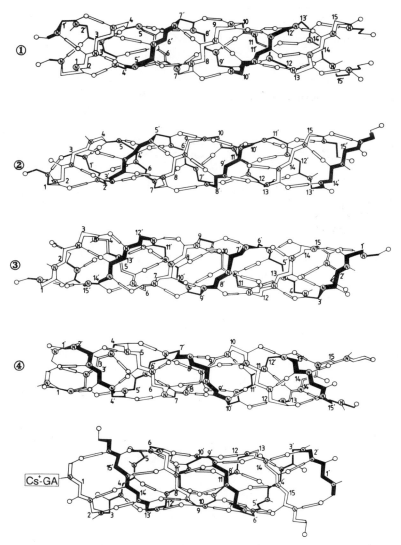

Figure 3. Schematic presentation of the gramicidin A backbone folding for double intertwined helices detected in ethanol solution by 2D NMR spectroscopy: species 1 - parallel left handed double helix, species 2 - parallel left handed helix with shifted backbone alignment, species 3 - antiparallel left handed double helix, species 4 - parallel right handed double helix, Cs^+-gramicidin A complex in methanol-chloroform and in methanol solutions - right handed antiparallel double helix $\uparrow\downarrow\pi\pi_{L,D}^{7.2}$.

conformations are in general agreement with predicted earlier by Veatch et al. (1974) structures of species 1, 2, 3 and 4 of gramicidin A on the basis of CD spectra (Figure 4) analysis, although the handedness of the helices was not specified.

Species 1 (Figure 3) is the parallel left handed double helix $\uparrow\uparrow\pi\pi_{L,D}^{5.6}$ with symmetrical alignment of the backbones; species 2 (Figure 3) - also parallel left handed double helix but with unsymmetrical, shifted backbone arrangement, each gives separate resonances in the NMR spectra (see Figure 2); species 3 (Figure 3) - antiparallel left handed double helix $\uparrow\downarrow\pi\pi_{L,D}^{5.6}$, and

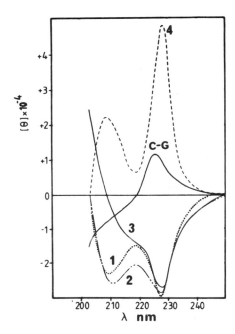

Figure 4. Circular dichroism spectra of gramicidin A isolated species 1, 2, 3 and 4 in dioxane
(Veatch et al., 1974) and of Cs$^+$-gramicidin A complex (C-G) in methanol-
chloroform solution (Barsukov, 1987).

species 4 (Figure 3) - parallel right handed double helix. The estimated relative proportion of
species 1, 2, 3 and 4 at equilibrium in ethanol accords with that evaluated by Veatch and Blout
(1974) - 0.16:0.23:0.46:0.14, respectively.

One of the conformers, namely, species 3 was isolated by crystallization from ethanol at
20°C as by Veatch et al. (1974) and subsequently dissolved in dioxane-d$_8$ to obtain high
resolution 2D NMR spectra of this species individually (Arseniev et al., 1984a, b). Its
conformation, analyzed in full extent by NOE connectivities, vicinal proton coupling constants
and NH deuterium exchange rates, perfectly coincides with one of the conformations in the
equilibrated set of species in ethanol (Figure 3). Later on Langs (1988) demonstrated by X-ray
analysis that precisely the same backbone folding also is formed in crystal. For instance, the φ
torsional angles for L-residues were estimated to be in the range from -150° to -155° by NMR
(Bystrov, 1985) and -155°±9° by X-ray analysis, and the φ angles for D-residues in the range
from 100° to 105° and 110°±22°, respectively.

At first glance the structures of the gramicidin A species in organic solvent due to their polar
axial cavities and lipophilic external surfaces might be considered as candidates for at least pores
for monovalent cations, not to mention transmembrane channel. However, direct addition of
corresponding salts to the gramicidin A solution demonstrates a remarkable change in
conformational situation (Barsukov, 1987; Bystrov et al., 1987). While lithium and sodium
disorganize the predominant regular structures and convert gramicidin A in a flexible random
coil, the addition of potassium, caesium and thallium salts promote generation of the new, single
structure of the complexes.

The caesium (CsCNS) complex formation was studied extensively in a methanol-chloroform solution by 2D NMR (Arseniev et al., 1985a) and the Cs+-gramicidin A conformation was established as right handed antiparallel double helix $\uparrow\downarrow\pi\pi_{L,D}^{7.2}$ (Figure 3). The antiparallel alignment of two polypeptide chains provides two symmetrical and thereby identical binding sites for polarizable monovalent cations, whose diameters fit the size of the lipophilic axial cavity of the double helix.

Quite unexpectedly Wallace and Ravikumar (1988) revealed recently by X-ray analysis that the Cs+-gramicidin A complex in a crystalline state has the principally different structure namely, left handed antiparallel double helix with 6.4 residues per turn. Sketched comparison of solution and crystal structures are shown on Figure 5. The specific differences between the right and left handed double helices in the NMR sense are demonstrated by characteristic NOE connectivity maps (Barsukov et al., 1987; Bystrov and Arseniev, 1988) and by peptide NH deuterium exchange. Fast exchange was observed by Arseniev et al. (1985a) for NH groups of the L-Val 1, Gly 2, L-Ala 3 and L-Ala 5 which are exposed in the NMR solution structure, while for the X-ray crystal structure the fast exchange has to be expected only for the Gly 2 and L-Trp 15 residues.

Initially we presumed that the divergence could be accounted to the difference in medium being an origin of the crystals. Accordingly we carried out NMR experiments in solution under the Wallace and Ravikumar (1988) conditions, that is in methanol with CsCl salt. The NMR samples were prepared by three procedures: 1) dissolution of gramicidin A in methanol and addition of CsCl, 2) solution of CsCl in methanol and then addition of gramicidin A, and 3) dissolution in methanol of Cs+-gramicidin crystals prepared according to Wallace and Ravikumar (1988).

The results of NMR structure elucidation of these samples were absolutely the same and identical with the original NMR solution structure in methanol-chloroform - right handed $\uparrow\downarrow\pi\pi_{L,D}^{7.2}$ double helix. The reason for the alternative crystal structure might be that during

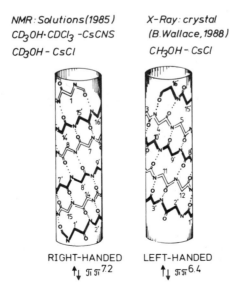

NMR: Solutions (1985)   X-Ray: crystal
CD$_3$OH·CDCl$_3$ -CsCNS   (B. Wallace, 1988)
CD$_3$OH - CsCl   CH$_3$OH - CsCl

RIGHT-HANDED   LEFT-HANDED
$\uparrow\downarrow$ $\pi\pi^{7.2}$ $\uparrow\downarrow$ $\pi\pi^{6.4}$

Figure 5.   Schematic comparison of the solution (Arseniev et al., 1985a) and crystal (Wallace and Ravikumar, 1988) structure of the Cs+-gramicidin A complex.

crystallization either one of the minor components, unobservable by NMR due to very low content, becomes the sole conformation, or a fundamentally new structure is formed, for instance, because of aggregation during solvent evaporation. Several examples of both cases could be recalled from NMR and X-ray analysis of other peptides and even proteins and oligonucleotides.

To finalize the discussion of the gramicidin A conformation in organic solvents it has to be specifically indicated that CD spectra adequately represent the handedness of the backbone folding. As depicted on Figure 4 left and right handed helices are clearly distinguished by the ellipticity sign of the most intensive band at 228 nm. In particular the mirror imaged CD spectra of species 1 and 4 correspond to similarly arranged backbones differing only in handedness.

Now when the problem of the gramicidin A conformations in organic solvents, which has presented a challenge for many years, is resolved one can speculate that one or several of the species, shown in Figure 3, are responsible for gramicidin A biological action as the endogeneous RNA-polymerase inhibitor regulating gene expression in relation to sporulation of producing organism Bacillus brevis (Paulus et al., 1979).

Gramicidin A Transmembrane Channel Structure

The CD spectrum (Figure 6) of gramicidin A in membranes (liposomes, vesicles) and micelles (Masotti et al., 1980; Wallace et al., 1981; Sychev and Ivanov, 1984; Wallace, 1987) totally differs from those obtained in organic solvents (Figure 4). The membrane spectrum cannot be produced by any linear combination of the spectra of individual conformers in solution (Wallace et al., 1981).

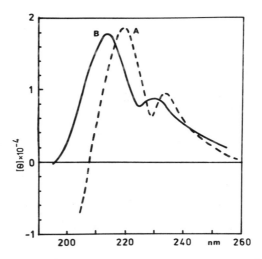

Figure 6. CD spectra of gramicidin A in liposome membranes (control) and in micelles (NMR sample): (A) 0.5 mM gramicidin A incorporated into dipalmitoylphosphotidyl-choline liposomes (peptide to lipid molar ratio 1:300) after heat incubation for 15 h at 70°C; (B) 5 mM gramicidin A incorporated into SDS micelles (peptide to detergent ratio 1:50) in water with trifluoroethanol (molar ratio 16:1).

Bearing in mind that gramicidin A adopts the same conformation in membrane and in micelles we chose detergent micelles as mimicing environment due to their smaller size (and molecular mass) as compared to liposomes and vesicles. To obtain the high resolution NMR spectrum of the gramicidin A transmembrane channel we tried several detergents, i.e. dodecylmethyl amine oxide, dodecyltrimethyl ammonium bromide, dodecylphosphocholine, etc. However the only successful result was attained by incorporation of gramicidin A into sodium dodecyl sulphate (SDS) micelles by dropwise addition of 0.05 M solution in 2,2,2-trifluoroethanol (TFE) into 0.5 M SDS dispersion in water (Arseniev et al., 1985b). The sample was then diluted to a final concentration of 5 mM gramicidin A and 0.25 M SDS in water/TFE (16:1 molar ratio). TFE was used to facilitate incorporation of gramicidin A into liposomes and micelles without effect on the peptide conformation (Masotti et al., 1980).

The experimental condition corresponds to incorporation of two gramicidin A molecules (a dimer) into single SDS micelle (~50-60 molecules of the detergent), as deduced from CD and NMR spectra recorded for different peptide to detergent molar ratios (Arseniev et al., 1986a) and from on Sephadex G-75 gel-chromatography, which shows that elution profiles correspond to the increasing in the molecular mass of SDS micelles by the value approximately equal to the molecular mass of the gramicidin A dimer. The total molecular mass of micelle with the gramicidin A dimer could be estimated as 20 kD, that is in the suitable range of high resolution NMR spectroscopy. The average diameter of the SDS micelle lipophilic core (~30 Å) in the range of exploited experimental conditions (Mazer et al., 1976) well corresponds with the thickness of bilayer membranes.

The validity of this preparation as a model of the genuine membrane spanning channel is proved by several approaches, including CD spectra (Figure 6), which are universally recognized by extensive experimental data as a direct and exhaustive "fingerprint" evidence of the transmembrane channel state of gramicidin A (Urry et. al., 1983). The CD curve of gramicidin A in dipalmytoylcholine liposomes is red shifted by 5 nm due to light scattering in the lipid suspension (Long and Urry, 1981), otherwise its shape is completely the same as for SDS micelles. It should be noted, that according to the NMR study there is only a single conformational form of gramicidin A in the SDS preparation, and thus the CD curves on Figure 6 correspond to the individual molecular structure.

The proof that gramicidin A, embedded into SDS micelles does manifest specific interaction with monovalent metal cations, as an important facet of the channel functioning, comes from ^{23}Na NMR (Arseniev et al., 1986a) (Figure 7.). Figure 7A corresponds to the SDS micelles preparation and demonstrates a relatively narrow [23]Na line. For SDS micelles with incorporated gramicidin A dimers the line is shifted and broadened (Figure 7B), indicating fast exchange of bulk sodium cations with bounded those in the gramicidin A channel. Specificity of interaction was confirmed by competitive binding of thallium cations, which are known to inhibit the sodium transport through gramicidin ion channels in membranes (Neher, 1975). Dissolving the thallium salt in the SDS-gramicidin A sample results in narrowing and shifting back of the [23]Na resonance (Figure 7D). Similar effects were observed for gramicidin A incorporated into lysophosphatidylcholine micelles (Urry et al., 1980) and were taken as evidence for the channel state formation. For instance, the single channel ion currents and the energy of activation for sodium transport through the gramicidin A channel calculated by using the [23]Na relaxation rates in micelles (Urry et al., 1980; Urry, 1987) are compared favorably with those obtained for the channel in lipid bilayer membranes.

Thus, the evidences obtained testify that gramicidin A embedded into SDS micelles forms the ion channel with a similar molecular structure and properties to that in lipid membranes and vesicles, as well in lysolecitin micelles.

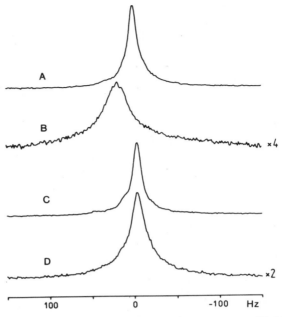

Figure 7. ^{23}Na NMR (132 MHz) at 55°C, pH 6.5 for (A) 250 mM SDS in water-TFE (16:1), (B) with incorporated 5 mM gramicidin A, (C) with 30 mM TlClO$_4$ added to the sample A, (D) with 30 mM TlClO$_4$ added to the sample B.

Proton NMR spectrum of gramicidin A incorporated into SDS-d$_{25}$ micelles displays well resolved signals, which makes it acceptable for 2D NMR study of the detailed molecular structure (Arseniev et al., 1985b; 1986a). Perdeuterated SDS is used so that the detergent does not obscure the peptide resonances. It should be specifically stressed, that the sample under investigation contains abundance of sodium cations and thus the results to be discussed correspond to the sodium complex of gramicidin A in micelles, that is the open state of the transmembrane channel.

Signal assignment was performed by a conventional procedure of sequential resonance assignment (Wieder et al., 1982) that is proton spin systems of individual amino acid residues were identified in COSY spectra via J-connectivities, and then by NOESY spectra via proton Overhauser effect connectivities between neighboring residues they were aligned into the amino acid sequence of gramicidin A (Figure 8).

Table 1 depicts the vicinal backbone H-NC$^\alpha$-H proton couplings, which, by their relatively large values, correspond to the extended backbone conformation, and the NH exchange rates estimated qualitatively by magnetization transfer upon water resonance irradiation - to screening of the largest part of the molecule from contacts with water medium.

It should be noted, that N-terminal ethanolamine, D-Leu 14 and D-Leu 12 amide groups are exposed to the medium. This observation is important for structure elucidation, as will be mentioned later on.

The analysis of phase sensitive NOESY spectrum reveals about 200 NOE connectivities between distinct protons (Arseniev et al., 1986a). Qualitatively, there are two sorts of the

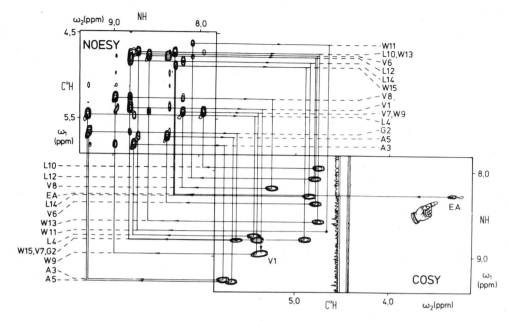

Figure 8. Combined phase sensitive COSY-NOESY diagram of 5 mM gramicidin A in SDS-d_{25} micelles (1:50) in H_2O-TFE-d_2 (16:1) at 50°C, pH 6.5. Straight lines with arrows indicate $^1d_{\alpha N}$ connectivities of neighboring amino acid residues and scalar $^3J(H\text{-}NC^\alpha\text{-}H)$ couplings within the residue, starting with NH/C-H COSY cross peaks of ethanolamine group and terminating on L-Val 1 NH/C$^\alpha$-H cross peak.

Table 1. Vicinal proton coupling constants H-NC$^\alpha$-H and amide NH exchange for gramicidin A embedded into SDS-d_{25} micelles (peptide-detergent molar ratio 1:50) in water-TFE-d_2 (molar ratio 16:1) at 50°C, pH 6.5.

Residue	$^3J(H\text{-}NC^\alpha\text{-}H)$, Hz	NH Exchange
L-Val 1	9.4	Slow
Gly 2		Slow
L-Ala 3	9.4	Slow
D-Leu 4	8.8	Slow
L-Ala 5	9.2	Slow
D-Val 6	8.8	Slow
L-Val 7	9.9	Slow
D-Val 8	8.7	Slow
L-Trp 9	9.1	Slow
D-Leu 10	9.7	Slow
L-Trp 11	9.0	Slow
D-Leu 12	9.8	Fast
L-Trp 13	9.8	Slow
D-Leu 14	9.8	Fast
L-Trp 15	9.4	Slow
Ethanolamine		Fast

connectivities fairly important for structure elucidation of the gramicidin A channel. Firstly, the observed NOE connectivities between the $N_{i+1}H$ and $C_i^{\alpha}H$ protons support the conclusion derived from $^3J(H\text{-}NC^{\alpha}\text{-}H)$ couplings on the extended backbone structure. On the other hand, the connectivities between the NiH and $C_{i+6}^{\alpha}H$ protons of D-residues and between the $C_i^{\alpha}H$ and $N_{i+6}H$ protons of L-residues as well as inhibited NH exchange for the majority of the residues suggest that the extended backbone is spiralled.

Taken together these NMR data inherently produce a right handed single stranded $\pi_{L,D}^{6.3}$ helix with 6.3 residues per turn. Some of the observed NOE connectivities and slow exchange rates of NH protons of Val 1, Ala 3 and Ala 5 cannot be consented with the monomer structure, so they fit the dimer conformation that is N-terminal to N-terminal ("head to head") right handed single stranded $\overrightarrow{\pi}_{L,D}^{6.3}\overleftarrow{\pi}_{L,D}^{6.3}$ helix, schematically shown on Figure 9. The dimer has the symmetry axis perpendicular to the channel axis in accordance with NMR observation of only one signal for each chemically equivalent proton. The structure is stabilized by 20 intramolecular and 6 intermolecular hydrogen bonds NH...O=C.

The corresponding wire model was constructed and side chains of amino acid residues were adjusted in accordance with the vicinal $H\text{-}C^{\alpha}C^{\beta}\text{-}H$ proton couplings and NOE connectivities of side chain protons. The torsion angles ϕ, ψ, χ^1 and χ^2 measured on the model were used as initial parameters for conformational energy optimization (Arseniev et al., 1986b). Calculations were performed by using fixed bond angles and bond lengths and planar trans-peptide units. Energy parameters and geometry of amino acid residues were identical to those used in ECEPP/2 (Momany et al., 1975; Nemethy et al., 1983).

The energy refined conformation was used as basis for exhaustive search of conformations which are in agreement with NMR data. Torsion angles ϕ, ψ, χ^1 and χ^2 of amino acid residue, one after the other, were modified in a systematic way and after energy minimization each conformation was compared with NMR data set. For this purpose, a special computer program CONFORNMR was developed (Arseniev et al., 1986b). The program uses as an input the list of NMR parameters - NOE cross peaks, spin-spin coupling constants of vicinal protons and solvent accessibility of NH groups. The output of the program is the NOE cross peaks assignment, stereochemical assignment of NMR signals and list of violated spectral parameters.

Analysis of the output leads us to conclude that backbone conformation of gramicidin A incorporated into micelles is unequivocally defined by NMR data, except that of the Gly 2-Ala 3 peptide bond orientation and highly mobile C-terminal ethanolamine group. Side chain

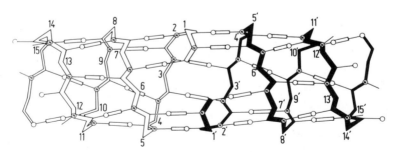

Figure 9. Schematic presentation of gramicidin A channel conformation - right handed $\overrightarrow{\pi}_{L,D}^{6.3}\overleftarrow{\pi}_{L,D}^{6.3}$ helical dimer. Side chains are omitted for clarity. Hydrogen bonds are shown by empty bars.

conformations of all amino acid residues, except for the angles χ^1 of Val 7 and χ^2 of all D-leucine residues, are located in corresponding single potential wells. One of the low energy conformations which is in excellent accordance with the NMR data set is shown in Figure 10. The atomic coordinates and torsional angles of the conformation are listed in Arseniev et al. (1986b).

The aforementioned flexibility of L-Val 7 (χ^1) and all D-leucines (χ^2) side chains and ethanolamine group can be the reason for dispersity of ion conductivity of the gramicidin A transmembrane channel originally observed by Hladky and Haydon (1972). Energy refinement demonstrates that changes in conformation of these fragments modify the orientation of the D-Leu 10, D-Leu 12 and D-Leu 14 carbonyl groups, located at the channel entrance forming a "funnel" (Figure 9 and 10) where cations should be partially dehydrated on their passage across the channel. Indeed, it was shown by Prasad et al. (1986), that replacement of the isopropyl group of L-Val 7, which might modulate the interaction of the carbonyl groups with cation, by methyl group in [L-Ala 7] gramicidin A, leads to a remarkable reduction in dispersion of conducting states of channel.

The derived structure of the sodium complex of gramicidin A, as an open, conductive state of the transmembrane channel, and revealed internal mobility of selective fragments are in accordance with available data on the cationic channel properties of gramicidin A and its numerous synthetic analogs in membranes, although this structure differs in handedness as well as in arrangement of polar NH and carbonyl groups at the entrance to the channel from the widely accepted similar, but left handed structure proposed by Urry et al. (1971 and 1983).

Taking this particular situation into account we feel it necessary to emphasize specific evidences for the right handed helicity of the dimer backbone.

First, the right and left handed $\overrightarrow{\pi}_{L,D}^{6.3}\ \overleftarrow{\pi}_{L,D}^{6.3}$ structures differ in the exposed, free of intramolecular hydrogen bond amide NH groups. The observed fast NH exchange for ethanolamine, D-Leu 14 and D-Leu 12 (Table 1) corresponds precisely to the right handed

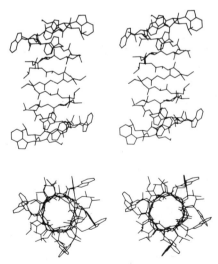

Figure 10. Stereo drawing of the energy refined conformation of the $\overrightarrow{\pi}_{L,D}^{6.3}\ \overleftarrow{\pi}_{L,D}^{6.3}$ head to head right handed single strained helical dimer of gramicidin A. Above: side view, below: channel view.

structure (Figure 9), while for the left handed structure fast exchange is expected for NH groups of L-Trp 11, 13, 15 residues (c.f. Figure 4C in Urry et al., 1983).

Second, the array of NOE connectivities is quite different for right and left handed structures (Bystrov and Arseniev, 1988). For instance, the long range $^6d_{N\alpha}$ and $^6d_{\alpha N}$ contacts (and, hence, short distances) in right handed structure are realized for D-residues and L-residues, respectively (Figure 11). Indeed, this is exactly the case observed in NOESY spectrum (Arseniev et al., 1985b and 1986a). The other way around is anticipated for the left handed structure (Figure 11). Similarly, the NOE connectivities revealing head to head junction could be correlated only with right handed structure of the gramicidin A dimer.

Finally, as mentioned above, the CD spectra of gramicidin A and its metal complexes in organic solvents (Figure 4) and membrane like environment (Figure 6) can be considered as specific fingerprints of the particular backbone folding motif. The right and left handed helices are clearly distinguished by the sign of ellipticity, the positive sign in the range of 210-230 nm corresponding to the right handed helicity. Thus, the CD spectra of the gramicidin A channel state, having positive ellipticity, fit better the right handed structure.

It is of interest to compare revealed by NMR structure of the gramicidin A ion channel in micelles (Figure 9 and 10) and the Cs+-gramicidin A complex in organic solvents (Figure 5). Both helices have hydrophobic external surfaces and polar internal axial cavities. What is more, the dimensions of these helices are the same - axial cavity diameter 4 Å and axial cavity length 26 Å. The question is why gramicidin A adopts $\uparrow\downarrow\pi_{L,D}^{7.2}$ dimer in isotropic organic solvent and $\overrightarrow{\pi}_{L,D}^{6.3} \overleftarrow{\pi}_{L,D}^{6.3}$ helix in anisotropic micelle milieu. The double helix has almost uniform distribution of different types of amino acid residues (valines, leucines, tryptophans) on the external surface. On the contrary, the $\overrightarrow{\pi}_{L,D}^{6.3} \overleftarrow{\pi}_{L,D}^{6.3}$ helix has obvious (Figure 10) anisotropy of its molecular surface which fits anisotropy of a micelle. Indole rings of tryptophan residues are located near the edges of the $\overrightarrow{\pi}_{L,D}^{6.3} \overleftarrow{\pi}_{L,D}^{6.3}$ dimer and are oriented to provide optimal interactions of their dipoles with the negatively charged water-micelle interface.

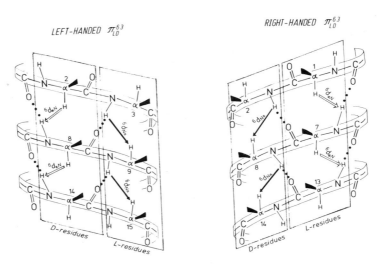

Figure 11. The long range NOE connectivities $^6d_{N\alpha}$ (between NH proton of i residue and $C^\alpha H$ proton of i+6 residue) and $^6d_{\alpha N}$ ($C^\alpha H_i$ and NH_{i+6}) expected for left handed and right handed $\pi_{L,D}^{6.3}$ helices.

The revealed suitable for the high resolution NMR spectroscopy artificial milieu, where the actual transmembrane ion channel conformation is accomplished, opens the way for discovery the structure-function relationships by, for instance, site directed structure modification. One of the most crucial structure element of the gramicidin A channel is the helical monomer junction. Obviously, sterical hindrances at the head to head connection, as introduced by substitution of the N-terminal formyl hydrogen by methyl group in N-acetyl(desformyl)gramicidin A (Figure 12) has to result in diminishing the dimer stability, while the covalent coupling of the monomer helices as in succinyl-bis(desformyl)gramicidin A (Figure 12) has to stabilize the dimer.

The spatial structures of these N-terminal modified gramicidin A analogs incorporated into SDS-d_{25} micelles were established by using the 2D NMR data in combination with energy minimization and turned out to be practically identical with the channel conformation (Barsukov et al., 1987b; Maslennikov et al., 1988). As to their thermal stability, the *a priori* expectations are fully fulfilled: the N-acetyl derivative demonstrates the decrease in the destructing temperature by ~20°C in comparison with the parent compound, while the covalently linked analog retains its structure up to 85°C (Figure 13). Similar results were deduced also from ^{23}Na NMR (Maslennikov et al., 1988). This sequence of the conformational thermal stability directly correlates with life time of the corresponding ion channels in bilayer membranes: for the N-acetyl derivative it is ~60 times shorter than for gramicidin A (Szabo and Urry, 1979), while for the succinyl analog - ~60 times longer (Fonina et al., 1982).

What is specifically essential in the temperature dependent proton spectra of gramicidin A and its N-acetyl analog shown in Figure 13, that above the transition temperature the whole particular spectrum is perturbed and all resonances are broadened. It means that not only the interface between the monomer helices is destroyed due to dissociation of the dimer, but at the same time the monomer helical structure itself is also demolished. These temperature transitions are completely reversible demonstrating fast exchange between the ordered and disordered structure of gramicidin molecules in micelles.

For the mechanism of the gramicidin A channel functioning these observations presumably are significant in indicating that the conversion from the open to closed state might be governed

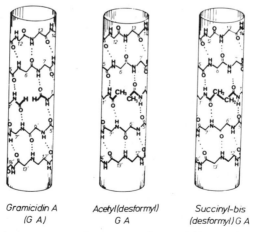

Gramicidin A (G A)   Acetyl(desformyl) G A   Succinyl-bis (desformyl)G A

Figure 12. Schematic presentation of the transmembrane dimer structures (N-terminal to N-terminal dimers of right handed single stranded helices with 6.3 residues per turn, $\overrightarrow{\pi}^{6.3}_{L,D} \overleftarrow{\pi}^{6.3}_{L,D}$) of gramicidin A and its analogs modified at the N-terminal site: Nacetyl(desformyl)gramicidin A and succinyl-bis(desformyl)gramicidin A.

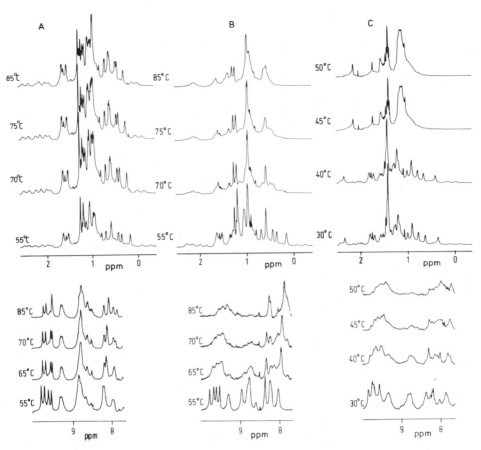

Figure 13. High (above) and low (below) field regions of 500 MHz proton NMR spectra of 5 mM (a) succinyl-bis(desformyl)gramicidin A, (b) gramicidin A and (c) N-acetyl(desformyl)gramicidin A incorporated into SDS-d_{25} (250 mM) micelles in H_2O-CF_3CD_2OH (16:1), pH 7 at different temperatures, indicated to the left side of the spectrum.

not only by dissociation of dimers, as was suggested by Elliot et al. (1983) from the channel life time dependence on surface tension of membranes, but also by cooperative disintegration of the whole molecular structure.

One of the fundamental questions of interest is the origin of ion selectivity. The gramicidin A channel is impermeable to anions and divalent cations and yet it exhibits relatively weak specificity for monovalent cations with permeability sequence H^+ > NH_4^+ > Cs^+ > Rb^+ > K^+ > Na^+ > Li^+ (Myers and Haydon, 1972). In principle the selectivity might be due to changes of the gramicidin A conformation. But the NMR results reveal no pronounced influence of a sort of a complexed monovalent cation (Li^+, Na^+ and Tl^+) on the spatial structure of gramicidin A incorporated into corresponding dodecyl sulphate micelles (Arseniev et al., 1986a). Hence, the same gramicidin A molecular model, right handed dimer $\overrightarrow{\pi}_{L,D}^{6.3} \overleftarrow{\pi}_{L,D}^{6.3}$, has to be used for theoretical computations of factors responsible for interactions of different monovalent ions with the ion channel.

The divalent cations, like Ca^{2+} and Ba^{2+}, cannot penetrate trough the channel although they interact with it and block the monovalent cation conductance (Bamberg and Lauger, 1977; Urry et al., 1983). To study the mechanism of interaction and blocking the paramagnetic Mn^{2+} cations were used to induce specific relaxation contribution for protons closely located to the ion binding sites. Before that it was shown by single channel current-voltage measurements that Mn^{2+} as other divalent cations does not penetrate through the gramicidin A channel and blocks the alkali ion transport (Golovanov et al., 1989b).

By adding varied amounts of Mn^{2+} salt to the gramicidin A preparation in SDS micelles and evaluating different sources of paramagnetic contribution to the proton longitudinal relaxation rates, the distances between the Mn^{2+} binding site and the channel protons were estimated (Golovanov et al., 1989a). Taking into account that distinct evidences indicate preservation of the gramicidin A dimer conformation in micelles on interaction with Mn^{2+} ions, the atomic coordinates of the channel (Arseniev et al., 1986b) were used to calculate the position of the binding site, as shown by the stereo drawing of the structure in Figure 14 and schematically in Figure 15.

It is evident that the Mn^{2+} cation resides at the channel entrance and that there are two symmetrical binding sites on both ends of the dimer located at the micelle surface (Figure 14). The binding site is shifted by ~3 Å from the inner pore axis to provide favorable interaction with exposed carbonyl groups of the D-Leu 10, D-Leu 12 and D-Leu 14 residues (Figure 15). The carbonyl groups of the L-Trp 11, L-Trp 13 and L-Trp 15 involved into intramolecular hydrogen

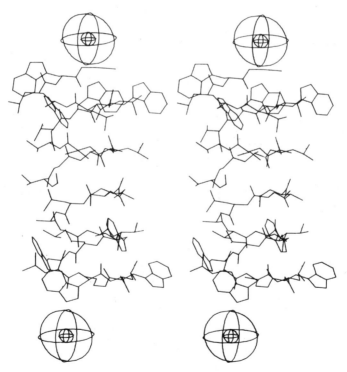

Figure 14. Stereo view of the gramicidin A channel structure (right handed $\overrightarrow{\pi}_{L,D}^{6.3}$ $\overleftarrow{\pi}_{L,D}^{6.3}$ helical dimer) with bound Mn^{2+} cations. There are two symmetrical binding sites per dimer. The outer sphere of the cations corresponds to the hydration shell 6 Å in diameter and the inner sphere - to the cation diameter 1.6 Å.

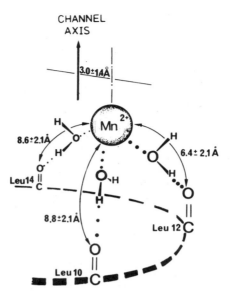

Figure 15. Schematic presentation of the Mn^{2+} binding site at the entrance of the gramicidin A transmembrane channel. The axis of the channel is indicated by dashed vertical line. Assumed positions of water molecules in the Mn^{2+} hydration shell are shown.

bonding, are also at a relatively close distance from the binding site (9-10 Å) that accords with previously observed ^{13}C chemical shifts of the carbonyl groups in the presence of divalent cations Ca^{2+} and Ba^{2+} (Urry et al., 1983).

To judge by the distances between bound Mn^{2+} and the nearest carbonyl oxygens (Figure 15), it is evident that the divalent cation preserves its hydration shell. Presumably the energy of divalent cation interaction with the channel carbonyl groups is too low for even partial dehydration and thus the solvated ion (6 Å in diameter) cannot penetrate trough the channel which is only 4 Å in diameter. The ion, strategically located at the channel entrance, creates electrostatic barrier that prevents the approach of monovalent cations and their penetration through the channel. Rapid exchange of Mn^{2+} between the bound and free states, which is evident from the NMR spectra, explains reduction of the effective conductance of the gramicidin A channel in the presence of divalent cations.

In summary, using 2D NMR spectroscopy a variety of double stranded intertwined helices of gramicidin A and its alkali metal complexes were revealed to exist in organic solvents. Thanks to developing an adequate artificial milieu the 2D NMR has proved that on the contrary, gramicidin A being incorporated into micelles, vesicles, liposomes and black lipid membranes adopts a conformation of a single stranded right handed $\overrightarrow{\pi}_{L,D}^{6.3} \overleftarrow{\pi}_{L,D}^{6.3}$ helical dimer and thus deeper insight into the intimate mechanism of ion channeling and blocking has been gained.

BACTERIORHODOPSIN

Bacteriorhodopsin, the sole protein component of the purple membrane of Halobacterium halobium, consists of a single polypeptide chain of 248 amino acid residues (Ovchinnikov et al., 1979; Khorana et al., 1979) and the retinal chromophore bound via the Shiff base to lysine 216

(Figure 16). Light absorption by the chromophore triggers a photocycle that leads to proton translocation across the cell membrane. By electron diffraction of the ordered two dimensional crystals of purple membranes Henderson and Unwin (1975) demonstrated that bacteriorhodopsin is folded in seven transmembrane segments. Additional results obtained by selective chemical and enzymatic modification, partial proteolytic cleavage (Ovchinnikov, 1982), cross linking (Huang et al., 1982), monoclonal antibody binding (Ovchinnikov et al., 1985), site directed mutagenesis (Khorana, 1988) and thermally activated tritium labeling (Tsetlin et al., 1988) provide further insight into the localization and interactions of the molecular transmembrane and surface loop fragments in the amino acid sequence. One of the derived secondary structure models is shown in Figure 16. However the precise folding and detailed conformation of bacteriorhodopsin remain to be disclosed.

The major problem in using high resolution NMR spectroscopy to solve the bacteriorhodopsin structure was to reveal a proper membrane mimicing milieu and to develop a suitable protocol for transferring the native conformation into the milieu. That was achieved by isolation of bacteriorhodopsin and its proteolytic fragments (Figure 16) by Sephadex LH-60 chromatography directly in the medium shown to be adequate to preserve the native like structure. Careful consideration demonstrated that the most convenient medium is the 1:1 mixture of methanol and chloroform with 0.1 M $LiClO_4$ (Arseniev et al., 1987). The CD spectrum (Figure 17) of bacteriorhodopsin solubilized in this medium has the same shape as that of the purple membrane (see Wallace and Mao, 1984) and is practically identical with the spectrum of bacteriorhodopsin in Triton X-100 micelles, where a monomeric molecule keeps spectral as well as many structural features of the native chromoprotein (Dencher and Heyn, 1979). It should be noted that in CH_3OH-$CHCl_3$ not only the CD pattern of the native bacteriorhodopsin is preserved but also the retinal-protein aldimine bond is retained. If small

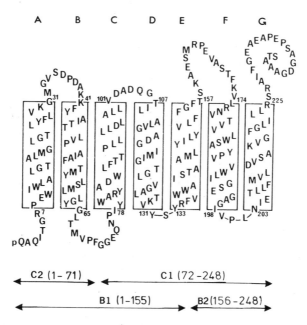

Figure 16. Primary structure of bacteriorhodopsin arranged according to the model of Khorana (1988). Seven transmembrane segments are designated by A-G; C1, C2 and B1, B2 denote the proteolytic peptides produced by α-chymotrypsin and NaBH$_4$ cleavage, respectively.

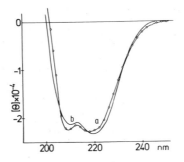

Figure 17. CD spectra of bacteriorhodopsin in Triton X-100 (a) and in methanol-chloroform (1:1), 0.1 M LiClO$_4$ (b).

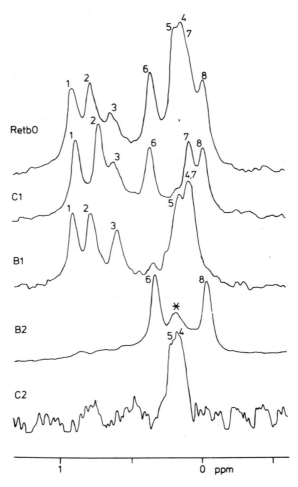

Figure 18. ^{19}F NMR spectra (470.6 MHz) of 0.1 mM [Trp(5F)]bacteriorhodopsin and its proteolytic fragments obtained by α-chymotrypsin (C1 and C2) and NaBH$_4$ (B1 and B2) cleavage in methanol-chloroform (1:1), 0.1 M LiClO$_4$ at 32°C. Origin of signal marked by asterisk is not known.

amount of formic acid is added to the solution, a typical bathochromic fluorescence shift of 370 -> 455 nm occurs due to Shiff base protonation.

Still bacteriorhodopsin molecule is too large to provide high resolution proton NMR spectrum appropriate for conformational analysis. Thus, in order to probe the spatial structure of the molecule in methanol-chloroform, ^{19}F NMR spectra were recorded for biosynthetically prepared 5-fluorotryptophan labeled bacterio-rhodopsin (Kuryatov et al., 1984), where all eight tryptophan residues were substituted by Trp(5F). The [Trp(5F)]bacteriorhodopsin in membrane suspensions retains the specific CD and laser Raman spectral patterns, as well as chromophore absorption maximum and the capacity for light driven proton transport of native bacteriorhodopsin.

^{19}F NMR spectrum of [Trp(5F)]bacteriorhodopsin in CD_3OH-$CHCl_3$ (deuterated methanol was used for field-frequency locking of the NMR spectrometer) demonstrates uniform incorporation of Trp(5F) residues into the polypeptide chain (Arseniev et al., 1987). Aldimine bond reduction and β-ionone ring modification induce changes of selected peak positions indicating that chromophore attached to the Lys 216 residue is located in close environment of distinct Trp(5F) residues although remoted in amino acid sequence from Lys 216. This clearly demonstrates the presence of a specific spatial structure of bacteriorhodopsin in CD_3OH-$CHCl_3$ (1:1), 0.1 M $LiClO_4$. Titration of [Trp(5F)]bacteriorhodopsin by formic acid reveals pronounced concentration dependence of two ^{19}F resonance chemical shifts associated with protonation of ionogenic groups spatially close to the corresponding Trp(5F) residues.

The ^{19}F NMR spectra of bacteriorhodopsin fragments (Figure 18) obtained by limited proteolytic cleavage indicate retention of the native tertiary structure. Thus they are eligible for signal assignment and for delineation of intramolecular interactions (Arseniev et al., 1987). The conclusion is supported by corresponding CD spectra study (Abdulaeva et al., 1987).

Figure 19 summarizes the partial ^{19}F signal assignment and depicts the chemical shift response to intramolecular interactions. For instance, the segment A proximity (Trp 10 or Trp 12) to aldimine bond of the F segment Lys 216 residue is obvious, as well as the close distance between β-ionone ring and the segment E (Trp 137 or Trp 138).

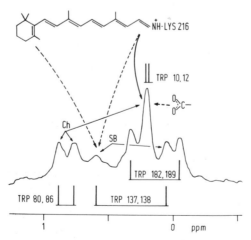

Figure 19. Schematic presentation of the ^{19}F signal assignment of bacteriorhodopsin solubilized in methanol-chloroform (1:1), 0.1 M $LiClO_4$. Ch and SB indicate effects of α-chymotrypsin and $NaBH_4$ cleavage, respectively.

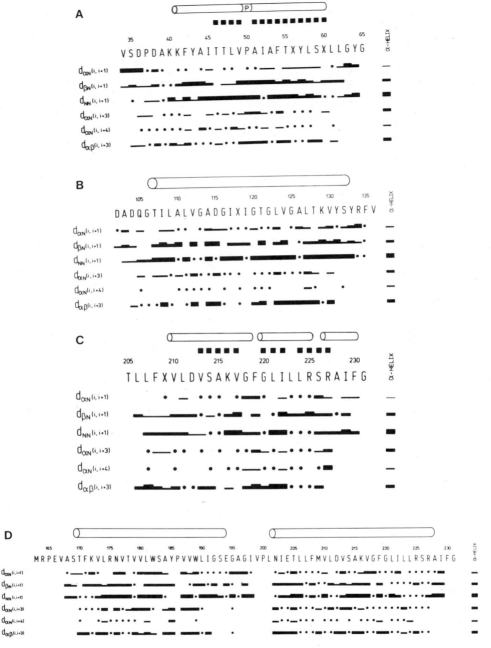

Figure 20. Amino acid sequences and survey of NOE connectivities involving NH, CαH and
CβH protons of bacteriorhodopsin segments B (34-65 residues), D (102-136
residues) and G (205-231 residues), and combined F-G segment (163-
231 residues - next page) obtained from B2 fragment. One letter code is used for
amino acid residues, X - norleucine residue. The NOE intensities classified as
strong, medium and weak are indicated by thick, medium and thin lines,
respectively. Filled circles correspond to so far doubtful assignment of
corresponding NOE cross peak. On the right hand side of the diagrams are shown

Proper choice of the membrane mimicing milieu provides opportunity to study the detailed conformation of the membrane spanning segments themselves. Up to now the proton 2D NMR spectra were recorded for the following fragments of bacteriorhodopsin: synthetic segments B (34-65 amino acid residues, see Figure 16), D (102-136) and G (205-231), and combined segments F-G (163-231) produced by papain treatment of the $NaBH_4$ cleaved purple membrane. The signal assignment was made by standard procedure. The NOE cross peak intensities (Figure 20) were classified as strong, medium and weak by counting the exponentially spaced contour levels in NOESY spectra of all these polypeptides definitely demonstrate the existence of the single predominant ordered structure.

Preliminary spatial structure elucidation allows us to delineate in all segments the extended right handed α-helical portions of 20-25 amino acid residues, long enough to penetrate the bilayer membrane. They are indicated in Figure 20 by horizontal cylinders. Most of the fragments are still under further refinement and conformational evaluation. Only the spatial structure of segment B was built up with some details (Arseniev et al., 1988). The remarkable feature of its α-helical part is the proline residue just in the center of the helix (Figure 21).

Figure 21. Schematic presentation of the right handed α-helical conformation of segment B in the Val 49-Pro 50 region. Arrows indicate NOE connectivities observed between protons of Pro 50 and other residues. Dots show intramolecular hydrogen bonds.

the NOE intensity pattern expected for i residue in regular right handed α-helix. The residues with slowly exchanged peptide NH protons are indicated by filled circles above the sequences.

Although usually proline is considered as a helix breaking residue, in this particular case it only introduces a slight bend in the helical axis.

The obtained, although at present tentative data could be nevertheless superimposed on the well known wheel model of Khorana (1988) (Figure 22). Orientation of the segment B, D, F and G wheels are chosen as to direct the hydrophobic surfaces of corresponding α-helices (Figure 20) into the periphery of the protein to face the lipid environment. Thus the functional groups are positioned in inner site of the molecule. Specifically there are shown the deduced from ^{19}F NMR spectra close proximity of aldimine bond to Trp 12 (or Trp 10) and of β-ionone ring to Trp 137 (or Trp 138). As soon as segments A and E are studied by NMR and the α-helical parts are established, the assignment of these Trp residues could be specified according to the proper orientation of the hydrophobic surface.

To summarize, the suitable membrane mimicing milieu for bacteriorhodopsin and its fragments is revealed. Further study is in progress to establish details of the spatial structure and interactions within the segments, as well as between the segments. We expect that some structure-functional relationships will be elucidated in this way.

CONCLUSION

Intrinsic membrane protein spatial structure presents a challenge to the NMR spectroscopy. Currently no universal technique exists to evaluate their conformation directly in native environment. Presumably for each particular system specific approach has to be developed. This paper considers two such cases with different protocols of transferring the native like conformation into membrane mimicing medium suitable for high resolution NMR spectroscopy. For gramicidin A the obtained results allowed to reach the final conclusion on the structure in the membrane and to disclose the route for further structure-functional investigations. The bacteriorhodopsin project is at the initial stage and much effort has to be involved to reach more conclusive outcome. In addition we strongly hope that further development of ever green NMR spectroscopy will discover entirely new extensive opportunities for the transmembrane structure elucidation.

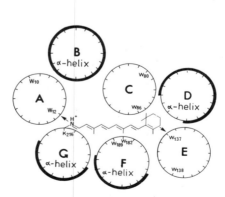

Figure 22.   Helical wheel model of bacteriorhodopsin (Khorana, 1988) with NMR results taken into account.

REFERENCES

Abdulaeva, G.V., Sychev, S.V. and Tsetlin, V.I., 1987, CD and IR study of the secondary structure of bacteriorhodopsin and its fragments in model systems, *Biol. Membrany (Sov. J. Biol. Membr.)* 4:1254.

Andersen, O.S., 1984, Gramicidin channels, *Ann. Rev. Physiol.* 46:531.

Arseniev, A.S., Barsukov, I.L., Sychev, S.V., Bystrov, V.F., Ivanov, V.T. and Ovchinnikov, Yu.A., 1984a, Double helical conformation of gramicidin, *Biol. Membrany (Sov. J. Biol. Membr.)* 1:5.

Arseniev, A.S., Bystrov, V.F., Ivanov, V.T. and Ovchinnikov, Yu.A., 1984b, NMR solution conformation of gramicidin A double helix, *FEBS Lett.* 165:51.

Arseniev, A.S., Barsukov, I.L. and Bystrov, V.F., 1985a, NMR solution structure of gramicidin A complex with a caesium cations, *FEBS Lett.* 180:33.

Arseniev, A.S., Barsukov, I.L., Bystrov, V.F., Lomize, A.L. and Ovchinnikov, Yu.A., 1985b, ^{1}H-NMR study of gramicidin A transmembrane ion channel. Head-to-head right-handed, single-stranded helices, *FEBS Lett.* 186:168.

Arseniev, A.S., Barsukov, I.L., Bystrov, V.F. and Ovchinnikov, Yu.A., 1986a, Spatial structure of gramicidin A transmembrane ion channel - NMR analysis in micelles, *Biol. Membrany (Sov. J. Biol. Membr.)* 3:437.

Arseniev, A.S., Lomize, A.L., Barsukov, I.L. and Bystrov, V.F., 1986b, Gramicidin A transmembrane ion-channel. Three-dimensional structure reconstruction based on NMR spectroscopy and energy refinement, *Biol. Membrany (Sov. J. Biol. Membr.)* 3:1077.

Arseniev, A.S., Kuryatov, A.B., Tsetlin, V.I., Bystrov, V.F., Ivanov, V.T. and Ovchinnikov, Yu.A., 1987, ^{19}F NMR study of 5-fluorotryptophan-labeled bacteriorhodopsin, *FEBS Lett.* 213:283.

Arseniev, A.S., Maslennikov, I.V., Bystrov, V.F., Kozhich, A.T., Ivanov, V.T. and Ovchinnikov, Yu.A., 1988, Two-dimensional ^{1}H-NMR study of bacteriorhodopsin-(34-65)-polypeptide conformation, *FEBS Lett.* 231:81.

Bamberg, E. and Lauger, P., 1977, Channel formation kinetics of gramicidin A in lipid bilayer membranes, *J. Membr. Biol.* 35:351.

Barsukov, I.L., 1987, "NMR study of conformational heterogeniety of gramicidin A in solutions," PhD thesis, Shemyakin Institute, Moscow.

Barsukov, I.L., Arseniev, A.S. and Bystrov, V.F., 1987a, Spatial structure of gramicidin A in organic solvents. ^{1}H-NMR analysis of four species in ethanol, *Bioorg. Khim. (Sov. J. Bioorg. Chem.)* 13:1501.

Barsukov, I.L., Lomize, A.L., Arseniev, A.S. and Bystrov, V.F., 1987b, Spatial structure of succinyl-bis(desformyl)gramicidin A in micelles. NMR conformational analysis, *Biol. Membrany (Sov. J. Biol. Membr.)* 4:171.

Braun, W., Bosch, Ch., Brown, L.R., Go, N. and Wüthrich, K., 1981, Combined use of proton-proton Overhauser enchancements and a geometry algorithm for determination of polypeptide conformation. Application to micelle-bound glucagon, *Biochim. Biophys. Acta* 667:377.

Braun, W., Wider, G., Lee, K.H. and Wüthrich K., 1983, Conformation of glucagon in a lipid-water interphase by ^{1}H nuclear magnetic resonance, *J. Mol. Biol.* 169:921.

Brown, L.R., Braun, W., Anil-Kumar, G. and Wüthrich K., 1982, High resolution nuclear magnetic resonance studies of the conformation and orientation of mellitin bound to a lipid-water interface, *Biophys J.* 37:319.

Bystrov, V.F., 1985, NMR and rotational angles in solution conformation of polypeptides, *J. Mol. Str.* 126:529.

Bystrov, V.F., Arseniev, A.S., Barsukov, I.L. and Lomize, A.L., 1987, 2D NMR of single and double stranded helices of gramicidin A in micelles and solutions, *Bull. Magn. Resonance* 8:84.

Bystrov, V.F. and Arseniev, A.S., 1988, Diversity of the gramicidin A spatial structure: two-dimensional ^{1}H NMR study in solution, *Tetrahedron* 44:925.

Dencher, N.A. and Heyn, M.P., 1979, Formation and properties of bacteriorhodopsin monomers in the non-ionic detergents octyl-β-D-glucoside and triton X-100, *FEBS Lett.* 96:322.

Dubos, R.J., 1939, Bacterial agent extracted from a soil bacillus. I. Preparation of the agent. Its activity *in vitro.*, *J. Exp. Med.* 70:1.

Elliot, J.R., Needham, D., Dilger, J.P. and Haydon, D.A., 1983, The effect of bilayer thickness and tension on gramicidin single-channel lifetime, *Biochim. Biophys. Acta* 735:95.

Fonina, L., Demina, A., Sychev, S., Irkin, A., Ivanov. V. and Hlavacek, J., 1982, Synthesis, structure and membrane properties of gramicidin A dimer analogs, *in:* "Chemistry of Peptides and Proteins", vol. 1, W. Voelter, J. Wünsh, Yu. Ovchinnikov and V. Ivanov, eds., Walter de Gruyer, West Berlin, p. 260.

Golovanov, A.P., Barsukov, L.I., Arseniev, A.S., Bystrov, V.F., Sukhanov, S.V. and Barsukov, I.L., 1989a, The divalent cation-binding sites of gramicidin A transmembrane ion-channel, submitted for publication.

Golovanov, A.P., Sukhanov, S.V. and Barsukov, L.I., 1989b, Mn^{2+} and Cd^{2+} influence on the ionic conductance of gramicidin A channels, *Biol. Membrany (Sov. J. Biol. Membr.)*, in press.

Gronenborn, A.M. and Clore, G.M., 1989, Three-dimensional structures of proteins in solution by nuclear magnetic resonance spectroscopy, *Protein Seq. Data Anal.* 2:1.

Henderson, R. and Unwin, P.N.T., 1975, Three dimensional model of purple membrane obtained by electron microscopy, *Nature* 257:28.

Higashijima, T., Wakamatsu, K., Milon, A., Okada, A. and Miyazawa, T., 1988, Conformational analysis of physiologically active peptides in membrane-bound state: correlation with physiological activity, *in:* "Peptide Chemistry 1987", T. Shiba and S. Sakakibara, eds., Protein Research Foundation, Osaka, p. 679.

Hladky, S.B. and Haydon, D.A., 1972, Ion transfer across lipid membrane in the presence of gramicidin A. I. Studies of the unit conductance channel, *Biochim. Biophys Acta* 274:294

Huang, K.-S., Ramachandran, R., Bayley, H. and Khorana, H.G., 1982, Orientation of retinal in bacteriorhodopsin as studied by cross-linking using photosensitive analog of retinal, *J. Biol. Chem.* 257:13616.

Khorana, H.G., Gerber, G.E., Herlihy, W.G., Gray, C.P., Anderegg, R.J., Niheu, K. and Biemann, K., 1979, Amino acid sequence of bacteriorhodopsin, *Proc. Natl. Acad. Sci. USA* 76:5046.

Khorana, H.G., 1988, Bacteriorhodopsin, a membrane protein that uses light to translocate proton, *J. Biol. Chem.* 263:7439.

Kuryatov, A.B., Ovechkina, G.V., Alyonycheva, T.N., Minaeva, L.P. and Tsetlin V.I., 1984, Byosynthetic ^{19}F- and ^{14}C-derivatives of bacteriorhodopsin, *Bioorg. Khim. (Sov. J. Bioorg. Chem.)* 10:333.

Langs, D.A., 1988, Three-dimensional structure at 0.86 Å of the uncomlexed form of the transmembrane ion channel peptide gramicidin A, *Science* 241:188.

Lee, K.H., Fitton, J.E. and Wüthrich, K., 1987, Nuclear magnetic resonance investigation of the conformation of δ-haemolysin bound to dodecylphosphocholine micelles, *Biochim. Biophys. Act* 911:144.

Long, M.M. and Urry, D.W., 1981, Absorption and circular dichroism spectroscopies, *in:* "Membrane Spectroscopy", E. Grell, ed., Springer, West Berlin, p. 143.

Maslennikov, I.V., Arseniev, A.S. and Bystrov, V.F., 1988, Spatial structure of N-acetyl(desformyl)gramicidin A in micelles, *Biol. Membrany (Sov. J. Biol. Membr.)* 5:459.

Masotti, L., Spisni A. and Urry, D.W., 1980, Conformational studies on the gramicidin A transmembrane channel in lipid micelles and liposomes, *Cell. Biophys.* 2:241.

Mazer, N.A., Benedek, B.B. and Carcy, M.C., 1976, An investigation of the micellar phase of sodium dodecyl sulfate in aqueous sodium chloride solutions using quasielastic light scattering spectroscopy, *J. Phys. Chem.* 80:1075.

McDonald, C.C. and Philips, W.D., 1967, Manifestation of the tertiary structures of proteins in high-frequency nuclear magnetic resonance, *J. Amer. Chem. Soc.* 89:6332.

Momany, F.A., McGuire, R., Burgess, A.W. and Sheraga, H.A., 1975, Energy parameters in polypeptides. 7. Geometric parameters, partial atomic charges, nonbonded interactions, hydrogen bond interactions, and intrinsic torsional potentials for the naturally occuring amino acids, *J. Phys. Chem.* 79:2631.

Myers, V.B. and Haydon, D.A., 1972, Ion transfer across lipid membranes in the presence of gramicidin A, *Biochim. Biophys. Acta* 274:313.

Neher, E., 1975, Ionic specificity of the gramicidin channel and the thallous ion, *Biochim. Biophys. Acta* 401:540.

Nemethy, G., Pottle, M.S. and Sheraga, H.A., 1983, Energy parameters in polypeptides. 9. Updating of geometrical parameters, nonbonding interactions and hydrogen bond interactions for naturally occuring amino acids, *J. Phys. Chem.* 87:1883.

Ovchinnikov, Yu.A. and Ivanov, V.T., 1975, Conformational states and biological activity of cyclic peptides, *Tetrahedron* 31:2177.

Ovchinnikov, Yu.A., Abdulaev N.G., Feigina, M.Yu., Kiselev, A.V. and Lobanov, N.A., 1979, The structural basis of the functioning of bacteriorhodopsin: an overview, *FEBS Lett.* 100:219

Ovchinnikov, Yu.A., 1982, Rhodopsin and bacteriorhodopsin: structure-function relationships, *FEBS Lett.* 148:179.

Ovchinnikov, Yu.A., Abdulaev N.G., Vasilov, R.G., Vturina, I.Yu., Kuryatov, A.B. and Kisilev, A.V., 1985, The antigenic structure and topography of bacteriorhodopsin in purple membranes as determined by interaction with monoclonal antibodies, *FEBS Lett.* 179:343.

Paulus, H., Sarkar, N., Mukherjee, P.K., Langley, D., Ivanov, V.T., Shepel, E.N. and Veatch W.R., 1979, Comparison of the effect of linear gramicidin A analogues on bacterial sporulation, membrane permeability and ribonucleic acid polymerase, *Biochemistry* 18:4532.

Prasad, K.U., Alonso-Romanowski, S., Venkatachalam, C.M., Trapane, T.L. and Urry D.W., 1986, Synthesis, characterization, and black lipid membrane studies of [7-L-alanine]gramicidin A, *Biochemistry* 25:456.

Ramachandran, G.N. and Chandrasekharan, R., 1972, Conformation of peptide chains containing both L- and D-residues: Part 1 - Helical structures with alternating L- and D-residues with special reference to LD-ribbon and LD-helices, *Indian J. Biochem. Biophys.* 9:1; Studies on dipeptide conformation and on peptide with sequences of alternating L and D residues with special reference to antibiotic and ion transport peptide, *Progr. Peptide Res.* 2:195.

Sarges, R. and Witkop, B., 1965, Gramicidin A. 5. The structure of valine and isoleucine-gramicidin A, *J. Amer, Chem. Soc.* 87:2011.

Sychev, S.V. and Ivanov, V.T., 1984, Conformational states of gramicidin A in phospholipid liposomes, *Biol. Membrany (Sov. J. Biol. Membr.)* 11:1109.

Szabo, G. and Urry, D.W., 1979, N-acetyl gramicidin: single channel properties and implications for channel structure, *Science* 203:55.

Tsetlin V.I., Alyonycheva T.N., Shemyakin V.V., Neiman L.A. and Ivanov V.T., 1988, Tritium thermal activation study of bacteriorhodopsin topography, *Eur. J. Biochem.* 178:123.

Urry, D.W., Goodall, M.C., Glickson, D. and Mayers, F., 1971, The gramicidin A transmembrane channel: characteristics of head-to-head dimerized π_{LD} helices, *Proc. Nat. Acad. Sci. USA* 68:1907.

Urry, D.W., Venkatachalam, C.M., Spisni, A., Lauger, P. and Khaled, M.A., 1980, Rate theory calculation of gramicidin single-channel currents using NMR-derived rate constants, *Proc. Nat. Acad. Sci. USA* 77:2028.

Urry, D.W., Trapane, T.L. and Prasad, K.U., 1983, Is the gramicidin A transmembrane channel single stranded or double-stranded helix? A simple unequivocall determination, *Science* 221:1064.

Urry, D.W., 1987, NMR relaxation studies of alkali metal ion interactions with the gramicidin A transmembrane channel, *Bull. Magn. Resonance* 9:109.

Veatch, W.R. and Blout E.R., 1974, The agregation of gramicidin A in solution, *Biochemistry* 13:5257.

Veatch, W.R., Fossel, E.T. and Blout, E.R., 1974, The conformation of gramicidin A, *Biochemistry* 13:5249.

Veatch, W.R. and Stryer, L., 1977, The dimeric nature of the gramicidin A transmembrane channel: conductance and fluorescence energy transfer studies of hybrid channels, *J. Mol. Biol.* 113:89.

Wakamatsu, K., Higashijima, T., Nakajima, T., Fujino, M. and Miyazawa, T., 1984, 1H-NMR analysis of conformation of mastoparans as bound to phospholipid bilayer, *in:* "Peptide Chemistry 1983", E. Munekatta, ed., Protein Research Foundation, Osaka, p.237.

Wakamatsu, K., Okada, A., Miyazawa, T., Masui, Y.,Sakakibara, S. and Higashijima, T., 1987, Conformation of yeast α-mating factor and analog peptides as bound to phospholipid bilayer. Correlation of membrane-bound conformation with physiological activity, *Eur J. Biochem.* 163:331.

Wallace, B.A., Veatch, W.R. and Blout, E.R., 1981, Conformation of gramicidin A in phospholipid vesicles: circular dichroism studies of effects of ion binding, chemical modification and lipid structure, *Biochemistry* 20:5754.

Wallace, B.A. and Mao, D., 1984, Circular dichroism analysis of membrane proteins: an examination of differential light scattering and absorption flattering effects in large membrane vesicles and membrane sheets, *Analyt. Biochem.* 142:317.

Wallace, B.A., 1987, The structure of gramicidin A, a transmembrane ion channel, *in:* "Ion transport through membranes", K. Yagi and B. Pullman, eds., Academic Press, New York, p. 255.

Wallace, B.A. and Ravikumar, K., 1988, The gramicidin pore, *Science* 241:182.

Wieder, G., Lee, K.H. and Wüthrich, K., 1982, Sequential resonance assignments in protein 1H-NMR spectra: glucagon bound to perdeuterated dodecylphosphocholine micelles, *J. Mol. Biol.* 155:367.

THE DYNAMICS OF OLIGONUCLEOTIDES AND PEPTIDES DETERMINED BY PROTON NMR

Patrice Koehl, Bruno Kieffer and Jean-François Lefèvre

Groupe de Cancerogénèse et de Mutagénèse Moléculaire et Structurale, Institut de Biologie Moleculaire et Cellulaire, 15 rue Descartes, 67000 Strasbourg France

INTRODUCTION

It is widely accepted that knowledge of the structure of a biological macromolecule is an important key to understanding its biological function. On the other hand, it is known that these macromolecules have adaptable structures that are specifically modified during their interactions with their partners. This adaptability is generally due to the flexibility of structural motifs in the macromolecule that are arranged in a specific way in space. The flexibility is possible because the potential energy around the average structure of the motif is quite flat, so that the structural motif is continuously moving between several isoenergetical structures, giving rise to the so called internal motion of the macromolecule. The knowledge of this internal motion is an important step toward understanding the biological function of a macromolecule and, consequently, toward engineering macromolecules.

The relationship between the structure and function of a biological macromolecule is often explored by mutations that alter the molecule function. Here again, we face the necessity of having a precise description of the motion and the structure of the molecule, since changing one amino acid in a protein or one nucleotide in an oligonucleotide often induces only a subtle change in the conformation or dynamics of the macromolecule.

NMR is a powerful method for studying the internal motions that agitate macromolecules in solution, as the relaxation phenomena are directly related to the correlation time and the amplitude of this motion. NMR is presently used for macromolecular structure determination; in fact, it has proven to be an efficient tool for finding secondary structure elements (e.g. alpha helix, beta sheet, turn) along the amino acid sequence of a protein and in elucidating the folding of these elements into a tertiary structure (Altman & Jardetzky, 1989; Braun & Go, 1985; Kaptein et al., 1985; Kline et al., 1986; Wüthrich, 1986). Current methods are based on the translation of observed nuclear Overhauser effects (NOE) in order to obtain a rough estimate of the distances between protons, thereby establishing spatial proximities between remote residues in the sequence.

However, the lack of precision prevents the application of these methods to certain kinds of macromolecules whose main structural features do not relate to a tertiary structure. Macromolecules of this kind include oligonucleotide and linear or cyclic polypeptides of limited

139

number of amino acids. Understanding the structure of these molecules depends on the determination of distances between immediate neighbors. Imprecise distances do not sufficiently constrain the conformational space available to the molecule. In these cases, there is a need to perform more accurate determinations of the structural parameters.

In the process of determining structure and dynamics, difficulties are encountered at three different levels:

1 - An experimental problem: even in a two dimensional experiment, overlap between cross peaks hinders the determination of all NOEs between all pairs of protons separated by a short distance.

2 - A problem in the interpretation of NOE: the values of the observed NOEs include indirect magnetization transfer (the so called spin diffusion) so that direct translation of NOE in distance is no longer trivial (Keepers & James, 1984, Lefèvre et al., 1987).

3 - A problem in the interpretation of relaxation parameters: internal motion has intricate effects on the correlation time and the averaging of distances.

We propose an approach in which the relaxation parameters, which are in fact the primary NMR parameters, are determined first. This procedure takes into account the spin diffusion phenomenon and circumvents the problem of underdetermining the set of NOEs. We also propose separating the determination of the dynamics from that of structure. The theme of this paper is therefore an approach in which the relaxation parameters are used first to study the internal motion of a molecule.

THE METHOD FOR RECONSTRUCTING THE RELAXATION MATRIX PARTIALLY OR TOTALLY

Nuclear Overhauser Effects are Non-Linear Functions of the Relaxation Parameters

The relaxation behavior of protons in a macromolecule is generally dominated by dipolar interactions. The Bloch equations (Bloch, 1946), adapted by Solomon (Solomon, 1955) for several dipolar coupled spins, describe the time dependence of the longitudinal magnetization of a spin i:

$$dM_i/dt = -\rho_i (M_i - M_i^0) - \sum_{i \neq j} \sigma_{ij} (M_j - M_j^0) \qquad [1]$$

where ρ_i is the longitudinal relaxation rate constant of the spin i, and σ_{ij} is the cross relaxation rate constant between the spin i and the spin j. In the case of homonuclear relaxation, the expression of these relaxation rate constants are well known (Abragam, 1986):

$$\rho_i = \alpha . \sum_j (1/r_{ij})^6.[J(0) + 3J(\omega) + 6J(2\omega)] \qquad [2]$$

$$\sigma_{ij} = \alpha . (1/r_{ij})^6.[6J(2\omega) - J(0)] \qquad [3]$$

For protons, the constant α is equal to $5.692 \cdot 10^{-38}$ $cm^6.s^{-2}$. $J(\omega)$ are the spectral density functions, which depend on the resonance frequency (ω) and the correlation time of the inter-proton vector motion (τ_c):

$$J(\omega) = \tau_c / (1 + \omega^2 \tau_c^2) \tag{4}$$

Equation [1] assumes that all the dipole-dipole interactions are fluctuating independently and does not include the relaxation through cross correlated fluctuations of the dipolar interactions (Bölhen et al, 1988). However, these cross correlation effects are generally neglected, as they are not observable in a traditional 1D or 2D experiment used to measure the Nuclear Overhauser Effect (NOE) (Bull, 1987).

The longitudinal relaxation rate constant can be measured alone through inversion or saturation recovery experiments, by inverting or saturating selectively the resonance of the proton i. However, this experiment provides only global information on the interactions of the spin i with its surrounding spins.

More interesting is the cross relaxation rate constant which involves two spins only. This constant can be deduced from the build up of NOE. This effect measures the change of magnetization of a spin i, when a neighboring spin j has been displaced from its equilibrium (by inversion or saturation):

$$NOE_{ij} (tm) = \{ M_i(t) - M_i^o \} / M_i^o \tag{5}$$

tm is the mixing time during which the magnetization transfer between the two spins is allowed.

Including [5] in [1] leads to an equation describing the kinetics of the NOE build up:

$$d\, NOE_{ij}(t)\, /\, dt = - \,\rho_i\, NOE_{ij}(t) - \sum_{k \neq i} \sigma_{ik}\, NOE_{ki}(t) \tag{6}$$

$NOE_{ki}(t)$ has the same definition as in [5] for the spin k. From the $1/r^6$ dependence of the relaxation rate constants, it should be noted that NOEs are short range effects (within 5Å).

At this stage, one dimensional and two dimensional experiments should be distinguished.

2D-NOE: The result of a NOESY experiment, performed at a mixing time tm, can be viewed as a matrix NOE(tm) whose elements (corresponding to the diagonal and the cross peaks volumes of the 2D NOESY) are $NOE_{ij}(tm)$ terms. By analogy to equation [6], the time dependence of this matrix can be written as:

$$d\, NOE(t)\, /\, dt = - \Gamma . NOE(t) \tag{7}$$

Γ is the so called relaxation matrix:

$$\Gamma = \begin{bmatrix} \rho1 & \sigma12 & \sigma13 & \ldots & \ldots \\ \sigma12 & \rho2 & \sigma23 & \ldots & \ldots \\ \sigma31 & \sigma32 & \rho3 & \ldots & \ldots \\ \ldots & \ldots & \ldots & \ldots & \ldots \\ \ldots & \ldots & \ldots & \ldots & \ldots \end{bmatrix}$$

The solution of the differential equation [7] can be symbolically written as:

$$NOE(tm) = exp(- \Gamma.tm) . NOE(0) \qquad\qquad [8]$$

where NOE(0) is the 2D-NOE matrix at zero mixing time, which is a diagonal matrix, because no magnetization has been transfered yet.

1D-NOE: In one dimensional driven truncated NOE experiment (Wagner & Wüthrich, 1979) for example, the magnetization of one spin is maintained to zero for the duration of the mixing time. The contribution of the magnetization of this spin (noted o here) to the transfer of magnetization recorded on other spins i, is constant in time ($= \sigma_{io}.Mo^o$). Equation [6] is no longer linear. For spins i, in the neighborhood of spin o:

$$d [NOE_{io}(t)] / dt = - \Gamma . [NOE_{io}(t)] + [\Sigma] \qquad\qquad [9]$$

$[NOE_{io}(t)]$ is now a vector formed by the NOE observed on the i spins, upon irradiation of the spin o. $[\Sigma]$ is the vector formed by the constant contribution of the magnetization of the irradiated spins (i.e. $\Sigma_{io} = \sigma_{io}$). The solution for such a vectorial equation is:

$$[NOE_{io}(tm)] = (I - exp (- \Gamma.tm)) (\Gamma)^{-1} [\Sigma] \qquad\qquad [10]$$

Direct Determination of the Relaxation Matrix from a 2D-NOE Experiment

For 2-D experiments, equation [8] is easily solved for Γ:

$$\Gamma = - (Ln [NOE(tm) . NOE(0)^{-1}]) / tm \qquad\qquad [11]$$

In fact, a 2D NOE experiment provides peak volumes which are proportional to the NOEs. Using the corresponding volume matrix V(t) , equation [11] becomes:

$$\Gamma = - (Ln [V(tm) . V(0)^{-1}]) / tm \qquad\qquad [12]$$

One measurement of the volume matrix at one mixing time should provide the full relaxation matrix. An application of this method is given below for an antigenic peptide. However, this approach has the very stringent requirement of knowing the complete V(tm) matrix. This ideal case is hardly encountered in macromolecules because of overlap between peaks in the 2D map. On the other hand, finding the relaxation matrix directly from equation [10] for a 1D experiment is not straightforward, since this equation is transcendental.

General Method for the Determination of the Relaxation Matrix

We have developed a method (described in detail in Koehl & Lefèvre, 1989) to extract the relaxation matrix from incomplete NOE matrix, obtained by one or two dimensional experiment. The goal of the method is to find the relaxation matrix which will minimize the deviation between observed (NOE_{obs}) and calculated (NOE_{cal}) NOEs. NOE_{cal} is calculated using equations [8] for 2D and [12] for 1D experiments. The risk function to be minimized is classically:

$$Ki^2 = \Sigma_i (NOE_{cal} - NOE_{obs})^2 / err(i)^2 \qquad\qquad [13]$$

The sum is made over the number of observed NOE. err(i) is the estimated error on NOE_{obs}.

Since Ki^2 is a non linear function of the relaxation parameters σ and ρ, the method of minimization is iterative. Its principle is to start from initial values of the relaxation parameters, which are calculated from a model, and to refine these values in order to reach the minimum of the Ki^2 function. As we know the analytical expression of the NOE as a function of the relaxation parameters (see eq. [8] and [10]), changes in the relaxation parameters between iterations are based on a gradient method that makes use of the derivative of the NOE expression according to these relaxation parameters, until the minimum of the risk function ki^2 is found (Koehl & Lefèvre, 1989). The gradient method is a modified Levenberg method (Levenberg, 1944; Press et al., 1986).

It should be noticed that the dynamics and structure of the molecule model are only utilized for calculating initial values of the relaxation rate constants, and are not used further during the refinement. Refinements can be started from several different model structures. They should end up with the same relaxation rate constants, and this can be used as a test for the reliability of the algorithm.

We should also emphasize the fact that the longitudinal relaxation rate constant is considered as a variable in the procedure, and is not directly linked to the cross relaxation rate constant values. This allows the relaxation caused by spins outside of the network of magnetization transfer between nuclei of interest (like the effect of the solvent molecules for example) or the relaxation mechanisms other than dipolar, to be taken in account. On the other hand, in a large molecule, it should be possible to isolate systems of spins within a limited volume and to calculate the relaxation parameters between spins within this volume. The longitudinal rate constants of the inner spins should account for the forwarding of the magnetization towards the surrounding nuclei.

Feasibility of the Method: a 2D Theoretical Case

Figure 1 shows the results of the application of the method to a theoretical case. 2D NOE values, at several mixing time (100, 200, 300, 400 and 500 ms) have been calculated for an isolated guanine mononucleotide (using eq. [8]). Only part of these "experimental" values have been incorporated in the NOE matrix (these are the NOE values whose calculation are not hindered by overlap) as shown in Table 1. The refinement leads to a set of sigma (Fig. 1) and rho values (not shown) that are very close to the real ones. This simulation demonstrates the capability of the method in reconstructing quite accurately the relaxation matrix.

Table 1.   2D NOE simulated matrix. 1 indicates that the NOE has been incorporated in the incomplete NOE matrix. The corresponding NOE have been calculated for 5 mixing time: 100, 200, 300, 400 and 500 ms.

	H1'	H2'	H2"	H3'	H4'	H8
H1'	1	1	1			1
H2'	1					1
H2"	1					1
H3'						
H4'						
H8	1	1	1			1

A real case is presented below, using 1D NOE values for refining the relaxation rate constants in an hexanucleotide.

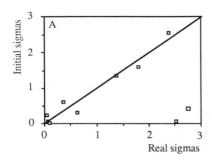 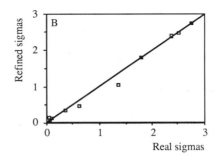

Figure 1. In the theoretical case of the isolated guanine, the reference structure So is defined from standard B DNA parameters; the corresponding sigma values are referred to as "real" sigmas. Panel A shows the dispersion of the initial sigmas used for the refinement procedure when they are plotted versus the real sigmas; this initial structure was calculated from standard Z DNA parameters and corresponds to a Ki^2 of 953 (based on NOE calculations) when compared to the reference structure. Panel B shows the result of the refinement, stopped when Ki^2 does not decrease significantly any longer ($Ki^2 = 0.01$); at this stage all the refined sigmas fall close to a straight line of slope 1 when plotted versus the real sigmas.

DETERMINATION OF THE DYNAMICS AND STRUCTURE FROM RELAXATION PARAMETERS

The cross relaxation rate constants are intimately linked to the dynamics (through the spectral density function), and structure (through its dependence on r^{-6}) of the molecule. Several modelization procedures have been proposed to characterize internal motion (King & Jardetzky, 1978; King et al., 1978; Ribeiro et al., 1980; Lipari & Szabo, 1984a,b; Woessner, 1962). Although they use different types of formalism, they all propose to characterize the internal motion by the value of its correlation time, and its contribution to the effective correlation time.

The simplest approach is the model free interpretation of nuclear magnetic relaxation of Lipari and Szabo (1984a,b). The total correlation function is approximated to the product of the correlation function of the overall motion by that of the internal motion. This gives rise to a modified spectral density function:

$$J(\omega) = S^2.\tau_m / (1 + \omega^2.\tau_m^2) + (1 - S^2).\tau / (1 + \omega^2.\tau^2) \qquad [14]$$

S is an order parameter which describes the spatial restriction of the motion. It is equal to 1 in the case of a rigid body, and 0 when the internal motion is isotropic. τ_m is the correlation time for the overall tumbling motion, and τ is obtained by adding the correlation rates for the overall (τ_m^{-1}) and the internal (τ_e^{-1}) motion:

$$\tau^{-1} = \tau_m^{-1} + \tau_e^{-1} \qquad [15]$$

The overall correlation time τ_m can be estimated by taking the longest correlation time calculated from the cross relaxation rate constants. This estimate can be confirmed by modelization of the shape of the molecule, or by using some other methods like fluorescence polarization measurement.

Some protons pairs are separated by a fixed distance. In proteins, these are some geminal ($r = 1.79$ Å) and aromatic ($r = 2.47$ Å) protons. In nucleic acids, the distances between the base protons of cytosines or thymines, between the geminal protons attached to the C2' and between the H1' and H2" protons ($r = 2.26 \pm 0.02$ Å) in deoxyriboses are fixed. The effective correlation time τ_c of the inter proton vector motion is obtained by solving Eq. [3] which is a 3rd degree polynome in τ_c.

For each proton pair separated by a fixed distance, we are then dealing with two parameters, S and τ_e, and only one experimental value to calculate them: the cross relaxation rate constant. One solution is to study the relaxation at different fields. However, it is often likely that the internal motion is very fast, so that $\tau_e \ll \tau_m$. The spectral density function of Eq. [14] becomes:

$$J(\omega) = S^2.\tau_m / (1 + \omega^2.\tau_m^2) + (1 - S^2).\tau_e / (1 + \omega^2.\tau_e^2) \qquad [16].$$

The cross relaxation rate constants σ_{ij} (Eq. [3]) simplifies to:

$$\sigma_{ij} = S^2.\alpha . (1/r_{ij})^6.[6J(2\omega) - J(0)] = S^2. \sigma_{ij}^0 \qquad [17].$$

σ_{ij}^0 is the cross relaxation rate constant which can be calculated for a rigid body. The σ_{ij} are scaled only by the order parameter S^2.

For another proton pair, whose interproton distance is to be determined, it is a prerequisite to understand the dynamics of the molecule, in order to integrate the averaging effect of the internal motion. Models of motion can be validated by comparing the value of the order parameter obtained experimentally to the one calculated from the model, using the expression of S^2 given by Lipari and Szabo (1984a,b). If several S^2 are available for different proton-proton vectors, this may sufficiently constrain the modeling of the motion and leaves only a small number of possible dynamic model. Then, the averaging effects of motion can be evaluated for the few validated models, and taken in account, whenever it is important.

In this paper, we only deal with the first part of the problem which is the determination of the effective correlation time induced by internal motion. Although this is only the first step of the structure determination procedure, it already gives a wealth of information for understanding the function of the molecule, as stated above.

THE DYNAMICS OF AN ANTIGENIC PEPTIDE

We are studying the dynamics and the conformation of a peptide that mimics an antigenic loop of hemaglutinin, a coat protein of the influenza virus. The sequence of the peptide is:

CYS1 - ARG2 - LYS3 - GLY4 - PRO5 - GLY6 - SER7 - ASP8 - PHE9 - ASP10 - TYR11

The peptide has been cyclisized by linking the side chain carboxyl group of the ASP10 to the amide of the CYS1. The synthesis of this peptide was performed by S. Plaué (Neosystem, Strasbourg).

145

Assignments

The NOESY spectrum of the peptide, obtained on a Bruker AM 400 spectrometer, with a mixing time of 250 ms in D2O, is presented in Figure 2. Surprisingly, all observed NOE are negative. At this mixing time, some spin diffusion is already observed, which increases when the duration of the mixing period in the NOESY experiment is increased. The NOESY spectrum alone allows a very straightforward assignment of non exchangeable protons, by identifying the different spin systems (Wüthrich, 1986).

The two glycines in the sequence are identified on the basis of NOE observed at long mixing time (500ms) between the alpha protons of these residues and the side chain protons of the proline. The aspartic spin systems are assigned by comparison with the spectrum of a similar peptide where one of the aspartic residues where replaced by a lysine. The position of the cross peaks between alpha and beta protons are indicated in Figure 2. The total assignment list will be published elsewhere.

Calculation of NOE Values

As can be observed in Figure 2, the spin systems are quite isolated, and no noticeable dipolar couplings are measured from one side chain to another, when the mixing time is kept under 300ms. It is then possible to treat each spin system alone, although the number of protons in the peptide leads to a total NOE matrix of reasonable size which could be easily solved through equation [12]. However, in rare cases, there are overlaps between diagonal or cross peaks, which hinder the determination of the complete relaxation matrix by the direct method (Eq. [12]).

We have integrated all the peaks in the 2D NOE map and converted them into NOE values. In equation [12], the matrix V(0) is diagonal as no magnetization transfer has given rise to cross peaks at a zero mixing time. V(0)$^{-1}$ is then formed by the inverse value of the diagonal peak volumes obtained in a zero mixing time NOESY.

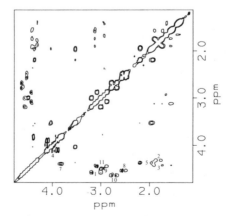

Figure 2.   Contour plot of a NOESY spectrum of the antigenic peptide. The peptide is dissolved in D2O, in a phosphate buffer, pD 7.0, at a concentration of about 10 mM. The temperature is 25 °C. The mixing time is 250 ms. 32 scans were accumulated for each of the 512 t1 values. The FIDs were multiplied by a sine bell function, shifted by 60°, prior to Fourier transformation in each direction. Cross peaks between alpha, alpha and beta, and delta protons are indicated by the number of the amino acid in the sequence: Cys-1 Lys-2 Arg-3 Gly-4 Pro-5 Gly-6 Ser-7 Asp-8 Phe-9 Asp-10 Tyr-11.

However, it is not necessary to perform this experiment. One should consider that the sum of the volumes of the cross peaks lying along a particular column is equal to the surface of the peak, corresponding to this column, in the first spectrum of the 2D experiment. This comes from the fact that the first point of the FID is equal to the total area of the peaks in the spectrum. It is then sufficient to run an experiment where only the first FID of NOESY spectra at different mixing times are recorded. Inspection of the intensities of the peaks in these "first spectra of NOESY" indicates how the net longitudinal magnetization of each spin has evoluated under T1 relaxation during the mixing time. Starting from the sum of the peak volumes in a column, it is then possible to recalculate the volume of the corresponding diagonal peak at zero mixing time. However, for the peptide, the intensities of the peaks in the first spectra of NOESY did not change significantly, up to a mixing time of 300 ms. Therefore, the volumes $V_{ii}(0)$ were taken equal to the sum of the cross peak volumes $V_{ij}(250ms)$.

Calculation of the Relaxation Parameters and Dynamics of the Peptide

Table 2 shows the sigma values between geminal protons as they result from the transformation of the NOE matrix into the relaxation matrix, using equation [12]. The longer correlation time is obtained for the vector joining the beta protons of ASP-8. This value of 2.8 ns was used to compute the order parameters S^2, reported along the sequence in Figure 3.

Table 2. Cross relaxation rate constants in the antigenic peptide, calculated from the NOE value calculated at 250 ms mixing time. σ are the cross relaxation rate constants between the alpha, beta and delta resolved geminal protons. τ_c are the correlation time (in ns) deduced from the σ values, solving Eq. [3].

residue	alpha		beta		delta	
	σ_a	τ_c	σ_b	τ_c	σ_d	τ_c
CYS-1	-	-	-0.83	0.8	-	-
LYS-2	-	-	-3.35	2.0	-	-
ARG-3	-	-	-1.08	0.9	-	-
GLY-4	-0.30	0.6	-	-	-	-
PRO-5	-	-	-0.67	0.7	-0.70	0.7
GLY-6	-1.92	1.3	-	-	-	-
SER-7	-	-	-4.03	2.4	-	-
ASP-8	-	-	-4.65	2.8	-	-
PHE-9	-	-	n.d.	n.d.	-	-
ASP-10	-	-	-3.98	2.4	-	-
TYR-11	-	-	-1.74	1.2	-	-

n.d.: not determined. The beta protons of PHE-9 are not resolved.

The value of the correlation time is unexpectedly high for a peptide of this size. In fact, analysis by HPLC methods (not shown) reveal that the peptide is in a dimeric form, resulting probably from bond formation between the cysteines. It may be also that, at the concentration used for the experiment (around 10mM), the dimeric peptide aggregates as well.

Although the order parameters are not measured between the same type of protons along the sequence, Figure 3 shows that there are zones of different mobilities along the sequence. It also indicates that the motion of the backbone and the side chains are locally bound up together, as correlation times measured either on alpha or beta protons, lie in the same range.

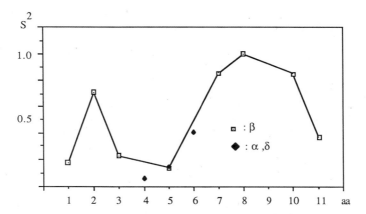

Figure 3. Plot of the order parameter S^2 as a function of the position in the sequence.
S^2 was calculated from the cross relaxation rate constants between the beta (points
joined by the lines), or alpha (Gly-4, Gly-6), or delta (Pro-5) protons.

The two vectors joining the alpha to the beta protons have correlated motions. If the
internal motion is mainly due to rotation around the $C\alpha$-$C\beta$ bond, it is expected that both vector
motions are characterized by similar correlation times and order parameters. The ratio of the
cross relaxation rates should then only depends on the sixth power ratio of the vector
magnitudes.

The coupling constants observed between the alpha and the beta protons are generally
interpreted in terms of populations of the three rotamers, corresponding to the three eclipsed
positions of the two tetrahedral carbons (Jardetzky & Roberts, 1981). The $C\alpha$ - $C\beta$ torsion
angle, X is equal to 60° for rotamer I, to 180° for rotamer II and 300° for rotamer III. However,
rotamer I and II cannot be distinguished unambiguously, as long as the stereospecific
assignment of the beta protons is not worked out. Indeed, the set of coupling constants are
identical for both rotamers. In the following, we will arbitrarily assign the population to each
rotamer, keeping in mind that populations of rotamers I and II can be inverted. This will not
alter the reasoning. Using the values of the coupling constants as they come from the Karplus
equation (Karplus, 1959; Wüthrich, 1986), each rotamer population f_I, f_{II}, and f_{III} are given by:

$$f_I = 0.11 \ (J_2 - 3.25) \qquad f_{II} = 0.11 \ (J_1 - 3.25) \qquad f_{III} = 1 - f_I - f_{II} \qquad [18]$$

The subscripts 1 and 2 refer arbitrarily to the $H\alpha$ - $H\beta$ and the $H\alpha$ - $H\beta'$ couple. J_1 and J_2 are
the coupling constants, in Hz, which are listed in Table 3 for the peptide. These values were
obtained from a J-resolved experiments (data not shown).

Assuming that the correlation time and the order parameter is the same for vectors 1 and 2
(same definition of the subscripts as above), the ratio between the cross relaxation rate constants
can be calculated by:

$$\sigma_1/\sigma_2 = (f_I.d_{1,I}^{-6} + f_{II}.d_{1,II}^{-6} + f_{III}.d_{1,III}^{-6})/(f_I.d_{2,I}^{-6} + f_{II}.d_{2,II}^{-6} + f_{III}.d_{2,III}^{-6}) \qquad [19]$$

where $d_{i,J}$ denotes the length of vector 1 or 2 in the rotamer J. Experimental and calculated ratios
are reported in Table 3.

If one admit a maximum 15% error on calculated sigma values, and then a 30% error on the
(σ_1/σ_2) ratios, half of the recalculated and experimental values of these ratios lie within

experimental errors (for CYS-1, SER-7 and TYR-11). However, for the other residues (LYS-2, ARG-3 and ASP-8), the discrepancy is larger than the expected maximum error. In these latter cases, the order parameter is certainly different for the two vectors, due to a differential effect of another motion than the rotation around the Cα - Cβ bond. Any modelization of the dynamics of the peptide should account for the S^2 values calculated along the sequence, and for the behavior of the sigma values of vectors 1 and 2.

There has been some connection made between the mobility of protein sequences and their ability to be an antigenic site (Westhof et al., 1984). The peptide alone is recognized by

Table 3. Dipolar and scalar interactions between the alpha and the two beta protons, in the antigenic peptide. As the stereospecific assignment is not yet done, the two cross relaxation rate constants are noted σ_1 and σ_2, and the ratio is calculated by dividing the largest σ by the smallest one $(\sigma_1/\sigma_2)_0$. The multiplicity (T stands for triplet, and Q for quadruplet) and the coupling constant values (J_1 and J_2, in Hz) are measured on the alpha protons resonances, in a J-resolved experiments. $(\sigma_1/\sigma_2)_c$ is the ratio recalculated using the population in each rotamer, obtained from the coupling constants (Eq. [19]).

Residue	σ_1	σ_2	$(\sigma_1/\sigma_2)_0$	multiplicity	J_1	J_2	$(\sigma_1/\sigma_2)_c$
CYS-1	-0.32	-0.23	1.4	T	7.4	7.4	1.0
LYS-2	-0.29	-0.22	1.3	Q	10.0	5.0	1.9
ARG-3	-0.23	-0.06	3.6	Q	10.4	5.0	2
PRO-5	-0.23	-0.16	1.4	Q	9.8	5.0	n.d.
SER-7	-0.20	-0.18	1.1	T	5.5	5.5	1
ASP-8	-0.49	-0.45	1.1	Q	9.1	5.0	1.6
ASP-10	-0.39	-0.08	4.6	n.d.	n.d.	n.d.	n.d.
TYR-11	-0.31	-0.14	2.13	Q	9.2	5.6	1.5

n.d.: not determined

monoclonal antibodies (S. Plaué, personal communication). The internal dynamics observed in this peptide, around the GLY-PRO-GLY motif, might be the key of its antigenic behavior. Obviously, the relation between the dynamics and the function of this peptide should be studied through an engineering of this loop aiming to change its internal motion. Incorporating amino acids with bulky side chains, in place of the glycines for example, would be one approach. Analogs of this peptide are actually studied to work out this problem.

THE DYNAMICS OF AN OLIGONUCLEOTIDE

In order to test the method of reconstructing the relaxation matrix, we have used a set of 1D driven truncated NOE obtained on an autocomplementary hexamer by A.N. Lane (MRC, London). The sequence of the oligonucleotide is :

C - G - T - A - C - G

For the purpose of this paper, we will only concentrate on the intranucleotide NOES. The set of NOE used (not shown) is not complete for all nucleotide and does not allow to fill all the NOE vectors described in equation [10]. Moreover, as stated above, equation [10] is not trivial in Γ, as compared to the 2D case (eq. [12]). These facts justify the use of the method for reconstructing the relaxation matrix described above.

Relaxation Parameters and Dynamics of the Oligonucleotide

The relaxation parameters of each nucleotide have been refined separately. The spin systems which have been considered to perform the refinement involved the following protons: H8 or H6, H5 or CH3 for pyrimidines, H1', H2', H2", H3' and H4'. The refined spin lattice relaxation rate constants take care of the magnetization loss to other protons of the same or neighboring nucleotides.

Figure 4 shows the evolution of some NOEs calculated by using equation [10] and the relaxation matrix obtained at different step of the refinement procedure. The first point of the NOE build up curve are first adjusted, then the points at longer mixing time are refined during the last iteration steps. This clearly indicates that it is necessary to record NOE at long mixing times to get right values of the relaxation parameters. Three different starting structures have been used to calculate initial values of the relaxation rate constants. These structures correspond to the three A, B and Z standard structures. Except those involving protons for which no NOE information at all is given (H3' and H4'), the refined values of the relaxation rate constants fall within a deviation of 5 to 10% with respect to the starting structure. Table 4 gives the averaged values of the cross relaxation rate constants for which a satisfying RMS was obtained.

Two distances in the sugars remain constant, no matter what the change in pseudorotation is. These are the H2' - H2" (1.79Å) and the H1' - H2" (2.25Å) vectors. They provide an insight into the dynamics of the riboses. The distances between the H5 and the H6 of cytosines are also constant. They allow to evaluate the overall tumbling correlation time of the molecule (Lane et al., 1986; Lefèvre et al., 1987). In the case of the hexamer, the experiments were performed at 15°C, and the correlation time is 2.4 ns (A.N. Lane, personal communication). The correlation time τ_c of the H2' - H2" and H1' - H2" vectors, calculated from the resolution of Eq. [3] are presented in Table 4.

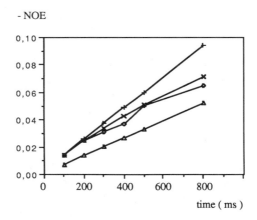

Figure 4. Buildup of NOE intensities for H6-H1' proton pair in the course of the refinement process for the C1 nucleotide in the hexamer d(CGTACG). ◊ experimental curve; Δ NOE for the starting structure; C1 was constructed in a A type conformation, assuming an isotropic correlation time of 2.4 ns (the corresponding Ki2 is 900); + NOE after 6 steps of the refinement (Ki2 = 30) ; x final NOE buildup, when the refinement reaches a minimum (Ki2 = 6).

Table 4. Refined cross relaxation rate constants (in s⁻¹) for the hexamer d(CGTACG). The correlation times (in ns) are calculated using Eq. [3]. The σ ratios are to be compared with the theoretical values in Figure 6.

Base	cross relaxation rate constant						correlation time		σ ratio
	H1'-H2'	H1'-H2"	H1'-H6,8	H2'-H2"	H2'-H6,8	H2'-H6,8	H2'-H2"	H1'-H2"	H1'-H2"/H1'-H2'
C1	-0.05	-0.55	-0.10	-2.45	-0.50	-0.05	1.6	1.4	10.0
G2	-0.32	-0.57	-0.06	-3.30	-0.90	-0.40	2.0	1.5	1.8
T3	-0.07	-0.72	-0.07	-3.10	-1.18	-0.10	1.9	1.8	10.3
A4	-0.20	-0.68	-0.09	-3.10	-1.20	-0.32	1.9	1.7	3.4
C5	-0.22	-0.50	-0.06	-2.80	-1.02	-0.07	1.8	1.3	2.3
G6	-0.12	-0.71	-0.08	-3.60	-1.25	-0.08	2.2	1.8	5.9

Figure 5 shows the variation of these correlation time along the sequence of the hexanucleotide. There is clearly an internal motion at the level of the riboses, which is characterized by a small correlation time at the 5' end of the molecule and a surprising large one at the 3' end. Between these two extreme riboses, the correlation times pass through a maximum for the central nucleotide. This suggest that, in this sequence, the internal motion increases when going from the center to the ends, which can be intuitively understood as the effect of fraying ends of the molecule. The rigidity of the 3' end sugar may be correlated which the general observation on crystal structure of oligonucleotides, that the base is better stacked at the 3' end than at the 5'end.

As the H2' and H2" protons are linked to the same carbon, their motion are correlated. It might be thought that the H1' - H2' and the H1' - H2" vectors have also correlated motions which are characterized by the same correlation time. The ratio between the cross relaxation rate constants of the two pairs of proton should only depends on the sixth power of the distances ratio:

$$\sigma (H1' - H2") / \sigma (H1' - H2') = [d(H1' - H2') / d(H1' - H2")]^6 \qquad [20]$$

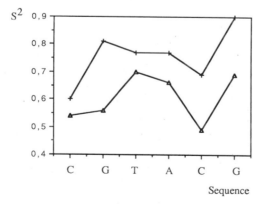

Figure 5. Order parameter S^2 reported along the sequence of the hexamer d(CGTACG) for: +, the H2'-H2" interproton vector and Δ, H1'-H2" interproton vector; S^2 is defined as σ/σ_r, where σ is the real sigma of the proton pair and σ_r is calculated with the overall correlation time (2.4 ns for the hexamer) for the same proton pair.

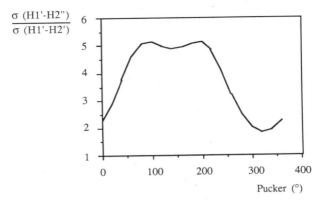

Figure 6. σ (H1'-H2") / σ (H1'-H2') versus the pseudorotation phase angle of a deoxyribose ring. This ratio was calculated assuming the same correlation time for the two interproton vectors (see equation [20]).

This ratio, plotted against the pseudorotation phase angle (P) of the sugar in Figure 6, appears to be constant in the 04'endo (P = 100°) to C2'endo (P = 180°). The puckers of the sugars, which were determined using the coupling constants between the H1' and the H2'/H2" protons (Lefèvre et al., 1987), are all in the C2'endo region (A.N. Lane, personal communication). The sigma ratio given in Table 4 never fall on the curve and are either bigger or smaller than the expected values. This suggests that the correlation time of the two vectors are different, indicating a strong anisotropy of the internal motion.

The motion of the sugar may well be the results of several contributions. The pseudorotation is likely to be one of these, but also libration of the base and oscillation of the torsion of the glycosidic link between the base and the sugar. Work is in progress to evaluate the effect of these motions on the value of the order parameter. This should guide us on the elucidation of the dynamics of the molecule, which is a prerequisite for the determination of the structure.

CONCLUSION

We have designed a method for reconstructing a relaxation matrix partially or totally. The power of this approach lies in the fact that one obtains the primary NMR parameters. These parameters can then be precisely interpreted in term of dynamics and used to guide the modeling of the internal motion. There are two major reasons why it is important to understand the dynamics of a biological macromolecule: the first deals with the function of the molecule, and the second with its precise structure determination.

The amplitude of the internal motion is a measure of the conformational space available to a molecule. Points of high mobility in the molecule are points of high flexibility. They easily absorb a constraint that is applied on the molecule during a specific interaction. A DNA molecule, for example, should have well placed flexibility points in order to wrap around the CRP protein (Gartenberg & Crothers, 1988) or the nucleosome (Travers & Klug, 1987). The peculiar dynamics of different types of nucleotide sequences can be studied by the method presented above. The same is true for proteins that have to interact with various ligands, in that the active site and the substrate mutually adapt their conformations.

Internal motion poses a problem for precise structure determinations. It affects the local correlation time and averages the fluctuating distances in a nontrivial manner. It is therefore important to appreciate the effects of the dynamics on the cross relaxation rate constants in order to interpret these parameters correctly.

ACKNOWLEDGEMENTS

We are very grateful to Dr. A. N. Lane for giving us the NOE data on the hexanucleotide, and for fruitful discussions. We thank S. Plaué for providing us with the antigenic peptide.

REFERENCES

Abragam, A., 1986 *in:* "The Principles of Nuclear Magnetism", Clarendon, Oxford, (latest edition).

Altman, R.B. & Jardetzky, O., (in press), The heuristic refinement method for the determination of the solution structure of proteins from NMR data, *in:* "Nuclear Magnetic Resonance, Part B: Structure and Mechanisms," Methods in Enzymology 177:218, Academic Press, NY.

Bloch, F., 1946, Nuclear induction, *Phys. Rev.* 70:460.

Böhlen, J.M., Wimperis, S. & Bodenhausen, G., 1988, Observation of longitudinal three-spin order for measuring dipole-dipole cross correlation, *J. Magn. Reson.* 77:589.

Borgias, B.A. & James, T.L., 1988, Comatose, a method for constrained refinement of macromolecular structure based on two dimensional nuclear Overhauser effect spectra, *J. Magn. Reson.* 79:493.

Braun, W. & Go, N., 1985, Calculation of protein conformations by proton-proton distance constraints, a new efficient algorithm, *J. Mol. Biol* . 186:611.

Bull, T.E., 1987, Cross correlation and 2D NOE spectra, *J. Magn. Reson.* 72:397.

Gartenberg, M. R. & Crothers, D. M., 1988, DNA sequence determinants of CAP-induced bending and protein binding affinity, *Nature* 333:824.

Jardetzky, O. & Roberts, G.C.K., 1981, *in:* "NMR in Molecular Biology," Academic Press, New York.

Kaptein, R., Zuiderweg, E.R.P., Scheek, R.M., Boelens, R. & Van Gunsteren, W.F., 1985, A protein structure from Nuclear Magnetic Resonance data: *Lac* repressor headpiece, *J. Mol. Biol.* 182:179.

Karplus, M., 1959, Contact electron-spin coupling of nuclear magnetic moments, *J. Chem. Phys.* 30:11.

Keepers, J.W. & James, T.L., 1984, A theoretical study of distance determinations from NMR two-dimensional nuclear Overhauser effect spectra, *J. Magn. Reson.* 54:404.

King, R. & Jardetzky, O., 1978, A general formalism for the analysis of NMR relaxation measurements on systems with multiple degrees of freedom. *Chem. Phys.Lett.* 55:15.

King, R., Maas, R., Gassner, M., Nanda, R. K., Conover, W.W. & Jardetzky, O., 1978, Magnetic relaxation analysis of dynamic processes in macromolecules in the pico- to microsecond range,*Biophys.* 6:103.

Kline, A.D., Braun, W. & Wüthrich, K., 1986, Studies by [1]H Nuclear Magnetic Resonance and Distance Geometry of the solution conformation of the α-amylase inhibitor Tendamistat, *J. Mol. Biol.* 189:377.

Koehl, P. & Lefèvre, J.F., 1989, The reconstruction of the relaxation matrix from an incomplete set of nuclear Overhauser effects, *J. Magn. Reson.*, in press.

Lane, A.N., Lefèvre, J.F. & Jardetzky, O., 1986, A method for evaluating correlation times from tumbling and internal motion in macromolecules using cross relaxation rate constants from proton NMR spectra, *J. Mag. Reson.* 66:201.

Lefèvre, J.F., Lane, A.N. & Jardetzky, O., 1987, Solution structure of the *Trp* operator of *Escherichia coli* determined by NMR, *Biochemistry* 26:5076.

Levenberg, K., 1944, A method for the solution of certain non-linear problems in least squares, *Quart. Appl. Math.* 2:164.

Lipari, G. & Szabo, A., 1984a, Model free approach to the interpretation of Nuclear Magnetic Resonance relaxation in macromolecules. 1. Theory and range of validity, *J. Am. Chem. Soc.* 104:4546.

Lipari, G. & Szabo, A., 1984b, Model free approach to the interpretation of Nuclear Magnetic Resonance relaxation in macromolecules. 2. Analysis of experimental results, *J. Am. Chem. Soc.* 104:4559.

Press, W.H., Flannery, B.L., Teukolsky , S.A.& Vetterling, W.T., 1986, *in:* "Numerical Recipes, The Art of Scientific Computing," Cambridge University Press, Cambridge.

Ribeiro, A.A., King, R., Restivo, C. & Jardetzky, O., 1980, An approach to the mapping of internal motions in proteins. Analysis of ^{13}C NMR relaxation in the Bovine Pancreatic Trypsin Inhibitor, *J. Am. Chem. Soc.* 102:4040.

Solomon, I., 1955, Relaxation processes in a system of two spins, *Phys. Rev.* 99:559.

Travers, A. A. & Klug, A., 1987, The bending of DNA in nucleosomes and its wider implications, *Phil. Trans. R. Soc. Lond.* B 317:537.

Wagner, G. & Wüthrich, K., 1979, Truncated driven nuclear Overhauser effect (TOE). A new technique for studies of selective ^{1}H-^{1}H Overhauser effects in the presence of spin diffusion, *J. Mag. Reson.* 33:675.

Westhof, E., Altschuh, D.,Moras, D., Bloomer, A. C., Mondragon, A., Klug, A. & Van Regenmortel, M. H. V., 1984, Correlation between segmental mobility and the location of antigenic determinants in proteins, *Nature* 311:123.

Woessner, E.D., 1962, Nuclear spin relaxation in ellipsoids undergoing rotational Brownian motion, *J. Chem. Phys.* 37:647.

Wüthrich, K., 1986, *in:* "NMR of Proteins and Nucleic Acids," John Wiley and Son, NY.

METHODS OF STABLE-ISOTOPE-ASSISTED PROTEIN NMR SPECTROSCOPY IN SOLUTION

Brian J. Stockman and John L. Markley

Department of Biochemistry, College of Agricultural and Life Sciences, 420 Henry Mall, University of Wisconsin-Madison, Madison, WI 53706

INTRODUCTION

The scope of this review, which is abbreviated from Stockman and Markley (1989) will be limited to ^{2}H-, ^{13}C-, and ^{15}N-assisted NMR spectroscopy of proteins in solution. Since hydrogen, carbon, and nitrogen are ubiquitous in proteins, the procedures presented here are applicable to any protein. Elegant reviews covering NMR spectroscopy of less predominant atoms in proteins have appeared recently: ^{19}F (Ho et al., 1985), ^{31}P (Gorenstein et al., 1989), ^{113}Cd (Summers, 1988). Techniques utilizing stable-isotope-assisted solid state NMR spectroscopy will not be discussed here.

It was realized from the earliest nuclear magnetic resonance (NMR) studies of proteins that spectral overlaps presented serious obstacles to the extraction of their full information content (Saunders et al., 1957). Studies of spectral envelopes held out tantalizing evidence for the conformation dependence of spectra (Bovey et al., 1959; Kowalsky, 1962), and investigations of those isolated resonances that could be resolved showed how quantitative results could be derived once problems of resolution and assignments could be overcome (Mandel, 1965; Bradbury and Sheraga, 1966; Meadows et al., 1967; McDonald and Phillips, 1967; Wüthrich et al., 1968; Kurland et al., 1968). Selective deuteration (Katz and Crespi, 1966; Crespi et al., 1968; Markley et al., 1968) was shown to provide a powerful approach to spectral simplification. Jardetzky (1965) presented a general strategy for spectral assignments based on selective deuteration that presaged many of the exciting developments of the past five years. Carbon-13 labeling also was developed as a general methodology for protein NMR studies (Browne et al., 1973). This approach, like nitrogen-15 labeling (Gust et al., 1975), which appeared somewhat later, led to a number of exciting applications, but these were overshadowed by the ^{1}H{^{1}H} two-dimensional methods that emerged in the early 1980's (Wüthrich, 1986; Ernst et al., 1987).

Recent advances in biotechnology have made it easier and more economical to introduce stable isotopes into proteins and a number of new heteronuclear two- and three-dimensional NMR experiments are permitting a much fuller exploitation of their inherent spectral information. The renaissance in interest in stable isotope labeling is leading to rapid technological advances that are permitting increasingly more detailed studies of small proteins and pushing back the molecular weight limitation on solution structural studies. This chapter attempts to review this accelerating field.

This chapter is divided into four sections. **Protein Sample Preparation** will be discussed first. Choice of enrichment level and pattern, enrichment methodologies, and requisite protein quantities will be considered. **Hardware Requirements** will then be discussed briefly. Advantages of **Deuterium-Assisted Protein NMR Spectroscopy** will be considered third. The merits of random fractional deuteration and β-chiral deuteration will be demonstrated. Finally, a long section will consider powerful techniques available with **Carbon-13- and Nitrogen-15-Assisted Protein NMR Spectroscopy**. Topics discussed include: approaches to first-order and sequential ^{1}H, ^{13}C, and ^{15}N resonance assignments, spreading ^{1}H-^{1}H correlations over a third, heteronucleus chemical shift dimension, simplification and enhancement of ^{1}H-^{1}H correlations, and extraction of structural and dynamic properties.

PROTEIN SAMPLE PREPARATION

Choice of Enrichment Level and Pattern

The optimal pattern and level of isotopic enrichment are determined by the NMR experiments that will be carried out on the protein (Table 1). *Pattern of enrichment* refers to whether the labeling is specific to a particular residue or residue type, or is uniform throughout the protein. *Level of enrichment* refers to the percentage of isotope incorporated into a given position in the protein.

Table 1.   Optimum enrichment levels for NMR experiments

Experiment	Enrichment Level
^{1}H{^{15}N}SBC	100% ^{15}N
^{1}H{^{15}N}MBC	100% ^{15}N
^{1}H{^{13}C}SBC	26% ^{13}C
^{1}H{^{13}C}MBC	26% ^{13}C
^{13}C{^{15}N}SBC	26% ^{13}C, 100% ^{15}N
^{13}C{^{13}C}DQC	44% ^{13}C
^{2}H observe	100% ^{2}H
^{1}H observe selective ^{2}H	100% ^{2}H
random fractional ^{2}H	50 - 80% ^{2}H

Three different patterns of deuterium labeling are commonly employed to simplify ^{1}H NMR spectra (LeMaster, 1989): (1) the incorporation of selected protonated amino acids into an otherwise deuterated protein (Markley et al., 1968; Markley, 1972); (2) the incorporation of selected deuterated amino acids into an otherwise protonated protein (LeMaster and Richards, 1988; LeMaster, 1987; Akasaka, 1988); and (3) random fractional deuteration (LeMaster and Richards, 1988; LeMaster, 1987; LeMaster, 1988). The first two are designed to permit rapid assignment of resonances to specific amino acid type. All hydrogens in a residue can be labeled with ^{2}H or, alternatively, certain protons in a given amino acid type can be (selectively) deuterated. In either case, the level of ^{2}H enrichment should be as high as possible. In the random fractional deuteration strategy, all carbon bound protons are enriched with deuterium to a level of 50-85%. This level of enrichment provides a compromise between the loss of intrinsic ^{1}H NMR sensitivity and the gain in resolution. Since proton-proton dipolar interactions dominate proton relaxation times, the dilution of protons with deuterons results in narrower linewidths (LeMaster, 1989; Kalbitzer et al., 1985). This, combined with more limited coupling, increases resolution. When all factors are considered, the reduction in signal-to-noise obtained in a 75% random fractionally deuterated protein spectrum is only two- to three-fold compared to the spectrum at natural abundance (LeMaster, 1989). Labeling patterns also may be combined, such as in the incorporation of chirally deuterated amino acids into a random fractionally deuterated background (LeMaster and Richards, 1988).

The optimum level of uniform ^{13}C enrichment depends on both the nucleus to be detected (^{1}H or ^{13}C) and the experiment (Table 1). Gains in sensitivity begin to be offset by increased complexity caused by carbon-carbon couplings at higher levels of uniform enrichment. For ^{1}H-detection as in ^{1}H{^{13}C}SBC, one must maximize the ratio of singlet to doublet (arising from one-bond carbon-carbon couplings) carbon resonances. This results in an optimum enrichment level of 26% ^{13}C (Wüthrich, 1976) (Figure 1). In practice, lower levels of enrichment (15-20%) are often used, since they represent a good compromise between the cost of enrichment and the gain in sensitivity over natural abundance ^{13}C. For ^{13}C-detection, the optimum enrichment level depends on the nucleus to be correlated. In the ^{13}C{^{15}N}SBC experiment, 26% ^{13}C is optimum, since one again must maximize the ratio of singlet to doublet carbon resonances. In contrast, the ^{13}C{^{13}C}DQC (double quantum correlation) experiment must be optimized to provide the maximum ratio of doublet to triplet carbon resonances. This corresponds to a level of 44% ^{13}C, as shown in Figure 1. Since it would be costly to produce two ^{13}C analogues of a protein in order to optimize independently for the ^{1}H{^{13}C}SBC and ^{13}C{^{13}C}DQC experiments, a good compromise should be to enrich to a level of 35-38% ^{13}C (Figure 1). Some of the ^{1}H{^{13}C}SBC sensitivity is sacrificed for improved ^{13}C{^{13}C}DQC sensitivity.

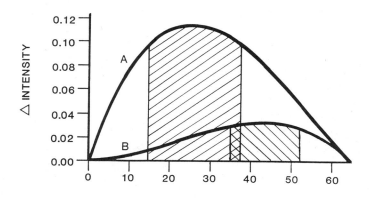

CARBON-13 ATOM % ENRICHMENT

Figure 1. Intensity difference between singlet and doublet (A) and doublet and triplet (B) carbon-13 signal mulitiplicities as a function of enrichment level. Shaded regions represent the enrichment range for the highest intensity difference. The region of overlap between the shaded areas represents a compromise level of enrichment that results in good intensity differences for observation of both singlet and doublet multiplicities. Curve A is a plot of $X(1-X)^2-0.5X^2(1-X)$ and curve B is a plot of $0.5X^2(1-X)-0.25X^3$, where X is ^{13}C isotopic enrichment.

Because the ^{1}H{^{13}C}SBC experiment is intrinsically much more sensitive that the ^{13}C{^{13}C}DQC experiment, this represents a good trade off.

Protein enrichment with ^{15}N is not complicated by increased ^{15}N-^{15}N coupling since adjacent ^{15}N atoms do not occur in the twenty common amino acids. Thus one should opt for the highest level of ^{15}N enrichment possible. Most ^{1}H{^{15}N}SBC and ^{13}C{^{15}N}SBC experiments are carried out with proteins enriched to greater than 95% isotopic purity, although lower levels (60%) also have been used successfully (Westler et al., 1988a).

Enrichment Methodologies

A variety of methods are used to enrich proteins, depending on the system being used to obtain the protein and the labeling pattern desired. Methodologies for ^{2}H (LeMaster, 1989), ^{13}C (Hilber et al., 1989; Wilde et al., 1989), and ^{15}N (Muchmore et al., 1989) protein enrichment have been reviewed recently; key points will be reiterated here.

Uniform labeling can be accomplished efficiently in cases where the protein is obtained from bacterial, cyanobacterial, or algal sources. This allows the use of relatively inexpensive labeled precursors, such as [^{13}C]carbon dioxide, [^{15}N]nitrate, [^{15}N]ammonia, [^{2}H]water, [^{2}H]succinate, and [^{2}H]alanine. After removal of the desired protein, the labeled by-products (such as other proteins, lipids, nucleic acids) can be recovered and utilized for other experiments.

It is most economical to obtain the desired protein from overproducing strains of bacteria or other micro-organisms or cultured cells, since one can thereby optimize the protein yield with respect to expensive labeled precursors. Enrichment of a particular amino acid type or types is accomplished by supplying the labeled amino acid or acids in the growth medium. Production in metabolically-deficient strains may be necessary in order to achieve the desired labeling patterns. Metabolic scrambling or dilution of the label (by *de novo* synthesis of the amino acid by the organism) can be reduced or eliminated by the use of transaminase-deficient or metabolically-deficient strains of bacteria.

In addition to the above strategies, selective labeling patterns can be obtained by taking advantage of the metabolism of the over-expressing organism. Specifically-labeled precursors can be chosen that, by means of their metabolic usage by the organism, introduce labels into specific types and/or positions of amino acids in the protein (Senn et al., 1989). Labeling of a particular residue in a protein also has been accomplished by incorporating the labeled amino acid during solid phase synthesis of the protein (Shon, 1989). This could also be accomplished by using suppressor tRNA's to incorporate labeled amino acids into specific positions of the protein sequence (Noren et al., 1989). Chemical exchange (Markley and Cheung, 1973; Markley and Kato, 1975; Markley, 1975) and enzymatic semi-synthesis (Baillargeon et al., 1980; Richarz et al., 1980) are two other methods of enrichment. Amide or histidine H$^{\varepsilon 1}$ (Markley and Cheung, 1973) protons can be chemically exchanged for deuterons, or vice versa. Enzymatic semi-synthesis involves the use of a specific enzyme to introduce a label into the appropriate position of the protein.

Protein Quantities Needed

In general, the amount of enriched protein needed for the NMR experiments described here is similar to that used routinely for ^{1}H NMR experiments. Typically, 3-10 mM protein (in 0.45 ml) is used in most experiments (Markley, 1989). Excellent signal-to-noise can be obtained with data accumulation times of 18-24 hours for ^{1}H{X}SBC and MBC (and their pseudo three-dimensional extensions), 24-72 hours for $^{13}C\{^{13}C\}DQC$ and $^{13}C\{^{15}N\}SBC$ experiments, and 72-180 hours for three-dimensional experiments.

The practical molecular weight limits for these experiments are not yet well defined. For uniform labeling, the limit will probably be around 30 kDa. With the ability to label selectively, proteins larger than 50 kDa (Kato et al., 1989) will be approachable in regions selected for by either the protein labeling pattern or enrichment of a bound ligand. The ability to simplify assignments by introducing labels into selective types of amino acids is crucial in the application of these experiments to larger systems.

HARDWARE REQUIREMENTS

High field NMR spectrometers (greater than 10.4 T) and appropriate peripherals are essential to extracting the optimum amount of information from labeled proteins. Spectrometers must be capable of both direct and indirect detection of ^{13}C and ^{15}N. Necessary peripherals include a full complement of probes, appropriate amplifiers, a frequency synthesizer in addition to the transmitter and proton decoupler, and a variety of bandpass filters, bandstop filters, and crossed diodes.

In addition to a proton-only probe, broad band probes are also required. One probe should have the broadband coil inside of the ^{1}H decoupler coil for higher sensitivity of ^{13}C and ^{15}N during direct observation, while another should have the inverse arrangement of coils in order to achieve the highest sensitivity for proton observation. A quadruple-band probe (^{1}H, ^{2}H, ^{13}C, ^{15}N) is necessary for the ^{1}H-decoupled $^{13}X\{^{15}N\}SBC$ experiment. The original $^{13}C\{^{15}N\}SBC$ experiments (Westler et al., 1988a) were carried out by using a 10 mm probe with the $^{13}C/^{15}N$ coil closest to the sample. Alternative carbon-nitrogen experiments utilize an inverted geometry and proton detection (Montelione, 1989). All probes should possess excellent water-suppression characteristics.

Sufficient range of transmitter pulse power must be available to provide both hard pulses and low-power decoupling pulses at the X-nucleus frequency. When a pulsed transmitter cannot be gated on long enough for composite-pulse-decoupling (i.e., WALTZ-16 [Shaka et al., 1983]) during acquisition, external amplifiers are useful to increase the output of a low power transmitter. Ideally, one should have the capability to use both high and low transmitter power levels in the same experiment. A third frequency channel is needed in the ^{1}H-decoupled $^{13}C\{^{15}N\}SBC$ experiment. We have employed an X-nucleus decoupler to supply this third frequency.

Finally, the success of many of the two- and three-dimensional experiments involving ^{13}C or ^{15}N depends on the elimination of extraneous noise arising from many components of the experimental setup. This can be accomplished by judicious use of bandpass and bandstop filters and crossed diodes.

DEUTERIUM-ASSISTED PROTEIN NMR SPECTROSCOPY

A convenient way to simplify strongly overlapping regions of ^{1}H NMR spectra is to deuterate the protein except for a few selected amino acid types (Crespi et al., 1968; Markley et al., 1968; Markley, 1972). This allows residue assignments based solely on spectra obtained from the protein analogue. Alternatively, selective deuteration may be used to eliminate resonances from particular amino acid types. This permits residue assignments in spectra of the fully protonated protein by observing the disappearance of the deuterated resonances in spectra of the protein analogue (LeMaster and Richards, 1988). Natural abundance proton minus deuterated analogue difference spectroscopy can be used to display such differences. Selective deuteration combined with difference spectroscopy has been used to determine the interactions between Fab antibody fragments and spin-labeled haptens (Anglister et al., 1984a; Anglister et al., 1984b; Frey et al., 1984; Anglister et al., 1987; Frey et al., 1988) and peptide antigens (Anglister et al., 1988; Anglister et al., 1989), as well as interactions between the Fab heavy and light chain components (Anglister et al., 1987; Anglister et al., 1985).

Selective deuteration can also be used to provide stereoselective resonance assignments. This method involves the incorporation into the protein of amino acids that are chirally deuterated in the beta position of the side chain (LeMaster, 1989; LeMaster and Richards, 1988; LeMaster, 1987). Chiral deuteration allows stereospecific assignment of the beta protons and extraction of the alpha-beta coupling constant by removing the passive coupling to the other beta proton. An example of this approach is shown in Figure 2, which displays the same region of the nuclear Overhauser enhancement spectroscopy (NOESY) spectrum from three different analogues of $E.$ $coli$ thioredoxin (LeMaster, 1987). Non-stereospecific β-proton assignments from Figure 2B can be stereospecifically assigned by comparison to NOESY spectra of pro-R (Figure 2A) and pro-S (Figure 2C) deuterated thioredoxin. This technique has allowed the χ_1 angle to be determined for four asparagine and eight aspartic acid residues in thioredoxin. Chiral assignments of beta-protons lead to more precise distance constraints, which then combined with dihedral angle information result in more accurate solution structures (Driscoll et al., 1989).

As discussed earlier, random fractional deuteration of proteins simplifies homonuclear two-dimensional ^1H spectra by narrowing proton resonance linewidths (LeMaster, 1989). An example of the increased resolution afforded by 50% random fractional deuteration is shown in Figure 3, which compares the COSY fingerprint regions recorded in ^1H_2O of [100% ^1H] staphylococcal nuclease and [50% ^1H] staphylococcal nuclease (Wang et al., 1989b). The increased resolution is readily apparent.

Random fractional deuteration (50%) also has been shown to remove some ambiguities in homonuclear Hartmann-Hahn (HOHAHA) experiments (LeMaster, 1988). When a group of resonances are identified as belonging to a single spin system, uncertainties may exist in the assignment of individual spin system members because the path of magnetization transfer may not be obvious from the peak intensities. In random fractionally deuterated proteins, progressive isotopic dilution readily identifies the magnetization transfer pathway (LeMaster, 1988).

The above techniques all use ^2H-labeling to simplify ^1H spectra. Deuterium-labeling also has been used widely in conjunction with ^2H NMR investigations of proteins. Akasaka et al. (1988), for example, have incorporated ^2H-labeled tryptophan into $Streptomyces$ subtilisin inhibitor to study the internal motion of the tryptophan residue.

CARBON-13- AND NITROGEN-15-ASSISTED PROTEIN NMR SPECTROSCOPY

A prerequisite to investigation of protein structure and dynamics is the sequence-specific assignment of the ^1H, ^{13}C, and ^{15}N resonances, or at least an interesting subset thereof. In this section we will demonstrate how enrichment with ^{13}C or ^{15}N both facilitates assignments of ^1H, ^{13}C, and ^{15}N resonances, and increases the number of spectral parameters that can be extracted and correlated with molecular structure and dynamics. Experiments that exploit ^{13}C and/or ^{15}N enrichment will be grouped into four broad classes: (1) chemical shift correlations between any two of the three heteronuclei (^1H, ^{13}C, ^{15}N); (2) combination of heteronuclear chemical shift

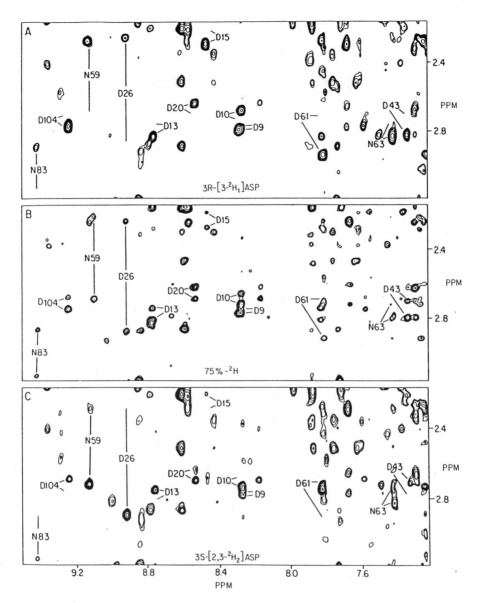

Figure 2. Chiral assignment of selected β-methylene protons in the NOESY spectrum (500 Mhz) of 8 mM *E. coli* thioredoxin. Assignments are made by comparing the spectra of pro-R deuterated (A) and pro-S deuterated (C) samples to the 75% random fractionally deuterated spectrum (B). From LeMaster, 1987.

correlation and [1]H-^{1}H correlation experiments; (3) simplification and enhancement of [1]H-^{1}H correlation spectra; and (4) isotope-assisted measurement of protein conformation and dynamics. The first class provides powerful tools for determining chemical shifts, elucidating spin systems, and making sequence-specific assignments. The second and third classes provide information on both assignments and, more importantly, secondary and tertiary structure. The fourth class provides information regarding main-chain and side-chain conformations, protein dynamics, and chemical exchange processes.

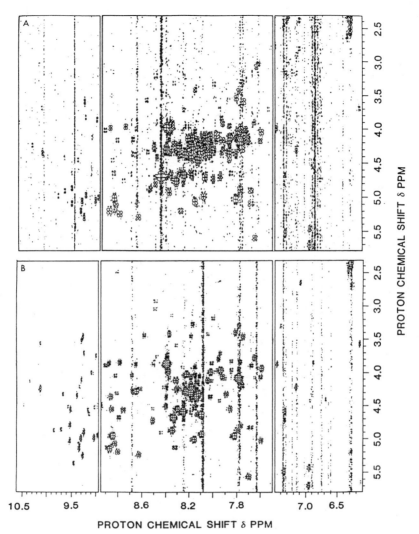

Figure 3.   Comparison of staphylococcal nuclease (H124L) COSY fingerprint regions of
A) natural abundance proton and B) 50% μl ^{2}H protein. The natural abundance
sample was 3.5 mM nuclease dissolved in 10.5 mM pdTp, 21 mM CaCl$_2$, and 300
mM KCl at pH 5.1 at 45°C. Deuterated nuclease was 5 mM in 15 mM pdTp, 30
mM CaCl$_2$, and 300 mM KCl at pH 5.1 at 45°C. J. Wang, D.M. LeMaster and J.L.
Markley, unpublished data.

Chemical Shift Correlations Between Heteronuclei

The classic approach to extensive peak assignments in small proteins has been to first
elucidate proton-proton coupling patterns (spin systems) by using a combination of two-
dimensional homonuclear ^{1}H NMR experiments (Wüthrich, 1986; Wüthrich et al., 1982;
Billeter, 1982), such as correlated spectroscopy (COSY) (Nagayama et al., 1980), NOESY
(Anil Kumar et al., 1980), relayed correlated spectroscopy (RELAY) (Bax and Drobny, 1985),
and HOHAHA (Davis and Bax, 1985). Sequence-specific assignments are then developed by

comparing interresidue NOE's from NOESY spectra with the protein sequence. The proton assignments can then be extended to carbon and nitrogen nuclei with heteronuclear correlation experiments (discussed below). Because of spectral complexity, this procedure becomes less efficient as the molecular weight increases above 10 kDa. In this section, scalar correlation experiments will be discussed that permit the identification of multinuclear spin systems of individual residues, as well as provide cross peptide bond connectivities. These approaches exploit heteronuclear coupling, and their efficiency depends on the magnitude of the coupling constant involved. Table 2 lists the observed magnitude ranges for such coupling constants in peptides.

Table 2. Coupling constant magnitude ranges in peptides/proteins[a]

Atoms coupled	Coupling constant range (Hz[b])
^{2J}HH	12 - 18
^{3J}HH	0 - 10
^{1J}NH	85 - 95
^{2J}NH	0 - 4
^{3J}NH	0 - 10
^{1J}CH	
aliphatic	120 - 130
aromatic	170 - 220
^{2J}CH	0 - 8
^{3J}CH	0 - 14
^{1J}CC	
aliphatic	35 - 50
aromatic	60 - 70
$^{13}C\alpha\text{-}^{13}C'$	50 - 60
$^{1J}C'N$	12 - 15
$^{1J}C\alpha N$	4 - 6
$^{2J}C\alpha N$	8 - 12

[a] Values are taken from Bystrov, 1976.
[b] Absolute value.

^{13}C-^{1}H and ^{15}N-^{1}H moieties in proteins can be assigned coordinately by using heteronuclear correlation experiments. The first heteronuclear correlation experiments directly detected the less sensitive nucleus rather than the proton (Chan and Markley, 1983; Kojiro and Markley, 1983) and greatly benefited from stable-isotope enrichment of the protein. More recently, proton detection has become the standard (Bax et al., 1983; Ortiz-Polo et al., 1986; Glushka and Cowburn, 1987; Sklenar and Bax, 1987). Isotopic enrichment, although not required in some cases, greatly improves the quality of spectra obtained from the proton-detected experiments. This class of experiments is known under a variety of names, including HETCOR, HMQC, and HSBC. This paper will use $^{1}H\{^{13}C\}SBC$ to refer to proton-carbon single-bond correlation experiments, and $^{1}H\{^{15}N\}SBC$ to refer to proton-nitrogen single-bond correlation experiments. A hypothetical $^1H\{^{13}C\}SBC$ spectrum of a threonine residue is shown in Figure 4A.

The heteronuclear multiple-bond correlation (HMBC) experiment can be used to correlate protons with distantly coupled (more than one-bond) carbon or nitrogen atoms. We will specify the nuclei involved and refer to these experiments as $^1H\{^{13}C\}MBC$ and $^1H\{^{15}N\}MBC$ (Bax and Summers, 1986). Typically, one observes only two- and three-bond correlations, although an occasional four-bond coupling has been observed, especially in aromatic ring systems. A hypothetical $^1H\{^{13}C\}MBC$ spectrum of threonine is shown in Figure 4B. Since the magnitudes of the hetero-nuclear two- and three-bond coupling constants have an angular dependence (Bystrov, 1976), they contain structural information (Bystrov, 1976; Clore et al., 1988; Karplus and Karplus, 1972).

Extraction of the information contained in the one-bond carbon-carbon scalar couplings was demonstrated several years ago with a ^{13}C-enriched amino acid (Markley et al., 1984). In this experiment, a $^{13}C\{^{13}C\}COSY$ pulse sequence was applied to a sample of [85% ^{13}C]lysine, resulting in the elucidation of the ^{13}C spin system. We have found that carbon-carbon correlations are more efficiently obtained from $^{13}C\{^{13}C\}DQC$ spectra. This results in greater resolution of cross peaks because of the dispersion along the double quantum axis. The pulse sequence used is a modification (Westler et al., 1988b) of a resonance-offset-compensated INADEQUATE (Bax et al., 1980; Levitt and Ernst, 1983) experiment. Figure 4C shows a hypothetical $^{13}C\{^{13}C\}DQC$ spectrum of threonine. When applied to a sample of *Streptomyces* subtilisin inhibitor that was enriched to 85% ^{13}C in the leucine residues, extension of previous carbonyl assignments to alpha carbons resulted. At the 85% ^{13}C-enrichment level, however, passive ^{13}C coupling results in complex line shapes. Lower levels of enrichment (Figure 1) are more optimal (Westler et al., 1988b). Application of the experiment to *Anabaena* 7120 flavodoxin (Stockman et al., 1988a), ferredoxin (Oh et al., 1988), and cytochrome c_{553} (Zehfus et al.,1989) uniformly enriched with 26% ^{13}C has demonstrated that entire carbon spin systems for amino acid residues and prosthetic groups can be identified. In fact, this single experiment allows one to identify in principle the carbon spin systems for 16 out of the 20 amino acid types (redundancies being Asp/Asn and Glu/Gln) (Oh et al., 1988; Oh and Markley, 1989) (Figure 5). This contrasts with unique identification of only 8 out of the 20 proton spin systems from any single $^1H\{^1H\}$ two-dimensional NMR experiment (Wüthrich, 1986).

A modification of the HSBC experiment discussed above for proton-heteronuclear correlations has also been applied to doubly-enriched (26% ^{13}C, 95% ^{15}N) proteins to correlate carbon and nitrogen chemical shifts (Westler et al., 1988a; Mooberry et al., 1989). The experiment is analogous to $^1H\{^{15}N\}SBC$ except that one detects carbons instead of protons. The experiment is referred to as $^{13}C\{^{15}N\}SBC$ (although some two-bond correlations are seen). Because the coupling constants between these nuclei are small (7-14 Hz), the second delay in the SBC experiment was removed to increase sensitivity.

The correlation experiments discussed above are most powerful when the results are analyzed in concert (Stockman et al., 1989), as shown in Figure 4. Examples using real data will now be given that demonstrate a concerted approach to first-order assignments, as well as several scalar-coupling approaches to sequential assignments in proteins. The foundation for this method of analysis is the carbon-carbon double quantum correlation ($^{13}C\{^{13}C\}DQC$) experiment. This experiment correlates single-bond coupled carbon atoms. Since each amino acid type has a unique carbon spin system (Figure 5), this single experiment allows (in principle) the carbon spin systems of each amino acid in the protein to be outlined, starting at the carbonyl carbon and tracing along carbon-carbon bonds to the terminal carbon(s) of the side chain. The

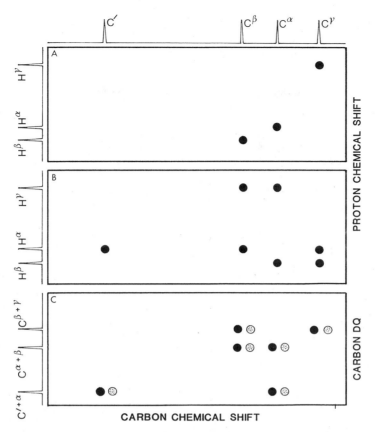

Figure 4. Hypothetical ^{1}H{^{13}C}SBC (A), ^{1}H{^{13}C}MBC (B), and ^{13}C{^{13}C}DQC (C) spectra of a threonine residue. Alignment of spectra in this way allows a concerted approach to resonance assignments.

^{13}C{^{13}C}DQC spectrum of uniformly enriched *Anabaena* 7120 flavodoxin is shown in Figure 6. Each pair of coupled carbon atoms is represented by a pair of antiphase doublets located at their double-quantum frequency on the vertical axis; each individual antiphase doublet is located at its single-quantum frequency on the horizontal axis. Horizontal lines in this spectrum connect resonances from single-bond coupled carbon atoms. In addition to amino acid resonances, the outlined correlations demonstrate the usefulness of this experiment to delineate assignment of cofactor resonances, in this case flavin mononucleotide. In practice, overlaps and ambiguities in the ^{13}C{^{13}C}DQC amino acid spin systems must be resolved by the use of proton-carbon single-bond (^{1}H{^{13}C}SBC), multiple-bond (^{1}H{^{13}C}MBC) correlations, or the pseudo 3D experiments discussed below. Each experiment complements the others by resolving spectral overlaps or by adding redundancy to well-resolved regions. The methodology works equally well for aliphatic and aromatic amino acids.

Aliphatic side chains. Comparisons of spectral regions illustrate how the information content of one type of spectrum can be used to sort out or confirm information available in the

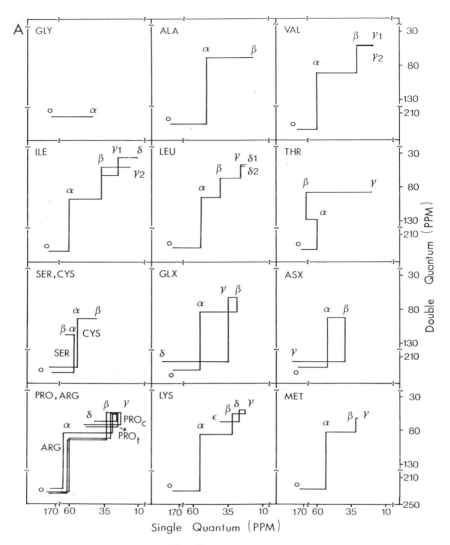

Figure 5.   (A) Two-dimensional $^{13}C\{^{13}C\}DQC$ linkage patterns for aliphatic amino acids. The patterns are based on ^{13}C chemical shifts in small peptides (Oh and Markley, 1989; Richarz and Wüthrich, 1978). From Oh et al., 1988, and Oh and Markley, 1989.

other types of spectra. Figure 7 shows corresponding aliphatic regions of the (A) $^{13}C\{^{13}C\}DQC$ and (B) $^1H\{^{13}C\}MBC$ spectra of *Anabaena* 7120 flavodoxin. The carbon dimension (vertical axis) is identical in each spectrum. Some regions of the $^{13}C\{^{13}C\}DQC$ spectrum are easily assigned based on Figure 5A, but others are too crowded for easy analysis. This overlap can be resolved by spreading the carbon chemical shifts out in the proton dimension by means of the $^1H\{^{13}C\}MBC$ coupling patterns characteristic for a given amino acid. Further redundancies can be added by reference to the $^1H\{^{13}C\}SBC$ spectrum shown in Figure 8.

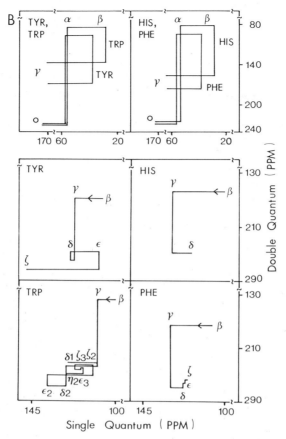

Figure 5.   (B) Two-dimensional $^{13}\text{C}\{^{13}\text{C}\}$DQC linkage patterns for aromatic amino acids.

The use of information from one spectrum to resolve another is exemplified by the outlined leucine connectivities. In Figure 7A, the leucine $^{13}\text{C}\{^{13}\text{C}\}$DQC correlations are not resolved. In Figure 7B, however, long-range proton-carbon coupling patterns characteristic of the pairs of leucine methyl $^1\text{H}^\delta$'s are observed. These connectivities assign the $^{13}\text{C}^\beta$, $^{13}\text{C}^\gamma$, $^{13}\text{C}^{\delta 1}$, and $^{13}\text{C}^{\delta 2}$ resonances in the $^{13}\text{C}\{^{13}\text{C}\}$DQC spectrum (Figure 7A). The position of the $^{13}\text{C}^{\delta 1}$ and $^{13}\text{C}^{\delta 2}$ and $^1\text{H}^{\delta 1}$ and $^1\text{H}^{\delta 2}$ resonances are confirmed in Figure 8, which shows the $^1\text{H}\{^{13}\text{C}\}$SBC connectivities for these and other methyl groups. The outlined alanine, threonine, and valine spin systems illustrate how redundant connectivities can be obtained from information in each type of 2D spectrum. Each carbon spin system is resolved in Figure 7A. The long-range coupling patterns of the methyl protons seen in Figure 7B confirm the carbon spin system assignments as well as assign the methyl proton chemical shifts. The methyl proton and carbon chemical shifts are also verified in Figure 8.

The extension of the assignments from side-chain to backbone atoms is critical. In particular, the redundancy of $^1\text{H}\{^{13}\text{C}\}$MBC and $^{13}\text{C}\{^{13}\text{C}\}$DQC in the assignment of carbonyl carbons permits the carbonyl carbon to be used as the focal point of several sequential assignment strategies (see below).

Aromatic side chains. The aromatic portions of the proton and carbon spin systems can be assigned by using a similar analysis. Figure 9 A illstrates the one-bond carbon-carbon correlations observed in the $^{13}\text{C}\{^{13}\text{C}\}$DQC spectrum for the $^{13}\text{C}^\beta$ and aromatic carbons of the

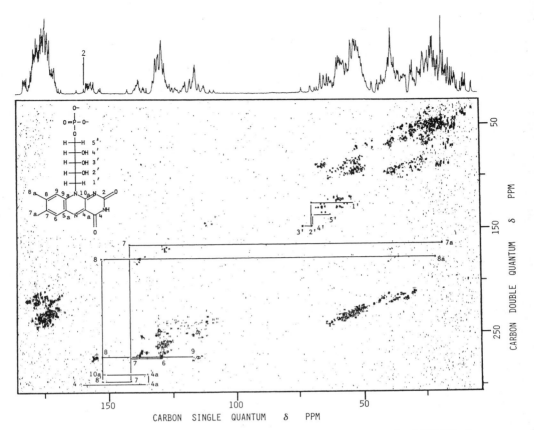

Figure 6. ^{13}C{^{13}C}DQC spectrum of flavodoxin. The sample was 7 mM [26% μl
^{13}C]flavodoxin in ^{2}H$_2$O containing 150 mM phosphate buffer at pH 7.5. The
single-quantum frequency is plotted along the vertical axis, and the double-quantum
frequency is plotted along the horizontal axis. Several correlations are shown for
the non-covalently bound flavin mononucleotide cofactor. B.J. Stockman and J.L.
Markley, unpublished data.

single tryptophan, one of the two tyrosine residues, and one of the two phenylalanine residues,
along with the one-bond carbon-carbon correlations observed for the ^{13}C$^{\beta}$, ^{13}C$^{\gamma}$ and ^{13}C$^{\delta 1}$ of
the iron-ligated histidine residue in *Anabaena* 7120 cytochrome c$_{553}$. The characteristic
^{13}C{^{13}C}DQC connectivity patterns (Figure 5B) allow easy identification of each carbon
resonance.

Extension of the carbon assignments obtained from the ^{13}C{^{13}C}DQC data to the
corresponding ^{1}H$^{\beta}$'s (Figure 9B) and aromatic ring protons (Figure 10A) was accomplished by
using ^{1}H{^{13}C}SBC data. Many of the carbon and proton resonance assignments were verified
by cross peaks in the ^{1}H{^{13}C}MBC spectrum (Figure 10B); characteristic long-range coupling
patterns were observed for several types of aromatic protons. Note that redundancy in the ^{1}H
and ^{13}C chemical shifts for the pair of symmetric δ- or ε-atoms of the phenylalanine and tyrosine
rings (presumably from ring flips that are rapid on the NMR time scale) results in identical cross
peaks for the δ- or ε-correlations in ^{1}H{^{13}C}SBC and ^{1}H{^{13}C}MBC spectra.

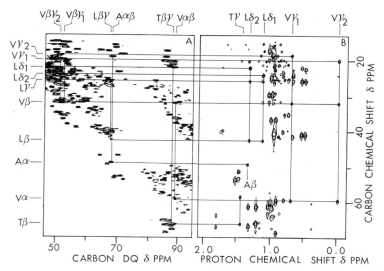

Figure 7.   Composite 11.7 Tesla NMR spectra of flavodoxin showing a portion of the aliphatic proton and carbon region. Sample conditions are described in the legend to Figure 6. The same region of the carbon spectrum (vertical axis) is shown in both contour plots. Labeled positions along the horizontal and vertical axes identify cross peaks located at their intersection. The notation used in the labels in "Pr," where "P" is the single-letter amino acid abbreviation and "r" is the Greek letter(s) identifying the atoms(s) involved in the correlation. Only selected spin systems are outlined in the figure. (A) A small region of the $^{13}C\{^{13}C\}$DQC spectrum. (B) A region of the $^1H\{^{13}C\}$MBC spectrum. From Stockman et al., 1989.

The connection of the aliphatic and aromatic portions of spectra from aromatic amino acids by using only scalar correlations presents an alternative to the use of through-space (NOESY) correlations (Billeter et al., 1982). The aromatic ring carbon spin systems can be entered via their aromatic proton resonances; then, by following carbon-carbon correlations around the ring,

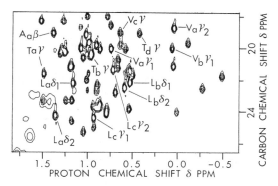

Figure 8.   Methyl region of the 11.7 Tesla NMR $^1H\{^{13}C\}$SBC spectrum of flavodoxin. Sample conditions were as described in the legend to Figure 6. The notation is "$P_q r$," where "P" is the single-letter amino acid abbreviation, "q" corresponds to the amino acid lettering in the data tables, and "r" is the Greek letter identifying the carbon and proton atoms involved in the correlation. From Stockman et al., 1989.

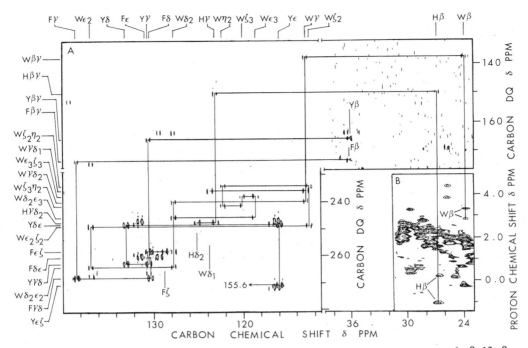

Figure 9. Composite 11.7 Tesla NMR spectra of cytochrome c_{553} showing the $^1H\beta$, $^{13}C\beta$, and ^{13}C aromatic regions. Peaks are labeled as in Figure 7. Only selected spin systems are outlined. The sample was 15 mM [26% μl ^{13}C]cytochrome c_{553} in ^2H_2O containing 50 mM phosphate buffer at pH 5.6 at 40°C; the total sample volume was 0.15 ml in a 5 mm (OD) cylindrical microcell. (A) Selected regions of the $^{13}C\{^{13}C\}$DQC spectrum. (B) β-region of the $^1H\{^{13}C\}$SBC spectrum. From Stockman et al., 1989.

the $^{13}C\beta$ can be assigned. Proton-carbon correlations to the $^{13}C\beta$ then can be used to exit the carbon spin system and assign the $^1H\beta$ resonances. In this way, aromatic ring proton resonances can be assigned unambiguously.

Figure 11 illustrates how $^1H\{^{15}N\}$MBC data were used to assign the aromatic nitrogen resonances in flavodoxin. The tryptophan $^{15}N\epsilon 1$ assignments were based on proton assignments made by using the method described in Figures 9 and 10. Tryptophan $^{15}N\epsilon 1$ resonances are difficult to assign by other methods because they are found near the amide backbone region (100-135 ppm). However, since no other cross peaks occur in this region of the $^1H\{^{15}N\}$MBC spectrum, the assignments here are unambiguous. Histidine $^1H\delta 2$ and $^1H\epsilon 1$ correlations to both $^{15}N\delta 1$ and $^{15}N\epsilon 2$ are observed as a characteristic box-shaped pattern. Although the histidine nitrogen resonances cannot be distinguished here (since both show correlations to both ring protons), they are assignable as long range correlations "leaking through" in ^{15}N-coupled $^1H\{^{15}N\}$SBC spectra (Oh et al., submitted). The relative magnitudes of the correlations provides a means of assigning the proton and nitrogen resonances since the weakest correlation is expected between $^1H\delta 2$ and $^{15}N\delta 1$ (Blomberg et al., 1977).

Previous assignments of quaternary carbons required indirect techniques, such as chemical modification, relaxation probes, or the use of variant proteins (Allerhand, 1979; Chan and Markley, 1983). They are assigned directly by the methods described here. Specific

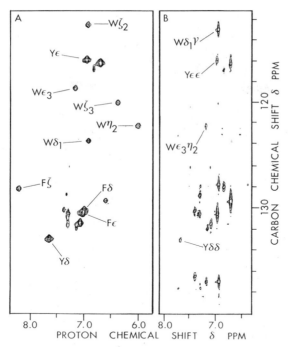

Figure 10. Composite 11.7 Tesla NMR spectra of cytochrome c_{553} showing the aromatic region. Sample conditions were as described in the legend to Figure 9. Each identified cross peak is associated with the outlined spin systems in Figure 10. (A) Region of the ^{1}H{^{13}C}SBC spectrum. Peak notation is "Pr," where "P" is the single-letter amino acid abbreviation, and "r" is the Greek letter identifying the type of carbon and proton atoms involved in the correlation. (B) Region of the ^{1}H{^{13}C}MBC spectrum. Peaks are labeled according to the notation "Prs," where "P" is the single letter amino acid abbreviation, and "r" and "s" are Greek letters identifying the atoms whose resonances along the horizontal and vertical frequency axes, respectively, give rise to the cross peak. From Stockman et al., 1989.

assignments of tyrosine ^{13}C$^\zeta$'s, which frequently can be resolved in the one-dimensional ^{13}C spectrum, are useful in investigations of tyrosine pK_a values or enzyme mechanisms thought to involve tyrosine (Grissom and Markley, 1989).

The concerted assignment techniques are also applicable to cofactors and prosthetic groups. In addition to the flavin mononucleotide cofactor of *Anabaena* 7120 flavodoxin, the complete assignment of ^{1}H, ^{13}C, and ^{15}N resonances of the heme group of *Anabaena* 7120 cytochrome c_{553} has been reported (Zehfus et al., 1989).

Sequential assignment strategies utilizing scalar correlations. The methods described above present several alternatives to the usual through-space approach to sequential assignments (Wüthrich, 1986; Wüthrich et al., 1982; Billeter et al., 1982). These alternative methods rely solely on scalar coupling pathways, are less ambiguous than the NOESY approach, and extend the applicability of NMR analysis beyond that possible with ^{1}H-^{1}H correlations alone. Two approaches use long-range heteronuclear correlations: ^{13}C$'_i$-N$^\alpha$-C$^\alpha$-^{1}H$^\alpha_{i+1}$ and ^{1}H$^\alpha_i$-C$^\alpha$-C'-^{15}N$^\alpha_{i+1}$. Since the three-bond coupling constants are angle

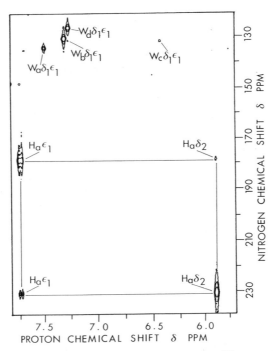

Figure 11. Aromatic correlations in the 11.7 Tesla NMR ^{1}H{^{15}N}MBC spectrum of flavodoxin. The sample was 10 mM [95% µl ^{15}N]flavodoxin in ^{2}H$_2$O containing 150 mM phosphate buffer at pH 7.5. Peaks are labeled according to the notation "P$_q$rs," where "P" is the single-letter amino acid abbreviation, "q" corresponds to the amino acid lettering in the data tables, and "r" and "s" are Greek letters identifying the atoms whose resonances along the horizontal and vertical frequency axes, respectively, give rise to the cross peak. Only the proton chemical shift is given for the histidine correlations. From Stockman et al., 1989.

dependent, some backbone conformations result in better correlations than others (Bystrov, 1976; Karplus and Karplus, 1972). This places limits on the overall applicability, but it is also useful in identifying regions of secondary structure and in measuring dihedral angles. A third approach, however, relies on one-bond correlations which have no angle dependence: ^{13}C'$_i$-^{15}N$^\alpha_{i+1}$.

^{1}H{^{13}C}MBC experiments can be used to correlate the carbonyl carbon of residue i to the alpha proton of residue i+1 (Westler et al., 1988b; Bax et al., 1988). An example of this is shown for [85% ^{13}C leucine]*Streptomyces* subtilisin inhibitor in Figure 12. Carbonyl assignments made previously were extended to the alpha proton of the following residue. Sequential three-bond correlations observed are between residues located in regions of β-sheet or extended conformations. This is consistent with model peptide data that predict larger three-bond coupling constants for β-sheet conformations (3-4 Hz) than for α-helical conformations (<2 Hz) (Bystrov, 1976; Karplus and Karplus, 1972). This sequential assignment strategy is thus best suited for β-sheet and random coil regions of proteins.

The ^{1}H{^{15}N}MBC experiment has been applied to the sequential assignment of protein *ner* from phage Mu (Clore et al., 1988). Many intraresidue two-bond couplings and interresidue (amide nitrogen of residue i+1 to alpha proton of residue i) three-bond couplings were observed.

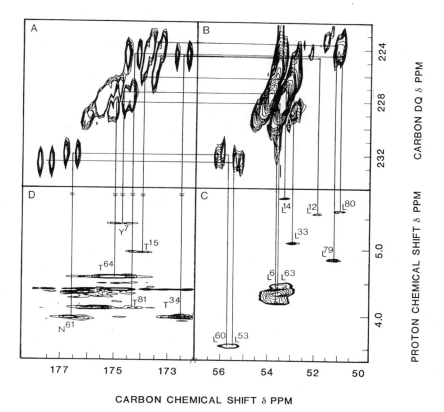

Figure 12. Composite NMR spectra of ([85% μl ^{13}C]leucine)SSI at 61°C. The sample was 100 mg of the labeled protein in 2 mL 0.05 M phosphate buffer, pH 7.3, in ^{2}H$_2$O. (A) and (B) show, respectively, the ^{13}C'/(^{13}C'+^{13}Cα) and ^{13}Cα/(^{13}C'+^{13}Cα) regions of the ^{13}C{^{13}C}DQC spectrum. (C) ^{1}Hα/^{13}Cα region of the ^{1}H{^{13}C}SBC spectrum. (D) ^{1}Hα/^{13}C' region of the ^{1}H{^{13}C}MBC spectrum. Carbonyl assignments are extended to intraresidue alpha carbon and proton resonances, and to the alpha proton of residue i+1. Note that the leucine-53 and leucine-60 spin systems are reversed from those of Westler et al., 1988a.

Intraresidue and interresidue couplings were distinguished with a ^{1}H{^{15}N}SBC-COSY experiment, which gives only intraresidue correlations. Nearly all three-bond correlations observed were between residues located in α-helical regions. This is consistent with model peptide data that predict larger three bond correlations for α-helical conformations (4-6 Hz) than for β-sheet conformations (0-2 Hz) (Bystrov, 1976; Karplus and Karplus, 1972). Similar results have been observed in ^{15}N-enriched staphylococcal nuclease (Bax et al., 1988). This sequential assignment strategy complements the ^{1}H{^{13}C}MBC strategy by being best suited for α-helical regions of proteins (Clore et al., 1988).

Transpeptide coupling provides a sequential assignment pathway unlimited by either through-space ambiguity or angular dependence. The double-labeling technique for making sequential assignments was first implemented by observing the splitting of carbonyl resonances (residue i) by the adjacent amide nitrogen (residue i+1) in one-dimensional ^{13}C spectra (Kainosho and Tsuji, 1982; Kainosho et al., 1987). Assignment of a methionyl-lysyl peptide bond in IgG (160 kDa) indicates that this approach to sequential assignments is feasible even with very large proteins (Kato et al., 1989). Recently, the same type of information has been obtained from $^1H\{^{15}N\}$SBC spectra that contain passive ^{13}C splittings arising from double-labeling (Griffey et al., 1986; Torchia et al., 1988). The two-dimensional $^{13}C\{^{15}N\}$SBC experiment (Westler et al., 1988a; Mooberry et al., 1989; Niemczura et al., 1989) provides direct intraresidue and interresidue carbon-nitrogen correlations and reduces the number of protein analogues required to make sequence-specific assignments. Figure 13 shows the $^{13}C\{^{15}N\}$SBC spectrum of uniformly ^{13}C (26%) and ^{15}N (95%) labeled *Anabaena* 7120 ferredoxin. Sequential and side-chain assignments are indicated. Since the one-bond coupling has no angle dependence, one expects to see all possible correlations.

Recently, several triple-resonance experiments have been used to make sequential assignments in peptides (Montelione and Wagner, 1989c). These results indicate that magnetization can be transferred efficiently from alpha (residue i) to amide (residue i+1) proton (or vice versa) through a series of scalar couplings. This type of experiment may prove fruitful for sequential assignment with doubly-enriched proteins.

Combination of heteronuclear chemical shift correlation and ^1H-^1H correlation experiments. Heteronuclear correlation pulse sequences can be combined with homonuclear ^1H correlation pulse sequences to create new kinds of 2D and 3D experiments (Lerner and Bax, 1986; Gronenborn et al., 1989a; Fesik and Zuiderweg, 1988; Marion et al., 1989b). In the resulting spectra, the information contained in the homonuclear ^1H spectrum is spread out by the chemical shifts of carbon or nitrogen atoms that are directly bonded to the protons involved in correlations with other protons. Heteronuclear RELAY experiments have also been used to spread ^1H-^1H correlations over a heteronuclear chemical shift range (Bolton, 1985; Bruhwiler and Wagner, 1986; Senn et al., 1987a), but these experiments are limited to ^1H-^1H scalar interactions. The experimental setup is analogous to $^1H\{^{13}C\}$SBC or $^1H\{^{15}N\}$SBC experiments, and the spectra obtained contain the same one-bond correlation peaks that are present in the $^1H\{^{13}C\}$SBC or $^1H\{^{15}N\}$SBC spectra. The difference is the

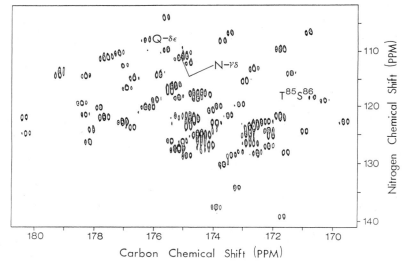

Figure 13. 2D $^{13}C\{^{15}N\}$SBC spectrum of ferredoxin. The sample was 4.5 mM [26% μl^{13}C, 95% μl ^{15}N]ferredoxin in 2.2 ml 40 mM phosphate buffer at pH 7.0 at 45°. The $^{13}C'$-$^{15}N^\alpha$ region is shown along with selected assignments. Several side-chain correlations are also indicated. E.S. Mooberry, B.-H. Oh and J.L. Markley, unpublished data.

addition of ^{1}H-^{1}H correlation data in the ^{1}H dimension. The basic pulse sequences used start with the usual ^{1}H{X}SBC pulses to create heteronuclear-chemical-shift-labeled ^{1}H magnetization. These are then followed by pulses on the proton channel that carry out a variety of ^{1}H{^{1}H} correlation experiments with the heteronuclear-chemical-shift-labeled ^{1}H magnetization (Gronenborn et al., 1989; Fesik and Zuiderberg, 1988; Marion et al., 1989b).

A single heteronuclear correlation with non-incrementation of the ^{1}H-^{1}H correlation part of the sequence is referred to as a pseudo three-dimensional experiment (one proton chemical shift axis, one heteronucleus chemical shift axis). A series of heteronuclear correlation experiments resulting from incrementation of the ^{1}H-^{1}H correlation part of the pulse sequence results in a true three-dimensional experiment (two proton chemical shift axes, one heteronucleus chemical shift axis). These experiments are referred to as ^{1}H{X}SBC-COSY, ^{1}H{X}SBC-NOESY, and ^{1}H{X}SBC-HOHAHA, depending on the ^{1}H correlations involved. Alternatively, three-dimensional spectra can be recorded with proton correlations first, followed by the heteronuclear correlations (i.e., NOESY-^{1}H{X}SBC). This sequence is preferable when correlations to protons that resonate near the water resonance are important (Kay et al., 1989b). The ^{1}H-^{1}H correlations obtained are the same as in the normal two-dimensional ^{1}H experiment, except now they are spread out over the chemical shift range of the heteronucleus.

A region of the ^{1}H{^{13}C}SBC-HOHAHA spectrum of *Anabaena* 7120 ferredoxin is shown in Figure 14B (Oh et al., 1989). This spectrum serves as the crucial link in a ^{13}C spin system directed approach to first-order assignment of the COSY fingerprint region (Oh et al., 1989). In this procedure, the ^{1}H$^\alpha$'s and ^{1}H$^\beta$'s of each amino acid residue are assigned by comparison of ^{13}C{^{13}C}DQC and ^{1}H{^{13}C}SBC-HOHAHA spectra, as demonstrated in Figure 14. One then compares these assignments with the ^{1}H{^{1}H}RELAY or ^{1}H{^{1}H}HOHAHA spectrum recorded in ^{1}H$_2$O to assign the fingerprint region. A similar pseudo three-dimensional experiment, ^{1}H{^{15}N}SBC-COSY, has been used to distinguish intraresidue and interresidue correlations in the ^{1}H{^{15}N}MBC spectrum of the 75-residue DNA-binding protein *ner* from phage Mu (Clore et al., 1988). Additional resolution afforded by the heteronuclear chemical shift aids both first-order and sequential assignment processes in protein NMR spectroscopy.

An important aspect of this class of experiments is in the resolution of ^{1}H{^{1}H}NOESY spectra, which contain information concerning interproton distances and thus are critical for solution structure determination (Wüthrich, 1986; Wüthrich, 1989). Since the NOESY spectrum of a protein is usually much more crowded than spectra containing scalar correlations (especially at higher molecular weights), the addition of the heteronucleus chemical shift dimension greatly assists assignment of the through-space correlations. Extraction of NOE build-up rates also may be more precise from this type of spectrum since one compares the intensity of NOE peaks to the intensity of the well-resolved one-bond correlation peaks, instead of the diagonal peaks of ^{1}H{^{1}H}NOESY spectra (which may be severely overlapping). In addition, secondary structural elements, which show characteristic through-space connectivity patterns in NOESY spectra, are easily identified in ^{1}H{^{13}C}SBC-NOESY (Wang et al., 1989a) and ^{1}H{^{15}N}SBC-NOESY spectra (Gronenborn et al., 1989a; Wang et al.,1989a; Gronenborn et al., 1989b; Shon and Opella, 1989).

Figure 15 shows the alpha region of ^{1}H{^{13}C}SBC-NOESY spectrum of staphylococcal nuclease (Wang et al., 1989a). Characteristic ^{1}H$^\alpha$-^{1}H$^\alpha$ NOE's, represented by the horizontal lines connecting two proton resonances at the same ^{13}C$^\alpha$ frequency delineate regions of β-sheet structure. Characteristic ^{1}H^N-^{1}H^N correlations indicative of an α-helix in staphylococcal nuclease are shown in Figure 16 (Wang et al., 1989a). Although this type of information can be obtained from ^{1}H{^{1}H}NOESY spectra, overlaps often prohibit a complete analysis. With the added heteronucleus chemical shift dimension, the patterns are more easily resolved.

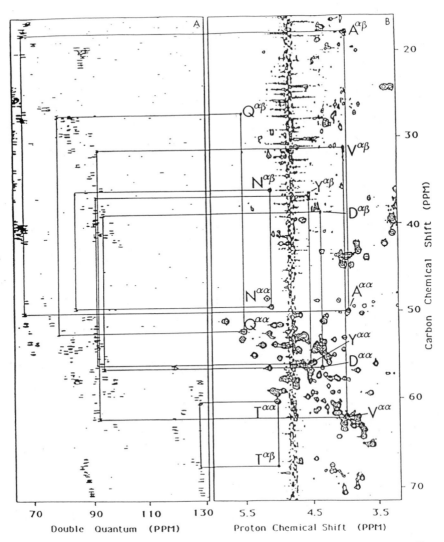

Figure 14.   Selected regions of (A) the $^{13}C\{^{13}C\}DQC$ spectrum and (B) the $^1H\{^{13}C\}SBC$-HOHAHA spectrum of oxidized [26% µl ^{13}C]ferredoxin from *Anabaena* 7120. The sample was 0.4 ml of 9.0 mM ferredoxin in ^2H_2O containing 50 mM phosphate buffer at pH 7.5. [$^1H^\alpha$,($^{13}C^\alpha$,$^{13}C^\beta$)] connectivities for seven different amino acids in the $^1H\{^{13}C\}SBC$-HOHAHA spectrum are matched to [$^{13}C^{\alpha+\beta}$,($^{13}C^\alpha$,$^{13}C^\beta$)] connectivities in the $^{13}C\{^{13}C\}DQC$ spectrum. From Oh et al., 1989.

Pseudo three-dimensional experiments can be easily extended to true three-dimensional experiments by incrementing the ^1H-^1H correlation part of the pulse sequence (Fesik and Zuiderweg, 1988; Marion et al., 1989b; Fesik et al., 1989b; Zuiderweg and Fesik, 1989). A series of heteronuclear correlation spectra are thus recorded, resulting in a three-dimensional data set. By recording each two-dimensional heteronuclear spectrum very rapidly (usually in less than two hours), three-dimensional data sets can be acquired in much less time than would be required to collect an equal number of "normal" two-dimensional heteronuclear correlation data

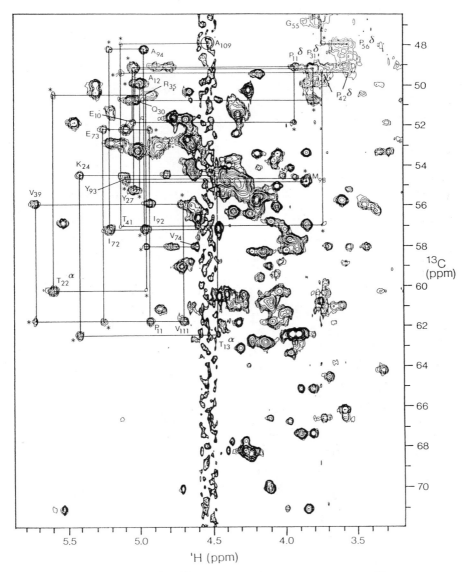

Figure 15. Alpha region of the ^{1}H{^{13}C}SBC-NOE spectrum of [26% μl ^{13}C]staphylococcal nuclease (H124L). Sample conditions were as described in the legend to Figure 3B. Interresidue ^{1}H$^\alpha$-^{1}H$^\alpha$ NOE's that correspond to regions of antiparallel β-sheet are indicated with asterisks. J. Wang, A.P. Hinck, S.N. Loh and J.L. Markley, 1989a.

sets. The decreased resolution in the rapidly-acquired heteronuclear correlation spectra is offset by spreading the correlations out into a third dimension. Typically, three-dimensional data sets are acquired over a period of 3-10 days.

Homonuclear three-dimensional NMR experiments (proton chemical shift on all axes) also have been reported (Vuister and Boelens, 1987; Griesinger et al., 1987a, b; Oschkinat et al.,

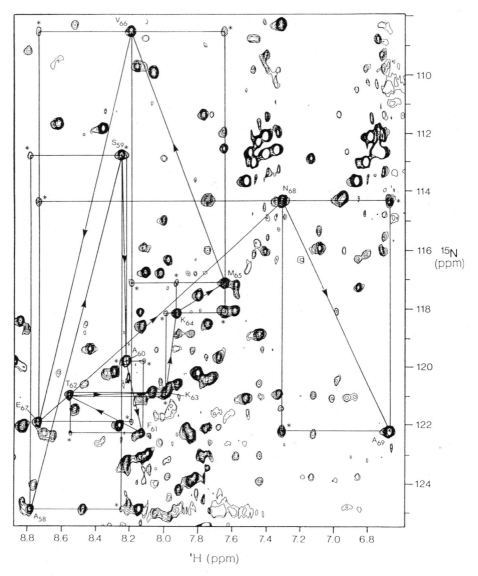

Figure 16. Amide proton region of the $^1H\{^{15}N\}$ SBC-NOE spectrum of [95% μl ^{15}N]staphylococcal nuclease (H124L). Sample conditions were as described in the legend to Figure 3B. Interresidue ^1HN_i-$^1HN_{i+1}$ NOE's diagnostic of an α-helical region from alanine-58 to alanine-69 are indicated with asterisks. J. Wang, A.P. Hinck, S.N. Loh and J.L. Markley, unpublished data.

1988; Vuister et al., 1988). With larger proteins, significant decay of proton magnetization occurs during the scalar magnetization transfer owing to J_{HH}'s (Table 2) that are on the same order as the T_2 relaxation times. This, coupled with spectral overlap in larger proteins, limits the sensitivity and resolution obtainable. In contrast, heteronuclear three-dimensional NMR utilizes the larger one-bond heteronuclear coupling to create the third dimension. A practical problem in heteronuclear three-dimensional spectroscopy is the long measuring time needed to obtain sufficient digital resolution in the first two dimensions. Digital resolution can be enhanced by minimizing the spectral widths of the first two dimensions by using either selective pulses (at the expense of lost information), or more elegantly by using folding and unfolding procedures (Marion et al., 1989b). Practical aspects of protein heteronuclear three-dimensional NMR regarding data acquisition and processing have been presented (Kay et al., 1989b).

Most heteronuclear three-dimensional experiments to date have employed ^{15}N-enriched samples and have exploited the role of the amide proton in making sequential assignments (Marion et al., 1989a). The same types of experiments also can be carried out with ^{13}C-enriched samples. Since much structural information is present in through-space correlations between carbon-bond protons, these experiments are likely to become valuable in the structural analysis of larger proteins. The natural abundance ^{1}H{^{13}C}SBC-COSY spectrum of an oligosaccharide (Fesik et al., 1989a) has been presented; it clearly demonstrates the advantages of dispersing [1]H-^{1}H correlations over the carbon chemical shift range.

Simplification and enhancement of [1]H-^{1}H correlation spectra. Carbon-13 and nitrogen-15 can serve as editors in one- and two-dimensional ^{1}H NMR spectroscopy (Bendall et al., 1981; Otting et al., 1986; Griffey et al., 1985a; Griffey and Redfield, 1987; Rance et al., 1987; McIntosh et al., 1987; Senn et al., 1987b; Griffey et al., 1985b). In this approach, the heteronuclei are used to simplify the spectrum by filtering out the protons of interest. The isotope-filtered NOESY experiment (Griffey et al., 1985a; Griffey and Redfield, 1987; Rance et al., 1987; McIntosh et al., 1987; Senn et al., 1987b; Weiss et al., 1986; Bax and Weiss, 1987; Fesik et al., 1987a; Fesik et al., 1988) has been used frequently to observe only that portion of the protein that has been enriched, either a particular residue type or a bound ligand or cofactor. The isotope-filtered NOESY experiment exploits the large heteronuclear scalar coupling to select for NOE's arising only from protons bound to the enriched nuclei (ω_1 filtered), or for NOE's from any proton to only heteronuclear-bonded protons (ω_2 filtered), or for NOE's between protons that are each bonded to heteronuclei (ω_1 and ω_2 filtered).

Results from isotope-filtered experiments can serve as starting points for sequential assignments (McIntosh et al., 1987; Senn et al., 1987b), or can be used to investigate the structure of one part of the protein without solving the entire structure. Figure 17 shows the isotope-filtered NOESY spectrum of *Anabaena* 7120 flavodoxin that was reconstituted with [^{13}C]flavin mononucleotide. The flavin methyl groups show NOE's to a tryptophan residue situated above the cofactor. Isotope-editing also has been used to measure amide proton exchange rates of specific amino acid residues or bound ligands (Griffey et al., 1985b; Fesik et al., 1987b), and to select for resonances from only one part of a molecular complex (Tsang et al., 1989; Tsang et al., 1988).

Nitrogen-15 has been used to increase the resolution of correlations involving amide protons in two-dimensional ^{1}H NMR spectra by narrowing the linewidths of the amide proton resonances (Bax et al., 1989). This technique employs the ^{15}N label to generate heteronuclear multiple-quantum coherences during the evolution period of the experiment. Normally, amide proton resonances are broadened by dipolar coupling to the nitrogen. However, since the multiple-quantum coherences have relaxation rates that are less affected by dipolar coupling, the resulting amide proton line widths are much narrower. Application of the "pseudo-single quantum COSY" experiment to staphylococcal nuclease resulted in a signal-to-noise increase of about 50% (Bax et al., 1989) for cross peaks involving amide protons.

Determination of Protein Conformation and Dynamic Properties

Protein isotopic enrichment permits the extraction of spectral parameters related to structure and dynamics which both complements and adds to those obtained from homonuclear ^{1}H NMR.

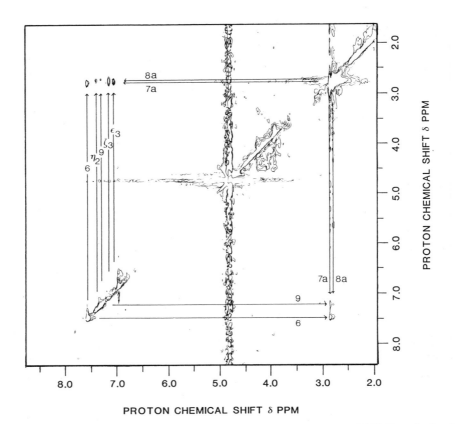

Figure 17. Isotope-edited NOESY spectrum (500 MHz) of *Anabaena* 7120 flavodoxin. The sample was 4 mM [26% ^{13}C FMN]flavodoxin in ^{2}H$_2$O containing 150 mM phosphate buffer at pH 7.5. Correlations between the flavin methyl groups and a tryptophan residue are indicated. B.J. Stockman and J.L. Markley, unpublished data.

Relaxation parameters (T_1, T_2, and nuclear Overhauser enhancements) of carbon and nitrogen can be used to investigate local protein mobilities. Enrichment allows evaluation of heteronuclear and homonuclear coupling constants, which have characteristic angle dependencies that can be correlated with molecular structure. In addition, enrichment can be used to investigate slow dynamic equilibria in proteins using two-dimensional exchange spectroscopy.

Relaxation measurements. One-dimensional NMR spectroscopy can be used to assess residue mobilities based on T_1, T_2, and NOE measurements. Dynamic parameters for side-chain atoms can reflect local mobility, and thus can give clues as to whether side chains are located on the surface or are buried within the protein (Jardetzky, 1981; Smith et al., 1987). Variations in main-chain mobility can be determined analogously. Dynamic properties have been assessed for natural abundance proteins (Gust et al., 1975; Allerhand et al., 1971), but the low sensitivity of ^{13}C and ^{15}N prohibits accurate direct observation measurements. ^{13}C-enrichment of glycine ^{13}C$^{\alpha}$'s in staphylococcal nuclease (McCain et al., 1988) and of alanine ^{13}C$^{\beta}$'s in M13 coat protein (Henry et al., 1986b; Weiner et al., 1987) has facilitated accurate measurement of relaxation parameters for these atoms. In addition, ^{15}N-enrichment has been used to qualitatively assess protein dynamics by measuring nuclear Overhauser enhancements of both main-chain and side-chain nitrogen resonances (Leighton and Lu, 1987; Bogusky et al., 1987;

Smith et al., 1987; Stockman et al., 1988b; Bogusky et al., 1989). Since the size of the NOE varies from 0.88 at long correlation times to -3.93 at short correlation times (Gust et al., 1975), nitrogen resonances arising from relatively mobile regions of the protein are easily determined.

Two-dimensional proton-detection of ^{13}C T_1's, however, has enabled the measurement of protein ^{13}C T_1's at natural abundance ^{13}C (Nirmala and Wagner, 1988; Wagner et al., 1989). Proton detection increases the sensitivity, while the added heteronucleus dimension provides spectral resolution. A two-dimensional double DEPT pulse sequence, modified from similar one-dimensional pulse sequences (Sklenar et al., 1987; Kay et al., 1987), has been used. The first DEPT sequence transfers proton magnetization to carbon, where it is allowed to relax for a series of variable delays and is also frequency labeled. The inverse DEPT returns the carbon magnetization to protons for detection. One thus obtains a $^{1}H\{^{13}C\}SBC$ spectrum for each variable delay time used. Relaxation times are then determined by plotting cross-peak intensities as a function of variable delay time. Results for the $^{13}C^{\alpha}$'s of bovine pancreatic trypsin inhibitor show a lack of variability indicating that internal motions (at least for the backbone region) are not biasing distances determined from nuclear Overhauser effects. Similar experiments could be performed to determine if this holds true for side chains. Differential mobilities determined in this way can then be incorporated when extracting distance information from NOE cross peak intensities.

A similar type of double DEPT pulse sequence has allowed proton detection of ^{13}C T_2's in ^{13}C-enriched small molecules (Wagner et al., 1989; Nirmala and Wagner, 1989). Comparison of this technique with direct detection of ^{13}C using the Carr-Purcell (Carr and Purcell, 1954) method yielded identical results, indicating that the resolution afforded in the two-dimensional spectrum should allow determination of accurate T_2 values for protein carbon resonances. As with T_1 measurements, ^{13}C enrichment is required for accurate T_2 measurements. Similar applications can be envisioned with ^{15}N-enriched proteins.

Amide hydrogen exchange rates. Protein amide proton lifetimes vary from less than a second to many months. Since hydrogen bonding interactions probably account for a great deal of this variation, amide proton exchange rates can be correlated with fluctuations in protein structure. Exchange rates can be measured by following the time course of signal disappearance upon transferring the protein from $^{1}H_2O$ to $^{2}H_2O$. As protein size increases, the resolution of the amide resonances becomes insufficient for obtaining exchange rates for each amide proton.

^{15}N-enrichment can be used to circumvent this by taking advantage of the inherently greater dispersion of amide ^{15}N resonances as opposed to amide ^{1}H resonances. Measurement of the disappearance of ^{15}N resonances, recorded with an INEPT pulse sequence, upon transferring the protein into fresh $^{2}H_2O$ allows the calculation of amide ^{1}H exchange rates; the intensity of the ^{15}N signal decreases as the number of protons involved in the polarization transfer diminishes. This type of experiment is useful only for slowly exchanging protons (Henry and Sykes, 1989).

Alternatively, one can monitor amide proton exchange by observing the two-bond isotope shift (0.08 ppm) (Feeney et al., 1974) of the carbonyl carbon resonance upon deuteration of the adjacent amide nitrogen (Kainosho and Tsuji, 1982; Henry and Sykes, 1989; Henry et al., 1986a; Henry et al., 1987). When dissolved in a solution of 50% $^{1}H_2O$: 50% $^{2}H_2O$, protein carbonyl resonance lineshapes are indicative of the amide exchange rate. In the slow exchange limit, two peaks are observed (protonated and deuterated amide nitrogen). In fast exchange, a single resonance is observed. The lines coalesce when the exchange rate is around 20 s^{-1} (Henry and Sykes, 1989). This technique is useful for measuring exchange rates that are faster than those measurable using ^{15}N NMR.

Coupling constant measurements. Recently, nitrogen enrichment has been exploited to provide accurate measurements of three-bond heteronuclear ($^3J_{NH}$) vicinal coupling constants (Wagner et al., 1989; Montelione et al., 1989), and homonuclear ($^3J_{HNH\alpha}$) vicinal coupling constants (Wagner et al., 1989; Montelione and Wagner, 1989b). Accurate determination of $^3J_{NH\beta}$ coupling constants, combined with measurements of homonuclear coupling constants between $^1H\alpha$ and $^1H\beta$, provide a measure of the side chain dihedral angle χ_1 and stereospecific assignment of β-methylene protons. For mobile side chains, distributions of rotomer populations also may be determined (Montelione et al., 1989). As an improvement over standard NOE and homonuclear coupling constant techniques (Hyberts et al., 1987) for stereospecific assignments, measurement of heteronuclear coupling constants will lead to better solution structure determinations (Driscoll et al., 1989; Montelione et al., 1989). In addition, interresidue $^3J_{NH\alpha}$ coupling constants can be used to determine the backbone dihedral angle ψ, and side-chain coupling constants can be used to provide Asn χ_2, Gln χ_3, and Arg χ_4 dihedral angle constraints (Montelione et al., 1989). $^3J_{HNH\alpha}$ coupling constants, often difficult to measure in two-dimensional ^1H NMR spectra, provide constraints on the backbone dihedral angle ϕ (Montelione and Wagner, 1989b). Figure 18 shows three-bond couplings used to evaluate ϕ, ψ, and χ_1 angles.

Heteronuclear coupling constants can be measured by recording homonuclear ^1H NMR spectra, such as NOESY, ROESY, RELAY, or HOHAHA, of ^{15}N-enriched proteins. Homonuclear cross peak patterns between protons coupled to a common nitrogen atom (one proton is single-bond coupled, the other is three-bond coupled) are then analyzed for passive heteronuclear couplings (Montelione et al., 1989).

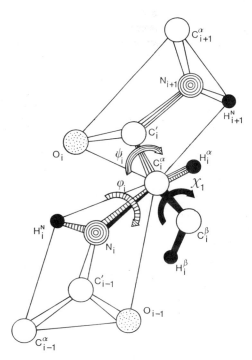

Figure 18.   Diagram of the peptide backbone. Measurement of three-bond couplings shown by solid, hatched, and striped bonds allows the correspondingly-labeled angles to be determined.

Homonuclear $^3J_{H_NH\alpha}$ coupling constants can be obtained from $^1H\{^{15}N\}$SBC experiments (Live et al., 1984; Kay et al., 1989a). They can also be measured by using a variation (Montelione and Wagner, 1989b) of the triple-resonance experiment (Montelione and Wagner, 1989c) discussed above. Polarization transfer from H^α to C^α is followed by an evolution period where the carbon coherence evolves decoupled from nitrogen but coupled to protons. The carbon-proton coupling results in the desired large separation of the multiplet components in ω_1. The remainder of the sequence involves the relay of magnetization to the amide proton via the amide nitrogen. One can also do the experiment in the opposite direction. Nitrogen-15 enrichment is required (Montelione and Wagner, 1989b).

Although heteronuclear coupling constants have been determined so far only for ^{15}N-enriched proteins, similar experiments are possible with ^{13}C-enriched proteins. $^3J_{CH}$ coupling constants should prove useful in delineating side-chain conformations. Heteronuclear coupling constants should also be useful for determining the conformation of enriched small molecules bound to proteins.

Exchange spectroscopy. One-dimensional investigations of chemical exchange processes in proteins often suffer from lack of resolution in the ^1H spectrum. Although two-dimensional chemical exchange spectroscopy (NOESY, ROESY) provides increased resolution, analysis is often complicated by exchange cross peaks obscured by the diagonal or cross peaks representing dipolar interactions.

Proton-detected heteronuclear correlations present an alternative approach that improves spectral resolution (Montelione and Wagner, 1989a; Alexandrescu et al., 1989). Spectra recorded by using the pulse sequence of Montelione and Wagner (Montelione and Wagner, 1989a) show $^1H\{^{13}C\}$SBC-type correlations (direct peaks) for each conformation in solution. In addition, exchange cross peaks are seen for interconverting conformations in slow dynamic equilibrium. Volumes of the direct peaks provide the relative populations of each conformer. Volumes of the exchange cross peaks can be correlated with the exchange rate by conducting a series of experiments with an incremented mixing time (Montelione and Wagner, 1989a). The sensitivity and selectivity of such experiments can be improved greatly by ^{13}C labeling. This has been illustrated recently by ^{13}C-directed 2D studies of exchange between native and thermally denatured staphylococcal nuclease (Alexandrescu et al., 1989). Analogous experiments can be performed on ^{15}N-enriched proteins.

FUTURE PERSPECTIVES

Rapid developments in the field of NMR spectroscopy have begun to exploit isotopic enrichment of proteins. Improvement in the quality and quantity of solution structural and dynamic information has resulted. Since the methodologies are applicable to any enriched protein, continued developments in the field of molecular biology very likely will enhance the usefulness of these techniques by creating the potential to enrich an increasing number of interesting proteins.

Newly emerging methods for (semi) automated data analysis are expected to facilitate the rapid assignment of protein spectra. Combined development of the techniques described here should result in more precise solution structures and better quantitation of protein dynamics. The molecular weight limitations of methods involving ^2H, ^{13}C, and ^{15}N labeling have yet to be evaluated fully, but it is clear that they will support detailed solution NMR studies of proteins in the 20 kDa range that have been inaccessible to study by 2D or 3D homonuclear ^1H NMR methods.

ACKNOWLEDGMENTS

Critical evaluation of the manuscript by Andrew P. Hinck, Stewart Loh, and W. Milo Westler is gratefully appreciated. We thank Byung-Ha Oh, Eddie S. Mooberry, and Jinfeng Wang for providing figures from their research. We also thank our many colleagues for providing preprints of their work. Supported by USDA Competitive Research Grant (88-37262-3406), NIH Grant RR02301, and the New Energy and Industrial Technology Development Organization (Tokyo, Japan). This work made use of the National Magnetic Resonance Facility at Madison, which is supported in part by NIH Grant RR02301 from the Biomedical Research Technology Program, Division of Research Resources. Equipment in the facility was purchased with funds from the University of Wisconsin, the NSF Biological Biomedical Research Technology Program (Grant DMB-8415048), NIH Shared Instrumentation Program (Grant RR02781), and the U.S. Department of Agriculture. Brian J. Stockman is supported by an NIH Training Grant in Cellular and Molecular Biology (GM07215).

REFERENCES

Akasaka, K., Inoue, T., Tamura, A., Watari, H., Abe, K. and Kainosho, M., 1988, Internal motion of a tryptophan residue in *Streptomyces* subtilsn inhibitor: deuterium nuclear magnetic resonance in solution, *Proteins: Structure, Function and Genetics* 4:131.

Alexandrescu, A.T., Loh, S.N. and Markley, J.L., 1989, ^{13}C NMR based chemical exchange spectroscopy: applications to enzymology and protein folding, *J. Magn. Res.*, in press.

Allerhand, A., 1979, Carbon-13 nuclear magnetic resonance: new techniques, *Methods Enzymol.* 61:458.

Allerhand, A., Doddrell, D., Glushko, V., Cochran, D.W., Wenkert, E., Lawson, P.J. and Gurd, F.R.N., 1971, Conformation and segmental motion of native and denatured ribonuclease A in solution. Application of natural-abundance carbon-13 partially relaxed Fourier transform nuclear magnetic resonance, *J. Am. Chem. Soc.* 93:544.

Anglister, J., Bond, M.W., Frey, T., Leahy, D., Levitt, M., McConnell, H.M., Rule, G.S., Tomasello, J. and Whittaker, M., 1987, Contribution of tryptophan residues to the combining site of a monoclonal anti dinitrophenyl spin-label antibody, *Biochemistry* 26:6058.

Anglister, J., Frey, T. and McConnell, H.M., 1984a, Distances of tyrosine residues from a spin-label hapten in the combining site of a specific monoclonal antibody, *Biochemistry* 23:1138.

Anglister, J., Frey, T. and McConnell, H.M., 1984b, Magnetic resonance of a monoclonal anti-spin-label antibody, *Biochemistry* 23:5372.

Anglister, J., Frey, T. and McConnell, H.M., 1985, NMR technique for assessing contributions of heavy and light chains to an antibody combining site, *Nature* 315:65.

Anglister, J., Jacob, C., Assulin, O., Ast, G., Pinker, R. and Arnon, R., 1988, NMR study of the complexes between a synthetic peptide derived from the B subunit of cholera toxin and three monoclonal antibodies against it, *Biochemistry* 27:717.

Anglister, J., Levy, R. and Scherf, T., 1989, Interactions of antibody aromatic residues with a peptide of cholera toxin observed by two-dimensional transferred nuclear Overhauser effect difference spectroscopy, *Biochemistry* 28:3360.

Anil Kumar, Ernst, R.R. and Wüthrich, K., 1980, A two-dimensional nuclear Overhauser enhancement (2D NOE) experiment for the elucidation of complete proton-proton cross-relaxation networks in biological macromolecules, *Biochem. Biophys. Res. Comm.* 95:1.

Baillargeon, M.W., Laskowski, Jr., M., Neves, D.E., Porubcan, M.A., Santini, R.E. and Markley, J.L., 1980, Soybean trypsin inhibitor (Kunitz) and its complex with trypsin. Carbon-13 nuclear magnetic resonance studies of the reactive site arginine, *Biochemistry* 19:5703.

Bax, A. and Drobny, G., 1985, Optimization of two-dimensional homonuclear relayed coherence transfer NMR spectroscopy, *J. Magn. Res.* 61:306.

Bax, A. and Summers, M.L., 1986, ^{1}H and ^{13}C assignments from sensitivity-enhanced detection of heteronuclear multiple-bond connectivity by 2D multiple quantum NMR, *J. Am. Chem. Soc.* 108:2093.

Bax, A. and Weiss, M.A., 1987, Simplification of two-dimensional NOE spectra of proteins by ^{13}C labeling, *J. Magn. Res.* 71:571.

Bax, A., Freeman, R. and Kempsell, S.P., 1980, Natural abundance [13]C-^{13}C coupling observed via double-quantum coherence, *J. Am. Chem. Soc.* 102:4849.

Bax, A., Griffey, R.H. and Hawkins, B.L., 1983, Correlation of proton and nitrogen-15 chemical shifts by multiple quantum NMR, *J. Magn. Res.* 55:301.

Bax, A., Kay, L.E., Sparks, S.W. and Torchia, D.A., Line narrowing of amide proton resonances in 2D NMR spectra of proteins, 1989, *J. Am. Chem. Soc.* 111:408.

Bax, A., Sparks, S.W. and Torchia, D.A., 1988, Long-range heteronuclear correlation: a powerful tool for the NMR analysis of medium-size proteins, *J. Am. Chem. Soc.* 110:7926.

Bendall, M.R., Pegg, D.T., Doddrell, D.M. and Field, J., 1981, NMR of protons coupled to ^{13}C nuclei only, *J. Am. Chem. Soc.* 103:934.

Billeter, M., Braun, W. and Wüthrich, K., 1982, Sequential resonance assignments in protein ^{1}H nuclear magnetic resonance spectra, *J. Mol. Biol.* 155: 321.

Blomberg, F., Maurer, W. and Ruterjans, H., 1977, Nuclear magnetic resonance investigation of ^{15}N labeled histidine in aqueous solution, *J. Am. Chem. Soc.* 99:8149.

Bogusky, M.J., Leighton, P., Schiksnis, R.A., Khoury, A., Lu, P. and Opella, S.J., 1989, ^{15}N NMR spectroscopy of proteins in solution, *J. Magn. Res.*, in press.

Bogusky, M.J., Schiksnis, R.A., Leo, G.C. and Opella, S.J., 1987, Protein backbone dynamics by solid-state and solution ^{15}N NMR spectroscopy, *J. Magn. Res.* 72:186.

Bolton, P.H., 1985, Heteronuclear relay transfer spectroscopy with proton detection, *J. Magn. Res.* 62:143.

Bovey, F.A., Tiers, G.V.D. and Filipovich, G., 1959, Polymer NMR spectroscopy. I. The motion and configuration of polymer chains in solution, *J. Polymer Sci.* 38:73.

Bradbury, J.H. and Sheraga, H.A., 1966, Structural studies of ribonuclease. XXIV. The application of nuclear magnetic resonance spectroscopy to distinguish between the histidine residues of ribonuclease, *J. Am. Chem. Soc.* 88:4240.

Browne, D.T., Kenyon, G.L., Packer, E.L., Sternlicht, H. and Wilson, D.M., 1973, Studies of macromolecular structure by ^{13}C nuclear magnetic resonance. II. A specific labeling approach to the study of histidine residues in proteins, *J. Am. Chem. Soc.* 95:1316.

Bruhwiler, D. and Wagner, G., 1986, Selective excitation of ^{1}H resonances coupled to [13]C. Hetero COSY and RELAY experiments with ^{1}H detection for a protein, *J. Magn. Res.* 69:546.

Bystrov, V.F., 1976, Spin-spin coupling and the conformational states of peptide systems, *Prog. NMR Spect.* 10:41.

Carr, H.Y. and Purcell, E.M., 1954, Effects of diffusion on free precession in nuclear magnetic resonance experiments, *Phys. Rev.* 94:630.

Chan, T.-M. and Markley, J.L., 1982, Heteronuclear (^{1}H, ^{13}C) two-dimensional chemical shift correlation NMR spectroscopy of a protein. Ferredoxin from *Anabaena variabilis*, *J. Am. Chem. Soc.* 104:4010.

Chan, T.-M. and Markley, J.L., 1983, Nuclear magnetic resonance studies of two-iron-two-sulphur ferredoxins. 3. Heteronuclear (^{13}C, ^{1}H) two-dimensional NMR spectra, ^{13}C peak assignments, and ^{13}C relaxation measurements, *Biochemistry* 22:5996.

Clore, G.M., Bax, A., Wingfield, P. and Gronenborn, A.M., 1988, Long-range ^{15}N-^{1}H correlation as an aid to sequential proton resonance assignment of proteins, *FEBS Lett.* 238:17.

Crespi, H.L., Rosenberg, R.M. and Katz, J.J., 1968, Proton magnetic resonance of proteins fully deuterated except for ^{1}H-leucine side chains, *Science (Washington, D.C.)* 161:795.

Davis, D.G. and Bax, A., 1985, Assignment of complex ^{1}H NMR spectra via two-dimensional homonuclear Hartmann-Hahn spectroscopy, *J. Am. Chem. Soc.* 107:2820.

Driscoll, P.C., Gronenborn, A.M. and Clore, G.M., 1989, The influence of sterospecific assignments on the determination of three-dimensional structures of proteins by nuclear magnetic resonance spectroscopy, *FEBS Lett.* 243:223.

Ernst, R.R., Bodenhausen, G. and Wokaun, A., 1987, "Principles of Nuclear Magnetic Resonance in One and Two Dimensions," Oxford University Press.

Feeney, J., Partington, P. and Roberts, G.C.K., 1974, The assignment of carbon-13 resonances from carbonyl groups in peptides, *J. Magn. Res.* 13:268.

Fesik, S.W. and Zuiderweg, E.R.P., 1988, Heteronuclear three-dimensional NMR spectroscopy. A strategy for the simplification of homonuclear two-dimensional NMR spectra, *J. Magn. Res.* 78:588.

Fesik, S.W., Gampe, R.T., Jr. and Rockway, T.W., 1987a, Application of isotope-filtered 2D NOE experiments in the conformational analysis of atrial natriuretic factor (7-23), *J. Magn. Res.* 74:366.

Fesik, S.W., Gampe, R.T., Jr. and Zuiderweg, E.R.P., 1989a, Heteronuclear three-dimensional NMR spectroscopy. Natural abundance ^{13}C chemical shift editing of ^{1}H-^{1}H COSY spectra, *J. Am. Chem. Soc.* 111:770.

Fesik, S.W., Gampe, R.T., Jr., Zuiderweg, E.R.P., Kohlbrenner, W.E. and Weigl, D., 1989b, Heteronuclear three-dimensional NMR spectroscopy applied to CMP-KDO synthetase, *Biochem. Biophys. Res. Comm.* 159:842.

Fesik, S.W., Luly, J.R., Erickson, J.W. and Abad-Zapatero, C., 1988, Isotope-edited proton NMR study on the structure of a pepsin/inhibitor complex, *Biochemistry* 27:8297.

Fesik, S.W., Luly, J.R., Stein, H.H. and BaMaung, N., 1987, Amide proton exchange rates of a bound pepsin inhibitor determined by isotope-edited proton NMR experiments, *Biochem. Biophys. Res. Comm.* 147:892.

Frey, T., Anglister, J. and McConnell, H.M., 1984, Nonaromatic amino acids in the combining site region of a monoclonal anti-spin-label antibody, *Biochemistry* 23:6470.

Frey, T., Anglister, J. and McConnell, H.M., 1988, Line-shape analysis of NMR difference spectra of an anti-spin-label antibody, *Biochemistry* 27:5161.

Glushka, J. and Cowburn, D., 1987, Assignment of ^{15}N NMR signals in bovine pancreatic trypsin inhibitor, *J. Am. Chem. Soc.* 109:7879.

Gorenstein, D., Meadows, R.P., Metz, J.T., Nikonowicz, E. and Post, C., 1989, *in* "Advances in Biophysical Chemistry," in press.

Griesinger, C., Sorensen, O.W. and Ernst, R.R., 1987a, Novel three-dimensional NMR techniues for studies of peptides and biological macromolecules, *J. Am. Chem. Soc.* 109:7227.

Griesinger, C., Sorensen, O.W. and Ernst, R.R., 1987b, A practical approach to three-dimensional NMR spectroscopy, *J. Magn. Res.* 73:574.

Griffey, R.H. and Redfield, A.G., 1987, Proton-detected heteronuclear edited and correlated nuclear magnetic resonance and nuclear Overhauser effect in solution, *Q. Rev. Biophys.* 19:51.

Griffey, R.H., Jarema, M.A., Kunz, S., Rosevear, P.R. and Redfield, A.G., 1985a, Isotopic-label-directed observation of the nuclear Overhauser effect in poorly resolved proton NMR spectra, *J. Am. Chem. Soc.* 107:711.

Griffey, R.H., Redfield, A.G., Loomis, R.E. and Dahlquist, F.W., 1985b, Nuclear magnetic resonance observation and dynamics of specific amide protons in T4 lysozyme, *Biochemistry* 24:817.

Griffey, R.H., Redfield, A.G., McIntosh, L.P., Oas, T.G. and Dahlquist, F.W., 1986, Assignment of proton amide resonances of T4 lysozyme by ^{13}C and ^{15}N multiple isotopic labeling, *J. Am. Chem. Soc.* 108:6816.

Grissom, C.B. and Markley, J.L., 1989, Staphylococcal nuclease active-site amino acids: pH dependence of tyrosines and arginines by ^{13}C NMR and correlation with kinetic studies, *Biochemistry* 28:2116.

Gronenborn, A.M., Bax, A., Wingfield, P.T. and Clore, G.M., 1989a, A powerful method of sequential proton resonance assignment in proteins using relayed [15]N-^{1}H multiple quantum coherence spectroscopy, *FEBS Lett.* 243:93.

Gronenborn, A.M., Wingfield, P.T. and Clore, G.M., 1989b, Determination of the secondary structure of the DNA binding protein *Ner* from phage μ using ^{1}H homonuclear and ^{15}N-^{1}H heteronuclear NMR spectroscopy, *Biochemistry*, in press.

Gust, D., Moon, R.B. and Roberts, J.D., 1975, Applications of natural-abundance nitrogen-15 nuclear magnetic resonance to biochemically important molecules, *Proc. Natl. Acad. Sci. USA* 72:4696.

Henry, G.D. and Sykes, B.D., 1989, Structure and dynamics of detergent-solubilised M13 coat protein (an integral membrane protein) determined by ^{13}C and ^{15}N NMR spectroscopy, *in:* "Biochemistry and Cell Biology," in press.

Henry, G.D., O'Neil, J.D.J., Weiner, J.H. and Sykes, B.D., 1986a, Hydrogen exchange in the hydrophilic regions of detergent-solubilized M13 coat protein detected by ^{13}C nuclear magnetic resonance isotope shifts, *Biophys. J.* 49:329.

Henry, G.D., Weiner, J.H. and Sykes, B.D., 1986b, Backbone dynamics of a model membrane protein: ^{13}C NMR spectroscopy of alanine methyl groups in detergent-solubilized M13 coat protein, *Biochemistry* 25:590.

Henry, G.D., Weiner, J.H. and Sykes, B.D., 1987, Backbone dynamics of a model membrane protein: measurement of individual amide hydrogen-exchange rates in detergent-solubilized M13 coat protein using ^{13}C NMR hydrogen/deuterium isotope shifts, *Biochemistry* 26:3626.

Hilber, D.W., Harpold, L., Dell-Acqua, M., Pourtabbed, T., Gerlt, J.A., Wilde, J.W. and Bolten, P.H., 1989, Isotopic labeling with hydrogen-2 and carbon-13 to compare conformations of proteins and mutants generated by site-directed mutagenesis: Part 1, *Methods Enzymol.*, 177:74.

Ho, C., Dowd, S.R. and Post, J.F.M., 1985, ^{19}F NMR investigations of membranes, *Curr. Top. Bioenerg.* 14:53.

Hyberts, S.G., Marki, W. and Wagner, G., 1987, Sterospecific assignments of side-chain protons and characterization of torsion angles in Eglin c, *E.J. Bioch* 164:625.

Jardetzky, O., 1965, An approach to the determination of the active site of an enzyme by nuclear magnetic resonance spectroscopy, *in* "Proceedings of the International Symposium on Nuclear Magnetic Resonance," Tokyo, Japan, N-3-14.

Jardetzky, O., 1981, NMR studies of macromolecular dynamics, *Acc. Chem. Res.* 14:291.

Kainosho, M. and Tsuji, T., 1982, Assignment of the three methionyl carbonyl carbon resonances in *Streptomyces* substilisin inhibitor by a carbon-13 and nitrogen-15 double-labeling technique. A new strategy for studies of proteins in solution, *Biochemistry* 21:6273.

Kainosho, M., Nagao, H. and Tsuji, T., 1987, Local structural features around the C-terminal segment of *Streptomyces* subtilisin inhibitor studied by carbonyl carbon nuclear magnetic resonances of three phenylalanyl residues, *Biochemistry* 26:1068.

Kalbitzer, H.R., Leberman, R. and Wittinghofer, A., 1985, ^{1}H NMR spectroscopy on elongation factor Tu from *Escherichia coli*, *FEBS Lett.* 180:40.

Karplus, S. and Karplus, M., 1972, Nuclear magnetic resonance determination of the angle ψ in peptides, *Proc. Natl. Acad. Sci. USA* 69:3204.

Kato, K., Matsunaga, C., Nishimura, Y., Waelchli, M., Kainosho, M. and Arata, Y., 1989, Application of ^{13}C nuclear magnetic resonance spectroscopy to molecular structural analyses of antibody molecules, *J. Biochem. (Tokyo)*, in press.

Katz, J.J. and Crespi, H.L., 1966, Deuterated organisms: cultivation and uses, *Science (Washington, D.C.)* 151:1187.

Kay, L.E., Brooks, B., Torchia, D., Sparks, S. and Bax, A., 1989a, Measurement of backbone J couplings in proteins by two-dimensional heteronuclear multiple quantum NMR, *J. Cell. Biochem.* Supplement 13A:31.

Kay, L.E., Jue, T.L., Bangerter, B. and Demou, P.C., 1987, Sensitivity enhancement of [13]C T1 measurements via polarization transfer, *J. Magn. Res.* 73:558.

Kay, L.E., Marion, D. and Bax, A., 1989b, Practical aspects of 3D heteronuclear NMR of proteins, *J. Magn. Res.*, in press.

Kojiro, C.L. and Markley, J.L., 1983, Connectivity of proton and carbon spectra of the blue copper protein, plastocyanin, established by two-dimensional nuclear magnetic resonance, *FEBS Lett.* 162:52.

Kowalsky, A., 1962, Nuclear magnetic resonance studies of proteins, *J. Biol. Chem.* 237:1807.

Kurland, R.J., Davis, D.G. and Ho, C., 1968, Paramagnetic proton nuclear magnetic resonance shifts of metmyoglobin, methemoglobin, and hemin derivatives, *J. Am. Chem. Soc.* 90:2700.

Leighton, P. and Lu, P., 1987, λ cro repressor complex with OR3 DNA: ^{15}N NMR observations, *Biochemistry* 26:7262.

LeMaster, D.M. and Richards, F.M., 1988, NMR sequential assignment of *Escherichia coli* thioredoxin utilizing random fractional deuteration, *Biochemistry* 27:142.

LeMaster, D.M., 1987, Chiral B and random fractional deuteration for the determination of protein sidechain conformation by NMR, *FEBS Lett.* 223:191.

LeMaster, D.M., 1988, Protein NMR resonance assignment by isotropic mixing experiments on random fractionally deuterated samples, *FEBS Lett.* 233:326.

LeMaster, D.M., 1989, Deuteration in ^{1}H protein NMR, *Methods Enzymol.*, in press.

Lerner, L. and Bax, A., 1986, Sensitivity-enhanced two-dimensional heteronuclear relayed coherence transfer NMR spectroscopy, *J. Magn. Res.* 69:375.

Levitt, M.H. and Ernst, R.R., 1983, Improvement of pulse performance in NMR coherence transfer experiments. A compensated INADEQUATE experiment, *Mol. Physics* 50:1109.

Live, D.H., Davis, D.G., Agosta, W.C. and Cowburn, D., 1984, Observation of 1000-fold enhancement of ^{15}N NMR via proton-detected multiquantum coherences: studies of large peptides, *J. Am. Chem. Soc.* 106:6104.

Mandel, M., 1965, Proton magnetic resonance spectra of some proteins, *J. Biol. Chem.* 240:1586.

Marion, D., Driscoll, P. C., Kay, L. E., Wingfield, P. T., Bax, A., Gronenborn, A. M., and Clore, G. M., 1989a, Overcoming the overlap problem in the assignment of ^{1}H NMR spectra of larger proteins by use of three-dimensional heteronuclear [1]H-[15]N Hartmann-Hahn-multiple quantum coherence and nuclear Overhauser-multiple quantum coherence spectroscopy: application to Interleukin 1B, Biochemistry 28, 6150.

Marion, D., Kay, L.E., Sparks, S.W., Torchia, D.A. and Bax, A., 1989b, Three-dimensional heteronuclear NMR of ^{15}N labeled proteins, *J. Am. Chem. Soc.* 111:1515.

Markley, J.L. and Cheung, S.M., 1973, Differential exchange of the C2-hydrogens of histidine side chains in native proteins: proposed general technique for the assignment of histidine NMR peaks in proteins, U.S. Atomic Energy Commission CONF-730525, 103.

Markley, J.L. and Kato, I., 1975, Assignment of the histidine proton magnetic resonance peaks of soybean trypsin inhibitor (Kunitz) by a differential deuterium exchange technique, *Biochemistry* 14:3234.

Markley, J.L., 1972, High-resolution proton magnetic resonance spectroscopy of selectively deuterated enzymes, *Methods Enzymol.* 26:605.

Markley, J.L., 1975, Correlation proton magnetic Resonanace studies at 250 MHz of bovine pancreatic ribonuclease. I. Reinvestigation of the histidine peak assignments, *Biochemistry* 14:3546.

Markley, J.L., 1989, Two-dimensional nuclear magnetic resonance spectroscopy of proteins: an overview, *Methods Enzymol.* 176:12.

Markley, J.L., Putter, I. and Jardetzky, O., 1968, High-resolution nuclear magnetic resonance spectra of selectively deuterated staphylococcal nuclease, *Science (Washington, D.C.)* 161:1249.

Markley, J.L., Westler, W.M., Chan, T.-M., Kojiro, C. and Ulrich, E.L., 1984, Two-dimensional NMR approaches to the study of protein structure and function, *Federation Proc.* 43:2648.

McCain, D.C., Ulrich, E.L. and Markley, J.L., 1988, NMR relaxation study of internal motions in staphylococcal nuclease, *J. Magn. Res.* 80:296.

McDonald, C.C. and Phillips, W.D., 1967, Manifestations of the tertiary structures of proteins in high-frequency nuclear magnetic resonance, *J. Am. Chem. Soc.* 89:6332.

McIntosh, L.P., Griffey, R.H., Muchmore, D.C., Nielson, C.P., Redfield, A.G. and Dahlquist, F.W., 1987, Proton NMR measurements of bacteriophage T4 lysozyme aided by ^{15}N isotopic labeling: structural and dynamic studies of larger proteins, *Proc. Natl. Acad. Sci. USA* 84:1244.

Meadows, D.H., Markley, J.L., Cohen, J.S. and Jardetzky, O., 1967, Nuclear magnetic resonance sudies of the structure and binding sites of enzymes. I. Histidine residues, *Proc. Natl. Acad. Sci. USA* 58:1307.

Montelione, G.T. and Wagner, G., 1989a, 2D chemical exchange NMR spectroscopy by proton-detected heteronuclear correlation, *J. Am. Chem. Soc.* 111:3096.

Montelione, G.T. and Wagner, G., 1989b, Accurate measurements of homonuclear H^N-H^α coupling constants in polypeptides using heteronuclear 2D NMR experiments, *J. Am. Chem. Soc.*, in press.

Montelione, G.T. and Wagner, G., 1989c, ^{1}H, ^{13}C, ^{15}N triple resonance experiments for accurate measurements of homonuclear H^N-H^α vicinal coupling constants and identification of sequential connections in polypeptides, Relaxation Times 6: 2.

Montelione, G.T., Winkler, M.E., Rauenbuehler, P. and Wagner, G., 1989, Accurate measurements of long-range heteronuclear coupling constants from homonuclear 2D NMR spectra of isotope-enriched proteins, *J. Magn. Res.* 82:198.

Mooberry, E.S., Oh, B.-H. and Markley, J.L., 1989, Improvement of [13]C-^{15}N chemical shift correlation spectroscopy by implementing time proportional phase incrementation, *J. Magn. Res.*, in press.

Muchmore, D.C., McIntosh, L.P., Russell, C.B., Anderson, D.E. and Dahlquist, F.W., 1989, Expression and ^{15}N labeling of proteins for proton and nitrogen-15 NMR, *Methods Enzymol.*, in press.

Nagayama, K., Anil Kumar, Wüthrich, K. and Ernst, R.R., 1980, Experimental techniques of two-dimensional correlated spectroscopy, *J. Magn. Res.* 40:321.

Niemczura, W.P., Helms, G.L., Chesnick, A.S., Moore, R.E. and Bornemann, V., 1989, Carbon-detected correlation of carbon-13-nitrogen-15 chemical shifts, *J. Magn. Res.* 81:635.

Nirmala, N.R. and Wagner, G., 1988, Measurement of ^{13}C relaxation times in proteins by two-dimensional heteronuclear ^{1}H-^{13}C correlation spectroscopy, *J. Am. Chem. Soc.* 110:7557.

Nirmala, N.R. and Wagner, G., 1989, Measurement of ^{13}C spin-spin relaxation times by two-dimensional heteronuclear ^{1}H-^{13}C correlation spectroscopy, *J. Magn. Res.* 82:659.

Noren, C.J., Anthony-Cahill, S.J., Griffith, M.C. and Schultz, P.G., 1989, A general method for site-specific incorporation of unnatural amino acids into proteins, *Science (Washington D.C.)* 244:182.

Oh, B.-H. and Markley, J.L., 1989, Complete carbon-13 resonance assignments of a tryptophan in L-lysyl-L-tryptophyl-L-lysine by single-bond and multiple-bond correlated hydrogen-1-carbon-13 two-dimensional NMR, *Biopolymers*, in press.

Oh, B.-H., Mooberry, E.S. and Markley, J.L., Multinuclear magnetic resonance studies of the 2Fe·2S* ferredoxin from *Anabaena* sp. strain PCC 7120. 2. Sequence-specific carbon-13 and nitrogen-15 resonance assignments of the oxidized form, submitted to *Biochemistry*.

Oh, B.-H., Westler, W.M., Darba, P. and Markley, J.L., 1988, Protein carbon-13 spin systems by a single two-dimensional nuclear magnetic resonance experiment, *Science* 240:908.

Oh, B.-H., Westler, W.M. and Markley, J.L., 1989, Carbon-13 spin system directed strategy for assigning cross peaks in the COSY fingerprint region of a protein, *J. Am. Chem. Soc.* 111:3083.

Ortiz-Polo, G., Krishnamoorthi, R., Markley, J.L., Live, D.H., Davis, D.G. and Cowburn, D., 1986, Natural-abundance ^{15}N NMR studies of turkey ovomucoid third domain. Assignment of peptide ^{15}N resonances to the residues at the reactive site region via proton-detected multiple-quantum coherence, *J. Magn. Res.* 68:303.

Oschkinat, H., Griesinger, C., Kraulis, P.J., Sorensen, O.W., Ernst, R.R., Gronenborn, A.M. and Clore, G.M., 1988, Three-dimensional NMR spectroscopy of a protein in solution, *Nature* 332:374.

Otting, G., Senn, H., Wagner, G. and Wüthrich, K., 1986, Editing of 2D ^{1}H NMR spectra using X half-dilters. Combined use with residue-selective ^{15}N labeling of proteins, *J. Magn. Res.* 70:500.

Rance, M., Wright, P.E., Messerle, B.A. and Field, L.D., 1987, "Site-selective observation of nuclear Overhauser effects in proteins via isotopic labeling, *J. Am. Chem. Soc.* 109:1591.

Richarz, R. and Wüthrich, K., 1978, Carbon-13 NMR chemical shifts of the common amino acid residues measured in aqueous solutions of the linear tetrapeptides H-Gly-Gly-X-l-Ala-OH, *Biopolymers* 17:2133.

Richarz, R., Tschesche, H. and Wüthrich, K., 1980, Carbon-13 nuclear magnetic resonance studies of the selectively isotope-labeled reactive site peptide bond of the basic pancreatic trypsin inhibitor in the complexes with trypsin, trypsinogen, and anhydrotrypsin, *Biochemistry* 19:5711.

Saunders, M., Wishnia, A. and Kirkwood, J.G., 1957, The nuclear magnetic resonance spectrum of ribonuclease, *J. Am. Chem. Soc.* 79:3289.

Senn, H., Eugster, A., Otting, G., Suter, F. and Wüthrich, K., 1987a, ^{15}N labeled P22 c2 repressor for nuclear magnetic resonance studies of protein-DNA interactions, *Eur. Biophys. J.* 14:301.

Senn, H., Otting, G. and Wüthrich, K., 1987b, Protein structure and interactions by combined use of sequential NMR assignments and isotope labeling, *J. Am. Chem. Soc.* 109:1090.

Senn, H., Werner, B., Messerle, B.A., Weber, C., Traber, R. and Wüthrich, K., 1989, Stereospecific assignment of the methyl ^{1}H NMR lines of valine and leucine in polypeptides by nonrandom ^{13}C labeling, *FEBS Lett.*, in press.

Shaka, A.J., Keeler, J., Frenkiel, T. and Freeman, R., 1983, An improved sequence for broadband decoupling: WALTZ-16, *J. Magn. Res.* 52:335.

Shon, K. and Opella, S.J., 1989, Detection of ^{1}H homonuclear NOE between amide sites in proteins with ^{1}H/^{15}N heteronuclear correlation spectroscopy, *J. Magn. Res.* 82:193.

Shon, R., Schrader, P., Opella, S., Richards, J. and Tomich, J., 1989, NMR spectra of synthetic membrane bound coat protein species, *in:* "Frontiers of NMR in Molecular Biology," UCLA Symposium, in press.

Sklenar, V. and Bax, A., 1987, Two-dimensional heteronuclear chemical-shift correlation in proteins at natural abundance ^{15}N and ^{13}C levels, *J. Magn. Res.* 71:379.

Sklenar, V., Torchia, D. and Bax, A., 1987, Measurement of carbon-13 longitudinal relaxation using ^1H detection, *J. Magn. Res.* 73:375.

Smith, G.M., Yu, L.P. and Domingues, D.J., 1987b, Directly observed ^{15}N NMR spectra of uniformly enriched proteins, *Biochemistry* 26:2202.

Stockman, B.J. and Markley, J.L., 1989, Stable-isotope-assisted protein NMR spectroscopy in solution, *in:* "Advances in Biophysical Chemistry," in press.

Stockman, B.J., Reily, M.D., Westler, W.M., Ulrich, E.L. and Markley, J.L., 1989, Concerted two-dimensional NMR approaches to hydrogen-1, carbon-13, and nitrogen-15 resonance assignments in proteins, *Biochemistry* 28:230.

Stockman, B.J., Westler, W.M., Darba, P. and Markley, J.L., 1988a, Detailed analysis of carbon-13 NMR spin systems in a uniformly carbon-13 enriched protein: flavodoxin from *Anabaena* 7120, *J. Am. Chem. Soc.* 100:4095.

Stockman, B.J., Westler, W.M., Mooberry, E.S. and Markley, J.L., 1988b, Flavodoxin from *Anabaena* 7120: uniform nitrogen-15 enrichment and hydrogen-1, nitrogen-15, and phosphorus-31 NMR investigations of the flavin mononucleotide binding site in the reduced and oxidized states, *Biochemistry* 27:136.

Summers, M.F., 1988, ^{113}Cd NMR spectroscopy of coordination compounds and proteins, *Coordination Chem. Rev.* 86:43.

Torchia, D.A., Sparks, S.W. and Bax, A., 1988, NMR signal assignments of amide protons in the α-helical domains of staphylococcal nuclease, *Biochemistry* 27:5135.

Tsang, P., Fieser, T.M., Ostresh, J.M., Houghten, R.A., Lerner, R.A. and Wright, P.E., 1989, Solution NMR studies of Fab'-peptide complexes, *in:* "Frontiers of NMR in Molecular Biology," UCLA Symposium, in press.

Tsang, P., Fieser, T.M., Ostresh, J.M., Lerner, R.A. and Wright, P.E., 1988, Isotope-edited NMR studies of Fab'-peptide complexes, *Peptide Research* 1:87.

Vuister, G.W. and Boelens, R., 1987, Three-dimensional J-resolved NMR spectroscopy, *J. Magn. Res.* 73:328.

Vuister, G.W., Boelens, R. and Kaptein, R., 1988, Nonselective three-dimensional NMR spectroscopy. The 3D NOE-HOHAHA experiment, *J. Magn. Res.* 80:176.

Wagner, G., Nirmala, N.R., Montelione, G.T. and Hyberts, S., 1989, Static and dynamic aspects of protein structure, *in:* "Frontiers of NMR in Molecular Biology," UCLA Symposium, in press.

Wang, J., Hinck, A.P., Loh, S.N. and Markley, J.L., 1989a Two-dimensional NMR studies of staphylococcal nuclease: 2. sequence-specific assignments of carbon-13 and nitrogen-15 signals from the (nuclease H124L).deoxythymidine-3',5'-bisphosphate.Ca2+ ternary complex, submitted.

Wang, J., LeMaster, D.M. and Markley, J.L., 1989b, Two-dimensional ^1H NMR studies of Staphylococcal nuclease: 1. Solution structure of the (nuclease H124L)·(Deoxythymidine-3',5'-bisphosphate)-Ca^{+2} ternary complex, submitted.

Weiner, J.H., Dettman, H.D., Henry, G.D., O'Neil, J.D.J. and Sykes, B.D., 1987, Nuclear magnetic resonance studies of a model membrane protein (M13 coat protein) reconstituted in detergent micelles and phospholipid vesicles, *Biochem. Soc. Trans.* 15:81.

Weiss, M.A., Redfield, A.G. and Griffey, R.H., 1986, Isotope-setected ^1H NMR studies of proteins: a general strategy for editing interproton nuclear Overhauser effects by heteronuclear decoupling, with application to phage λ repressor, *Proc. Natl. Acad. Sci. USA* 83:1325.

Westler, W.M., Kainosho, M., Nagao, H., Tomonaga, N. and Markley, J.L., 1988a, Two-dimensional NMR straegies for carbon-carbon correlations and sequence-specific assignments in carbon-13 labeled proteins, *J. Am. Chem. Soc.* 110:4093.

Westler, W.M., Stockman, B.J., Hosoya, Y., Miyake, Y., Kainosho, M. and Markley, J.L., 1988b, Correlation of carbon-13 and nitrogen-15 chemical shifts in selectively and uniformly labeled proteins by heteronuclear two-dimensional NMR spectroscopy, *J. Am. Chem. Soc.* 110:6256.

Wilde, J.A., Bolton, P.H., Hilber, D.A., Harpold, L., Pourtabbed, T., Dell'Acqua, M. and Gerlt, J.A., 1989, Isotopic labeling with hydrogen-2 and carbon-13 to compare conformations of proteins and mutants generated by site-directed mutagenesis: Part 2, *Methods Enzym.* 177:282.

Wüthrich, K., 1976, "NMR in Biological Research: Peptides and Proteins," North-Holland/American Elsevier, 164.

Wüthrich, K., 1986, "NMR of Proteins and Nucleic Acids," Wiley, New York.

Wüthrich, K., 1989, Protein structure determination in solution by nuclear magnetic resonance spectroscopy, *Science (Washington, D.C.)* 243:45.

Wüthrich, K., Shulman, R.G. and Peisach, J., 1968, High-resolution proton magnetic resonance spectra of sperm whale cyanometmyoglobin, *Proc. Natl. Acad. Sci. USA* 60:373.

Wüthrich, K., Wider, G., Wagner, G. and Braun, W., 1982, Sequential resonance assignments as a basis for determination of spatial protein structures by high resolution proton nuclear magnetic resonance, *J. Mol. Biol.* 155:311.

Zehfus, M.H., Reily, M.D., Ulrich, E.L., Westler, W.M. and Markley, J.L., 1989, Complete ^{1}H, ^{13}C, and ^{15}N resonance assignments for a ferrocytochrome c533 heme by multinuclear NMR spectroscopy, *Arch. Biochem. Biophys*, in press.

Zuiderweg, E.R.P. and Fesik, S.W., 1989, Heteronuclear three-dimensional NMR spectroscopy of the inflammatory protein C5a, *Biochemistry* 28:2387.

NMR STUDIES OF PROTEIN DYNAMICS AND FOLDING

Christopher M. Dobson

Inorganic Chemistry Laboratory and Centre for Molecular Sciences, University of Oxford, South Parks Road, Oxford OX1 3QR, UK

INTRODUCTION

We have become familiar over the last 20 years with the nature of the compact globular states of proteins, initially from studies of protein crystals by diffraction methods (Richardson, 1981) and more recently also from direct structure determination in solution by NMR techniques (Wüthrich, 1989). In addition, from both these experimental techniques, and from theoretical studies, many details of the dynamical properties of these folded states have been revealed (McCammon and Karplus 1983). By contrast, however, very little is known about the unfolded or partially folded states of proteins, primarily because of the intrinsic difficulties inherent in their study. Crystallization of proteins in such states has not been achieved, and indeed seems unlikely to be possible in general. NMR spectroscopy, however, seems ideally suited to their characterization, as it is able to provide both structural and dynamical information about molecules in solution. The methods required for such studies may, however, be very different from those now becoming familiar from studies of proteins in their globular states. Further, the nature of any conformational description may need to be significantly different from that used for the globular state, because of the much greater conformational freedom likely to be characteristic of the unfolded or partially folded states.

We have been increasingly interested in the development and exploitation of suitable NMR methods for studying these states of proteins, as well as the more familiar globular states. The reasons for this are as follows. First, establishing the degree of stability of particular elements of protein structure will be of substantial value in understanding the nature of the factors important in stabilizing folded proteins, and be of value in the design and modification of proteins for specific purposes. Second, it provides the basis for experimental studies concerned with the pathways of protein folding. Even for partly folded states of proteins which are not directly implicated in folding pathways, their characterization will help to establish some fundamental understanding of such states. Third, it has become apparent that many proteins and protein complexes are not fully structured or folded in their functional states. The methods developed to characterize partly folded proteins could be invaluable in these cases, which may well be systems not amenable to study by more conventional methods, such as crystallographic techniques.

In order to study these states of proteins by NMR it is necessary to generate the appropriate conditions for their existence. One approach is to attempt to study them indirectly, and to infer their nature from the measurement of spectral features determined by their presence. An example of this is an approach which detects different rates of hydrogen exchange in states with different degrees of folding (Schmidt and Baldwin 1979, Roder and Wüthrich 1986). This has recently provided important clues as to the order in which regions of secondary structures form in ribonuclease and cytochrome **c** (Udgaonkar and Baldwin, 1988; Roder et al., 1988). A variation on the trapping theme involves studies where intermediate states of proteins are generated by stopping the folding process chemically. The clearest example of such an approach involves the study of proteins containing disulphide bridges, in which reformation of the bridges is prevented at different stages in the folding process by addition of reagents which react with free sulphydryl groups. This approach has the advantage that intermediate states of proteins can be isolated and studied at leisure. This has been achieved for the bovine pancreatic trypsin inhibitor protein BPTI (Creighton, 1988), and NMR studies have been of considerable importance in identifying structural features of the intermediate states (States et al., 1987). In particular, clear evidence for native like structure in the intermediates even early in the folding pathway was obtained. It is particularly pleasing that a recent report has demonstrated that a synthetic peptide fragment of BPTI contains the major structural elements identified in the early folding intermediates generated by the trapping experiment (Oas and Kim, 1988).

One of the important features of the existence of substantial elements of persistent structure is the identification of cooperativity in the folding interactions. This can be seen, for example, in the BPTI intermediates where increase in the temperature results in the reversible change of the NMR spectra from those characteristic of folded structured proteins to those characteristic of highly unstructured states (States et al., 1987). This aspect of the exploration of protein structure, in which the stability of the folded state is altered by changes in temperature, pH or the concentration of added denaturants, or by chemical modification or mutagenesis, to provide an unfolded or partly folded state stable under equilibrium conditions, will be the theme of much of this article. Studies under these conditions can provide thermodynamic and kinetic data about protein structures at the level of individual residues.

We shall begin by describing studies of two very closely related proteins, lysozyme and α-lactalbumin, which have proved invaluable in developing some of our basic approaches. Then, we shall describe studies of two other proteins which extend these approaches, and which point the way to future applications of the methods.

STUDIES OF INDIVIDUAL PROTEINS

Lysozyme and α-Lactalbumin

Lysozyme from hen egg white is a small, extremely well characterized protein containing 129 residues. The ^{1}H NMR spectrum of lysozyme in its native state has been analyzed in detail (Redfield and Dobson, 1988). Resonances have been assigned virtually completely to individual residues in the structure, enabling extensive studies of the structure, dynamics and binding properties of the protein in solution to be made.

An early observation in the NMR spectroscopy of lysozyme was that, under certain conditions, the protein can be taken reversibly through the thermal unfolding transition and the NMR spectrum reflects dramatically the change in the conformational state (McDonald et al., 1971). Further, the rate of unfolding and refolding are such that at conditions close to the mid-point of the transition, where both folded and unfolded states are present in significant concentrations, the spectra of the folded and unfolded states of the protein are superimposed, rather than averaged as would be the case if the kinetics were fast. Under these conditions the relative intensities of the resonances of individual protons in both the folded and unfolded states provide a direct measure of the equilibrium constant reflecting the unfolding process for individual residues in the protein. By studying the temperature dependence of the equilibrium constant, the enthalpy and entropy changes associated with the transition can be obtained. The data obtained in this way for lysozyme are in excellent agreement with calorimetric and other studies, and the measurements for individual residues show directly the high cooperativity (two-state nature) of the unfolding transition (Dobson and Evans, 1984).

Although the kinetics of the unfolding and folding transition are sufficiently slow for the resonances in the two conformational states to be observed separately, the equilibrium is of course a dynamic one; protein molecules do interconvert between the folded and unfolded states. Under these conditions, provided that the rates of interconversion are not slow compared with the nuclear relaxation rates, they can be probed by magnetization transfer methods. If, for example, saturation of the resonance of a residue in its folded state is carried out, this saturation may be transferred to the resonance of the same residue in its unfolded state. The steady state degree of saturation of this resonance will depend simply on the rate at which magnetization is lost from the unfolded state by the transition to the folded state, relative to the rate at which magnetization returns by the normal relaxation methods. Because the time course of achieving the steady state effect also depends on these two parameters, they can be separated and the rate of folding determined. By reversing the experiment and saturating the unfolded state and observing the folded state resonance, the rate of unfolding may similarly be determined for individual residues. For a two-state system, the ratio of the rates should simply equal the equilibrium constant, determined separately as described above. These experiments have been carried out for lysozyme at a variety of temperatures, demonstrating the two state nature of the kinetics of the folding; the kinetic data are the same when measured for different residues in the protein and are fully consistent with the equilibrium constant measurements (Dobson and Evans, 1984).

The exchange of magnetization between the folded and unfolded states can be detected by 2-D as well as 1-D experiments; here it is manifest in cross-peaks linking the resonances of exchanging nuclei (see Figure 1). The resolution of the 2-D experiments enables the exchange process to be studied for many resonances not resolvable in 1-D experiments. Further, because the cross-peaks arising from the exchange process link resonances which arise from the same residue in the different states of the protein, it enables assignments made in the spectrum of one state to be transferred to that of the other state (Dobson et al., 1984; 1986). This is of major importance because it provides a method of assigning the spectra of the unfolded state. Because of the very limited resolution of resonances in this state this process is largely impracticable by direct methods such as those applied to folded states of proteins. In the case of lysozyme, many resonances can be identified in this way, and this opens the door for a conformational analysis of an unfolded state of a protein in a manner not previously possible.

The conclusions from such studies of lysozyme, albeit under conditions where unfolding takes place at temperatures in excess of 70°C, are that little persistent structure is evident in the thermally unfolded state (Dobson et al., 1984; 1987). A very different situation is, however, found for α-lactalbumin where the 2-D exchange experiment can be seen to be of major significance. α-lactalbumins are proteins with high sequence homology to lysozymes (although with quite different functional roles) and with very closely similar folded structures. It turns out, however, that their unfolding behavior is quite different. In particular, under certain circumstances cooperative unfolding of the native structure leads to a state, called a molten globule state, in which a variety of techniques have indicated a degree of compactness little

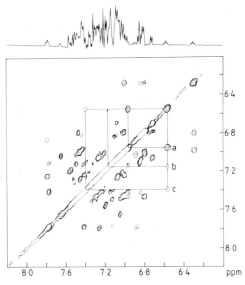

Figure 1. Aromatic region of a 500 MHz 2-D exchange (NOESY) spectrum of hen lysozyme in ^2H_2O, pH 3.8, at the mid-point of the thermal unfolding transition (77°C). The cross-peak labelled (a) is an example of an exchange peak enabling correlation of resonances of the same residue (here Trp 108) in the folded and unfolded states. The cross-peaks labelled (b) and (c) correspond to NOE effects in the folded protein. (From Dobson et al., 1987).

different from that of the native state, and with clear evidence of helical secondary structure (Dolgikh et al., 1985). Only in the presence of high concentrations of chemical denaturants do these characteristics disappear, and the state resembles a "normal" unfolded one. The NMR spectrum of α-lactalbumin in the molten globule state is clearly very different from that of both the native and the fully unfolded protein (Figure 2). It is, however, broad and poorly dispersed, and hence difficult to study directly by NMR. The 2-D exchange experiment enables direct correlation with the native state to be made, and hence the identity of the resonances most perturbed from their position in the unfolded protein to be established (Baum et al., 1989). These resonances are amongst those most perturbed in the native state, implying that native like structural features could be present in the molten globule state.

Considerably more insight into the detailed nature of this state arises from a different experiment, but one which shares the philosophy of the correlation of resonances in the molten globule state with those of the fully folded state. One of the most striking observations in the spectrum of α-lactalbumin in its molten globule state is that there are several amide protons whose resonances persist in the spectrum hours or even days after the protein has been dissolved in ^2H_2O. Such resonances are completely absent in the spectra of unfolded lysozyme, or indeed of α-lactalbumin in the presence of high concentrations of denaturants. Slowly exchanging amides are, however, common in the spectra of folded proteins where they imply the existence of persistent secondary structure in a protected environment. The 2-D exchange experiment proved difficult to utilize for the identification of these resonances, not least because hydrogen exchange was relatively fast under the conditions where the experiment worked efficiently. To identify the resonances, however, the following procedure was adopted (Baum et al., 1989). After dissolution of the protein in ^2H_2O in the molten globule state at pH 2 for several hours, the pH was changed to pH 5.5. At this pH, the protein refolds rapidly to its native state. Observation of the NMR spectrum reveals the presence of several amide proton resonances. As

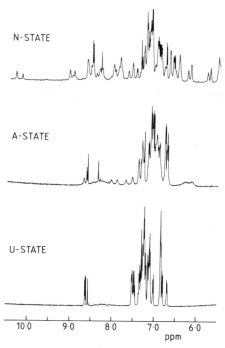

Figure 2. Low-field regions of the 500 MHz ^{1}H NMR spectra recorded at 52°C of guinea pig α-lactalbumin at (a) pH 5.4 (the native state), (b) pH 2.0 (the molten globule or A- state) and (c) pH 2.0 in the presence of 9M urea (the unfolded state). In each case the protein was dissolved in ^{2}H$_2$O for 6 h prior to recording the spectrum. The resonances in the spectra shown correspond to protons from aromatic residues and from amide protons that have not exchanged with deuterons from the ^{2}H$_2$O solvent. (From Baum et al., 1989).

the whole operation was carried out in ^{2}H$_2$O, the origin of these protons can only have been that they were present in the molten globule state. Conventional 2-D NMR experiments allow clear identification of the resonances (Figure 3). They arise predominantly from one of the α-helical regions of the structure, which includes residues 89 to 97, and therefore indicates that this helix is stable in the molten globule state. It is likely from the spectra that other regions of secondary structure, with less well protected amide hydrogens, also exist; these have not yet been firmly identified. The differences in the spectra of the molten globule and the native state, however, indicate that the molten globule state has much greater conformational flexibility than the native state does. The NMR experiments, like those of the BPTI intermediates, do, however, provide clear evidence for elements of native like structure in a partly folded protein.

The NMR experiments described here, which depend on establishing conditions whereby non-native states of a protein are present in significant concentrations in equilibrium with the native state, are able to provide, at the individual residue level, thermodynamic, kinetic and structural information about the folding/unfolding process. The lysozyme/α-lactalbumin system offers opportunities in relatively simple proteins to probe the influence on all these parameters of features influencing protein structure and folding. The difference in the unfolded states of lysozymes and α-lactalbumin, which have the same folded structure, indicate the importance of sequence effects on the stability of individual folding elements. Site directed mutagenesis should

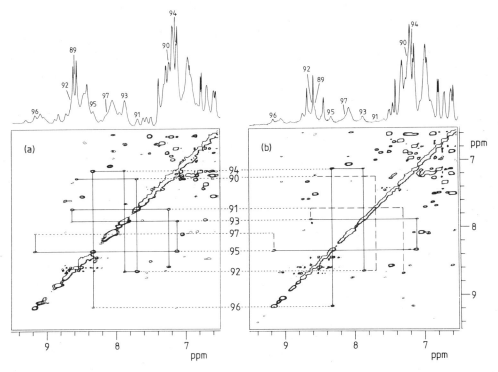

Figure 3. Low-field regions of the 500 MHz NOESY spectra of α-lactalbumin at 35° freshly dissolved (a) in ^2H_2O at pH 5.4 and (b) in ^2H_2O at pH 5.4 following a pH jump from pH 2, where the protein had been allowed to remain for 10 h. The assignments for the resonance of residues 90-97 are indicated (From Baum et al., 1989).

therefore be of particular interest in these proteins, and offers the opportunity to examine the importance of specific residues in the stability and folding of a protein.

Staphylococcal Nuclease

Lysozyme provides a straightforward example of a protein which behaves closely in accord with a two-state model of folding/unfolding. In this section we describe experiments which illustrate the value of the NMR methods described for lysozyme in a system with more complex behavior. The protein involved is staphylococcal nuclease which, like lysozyme, undergoes reversible thermal unfolding under appropriate conditions. Staphylococcal nuclease is a protein of nearly 150 amino acid residues, which contains four histidine residues. In the region of the spectrum where the characteristic $H^{\epsilon 1}$ resonances are observed it was noted in an early study by NMR that several minor resonances exist in addition to the resonances attributable to these protons in the major folded state of the protein (Markley and Jardetzky, 1970). There could in principle be several different origins of such resonances, including degradation, sequence differences, or multiple folded conformations. That the latter was the correct explanation is clearly demonstrated by a 2-D exchange experiment of the type described for lysozyme (Dobson et al., 1987; Evans et al., 1989). Exchange cross-peaks between a minor and a major peak for 3 of the 4 histidine resonances are clearly evident in the spectra obtained, see Figure 4, indicating that interconversion of the two states takes place. In addition, exchange cross-peaks were observed in the region of the mid-point of unfolding between both major and minor folded state resonances and resonances arising from the unfolded protein. In the case of one of the histidines, two resonances could be identified in the region of the resonances of the unfolded

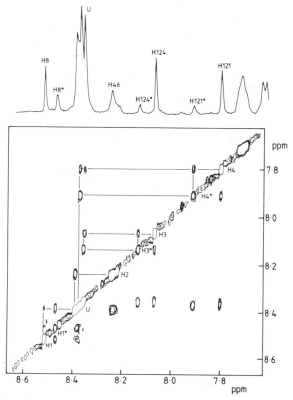

Figure 4. Part of a 500 MHz 2-D exchange spectrum of staphylococcal nuclease recorded at 48°C, close to the midpoint of thermal unfolding. The region of the histidine $H^{\varepsilon 1}$ resonances is displayed. (From Evans et al., 1989).

state. This, along with data from a series of 1-D saturation experiments, indicated the presence of four states of the protein in the region of the unfolding transitions, two of which are folded and two unfolded, all of which interconvert sufficiently slowly on the NMR timescale for separate resonances to be resolved.

There are several possible explanations for slowly interconverting folded structures. For unfolded states far fewer possibilities exist, and prominent amongst these is the possibility of *cis-trans* isomerism about an X-proline peptide bound. Examination of the crystal structure of staphylococcal nuclease indicated at once a possible candidate for the distinctive proline residue; this is Pro 117, which is in the less common *cis* conformation in the folded structure (Cotton et al., 1979). Further, the observation that in solution the equilibrium shifts in the presence of inhibitor to show only the major folded resonances, coupled with the fact that the original crystal structure was carried out in the presence of the inhibitor, suggested that the major folded state of staphylococcal nuclease in solution has Pro 117 in the *cis* conformation. This is also consistent with the finding from magnetization transfer experiments that the major folded state interconverts with the minor unfolded state; *cis* proline conformations are found to be less stable than the *trans* in simple peptides. In order to test the hypothesis that structural heterogeneity arose from Pro 117, a mutant protein was constructed in which this residue were replaced by glycine. The spectra of the mutant indicated (see Figure 5) only a single set of histidine $H^{\varepsilon 1}$ resonances, and magnetization transfer experiments showed a single resonance of His 121 in the unfolded state (Evans et al., 1987).

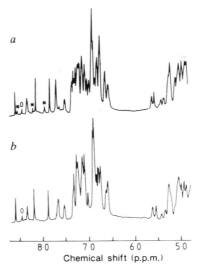

a

b

8·0	7·0	6·0	5·0

Chemical shift (p.p.m.)

Figure 5. 500 MHz NMR spectra at pH 5.3, 40°C in ^2H_2O of (a) wild-type staphylococcal nuclease and (b) a mutant protein with Pro 117 replaced by Gly 117. The resonances labelled with an asterisk correspond to the histidine residues in the minor folded state. (From Evans et al., 1987).

By means of the same approach as described for lysozyme, information on the thermodynamic properties of these different states was gained from the intensities of the various resonances as a function of temperature. From the relative intensities of the minor and major folded states the equilibrium constant and hence the difference in free energy between the two states could be determined. The temperature dependence of this then gave the enthalpy and entropy differences. A similar thermodynamic analysis was possible for the unfolded states, except that the enthalpy difference between the *cis* and *trans* forms was assumed to be zero (by analogy with simple peptides); a detailed temperature analysis of the spectra of the unfolded state was not practicable. Subtraction of the thermodynamic parameters for the folded and unfolded state then allows the "conformational" contribution to the thermodynamics to be calculated (Evans et al., 1989). This represents the influence of the folded protein structure on the relative stability of the *cis* and *trans* isomers. These data prove to be of considerable interest. The *cis* isomer is estimated to be stabilized by a substantial ΔH term (-45 ± 5kJ mol^{-1}). This suggests that the bonding interactions within the folded protein are considerably more favorable with the *cis* than the *trans* isomer. This can be rationalized when the crystal structure of the protein in this region of the protein is examined, as Pro 117 is in a type VI reverse turn. Such a turn can only accommodate proline in its *cis* configuration. Presumably, the structural change which is inevitable to accommodate the *trans* conformational state results in a less satisfactory fold in this region of the protein. This is also likely to be the origin of the shifts of the various histidine (and indeed other) resonances in the different states of the protein. The ΔS term, however, is estimated to be -99 ± 24 JK^{-1} mol^{-1} and so favours the *trans* protein in the folded state, but not to an extent sufficient to outweigh the ΔH term. Perhaps the less satisfactory fold of the folded state with *trans* Pro-117 has greater dynamic freedom, hence giving an increased entropy. Regardless of the validity of such speculation, this study indicates the detailed understanding of specific structural features of a protein that can be achieved by these approaches.

So far we have discussed only the thermodynamics of the nuclease system. The magnetization transfer experiments, as well as showing the interconversion of the various species and correlating the resonances of the states, indicates the potential availability of the kinetic parameters linking the different states. The analysis of the experimental data is, however, substantially more involved than that of lysozyme. For example, the existence of a cross-peak between the two folded states of nuclease does not by itself show that they interconvert directly. Magnetization transfer could, in principle, occur indirectly by interconversion of the two states via the unfolded states. By carrying out a series of 1-D multiple saturation experiments it has proved possible with certain assumptions to obtain values for the rate constants for all 8 different transitions (Evans et al., 1989).

Examination of the various rate constants determined in this way allows a number of important conclusions to be drawn. For example, the data provide insight into the nature of the folded conformation and show clearly that isomerization of the proline residue can take place within the folded state without the need for complete unfolding of the protein. Further, they enable insight into the folding process to be gained. As an example of this, the kinetic data show that the difference between the *cis* and the *trans* forms is much greater in the unfolding kinetics than in the folding kinetics. This strongly suggests that in the pathway of folding, the turn that contains Pro 117 is formed after the major rate determining step. The implication of this conclusion is that the turn is incorporated into the folded structure at a late stage along the folding pathway (Evans et al., 1989).

This study of staphylococcal nuclease shows how the fortuitous observation of multiple states in a protein has enabled information to be obtained about the stability and folding kinetics of specific structural features of a protein. It will be of great interest to probe these features in more depth, for example by studying mutant proteins and examining the consequence of specific mutations on the thermodynamics and kinetics of the various interconversions. In this way it should be possible to gain considerable insight into fundamental factors influencing these properties of the protein.

Urokinase

The ability to observe resonances of individual residues in a protein provides a unique opportunity to probe systems where, unlike the previous systems, the unfolding behavior of the protein is not fully cooperative. This aspect of the NMR experiments can be illustrated with some preliminary studies of the multidomain proteins of the fibrinolytic system. These proteins are substantially larger than those commonly considered suitable for detailed study by NMR, but our results indicate that they are surprisingly amenable to the types of study described already in this article (Oswald et al., 1989).

We have so far investigated three of the family of proteins involved in blood clotting and fibrinolysis. One of these three proteins, plasminogen is the protein that, on activation, is converted to plasmin which is responsible for the degradation of fibrin in the blood clot. Urokinase and tissue plasminogen activator (t-PA) are two proteins involved in the activation of plasminogen. All three of these proteins have been of considerable medical and pharmaceutical interest recently following the demonstration that administration of them, or their derivatives, has the potential to reduce substantially mortality following coronary thrombosis (Wilcox et al., 1988). Despite the substantial interest in these molecules, understanding of their biological properties has been limited by the lack of structural information about them. The sequences of the three proteins, like those of others involved in haemostasis, provides clear evidence that they have a domain or mosaic structure. Regions of the polypeptide chain show sequence homologies, and disulphide - bond linkages, characteristics of individual structured regions. None of the proteins, or their near relatives, has, however, been crystallized in a form suitable for X-ray diffraction analysis. Some information about the structures of the individual domains has, nevertheless, been deduced by comparison with known structures, and models of the proteins have been constructed based on this information (Holland and Blake, 1989). The main problems in defining the structures are undoubtedly concerned with the linker regions between domains, and the interactions between the different domains, which determine the overall architecture of the proteins.

Our initial studies focussed on the smallest protein, urokinase, which consists of three domains. The N-terminal domain, of nearly 50 residues, is called an EGF (epidermal growth factor) domain, because of its homology with the family of growth factors. The next region of the sequence, approximately 90 residues, is one of a domain type called a kringle. The C-terminal domain is the catalytic portion of the molecule, and has high homology with the digestive serine protease family of proteins; this region of the molecule also has carbohydrate attached at one site in the sequence. The NMR spectrum of urokinase at room temperature shows features characteristic of globular structure, with a number of resonances shifted significantly from the positions expected in unstructured molecules, Figure 6. Following the approach described in the earlier section the temperature was increased to explore the unfolding of the molecule. By performing these experiments at pH values below about 6, it was evident that fully reversible folding and unfolding could be achieved reproducibly despite the complexity of the molecule (Bogusky et al., 1989). As the temperature is increased, the spectrum changes towards that of an unstructured protein. The resonances characteristic of structured regions of the protein disappear from the spectrum in groups at characteristic temperatures. Examination of the region of the spectrum to high field shows this clearly, Figure 7. For the protein at pH 3.8, the resonance at -1 ppm broadens and reduces in intensity between 40 and 50°C. Another group of resonances, between -0.5 and 0.2 ppm, disappears at 55-60°C. Even at 85°C, a small subset of resonances remains, and only disappears from this region of the spectra at lower pH values; at pH 2.5 and 70°C, for example, the spectrum resembles that of a fully unfolded protein.

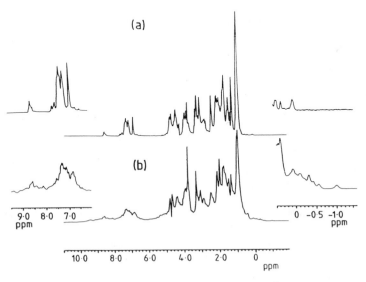

Figure 6. 600 MHz ^{1}H NMR spectrum of human urokinase in ^{2}H$_2$O, pH 3.8 at (a) 85°C and (b) 35°C. The inserts show expansions of the upfield and downfield regions of the spectrum (From Bogusky et al., 1989).

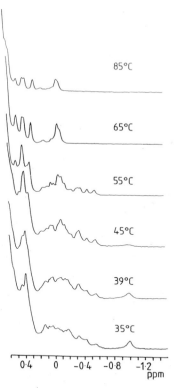

Figure 7. Upfield region of the 600 MHz ^{1}H NMR spectrum of urokinase at the temperatures indicated. (From Bogusky et al., 1989).

The obvious interpretation of this behavior is that the protein is unfolding in a series of steps, rather than in a single totally cooperative event. That this is indeed the case is clearly demonstrated by magnetization transfer experiments of the type already described in this article. Saturation, for example, of the resonance at -1ppm in the region of mid-point of its disappearance from the spectrum resulted in magnetization transfer to a resonance at +1.0ppm, exactly the position expected for a methyl group in an unstructured region of a protein. Similar experiments have been carried out for other resonances characteristic of other transitions; in all cases transfer occurred to positions consistent with an unfolded state.

In order to identify the nature of the structural transitions, assignment of the individual resonances must be achieved. This is a much more difficult task than for small proteins, but assignments of resonances to difference domains of the protein has been initiated for urokinase by comparison of the spectra of the intact protein with the spectra of fragments of the protein obtained by limited proteolysis (Bogusky et al., 1989; Oswald et al., 1989). These experiments also provide direct evidence that the structures of the domains isolated as proteolytic fragments are at least similar to their structures in the intact protein. On the basis of the assignments, it has been possible to probe a number of features of the unfolding of the urokinase molecule. This by itself has provided interesting information about a several aspects of the structure of the protein. First, it is apparent that the most stable region of the structure is within the protease domain. Indeed, it is evident from the results that the unfolding of the protease domain itself (whether in the intact protein or as a fragment) occurs in two distinct steps. This would be consistent with a model of the domain based on the digestive serine proteases since these proteins have two segments (Richardson, 1981), of similar structure, and it is possible that the two steps correspond to the separate unfolding of the two segments of the molecule. If future experiments show this to be the case, the origin of the very high stability of one of the segments relative to the other is of considerable interest. It could, perhaps, be associated with the disulphide linkages (none of the six disulphides in the protease domain of urokinase link the two segments of the molecule) or with other features of the sequence. Further investigations may provide important clues as to factors stabilizing protein structural elements.

The second particularly interesting feature of the unfolding transitions comes from a comparison of these results with studies of the unfolding of the fragments. At least under the conditions so far explored, the unfolding behavior of the individual domains in fragments appears closely similar to that of the same domains in the intact protein. Of particular interest here is the kringle domain, because of the lack of significant perturbation of its unfolding transition temperature by the presence of the protease domain, which itself remains folded during the unfolding transition of the kringle. As a strong interdomain interaction, if it persisted to temperatures in the region of the unfolding transition, might be expected to perturb the thermodynamics of the kringle unfolding transition, this result implies that the interactions between these two domains are relatively weak. More detailed experiments, involving the measurement of kinetic and thermodynamic parameters, along the lines outlined for lysozyme should shed light on this aspect of the structural properties of the molecule.

Further evidence as to the nature of this interdomain interaction comes, however, from closer examination of the spectrum of urokinase in its folded state. The molecular weight of this protein, 54,000, is large compared to that of proteins which conventionally give rise to well resolved NMR spectra. Indeed, the linewidths of the resolved resonances do not appear markedly greater in the spectra of the intact protein than in those of the fragments. In addition, excellent 2-D NMR spectra are obtained from the intact protein, showing a multitude of cross-peaks, Figure 8 (Oswald et al., 1989). As the intensity of these cross-peaks is a sensitive function of the linewidth of the contributing resonances, this implies that many of the unresolved resonances in the spectra of the protein are relatively narrow. The clue to the possible origin of this phenomenon arises from a comparison of the 2-D spectra of urokinase with those of its proteolytic fragments. This reveals that the overwhelming majority of the cross-peaks obtained in the spectra of the fully folded protein arise from residues in the kringle and EGF domains; the protease domain gives rise to few cross-peaks, and these are generally of lower intensity. The implication of this result is that these two sections of the molecule have different dynamic properties. This would be consistent with significant freedom of motion about the long linker region between the kringle and protease domains; under these conditions the smaller domains (EGF and kringle) will reorient faster than the larger protease domain (which also has carbohydrate attached) to give greater motional narrowing of the resonances, that is narrower linewidths. At elevated temperatures, however, it is possible to observe the resonances of the protease domains because faster reorientation of even the larger part of the molecule narrows lines sufficiently for them to give rise to well resolved cross-peaks, Figure 8. These observations are consistent with the proposal for the absence of strong specific interactions between the domains.

The ability to obtain well resolved 2-D spectra containing resonances of residues within the kringle and EGF domains provides an opportunity to carry out structural comparisons with the isolated domains at a level of detail much greater than that possible with 1-D experiments (Mabbut & Williams, 1988; Thewes et al., 1989). As these domains are within the range of molecular sizes for which complete structures can now be determined by NMR methods, this offers the opportunity to define the conformations of these two domains within the intact molecule. As under appropriate conditions resonances from the protease domain can be

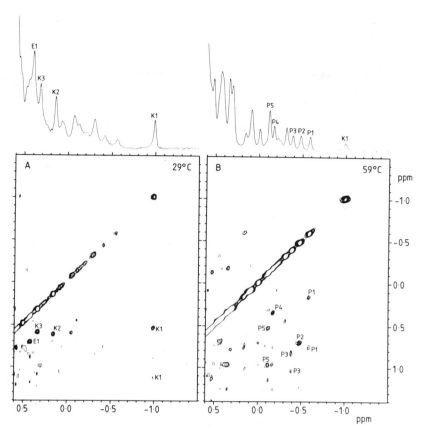

Figure 8. Comparison of the upfield region of the 600 MHz NOESY spectrum of human urokinase in ^2H_2O, pH 4.5, at (a) 29°C and (b) 59°C. At the higher temperature the EGF and kringle domains are partially unfolded. The resonances labelled K1-K3 and E1 arise from the kringle and EGF domains respectively, and those labelled P1-P5 are from the protease domain. This illustrates the selective observation of resonances from the different domains as a function of experimental conditions. (From Oswald et al., 1989).

resolved, some detailed study of even this part of the molecule is feasible. Most importantly, however, once resonances can be resolved and assigned in such experiments, the opportunity exists to explore domain-domain contacts in detail, and hence to attempt to define the overall architecture of the molecule. An important feature in all these methods will be the ability to compare spectra of the protein in its various partially folded states. There seems little doubt that existing NMR methods have the capability to describe at least in outline, the structural properties of the protein, although because of the dynamic properties of the molecule the nature of the description may differ from that of a simple globular protein.

The prospects for a detailed description of the structures, dynamics and folding properties of urokinase look very promising. But is this a unique and particularly favorable situation for an individual protein? Preliminary results show firmly that the answer is no. Excellent NMR spectra have been obtained from t-PA and from plasminogen. Further, in agreement with calorimetric data, reversible and, most clearly in the case of plasminogen, multi-step unfolding behavior can be achieved under suitably chosen conditions. This, as in the case of urokinase, implies significant independence of at least some of the domains in these proteins. As well as generating well resolved NMR spectra, such motion may well be of considerable importance in terms of its significance for functional properties of the proteins. This suggests that multi-domain proteins, though resistant to crystallography methods, may be particularly amenable to NMR studies of the type outlined here. This offers the opportunity not just for structure determination, but for exploration of the effects on the structure and dynamics of, for example, binding of either small molecules or other proteins, or of modification, mutagenesis, or domain swapping experiments. This could well be of real significance in the rational design of improved therapeutic agents based on these molecules.

CONCLUSIONS

NMR studies of protein stability and folding are still in their infancy. In this article we have chosen to illustrate some of the potential rewards of research activities in this field by examples chosen from our recent work. By developing these and other approaches we believe that NMR methods, in conjunction with other techniques, particularly those of protein engineering, will be capable of providing major insight into fundamental factors influencing protein structures, and of enabling detailed descriptions to be given of molecular events associated with the transitions of proteins from unfolded to folded states.

ACKNOWLEDGEMENTS

I should like to acknowledge my research associates whose work is described in this article and who contributed so many of the ideas on which it is based. These include Jean Baum, Michael J. Bogusky, Philip A. Evans, Robert O. Fox, Claire Hanley, Robert E. Oswald, Sheena E. Radford, Christina Redfield, Richard A.G. Smith and Karen Topping.

I should also like to thank my colleagues in the Oxford Centre for Molecular Sciences, particularly Colin C.F. Blake, Iain D. Campbell, Sir David Phillips and Robert J.P. Williams, for many stimulating discussions.

REFERENCES

Baum, J., Dobson, C. M., Evans, P. A. and Hanley, C., 1989, Characterization of a partly folded protein by NMR methods: studies on the molten globule state of guinea-pig α-lactalbumin, *Biochemistry* 18:7 .

Bogusky, M. J., Dobson, C. M. and Smith, R. A. G., 1989, Reversible independent unfolding of the domains of urokinase monitored by ^{1}H NMR, *Biochemistry* 28:6728.

Cotton, F. A., Hazen, E. E. & Legg, M. J., 1979, Staphylococcal nuclease - proposed mechanism of action based on the structure of enzyme-thymidine 3'-5'*bis* phosphate-calcium ion complex at 1.5 Å resolution, *Proc. Natl. Acad. Sci. U.S.A.* 76:2551.

Creighton, T. E., 1988, Towards a better understanding of protein folding pathways, *Proc. Natl. Acad. Sci. U.S.A.* 85:5082.

Dobson, C. M. and Evans, P. A., 1984, Protein folding kinetics from magnetization transfer NMR, *Biochemistry* 23:4267.

Dobson, C. M., Evans, P. A. and Williamson, K. L., 1984, Proton NMR studies of denatured lysozyme, *FEBS Letters* 168:331.

Dobson, C. M., Evans, P. A., Fox, R. O., Redfield C. and Topping, K. D., 1987, Two-dimensional exchange experiments in NMR studies of protein dynamics and folding, *Prot. Biol. Fluids* 35:433.

Dolgikh, D. A., Abaturov, L. V., Bolotina, I. A., Brazhnikov, E. V., Bychkova, V. E., Bushuev, V. N ., Gilmanshin, R. I., Lebedev, Y.O., Semisotnov, G. V., Tiktopulo, E. I., and Ptitsyn, O. B., 1985, Compact state of a protein molecule with pronounced samll-scale mobility: bovine α-lactalbumin, *Eur. Biophys. J.* 13:109.

Evans, P. A., Kautz, R. A., Fox, R. O. and Dobson, C. M., 1987, Proline isomerism in staphylococcal nuclease characterised by NMR and site directed mutagenesis, *Nature* 329:6136.

Evans, P. A., Kautz, R. A., Fox, R. O. and Dobson, C. M., 1989, A magnetisation transfer NMR study of the folding of staphylococcal nuclease, *Biochemistry* 28:362.

Holland, S. K. and Blake, C. C. F., Reconstructing the plasminogen molecule from its known domain structures, *EMBO J.*, in press.

Mabbut, B. C. and Williams, R. J. P., 1988, Two-dimensional ^{1}H NMR studies of the solution structure of plasminogen kringle 5, *Eur. J. Biochem.* 170:539.

Markley, J. L. and Jardetzky, O., 1970, Nuclear magnetic resonance studies of the structure and binding sites of enzymes, *J. Mol. Biol.* 50:223.

McCammon, J. A. and Karplus, M., 1983, The dynamic picture of protein structure, *Accts. of Chem. Res.* 16:187.

McDonald, C. C., Phillips, W. D. and Glickson, J. D., 1971, Nuclear magnetic resonance study of the mechanism of reversible denaturation of lysozyme, *J. Am. Chem.Soc.* 93:235.

Oas, T. G. and Kim, P. S., A peptide model of a protein folding intermediate, 1988, *Nature* 336:42.

Oswald, R. E., Bogusky, M. J., Bamberger, M., Smith, R. A. G. and Dobson. C. M., 1989, Dynamics of a multidomain fibrinolytic protein: a 2D NMR study of urokinase,*Nature* 337:579.

Redfield, C. and Dobson, C. M., 1988, Sequential ^{1}H NMR assignments and secondary structure of hen egg white lysozyme in solution, *Biochemistry* 27:122.

Richardson, J. S., 1981, The anatomy and taxonomy of protein structure, *Adv. Protein Chem.* 23:167.

Roder, H., Elove, G. A. and Englander, S. W., 1988, Structural chasracterisation of folding intermediates in cytochrome **c** by H-exchange labelling and proton NMR, *Nature* 335:700.

Roder, H., and Wüthrich, K., 1986, *Proteins* 1:34.

Schmidt, F. X. and Baldwin, R. L., 1979, Detection of an early intermediate in the folding of ribonuclease A by protection of amide protons against exchange, *J. Molec. Biol.* 135:199.

States, D. J., Creighton, T. E., Dobson, C. M. and Karplus, M., 1987, Conformations of intermediates in the folding of the pancreatic trypsin inhibitor, *J.Mol. Biol.* 195:731.

Thewes, T., Ramesh, V., Simplaceanu, E., and Llinas, M., 1988, Analysis of the aromatic ^{1}H NMR spectrum of the kringle 5 domain from human plasminogen, *Eur. J .Biochem.* 175:237.

Udgaonkar, J. B. and Baldwin, R. L., 1988, NMR evidence for an early framework intermediate in the folding pathway of ribonuclease A, *Nature* 335:694.

Wilcox, R. G., Olsson, C. G., Skene, A. M., von der Lippe, G., Jensen, G. and Hampton, J. R., 1988, Trial of tissue plasminogen activator for mortality reduction in acute myocardial infarction, *Lancet* 8610:525.

Wüthrich, K., 1989, Protein structure determination in colution by nuclear magnetic resonance spectroscopy, *Science* 243:45.

UNDERSTANDING THE SPECIFICITY OF THE DIHYDROFOLATE REDUCTASE BINDING SITE

Gordon C. K. Roberts

Department of Biochemistry and Biological NMR Centre, University of Leicester, Leicester LE1 7RH, UK

INTRODUCTION

Recent developments in molecular and structural biology have transformed our ability to explore the structure-function relationships of biological macromolecules. This applies escpecially to the intermolecular interactions which form the basis of biological specificity. It is now possible to make precisely defined structural changes to each partner in these interactions, by genetic or chemical means, and the spectroscopic and crystallographic methods are available to assess the consequences of these changes in detail. As a result, it is now beginning to become possible to make quantitative estimates of the contributions of individual interactions to the overall molecular recognition process.

Dihydrofolate reductase (dhfr) is a small (M_r 18-25,000) dehydrogenase which catalyses the reduction of dihydrofolate (and, less effectively, folate) to tetrahydrofolate, using NADPH as coenzyme. It is a ubiquitous enzyme which plays a crucial role in one-carbon metabolism in the cell by maintaining the levels of tetrahydrofolate coenzymes, and is pharmacologically important as the 'target' for the important 'anti-folate' group of drugs such as methotrexate, trimethoprim and pyrimethamine. As a result, there have been intensive studies of this enzyme over the last decade, including detailed structural studies by both crystallography and NMR. High-resolution crystal structures have been determined for dhfr from *Lactobacillus casei* and other organisms in a number of ligand complexes (Bolin et al., 1982; Filman et al., 1982; Volz et al., 1982; Matthews et al., 1985; Oefner et al., 1988). In the NMR spectrum, ^1H resonances from more than 30% of the residues of *L. casei* dhfr have been assigned (Hammond et al., 1986; Feeney et al., 1989; Arnold et al., 1989a), and as a result 'reporter groups' are available throughout the protein structure; in addition, ^1H, ^{13}C, ^{15}N, ^{19}F and ^{31}P resonances of bound ligands can be used to characterise their modes of binding (e.g., Hyde et al., 1980a,b; Clore et al., 1984; Cheung et al., 1986; Hammond et al., 1987; Tendler et al., 1988; Birdsall et al., 1989; Jimenez et al., 1989a; Arnold et al., 1989). This considerable background of structural information allows us to begin to ask precise questions about the roles of individual amino-acid residues in substrate, inhibitor and coenzyme binding to the enzyme.

Methotrexate binds to the enzyme with its 2,4-diaminopteridine ring in a predominantly hydrophobic 'slot' in the protein. The ring is protonated on N_1 when bound to the enzyme (as shown by NMR; Cocco et al., 1981) and forms an ion-pair with the carboxylate of Asp26 (*L.*

casei numbering). The *p*-aminobenzoyl-L-glutamate moiety of methotrexate lies in a relatively shallow groove in the surface of the protein, with the benzoyl ring close to Phe49 and the α- and γ-carboxylate groups forming ion-pairs to Arg57 and His28, respectively. The contributions of these individual intermolecular interactions to the overall binding energy can only be assessed by making structural analogues of the small molecule or the protein in which one attempts to affect only one of the interactions at a time. An enormous range of dhfr inhibitors has been synthesized over the last 40 years (Roth and Cheng, 1982), while much more recently reliable protocols for making specific residue changes in the protein by oligonucleotide-directed mutagenesis have been developed (e.g., for dhfr, Andrews et al., 1989a). It is thus reasonably straightforward to make defined changes in the chemical structure of either partner in the interaction, and to measure the consequent changes in strength of binding. However, if these changes are to be reliably interpreted, the three-dimensional structures of the complexes must be compared in detail. Crystal structures are available for selected mutants of *Eschericia coli* and human dhfrs (Howell et al., 1986; A. J. Geddes, personal communication). In the case of *L. casei* dhfr, we have used NMR to make structural comparisons between complexes with different ligands (Hyde et al., 1980a,b; Antonjuk et al., 1984; Cheung et al., 1986; Hammond et al., 1987; Birdsall et al., 1989b) and between wild-type and mutant enzyme (Birdsall et al., 1989a; Jimenez et al., 1989a,b; Thomas et al., 1989), and some of our recent work is described here.

COMPARISON OF THE BINDING OF STRUCTURALLY RELATED LIGANDS

As outlined above, the binding of methotrexate is seen in the crystal structure (Bolin et al., 1982) to involve a number of important ion-pairs and a network of hydrogen bonds between enzyme and ligand which must contribute to binding and specificity. We have studied the His28-γ-carboxylate and Arg57-α-carboxylate interactions using appropriate methotrexate analogues (Antonjuk et al., 1984; Hammond et al., 1987). The γ-amide of methotrexate binds nine times less tightly to the enzyme than methotrexate itself. Of the assigned resonances from the protein, only those of His28 and Leu27 differ between the γ-amide and methotrexate complexes, indicating that these complexes are essentially isostructural, and that the difference in binding constant is a reasonable measure of the contribution of the His28-γ-carboxylate ion pair to the overall interaction energy. The α-amide of methotrexate binds 100-fold less tightly than methotrexate, but the differences in chemical shifts of the assigned resonances indicate that in this case the whole of the *p*-amino-benzoyl-glutamate moiety binds differently in the analogue, and the His28-γ-carboxylate ion-pair is also broken, so that the difference in binding energy cannot be simply interpreted (Antonjuk et al., 1984; Hammond et al., 1987).

The electrostatic interactions between protein and ligand also involve hydrogen bonds, and NMR offers an excellent means of monitoring these. For example, the carboxylate of Asp26 accepts hydrogen bonds from N_1-H and the 2-NH_2 of the bound inhibitor, while the second proton of the 2-amino group hydrogen bonds through a water molecule to Thr116. By using specifically ^{15}N-labelled trimethoprim (Bevan et al., 1985) and methotrexate (Arnold et al., 1989b), together with indirect detection experiments, we have located the resonances of the protons involved in all three of these hydrogen bonds, allowing us to monitor them very directly.

The packing of non-polar residues around the inhibitors, notably around the pteridine and benzoyl rings, makes an important contribution to binding. Among the residues making hydrophobic contacts with methotrexate are Leu4, Leu19, Leu27, Phe30, Phe49 and Leu54. Resonances of all these residues have been assigned, and their packing around the inhibitor is well defined by a network of NOEs. These NOEs, together with chemical shift changes, provide sensitive measures of changes in this packing produced by mutagenesis or by changes in inhibitor structure. For example, 3',5'-difluoro- and 3',5'-dichloro-methotrexate bind about 100-fold less tightly to the enzyme than methotrexate itself. Structural comparison of their complexes shows that the binding of the pteridine ring and of the glutamate moiety is unaffected, but the orientation of the benzoyl ring about its symmetry axis in the complex changes by up to 25° to accomodate the bulky substituents (Hammond et al., 1987). The change in binding constant can thus be attributed to changes in packing of non-polar residues around the benzoyl ring.

In addition to the comparison of structurally related ligands, NMR can be used to compare the binding of different conformational isomers of the same molecule. For example, we have studied the binding of a series of pyrimethamine analogues synthesised by Stevens and his colleagues at Aston. Rotation about the phenyl-pyrimidine bond of pyrimethamine is (as expected for an o-substituted biphenyl-like compound) slow, allowing the two conformational isomers of the inhibitor, related by a rotation of approx. 180° about the phenyl-pyrimidine bond, to be distinguished by NMR. In unsymmetrically substituted compounds such as 2,4-diamino-5-(3-nitro, 4-fluorophenyl)-6-ethyl pyrimidine we have found that the two conformational isomers of the inhibitor both bind to the enzyme. They have similar but not equal affinities for the enzyme, which are reversed by addition of $NADP^+$ (Tendler et al., 1988).

COMPARISON OF WILD-TYPE AND MUTANT ENZYMES

Comparison of the structures of mutant and wild-type dhfr both by X-ray crystallography and by NMR has revealed considerable variation in the structural consequences of amino-acid substitutions. Crystallographic studies of the Asp27 -> Asn (D27N) mutant[1] of *E. coli* dhfr (Howell et al., 1986) and NMR studies of the corresponding *L. casei* mutant, D26N (Jimenez et al., 1989a, b), show that any structural changes in the enzyme-methotrexate complex are very small and local (and may result from the displacement of a bound water molecule). Close structural similarity to the wild-type enzyme has also been observed for the D26E mutant (Birdsall et al., 1989a), although here the differences, while still local, are somewhat larger. In dhfr W21L, the majority of the assigned resonances are again unaffected. However, in this case small chemical shift differences are seen for resonances of seven residues which are not close to the site of substitution - some as much as 10 Å away (Birdsall et al., 1989a). Finally, the T63Q substitution involves a residue which makes a hydrogen bond to the 2'-phosphate of the co-enzyme. NMR analysis shows that the conformational effects are again slight, but, surprisingly, they are concentrated in the active site and *not* around the adenine binding site (Thomas et al., 1989). This is thus a striking example of the effects of a mutation being transmitted a substantial distance through a protein.

In collaboration with Dr. C. Fierke and Prof. S. Benkovic we have undertaken detailed studies of the kinetic mechanism of *L. casei* dhfr (Andrews et al., 1989b). At pH 6.5, the major contribution to the rate-limiting step is the dissociation of the product FH_4, and there is thus only a small ^2H isotope effect with $NADP^2H$; the contribution of hydride transfer to the rate-limiting step, and hence the magnitude of this isotope effect, increases with increasing pH. With two contributions to k_{cat} of similar magnitude, a detailed kinetic analysis of the mutant enzymes is important if the functional consequences are to be properly characterised. For example, dhfr W21L has a k_{cat} value 40-fold lower than that of the wild-type at pH 6.5. However, there is an increased $NADP^2H$ isotope effect on k_{cat} in this mutant, indicating a change in the contributions

---

[1] The nomenclature used to describe mutants with single amino-acid substitutions is based on the single-letter code for the amino-acids; thus a mutant in which aspartic acid at position 27 of the wild-type sequence has been replaced by asparagine is denoted D27N.

to the rate-limiting step. Detailed kinetic studies (Andrews et al., 1989b) reveal that the rate of FH_4 dissociation is unaffected in dhfr W21L, but the rate of hydride transfer has decreased by a factor of 100. An even more striking example is the L54G mutant of *E. coli* dhfr (Mayer et al., 1986), which has a k_{cat} essentially identical to that of the wild-type enzyme, but with hydride ion transfer rather than product dissociation as the rate-limiting step.

A significant role for non-polar contacts with the *p*-aminobenzoyl moiety of methotrexate in the overall binding of the inhibitor was implied by the comparison with the 3',5'-dihalo-methotrexates discussed above, and this is supported by the observation (Mayer et al., 1986) that methotrexate binds 300-fold less tightly to *E. coli* dhfr L54G, in which one of the side-chains involved in these non-polar protein-ligand contacts has been removed. A similar convergence of the effects of ligand and protein modification is seen in studies of *L. casei* dhfr W21L. The major effect of this substitution in terms of binding is on the coenzyme; the binding constant of NADPH is decreased 400-fold but that of NADP⁺ only 2-fold (Birdsall et al., 1989a). The indole ring of Trp21 makes contact with the amide substituent on the nicotinamide ring of the bound coenzyme, but a leucine side-chain at this position is too small to do so. This difference appears to lead to a change in the orientation of the nicotinamide ring in the site, as indicated by differences in chemical shift for the protons of this ring in the complex (Birdsall et al., 1989a), and hence to the observed decreases in k_{cat} and binding constant. There is a qualitative parallel between the effects of the W21L substitution and those of replacing the nicotinamide carboxamide by a thioamide to give thioNADPH (Birdsall et al., 1980a). In both cases there is a marked decrease in K_a and k_{cat}, and an apparent change in the mode of binding of the nicotinamide ring; in both cases these effects probably arise from a perturbation of the interaction between Trp 21 and the nicotinamide. The marked difference in the effect of the amino-acid substitution on the binding of NADPH and NADP⁺ (Birdsall et al., 1989a), which is paralleled by the much smaller difference in binding constant between thioNADP⁺ and NADP⁺ (Birdsall et al., 1980a), may be explained by our observation (e.g., Birdsall et al., 1980a,b) that the oxidised nicotinamide ring makes little net contribution to the binding of NADP⁺.

A particularly important interaction between methotrexate and the protein is that involving Asp26 and the protonated pteridine ring of methotrexate (Cocco et al., 1981) or the corresponding protonated pyrimidine ring of trimethoprim (Roberts et al., 1981). By contrast, the pteridine ring of the substrate, folate, binds in a different orientation (see below), and in the uncharged state (Hood and Roberts, 1978). For both inhibitors, the pK of N_1 is markedly increased on binding (Cocco et al., 1981; Roberts et al., 1981), showing that the charged form binds much more tightly (as much as 10^5-fold) to the enzyme than does the neutral form. The D27N mutant of *E. coli* dhfr binds methotrexate only 27-fold less tightly than the wild-type at neutral pH (Howell et al., 1986), but it is difficult to use this comparison to assess the contribution of the ion-pair, since methotrexate is in the *neutral* (high pH) state in the complex (Howell et al., 1986), but the asparagine residue 'mimics' the *protonated* (low pH) form of aspartate (see below).

Up to now, the structural comparisons of mutant dhfrs and of complexes with different ligands have largely been made on the basis of chemical shift comparisons. These provide a simple and sensitive method for the detection of structural differences, but it is generally very difficult to use observed shift differences to describe the precise *nature* of the difference in conformation. As is widely recognised, the single NMR parameter which can most easily be interpreted precisely in structural terms is the nuclear Overhauser effect (NOE). Our preliminary results (Thomas et al., 1989) suggest that NOEs can be quantitated with sufficient accuracy to permit a detailed description of these (generally) small structural differences. The prospect of describing changes in protein structure in solution with a precision of ±0.2-0.3 Å is an exciting one, and work is in progress to develop the necessary methods.

STRUCTURAL DYNAMICS

Proteins and their complexes have structures which fluctuate at a wide range of rates; since molecular recognition is, by its nature, a dynamic process, it is important to characterise these fluctuations as well as the time-average structures of the complexes. NMR has permitted the identification and partial characterisation of structural fluctuations in dhfr at rates ranging from 1 s^{-1} to 10^9 s^{-1}. Slow conformational equilibria, discussed below, have been observed in the enzyme-trimethoprim-NADP$^+$ (Gronenborn et al., 1981; Birdsall et al., 1984), enzyme-folate-NADP$^+$ (Birdsall et al., 1982, 1987, 1989b) and enzyme-folinic acid (Birdsall et al., 1981) complexes.

Fluctuations in the conformations of bound ligands have been characterised for 3',5'-difluoromethotrexate (Clore et al., 1984) and trimethoprim (Searle et al., 1988). Lineshape analysis shows that the benzoyl ring of 3',5'-difluoromethotrexate rotates (or rather, more probably, 'flips') about its symmetry axis by 180° with a frequency of approximately 10^3 s^{-1} at room temperature (Clore et al., 1984). This frequency is determined, in large measure, by the packing of neighbouring amino-acid side-chains around the ring and their ability to move so as to allow the ring to flip, and it thus provides a sensitive measure of changes in non-polar inter-actions around the ring. For example, the binding of the coenzyme NADPH leads to conform-ational changes, one of which involves a movement of helix C (Hammond et al., 1986), residues 42-49, whose C-terminal residue, Phe49, is in contact with the benzoyl ring; the changes in packing around the ring resulting from this coenzyme-induced conformational change are reflected in a 3-fold *increase* in the rate of ring flipping (Clore et al., 1984).
Similar studies of the symmetrically substituted benzyl ring of trimethoprim bound to dhfr have been made using [*m*-methoxy ^{13}C]-trimethoprim (Searle et al., 1988). Relaxation measurements reveal fluctuations of ±25-35° at rates of about 10^9 s^{-1}, comparable to those seen for aromatic rings of phenylalanine and tyrosine residues in proteins, both experimentally and in molecular dynamics simulations. Again as for the aminoacid residues, 180° 'flips' about the symmetry axis of the ring are much rarer events; lineshape analysis of the *m*-methoxy ^{13}C resonances shows that they occur at 250 s^{-1} at room temperature (Searle et al., 1988). In trimethoprim, a major contribution to the barrier to this 'flipping' comes from steric interactions *within* the trimethoprim molecule, between the benzyl and pyrimidine rings. Molecular mechanics calculations show that the 'flipping' of the benzyl ring must be accompanied by a significant rotation (of about 20°) about the C5-C7 bond so as to change the relative orientation of the two rings in bound trimethoprim; this in turn requires significant movements of the surrounding protein structure. The use of ^{15}N-labelled trimethoprim to identify the resonance from the proton in the N$_1$H-Asp26 hydrogen bond has been mentioned above; lineshape analysis of this resonance yields an estimate of the rate of exchange of this proton with the solvent, and hence of the rate of making and breaking of this crucial hydrogen-bond between inhibitor and enzyme. This occurs at a rate of 34 s^{-1} at room temperature (Searle et al., 1988), and clearly results from a fluctuation in the protein structure distinct from that involved in the ring flipping. It seems likely that when *both* fluctuations happen to occur simultaneously then the trimethoprim will dissociate from the enzyme; the measured dissociation rate constant at 298K is 1.7 s^{-1} (Cayley et al., 1979).

The rates of all these fluctuations in the structure of the enzyme-trimethoprim complex are affected by coenzyme binding (Searle et al., 1988), and also by amino-acid substitutions. The mutant dhfr D26E shows close structural and functional similarity to the wild-type enzyme (Birdsall et al., 1989a; Andrews et al., 1989b), but this is accompanied by clear differences in the *dynamics* of bound trimethoprim. The rates of the conformational fluctuations discussed above are increased by as much as 30-fold in this mutant (Birdsall et al., 1989a). It appears that structural fluctuations are significantly more sensitive to the effects of amino-acid substitutions than is the time-average structure.

COOPERATIVITY BETWEEN COENZYME AND SUBSTRATE ANALOGUE BINDING

Many substrates and substrate analogues bind more tightly (by up to 600-fold) to dhfr in the presence of coenzyme than in its absence (Birdsall et al., 1980b); the magnitude (and indeed the sign; see below) of this effect depends critically on the structure of both coenzyme and substrate analogue. Some contribution to this cooperativity no doubt comes from the van der Waals' contact between the nicotinamide ring of the coenzyme and the pteridine ring of the substrate or inhibitor in the ternary complex (Bolin et al., 1982; Filman et al., 1982). For example, it is interesting that the W21L substitution which, as noted above, leads to a change in the orientation of the nicotinamide ring also leads to abolition of the cooperativity between NADP+ and folate binding (Andrews et al., 1989b). However, the observations that cooperativity is also shown by the coenzyme analogue PADPR-OMe (the methyl ß-riboside of 2'-phosphoadenosine-5'-diphosphoribose), which lacks a nicotinamide ring (Birdsall et al., 1980b), and that resonances of inhibitor nuclei remote from the coenzyme are affected by coenzyme binding (Cheung et al., 1986), demonstrate that there is also a contribution from effects transmitted through the protein. Clear evidence for coenzyme-induced conformational changes has come from studies of the effects of coenzyme binding on individual assigned resonances from the protein (Hammond et al., 1986), and these have also allowed a preliminary description of some of the conformational changes. For example, NADPH binding appears to lead to an axial movement of helix C, which runs from the binding site for the adenine ring of the coenzyme to that for the p-aminobenzoyl ring of the substrates and inhibitors (Hammond et al., 1986). The mutant dhfr T63Q, with a substitution in the adenosine binding site, shows increased cooperativity between coenzyme and substrate analogues (Thomas et al., 1989). Quantitative understanding of the role of these conformational changes will require detailed structural and energetic analysis of carefully chosen mutant dhfrs.

Negative cooperativity also plays a role in the mechanism of the enzyme. The most striking feature of the kinetic mechanism of dhfr (Andrews et al., 1989b) is that, in the kinetically preferred pathway for product release, tetrahydrofolate (FH_4) dissociates only *after* NADPH has bound. This originates from the marked negative cooperativity in binding between FH_4 and NADPH; FH_4 dissociates 300 times faster from the E.NADPH.FH_4 complex than from E.FH_4. An interesting consequence of this mechanism is that the free enzyme is not on the kinetically preferred pathway. Very similar negative cooperativity between NADPH and the stable product analogue 5-formyl-FH_4 was earlier demonstrated by NMR and binding constant measurements (Birdsall et al., 1981). The natural $6R,\alpha S$ diastereoisomer of 5-formyl-FH_4 has a binding constant of 1.3×10^8 M^{-1} for the enzyme alone, but it binds 3 times more weakly in the presence of NADP+ and 600 times more weakly in the presence of NADPH. It is notable that the single atom change between NADP+ and thioNADP+ converts negative into positive cooperativity. Negative cooperativity was unequivocally distinguished from competition by measurements of dissociation rate constants (Birdsall et al., 1981). The negative cooperativity must result from ligand-induced conformational changes, and NMR studies of coenzyme signals, for example, show that these changes are qualitatively distinct from those associated with the positive cooperativity between NADPH and methotrexate, and that the enzyme-coenzyme-5-formyl-FH_4 complex exists in more than one conformational state (Birdsall et al., 1981). These differences are presumably related to the different mode of binding of substrates and inhibitors (see below).

Thus, notwithstanding the small size and apparent simplicity of dihydrofolate reductase, both large positive and large negative free-energy coupling between coenzyme and substrate analogue binding has been observed. For NADPH and methotrexate, ΔG_{coop} = -3.84 kcal/mole, while for NADPH and 5-formyl-FH_4, ΔG_{coop} = +3.76 kcal/mole (Birdsall et al., 1980b, 1981). Couplings of this magnitude lead to substantial effects on the distribution of ligands between binary and ternary complexes. Consider solutions containing concentrations of NADPH and either methotrexate or 5-formyl-FH_4 sufficient to half-saturate the enzyme with each ligand. Under these conditions, if binding of coenzyme and substrate analogue were independent (ΔG_{coop} = 0), the population of the ternary complex would be 0.25. However, with the free energy couplings observed for NADPH and methotrexate the population of the ternary complex would be 0.48, while for NADPH and 5-formyl-FH_4 it would be 0.02.

CONFORMATIONAL EQUILIBRIA

The inhibitor methotrexate differs structurally from the substrate folate in only two ways: in having a methyl group on N_{10} (which does not affect binding to dhfr) and, crucially, in having a 4-amino rather than a 4-oxo substituent on the pteridine ring. In spite of this close structural similarity, not only does methotrexate bind 10^5 times more tightly to the enzyme, but the stereochemistry of reduction implies a difference of 180° in the orientation of the pteridine ring on the enzyme between substrate and inhibitor (Hitchings and Roth, 1980; Charlton et al., 1979, 1985). We have shown that both the enzyme-folate and the enzyme-folate-NADP+ complexes exist in solution as a mixture of two or three conformations which interconvert sufficiently slowly for separate NMR signals to be observed, and whose proportions are pH-dependent (Birdsall et al., 1982, 1987, 1989b). Comparisons of chemical shifts between the three conformations of the dhfr-folate-NADP+ complex indicate that the structural differences are localised to the active site region, close to the pteridine and nicotinamide rings. As noted above in another context, these chemical shift differences cannot be interpreted unambiguously to define the structural differences between the three conformations. Two-dimensional NOE experiments (Figure 1; Birdsall et al., 1989b) have recently allowed us to identify one crucial structural difference between these conformations: the orientation of the pteridine ring. The first step is to identify the resonance position of the pteridine 7-proton in each of the three conformations of the complex; this is done by means of a two-dimensional exchange experiment, relying on magnetization transfer by chemical exchange of folate molecules between the free state and the three bound states (Figure 1b). Having identified these resonance positions, one can then seek NOEs between the 7-proton and protons of the protein in each of the three conformations. In the conformations denoted I and IIa, NOEs are observed between the pteridine 7-proton of bound folate and the two methyl groups of Leu27. Similar NOEs are observed for the 7-proton of methotrexate in the enzyme-NADP+-methotrexae complex (although in the latter case, but not the former, NOEs are also observed to the methyl protons of Leu19). In conformations I and IIa of the enzyme-folate-NADP+ complex, the pteridine ring of folate therefore has an orientation in the binding site similar (but not identical) to that seen for methotrexate in the crystal structure. In conformation IIb, by contrast, no NOEs are observed between the pteridine 7-proton and these methyl groups; given the structural constraints of the pteridine ring binding site, this can only be explained if the ring has turned over by approximately 180° about an axis along the C2-NH_2 bond (Birdsall et al., 1989b).

The stereochemistry of reduction indicates that only the latter is a catalytically productive mode of binding for the substrate. The substrate is thus able to bind to dhfr with very similar affinity in both productive and non-productive orientations; the inhibitor methotrexate binds only in the non-productive orientation, in which it can form the ion-pair with Asp26 discussed above.

By contrast, the complex of folate and NADP+ with the mutant dhfr D26N exists in only a single conformation over the pH range 5.5-7.5 (Jimenez et al., 1989a); this strongly suggests

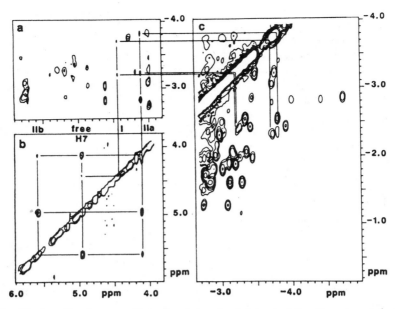

Figure 1.   Nuclear Overhauser effects involving the folate pteridine 7-proton and methyl groups of the protein in the three conformations of the dhfr-folate-NADP+ complex. Part b shows the aromatic region of the 2D exchange spectrum obtained from a sample containing the complex and a 2-fold excess of free folate. The cross-peaks linking the 7-proton of free folate at 4.95 ppm to the resonances of the same proton in the three conformations of the complex are connected by lines. Part a shows the region of a 2D NOESY spectrum containing cross-peaks between aromatic and methyl proton resonances. Cross-peaks involving the 7-proton resonance in conformations I and IIa are indicated; there are no cross-peaks in this region of the spectrum involving the 7-proton resonance in conformation IIb. Part c shows the high-field region of the COSY spectrum of the complex. The lines connect the resonances of the 7-proton in conformations I and IIa (in part b) with their NOESY cross-peaks in part a and with the corresponding COSY cross-peaks in part c. The latter help to identify the methyl protons giving NOEs to the 7-proton as those of Leu27. The spectra are referenced to internal dioxan. From Birdsall et al., 1989b with permission.

that Asp26 is the group responsible for the pH-dependence of the conformational equilibrium of this complex. The single conformation seen for this mutant corresponds to the low-pH conformer of the wild-type complex (Jimenez et al., 1989a), as one would expect from a comparison of the aspartate and asparagine sidechains. The very low k_{cat} for dhfr D26N (ca. 1% of wild-type) thus arises not only because Asp26 is, most probably, a proton donor in the reaction (Howell et al., 1986), but also because the mutant is 'locked' in a catalytically non-functional conformation.

The complex between dhfr, trimethoprim and NADP+ also exists in a conformational equilibrium; this is independent of pH, and involves primarily the nicotinamide ring of the coenzyme (Gronenborn et al., 1981; Birdsall et al., 1984; Cheung et al., 1986). In one conformation this ring is bound specifically to its site on the enzyme, but in the other, rotations about the nicotinamide ribose C5'-O and pyrophosphate O-P bonds have altered the conformation of the coenzyme so that the nicotinamide ring is no longer in contact with the enzyme, but is hanging free in solution, the coenzyme being bound solely through its adenosine moiety. The fact that these two conformations are almost equally populated supports the idea (Birdsall et al., 1980a) that the oxidised nicotinamide ring contributes very little to the overall binding energy. It seems likely that the favourable interactions, notably two hydrogen bonds to the protein formed by the carboxamide, are balanced by the energetically unfavourable desolvation of the charged ring. Due to this fine balance, the relative populations of the two conformations are readily altered by changes in coenzyme or inhibitor structure or by amino-acid substitutions in the protein. For example, in the dhfr-trimethoprim-thioNADP+ complex or the dhfr W21L-trimethoprim-NADP+ complex only the conformation in which the nicotinamide ring is not in contact with the enzyme is detectable (Birdsall et al., 1984, 1989a), reflecting the loss or perturbation of the Trp21-amide interaction.

CONCLUSIONS

The specificity of dihydrofolate reductase has been investigated by the full range of tools presently available to the protein scientist, from X-ray crystallography and NMR spectroscopy to site-directed mutagenesis. Our understanding of the system has reached the point where we can ask very precise questions about the details of the recognition process. Nonetheless, a truly quantitative understanding of protein specificity remains elusive, even in a system as well-characterised as this. The conformational flexibility of proteins, and the subtlety of their responses to amino-acid substitutions and changes in ligand structure, continue to present substantial challenges.

It is clear that NMR has much to contribute to the solution of these problems. It allows one to monitor individual protein-ligand interactions, such as hydrogen bonds, it provides a convenient and sensitive means for comparing the structure of wild-type and mutant proteins in solution, and it permits the characterisation of dynamic processes in proteins and their complexes over a wide range of rates. The limitations of the technique, for example in the size of protein that can be tackled and in the quantitative structural interpretation of the data, are gradually becoming less restrictive, with improvements in relaxation and NOE analysis, the continuing development of higher-field spectrometers, and the use of ^{13}C and ^{15}N labelling in conjunction with indirect detection experiments. The crucial strength of NMR, that it provides dynamic and structural information simultaneously, under conditions where equilibrium constants can also be readily measured, assures it of a continuing role in studies of protein specificity.

ACKNOWLEDGEMENTS

The work described here is part of a continuing collaboration with Jim Feeney and Berry Birdsall (Mill Hill) and Julie Andrews (Manchester). Many people, whose names appear in the reference list, have made essential contributions; my thanks to all of them, and particularly to John Arnold, Janette Thomas, Maria Jimenez and Lu-Yun Lian at Leicester. The work at Leicester has been supported by SERC (Protein Engineering Club) and the Wellcome Trust.

REFERENCES

Andrews, J., Sims, P.F.G., Minter, S., and Davies, R. W., 1989a, submitted for publication.
Andrews, J., Fierke, C.A., Birdsall, B., Ostler, G., Feeney, J., Roberts, G.C.K., and Benkovic, S.J., 1989b, A kinetic study of wild-type and mutant dihydrofolate reductases from *Lactobacillus casei, Biochemistry* 28:5743.

Antonjuk, D.J., Birdsall, B., Cheung, H.T.A., Clore, G.M., Feeney, J., Gronenborn, A.M., Roberts, G.C.K., and Tran, T.Q., 1984, A ^{1}H NMR study of the role of the glutamate moiety in the binding of methotrexate to dihydrofolate reductase, *Br. J. Pharmacol.* 81:309.

Arnold, J.R.P., Tendler, S.J.B., Thomas, J.A., Birdsall, B., Feeney, J., and Roberts, G.C.K., 1989a, in preparation.

Arnold, J.R.P., De Graw, J., Lian, L.-Y., and Roberts, G.C.K., 1989b, unpublished work.

Bevan, A.W., Roberts, G.C.K., Feeney, J., and Kuyper, L., 1985, ^{1}H and ^{15}N NMR studies of protonation and hydrogen-bonding in the binding of trimethoprim to dihydrofolate reductase, *Eur. J. Biophys.* 11:211.

Birdsall, B., Burgen, A.S.V., and Roberts, G.C.K., 1980a, Binding of coenzyme analogues to *Lactobacillus casei* dihydrofolate reductase: Binary and ternary complexes, *Biochemistry* 19:3723.

Birdsall, B., Burgen, A.S.V., and Roberts, G.C.K. 1980b, Effects of coenzyme analogues on the binding of p-aminobenzoyl-L-glutamate and 2,4-diaminopyrimidine to *Lactobacillus casei* dihydrofolate reductase, *Biochemistry* 19:3732.

Birdsall, B., Burgen, A.S.V., Hyde, E.I., Roberts, G.C.K., and Feeney, J., 1981, Negative cooperativity between folinic acid and coenzyme in their binding to *Lactobacillus casei* dihydrofolate reductase, *Biochemistry* 20:7186.

Birdsall, B., Gronenborn, A.M., Hyde, E.I., Clore, G.M., Roberts, G.C.K., Feeney, J., and Burgen, A.S.V., 1982, ^{1}H, ^{13}C and ^{31}P NMR studies of the dihydrofolate reductase-NADP$^+$-folate complex: Characterisation of three coexisting conformational states, *Biochemistry* 21:5831.

Birdsall, B., Bevan, A.W., Pascual, C., Roberts, G.C.K., Feeney, J., Gronenborn, A.M., and Clore, G.M., 1984, Multinuclear NMR characterisation of two coexisting conformational states of the *Lactobacillus casei* dihydrofolate reductase-trimethoprim-NADP$^+$ complex,*Biochemistry* 23:4733.

Birdsall, B., De Graw, J., Feeney, J., Hammond, S.J., Searle, M.S., Roberts, G.C.K., Colwell, W.T., and Crase, J., 1987, ^{15}N and ^{1}H NMR evidence for multiple conformations of the complex of dihydrofolate reductase with its substrate, folate, *FEBS Lett.* 217:106.

Birdsall, B., Andrews, J., Ostler, G., Tendler, S.J.B., Feeney, J., Roberts, G.C.K., Davies, R.W., and Cheung, H.T.A., 1989a, NMR studies of differences in the conformations and dynamics of ligand complexes formed with mutant dihydrofolate reductases, *Biochemistry* 28:1353.

Birdsall, B., Feeney, J., Tendler, S.J.B., Hammond, S.J., and Roberts, G.C.K., 1989b, Dihydrofolate reductase: multiple conformations and alternative modes of substrate binding, *Biochemistry* 28:2297.

Bolin, J.T., Filman, D.J., Matthews, D.A., Hamlin, R.C., and Kraut, J., 1982, Crystal structures of *Escherichia coli* and *Lactobacillus casei* dihydrofolate reductase refined at 1.7 Å resolution, *J. Biol. Chem.* 257:13650.

Cayley, P.J., Albrand, J.P., Feeney, J., Roberts, G.C.K., Piper, E.A., and Burgen, A.S.V., 1979, Nuclear magnetic resonance studies of the binding of trimethoprim to dihydrofolate reductase, *Biochemistry* 18:3886.

Charlton, P.A., Young, D.W., Birdsall, B., Feeney, J., and Roberts, G.C.K., 1979, Stereochemistry of reduction of folic acid using dihydrofolate reductase, *J. Chem. Soc. Chem. Comm.* 922.

Charlton, P.A., Young, D.W., Birdsall, B., Feeney, J., and Roberts, G.C.K., 1985, Stereochemistry of the reduction of the vitamin folic acid by dihydrofolate reductase, *J. Chem. Soc. Perkin I* 1349.

Cheung, H.T.A., Searle, M.S., Feeney, J., Birdsall, B., Roberts, G.C.K., Kompis, I., and Hammond, S.J., 1986, Trimethoprim binding to *Lactobacillus casei* dihydrofolate reductase: a ^{13}C NMR study using selectively ^{13}C-enriched trimethoprim, *Biochemistry* 25:1925.

Clore, G.M., Gronenborn, A.M., Birdsall, B., Feeney, J., and Roberts, G.C.K., 1984, ^{19}F NMR studies of 3'5'-difluoremethotrexate biding to *Lactobacillus casei* dihydrofolate reductase. Molecular motion and coenzyme induced conformational change, *Biochem. J.* 217:659.

Cocco, L., Groff, J.P., Temple, jr., C., Montgomery, J.A., London, R.E., Matwiyoff, N.A., and Blakley, R.A., 1981, ^{13}C NMR study of protonation of methotrexate and aminopterin bound to dihydrofolate reductase, *Biochemistry* 20:3972.

Feeney, J., Birdsall, B., Akiboye, J., Tendler, S.J.B., Barbero, J.J., Ostler, G., Arnold, J.R.P., Roberts, G.C.K., Kuhn, A., and Roth, K., 1989, Optimising selective deuteration of proteins for 2D ^{1}H NMR detection and assignment studies: Application to the phenylalanine residues of dihydrofolate reductase, *FEBS Letts.*, in press.

Filman, D.J., Bolin, J.T., Matthews, D.A., and Kraut, J., 1982, Crystal structures of *Escherichia coli* and *Lactobacillus casei* dihydrofolate reductase refined at 1.7 Å resolution II. *J. Biol. Chem.* 257:13663.

Hammond, S.J., Birdsall, B., Searle, M.S., Roberts, G.C.K., and Feeney, J., 1986, Dihydrofolate reductase: ^{1}H resonance assignments and coenzyme-induced conformational changes, *J. Mol. Biol.* 188:81.

Hammond, S.J., Birdsall, B., Feeney, J., Searle, M.S., Roberts, G.C.K., and Cheung, H.T.A., 1987, Structural comparisons of complexes of methotrexate analogues with *L. casei* dihydrofolate reductase by 2D ^{1}H NMR at 500 MHz, *Biochemistry* 26:8585.

Hitchings, G.H., and Roth, B., 1980, Inhibition of dihydrofolate reductase, *in:* "Enzyme Inhibitors as Drugs," M. Sandler, ed., Macmillan, London.

Hood, K., and Roberts, G.C.K., 1978, Ultraviolet difference spectroscopic studies of substrate and inhibitor binding to *Lactobacillus casei* dihydrofolate reductase, *Biochem. J.* 171:357.

Howell, E.E., Villafranca, J.E., Warren, M.S., Oatley, S.J., and Kraut, J., 1986, Functional role of aspartic acid 27 in dihydrofolate reductase revealed by mutagenesis,*Science* 231:1123.

Hyde, E.I., Birdsall, B., Roberts, G.C.K., Feeney, J., and Burgen, A.S.V., 1980a, Proton nuclear magnetic resonance saturation transfer studies of coenzyme binding to *Lactobacillus casei* dihydrofolate reductase, *Biochemistry* 19:3738.

Hyde, E.I., Birdsall, B., Roberts, G.C.K., Feeney, J., and Burgen, A.S.V., 1980b, Phosphorus-31 nuclear magnetic resonance studies of the binding of oxidised coenzymes to *Lactobacillus casei* dihydrofolate reductase, *Biochemistry* 19:3747.

Gronenborn, A.M., Birdsall, B., Hyde, E.I., Roberts, G.C.K., Feeney, J., and Burgen, A.S.V., 1981, ^{1}H and ^{31}P NMR characterisation of two conformations of the trimethoprim-NADP$^+$-dihydrofolate reductase complex, *Mol. Pharmacol.* 20:145.

Jimenez, M.A., Arnold, J.R.P., Thomas, J.A., Roberts, G.C.K., Feeney, J., and Birdsall, B., 1989a, Dihydrofolate reductase: Control of the mode of substrate binding by aspartate 26, *Protein Engineering* 2:627.

Jimenez, M.A., Arnold, J.R.P., Thomas, J.A., G.C.K. Roberts, J. Feeney & B. Birdsall, 1989b, in preparation.

Matthews, D.A., Bolin, J.T., Burridge, J.M., Filman, D.J., Volz, K.W., Kaufman, B.T., Beddell, C.R., Champness, J.N., Stammers, D.K., and Kraut, J., 1985, Refined crystal structures of *Escherichia coli* and chicken liver dihydrofolate reductase containing bound trimethoprim, *J. Biol. Chem.* 260:381.

Mayer, R.J., Chen, J.-T., Fierke, C.A., and Benkovic, S.J., 1986, Importance of a hydrophobic residue in binding and catalysis by dihydrofolate reductase, *Proc. Natl. Acad. Sci. USA* 83:7718.

Oefner, C., D'Arcy, A., and Winkler, F.K., 1988, Crystal structure of human dihydrofolate reductase complexed with folate, *Eur. J. Biochem.* 174:377.

Roberts, G.C.K., Feeney, J., Burgen, A.S.V., and Daluge, S., 1981, The charge state of trimethoprim bound to *Lactobacillus casei* dihydrofolate reductase, *FEBS Lett.* 131:85.

Roth, B., and Cheng, C.C., 1982, Recent progress in the medicinal chemistry of 2,4-diaminopyrimidines, *Prog. Med Chem.* 19:270.

Searle, M.S., Forster, M.J., Birdsall, B., Roberts, G.C.K., Feeney, J., Cheung, H.T.A., Kompis, I., and Geddes, A.J., 1988, The dynamics of trimethoprim bound to dihydrofolate reductase, *Proc. Natl. Acad. Sci. USA* 85:3787.

Tendler, S.J.B., Griffin, R.J., Birdsall, B., Stevens, M.F.G., Roberts, G.C.K., and Feeney, J., 1988, Direct ^{19}F NMR observation of the conformational selection of optically active rotamers of the antifolate compound fluoronitrophyrimethamine bound to the enzyme dihydrofolate reductase, *FEBS Lett.* 240:201.

Thomas, J.A., Andrews, J., Arnold, J.R.P., Roberts, G.C.K., Birdsall, B., and Feeney, J., 1989, unpublished work.

Volz, K.W., Matthews, D.A., Alden, R.A., Freer, S.T., Hansch, C.T., Kaufman, B.T., and Kraut, J., 1982, Crystal structure of avian dihydrofolate reductase containing phenyltriazine and NADPH, *J. Biol. Chem.* 257:2528.

HISTONE H1 SOLUTION STRUCTURE AND THE SEALING OF MAMMALIAN NUCLEOSOME

Claudio Nicolini, Paolo Catasti, Mario Nizzari and Enrico Carrara

Institute of Biophysics, University of Genova School of Medicine, Italy

INTRODUCTION

The lysine-rich chromosomal protein histone H1 has been shown to be critically related to the control of the tertiary (nucleosome) and quaternary (solenoid or rope-like) structure of mammalian chromosomes (Nicolini, 1983; Dolby et al., 1981; Bradbury and Baldwin, 1986; Nicolini, 1986).

In solution and under physiological conditions of pH and ionic strength, histone H1 (Figure 1) is well known to consists of three structural regions, two random-coil N-terminal and C-terminal regions and a central globular region which is highly structured and conserved (Chapman et al., 1975). This globular part or "core", which extends from residues 39 (±4) to 116 (±4), with a molecular weight of about 9000 Daltons, has an unknown function, although it has been suggested that it seals the DNA on the nucleosome surface (Bradbury and Baldwin, 1986).

Little is known at the atomic resolution about the secondary and tertiary structure of the globular domain of H1 histone, and so far no high quality crystals (i. e. diffracting to better than 5 Å resolution for X-ray crystallographic studies) have been reported.

In this report we present an overview of our present understanding of the structure of the globular part of histone H1 up to the atomic resolution, as recently obtained by a variety of complementary experimental probes (Pepe et al., 1989) and by a combination of statistical methods (Rauch et al., 1989) and molecular modeling.

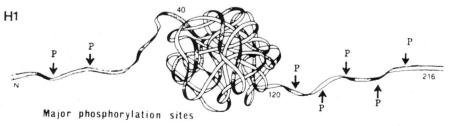

Figure 1.   Model of H1 histone structure, showing the globular "core" and the two random-coil regions.

Structural details of the globular domain of H1 histone were sought by 2DFT protein NMR spectra taken at high resolution (Nicolini et al., 1989), while its alterations due to changes in pH and temperature were monitored by Circular Dichroism (CD) spectra, differential scanning calorimetry and 1DFT proton NMR spectroscopy (Pepe et al., 1989).

^{31}P and ^1H NMR relaxation parameters of nucleosomes isolated from mammalian cells (Nicolini et al., 1989), as a function of the selective progressive removal of the bound chromosomal proteins (HMG, H1, H2A, H2B, H3 and H4), confirm the critical role of H1 histone core in the control of DNA internal motion within nucleosomes.

This observation, combined with earlier experimental evidences on the role of the two random coil tails and their enzymatic modifications in determining the quaternary polynucleosome superfolding and cell function (Dolby et al., 1981; Bradbury and Baldwin, 1986), suggests that the engineering of H1 histone might constitute an unique model system to study structure-function relationships towards the control of nucleosome sealing and ultimately of human chromosome structure and expression.

METHODS

Preparation of H1 Histone

The H1 histone was extracted from calf liver following the method of Sanders with minor modifications (Sanders, 1977).

The globular part of histone H1 was obtained by digestion, with trypsin (Barbero et al., 1980) added to give an enzyme/substrate ratio of 1:1000 (w/w).

After incubation at 24°C for 90 min., the purity of H1 histone core obtained with this procedure was controlled on SDS polyacrylamide gel. The three H1 subfractions, H1a, H1b and H1$_0$ (up to 12%), reduce to one broad band after trypsin digestion. The reaction was terminated adding soybean trypsin inhibitor to achieve an inhibitor/trypsin ratio of 2:1 (w/w). Cold trichloracetic acid was added up to a final concentration of 18% and the precipitate was centrifuged at 10,000g for 15 min., washed once with 18% trichloracetic acid, once with acidified acetone (1N HCl in acetone) and 2 times with acetone. The final precipitate containing the H1 core was then dried under vacuum and stored at -20°C (Barbero et al., 1980).

Circular Dichroism Spectra

Circular Dichroism (CD) measurements were made on a Spectropolarimeter (Jasco model J-500A) equipped with a DP-500N data processor, and recorded at 25°C.

Samples of 3 µM solution of H1 core were prepared, dissolving the protein in 0.3 M NaF and 10 mM buffer at different pH.

To correct for inactive sample artifacts, the instrument was calibrated by using a solution of microspheres of sizes and concentrations (total absorbance) similar to those in H1 solutions; difference spectra were recorded. Results are expressed in molar ellipticities with the dimension of deg. x cm^2/dmol. The percentage of α-helix (A), β-sheet (B), turn (T) and random-coil (C) were determined by solving the coefficients of a series of simultaneous equations (Yang et al., 1986):

$$X(\lambda) = f_A X_A^N(\lambda) + f_B X_B(\lambda) + f_T X_T(\lambda) + f_C X_C(\lambda)$$

where $X(\lambda)$ is the experimental CD spectrum of the protein at each wavelength; f_A, f_B, f_C, f_T, are the unknown coefficients representing the percentage of A,B,T and C conformations; X_A^N, X_B, X_C, X_T are the corresponding reference CD spectra based on a standard of known proteins (Cheng et al., 1974) and N is the mean number of residues per α-helix. $X_A^N(\lambda)$ takes into account the chain-length dependence of α-helices (Chang et al., 1978).

Differential Scanning Calorimetry (DSC)

A Setaram micro DSC model was used for measuring the heat connected with intramolecular transition of the protein fragment. Samples of up to 20 mg of histone H1 core were dissolved in 1 ml of 0.01 M Tris-HCl pH 7 and 0.35 M K_2SO_4. Cuvettes of 1 ml of volume were used. Temperature scanning was performed from 25°C to 50°C.

2DFT-NMR Spectroscopy

Proton NMR spectra were recorded on a Bruker AM-500 spectrometer at the Stanford Magnetic Resonance Laboratory (Nicolini et al., 1989) using solution of 2.0 mM of H1 core in D_2O or 90% H_2O - 10% D_2O at 25°C, 0.35 M K_2SO_4, pH=4.1. Double-quantum-filtered

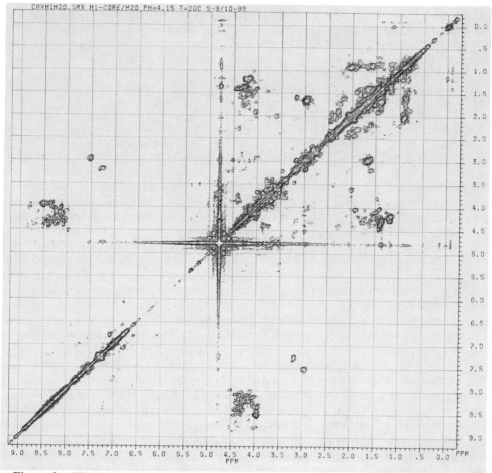

Figure 2.   H1 histone absolute-value COSY in H_2O as described on the following page.

COSY (Rance et al., 1984) and NOESY (Jeener et al., 1979) spectra with 100, 200 and 300 ms mixing times were acquired in the pure-phase mode with time-proportional phase incrementation TPPI (Marion and Wüthrich, 1983).

Typical acquisition parameters were: 2048 FID data points, 512 t_1-increments, spectral widths of 7042 and 5495 Hz for H_2O and D_2O-solutions respectively, 1.8-2.5s relaxation delay. The COSY experiments were recorded either in phase-sensitive or in absolute-value mode (Figure 2). The NOESY experiments were recorded in phase-sensitive mode (Jeener et al., 1979) using the TPPI method (Marion and Wüthrich, 1983). The carrier was placed at the position of the solvent. The residual HDO-resonance was suppressed by low-power CW-preirradiation during the relaxation delay (DQF-COSY) and the relaxation and mixing time periods (NOESY). Chemical shifts are referenced to internal sodium trimethylsilyl-propionate (TSP). The data were processed using one and no zero filling in the t_1 and t_2-dimensions, respectively, and applying shifted sine-bell or shifted squared sine-bell weighting functions in both dimensions before 2D Fourier-transform. The NOESY spectra were baseline-corrected, if necessary, in the F2-dimension using a polynomial fitting routine (ABS in the Bruker DISNMR software).

Alternatively, the COSY experiment was processed with a Lorentz-Gauss transformation in each dimension convoluted with a trapezoidal multiplication in the t2 dimension in order to avoid baseline artifacts due to incorrect first points of the FIDs. The NOESY experiment was processed with a Lorentz-Gauss transformation in the t2 dimension and a 45° shifted square sine-bell multiplication in the t1. All the experiments were also zero-filled in the t1 dimension in order to obtain a final square matrix of 2k per 2k data points with a digital resolution in each dimension of 3.44 Hz/pt. for H_2O and 2.68 Hz/pt. for D_2O.

PEPTO

For the peak assignment of COSY and NOESY spectra and the determination of subsequent interatomic distances, an expert system named PEPTO was utilized (Catasti et al., 1989). PEPTO is written in Borland Turbo Prolog and runs on a PC with 640 Kb RAM and 20 Mb hard disk (see Figure 3). Systematic peak-picking was performed on the NOESY spectra in D_2O at 100, 200 and 300 msec. of mixing time. In order to avoid t1-ridges and noise, symmetrization was performed on each spectrum. The presence of artifacts after symmetrization was also reduced by considering only the peaks appearing a minimum of in two experiments. This led to a list of 156 D_2O peaks.

In Figure 4a a D_2O NOESY spectrum with 300 msec. of mixing time is shown, and the consequent peak-picking is shown in Figure 4b. The same peak-picking was performed in the H_2O NOESY spectrum, neglecting the peaks already present in the D_2O spectra, yielding a list of 105 additional cross-peaks. The last input for PEPTO was the primary sequence of an 84 residues fragment of the H1 histone globular region (from residue 35 to 118) typically resulting from Trypsin digestion that always cuts on the carboxy-terminal of a LYS or an ARG.

The side-chain spin-system identification of PEPTO found 51 spin-systems out of 84. After a comparison with a manual inspection of the COSY D_2O spectrum we re-identified three of them plus 9 more spin-systems, reaching a total number of 60 residues. Of the 24 missing spin-systems, 9 are likely to be GLY with equivalent α protons; after an inspection of the COSY "fingerprint" (Figure 5) showing the α-N through-bond connectivities, we can identify all the 9 GLY in the NOESY spectrum (Figure 6) arriving at a global number of 68 residues currently identified. For 50 of the residues we have also identified the amide proton chemical shifts (Table 1). At this stage we can't say anything about the remaining 15 residues, which are: 3 THR, 3 SER, 3 VAL, 2 (LYS or ARG), 3 (ASP, TYR or ASN), 1 PRO.

Table 1. List of the 68 residues spin-system identified on the H1 histone globular region.

Residue	NH	CαH	CβH	CγH	CδH
3 (6) ASN+ASP+PHE+TYR					
	8.26	4.39	3.07/2.87		
	8.16	4.18	2.47		
	8.14	3.85	3.29		
14 (14) ALA					
	4.21	1.36			
	4.20	1.14			
	4.33	1.2			
	4.06	1.1			
	4.33	1.09			
	8.27	4.01	1.18		
	8.21	4.15	0.92		
	8.22	4.27	1.46		
	8.32	4.26	1.22		
	8.59	4.05	0.96		
	8.08	4.22	0.96		
	8.31	4.35	1.18		
	8.09	3.96	1.57		
	8.04	4.30	1.03		
13 (15) LYS+ARG					
	4.35	2.11			
	4.26	1.85			
	3.81	2.02			
	8.31	4.49	1.56/1.46		
	8.37	4.27	1.46/1.40		
	8.18	4.22	1.55		
	9.04	4.16	1.74		
	8.38	4.12	1.31/1.03		
	8.28	4.41	2.30/1.55		
	8.13	4.31	1.66		
	8.28	4.19	1.80		
	8.71	4.37	1.79		
	8.25	3.95	1.24/1.10		
8 (11) SER					
	4.06	3.87/3.79			
	4.08	3.96/3.96			
	7.98	4.17	3.97		
	8.38	4.58	3.82		
	8.24	4.27	4.05		
	8.57	4.53	3.97		
	8.09	4.39	3.95/3.87		
	8.16	4.34	3.91		
5 (5) GLU					
	4.32	2.12/1.97	2.47		
	8.38	4.48	2.15/1.99	2.52/2.37	
	8.42	4.29	1.77/1.57	2.36	
	8.20	4.45	1.95/1.88	2.32	
	8.52	4.31	1.91/1.75	2.35/2.27	
9 (9) LEU					
	4.08	1.95	1.62	1.13	
	3.95	1.97	0.83	0.61	
	3.96	1.74	1.09	0.88	
	4.07	1.68	1.46	0.59	

Table 1. continued...

Residue	NH	CαH	CβH	CγH	CδH
	4.37	1.44/1.16	1.15	1.00/0.80	
	3.81	1.90	1.13	0.57/0.47	
	8.20	3.94	1.79/1.50	1.49	0.96/0.81
9 (9) LEU					
	8.24	4.11	1.60	0.91	0.64/0.35
	8.33	4.25	1.76	1.57	0.72
3 (6) VAL					
	3.63	2.01	0.69/0.81		
	8.46	4.40	2.00	1.11/0.99	
	8.45	4.19	2.14	1.14	
9 (9) GLY					
	8.52	4.01			
	8.36	3.96			
	8.31	4.07/3.93			
	8.19	4.02			
	8.83	4.03			
	8.56	3.97			
	8.41	3.96			
	8.37	4.00			
	8.51	4.08			
2 (2) ILE					
	8.27	4.27	1.39	0.89	0.47
	3.80	1.41	1.49/1.13	0.87/0.69	
1 (2) PRO					
	4.40	2.50	2.12	3.82/3.63	
1 (4) THR					
	7.98	4.08	3.95	1.24	

Sequential resonance assignments were successful in individuating the 40 residues listed in Table 2. The primary structure is assumed to be equal to the calf thymus H1 histone "core" amino acid sequence (Chapman et al., 1978); verification on the actual amino acid sequence of our calf liver "H1 core" is, however, in progress.

Table 2. Partial sequential resonance assignments of the H1 histone globular region.

Residue	NH	CαH	CβH	CγH	CδH	others
PRO40	--	4.40	2.50	2.12	3.82/3.63	.
VAL41	8.46	4.40	2.00	1.11/0.99		
SER42	8.24	4.27	4.05			
GLU43	8.42	4.29	1.77/1.57	2.36		
LEU44	8.24	4.11	1.60	0.91	0.64/0.35	
ILE45	--	3.80	1.41	1.49/1.13	0.87/0.69	
THR46	7.98	4.08	3.95	1.24		
LYS47	8.25	3.95	1.24/1.10			
ALA48	8.04	4.30	1.03			
ALA50	8.31	4.35	1.18			
ALA51	8.08	4.22	0.96			
SER52	8.09	4.39	3.95/3.87			

Table 2. continued...

Residue	NH	CαH	CβH	CγH	CδH	others
LYS53	9.04	4.16	1.74			
GLU54	8.20	4.45	1.95/1.88	2.32		
ARG55	8.38	4.12	1.31/1.03			
SER56	7.98	4.17	3.97			
GLY57	8.37	4.00				
-						
LYS76	8.18	4.22	1.55			
ASN77	8.16	4.18	2.47			
ASN78	8.26	4.39	3.07/2.87			
SER79	8.38	4.58	3.82			
ARG80	8.37	4.27	1.46/1.40			
ILE81	8.27	4.27	1.39	0.89	0.47	
LYS82	8.31	4.49	1.56			
LEU83	8.20	3.94	1.79/1.50	1.49	0.96/0.81	
-						
VAL89	8.45	4.19	2.14	1.14		
SER90	8.16	4.34	3.91			
LYS91	8.28	4.19	1.80			
GLY92	8.41	3.96				
-						
LYS98	8.71	4.37	1.79			
GLY99	8.56	3.97				
.						
GLY104	8.83	4.03				
SER105	8.57	4.53	3.97			
PHE106	8.14	3.85	3.29	-	6.81	CεH 7.11
						CεH 7.26
LYS107	8.13	4.31	1.66			
LEU108	8.33	4.25	1.76	1.57	0.72	
-						
GLY115	8.51	4.08				
GLU116	8.52	4.31	1.91/1.75	2.35/2.27		
ALA117	8.09	3.96	1.57			
LYS118	8.28	4.41	2.30/1.55			

The last module of PEPTO has determined up to now 61 interatomic long (16), short (31) and medium (14) range NOEs, summarized in Table 3.

Table 3. Interatomic NOEs found by means of PEPTO.

	SHORT	MEDIUM	LONG
40 PRO	δH 41 NH		
41 VAL	αH 42 NH	αH 43 NH	βH 82 NH
			βH 45 NH
42 SER	αH 43 NH		
43 SER	αH 44 NH		αH 50 NH
			αH 79 NH
45 ILE		βH 48 NH	
46 THR	αH 47 NH		
47 LYS	αH 48 NH		

Table 3. continued...

	SHORT	MEDIUM	LONG
48 ALA			αH 107 NH
50 ALA	αH 51 NH	αH 54 NH	αH 82 NH
51 ALA	αH 52 NH		
52 SER	αH 53 NH		βH 80 NH
53 LYS	αH 54 NH	αH 56 NH	βH 81 NH
54 GLU	αH 55 NH	αH 57 βH	
	βH 55 NH		
55 ARG	αH 56 NH		
56 SER	αH 57 NH		αH 77 NH
			βH 83 NH
76 LYS	αH 77 NH	αH 79 βH	αH 82 NH
77 ASN	αH 78 NH		αH 98 NH
78 ASN	αH 79 NH		αH 98 NH
79 SER	αH 80 NH	αH 81 NH	αH 104 NH
		αH 83 NH	αH 105 NH
80 ARG	αH 81 NH		
81 ILE	αH 82 NH	αH 83 NH	
		βH 83 NH	
82 LYS	αH 83 NH		
89 VAL	αH 90 NH	βH 92 NH	
90 SER	αH 91 NH		βH 116 NH
91 LYS	αH 92 NH		
98 LYS	αH 99 NH		
99 GLY			αH 105 NH
104 GLY	αH 105 NH		
105 SER	αH 106 NH	βH 108 NH	
107 LYS	αH 108 NH		
115 GLY	αH 116 NH		
116 GLU	αH 117 NH	αH 118 NH	
117 ALA	αH 118 NH		

Computer-Generated Hypothesis of Secondary Structure

The algorithm used was based on a modified GOR method (Rauch et al., 1989; Robson and Garnier, 1986) which uses a data base containing the statistical "propensity", defined as the natural logarithm of the probability for each amino acid to belong to a particular conformation (derived from proteins of known crystal structure). As for Circular Dichroism, four kinds of conformation patterns are considered: α-helix, β-sheet, random-coil and turn. The conformational propensity for an amino acid in the sequence is computed by summation of the propensities assigned to each of the neighboring residues. The molecular organization is in fact strictly dependent on local boundary conditions and short-range interactions.

The decisional rules for the inclusion of an amino acid into a structural pattern had the following differences (Rauch et al., 1989) from those suggested by the GOR methods: a) the local forward gradient of the propensity, looking at the chain from N-terminus toward C-terminus, was taken into account, b) in addition to the helix percentage of the protein fragment, the mean length N of a helix (statistically estimated) and helical percentage from CD data were taken into account. The detailed algorithm was developed by Rauch et al. (1989). This procedure resulted in an assignment of conformational regions to the amino acid sequence; based on this, a tentative computation of the coordinates and relative distances of the residues was

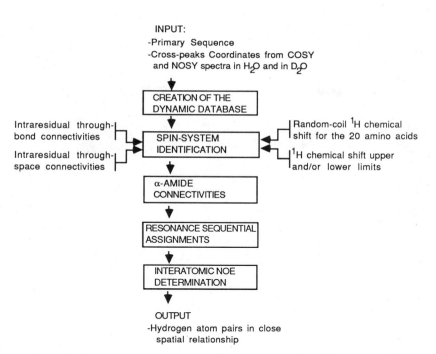

INPUT:
-Primary Sequence
-Cross-peaks Coordinates from COSY
 and NOSY spectra in H_2O and in D_2O

CREATION OF THE
DYNAMIC DATABASE

Intraresidual through-
bond connectivities

Intraresidual through-
space connectivities

SPIN-SYSTEM
IDENTIFICATION

Random-coil ^1H chemical
shift for the 20 amino acids

^1H chemical shift upper
and/or lower limits

α-AMIDE
CONNECTIVITIES

RESONANCE SEQUENTIAL
ASSIGNMENTS

INTERATOMIC NOE
DETERMINATION

OUTPUT
-Hydrogen atom pairs in close
 spatial relationship

Figure 3. Flow-chart of PEPTO. The rectangular boxes in the central column represent the functional modules. The four boxes at the sides display the information coming from the common database used by PEPTO.

performed utilizing well known geometrical properties of peptides (Robson and Garnier, 1986; Sanders, 1977).

^{31}P and ^1H 1DFT NMR of Nucleosomes

Trinucleosomes and pentanucleosomes were isolated from logarithmically growing He La cells using standard procedures (Yau et al., 1982). Increasing concentrations of NaCl were added to the nucleosome solution in order to selectively and successively extract HMG non-histone (0.35 M NaCl), H1 histone (0.60 M NaCl), H2A-H2B (1.2 M NaCl) and H3-H4 (2.0 NaCl). The molar concentration of protein and DNA are shown in Table 4. 1DFT proton and phosphorus NMR spectra were taken on an AM-500 Bruker spectrometer at the Stanford Magnetic Resonance Laboratory as described in the corresponding figure legends (Nicolini et al., 1989).

Table 4. Molar concentration of protein and DNA versus salt extraction of Nucleosomes.

Salt conc.	HMG	H1	H2A	H2B	H3	H4	DNA
10 mM	0.1	1.2	2.0	2.0	2.0	2.0	1.0
350 mM	0.0	1.2	2.0	2.0	2.0	2.0	1.0
600 mM	0.0	0.0	2.0	2.0	2.0	2.0	1.0
1.2 M	0.0	0.0	0.0	0.0	2.0	2.0	1.0
2.0 M	0.0	0.0	0.0	0.0	0.0	0.0	1.0

Nucleosomal protein and DNA molar concentration were computed assuming molecular weight of 21000 for H1, 14800 for H2A, 14000 for HMG 1-2, 14000 for H2B, 15000 for H3, 11400 for H4 and 126000 for nucleosomal DNA (equivalent to 180 bp).

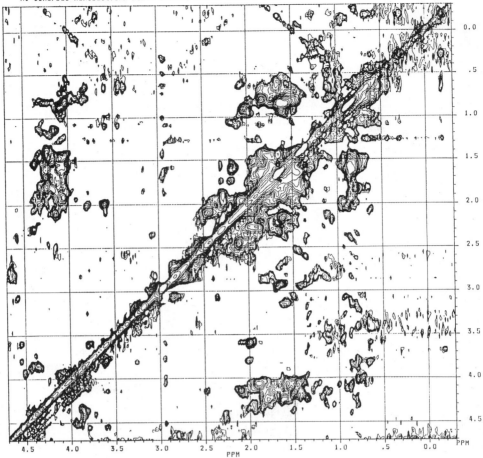

Figure 4.   (a) NOESY spectrum in D$_2$O at 300 msec of mixing time and elaborated as described in the text.

RESULTS

Histone H1 Core Secondary Structure

The CD spectra of the H1 core (Figure 7) show no noticeable variations between pH 1.9 and 2.6 and are very similar to those of typical random-coil polypeptides. A dramatic variation of the spectra is observed between pH 2.6 and 4.4, suggesting a conformational transition near pH 4.   At pH 4.4 and above, the best fitting of the experimental CD spectra of H1 core with the reference spectra of proteins of known structures suggests that a considerable percentage of α-helix is present (about 35 % to 38%, as shown in Table 5). Considering the dependence of CD spectra on the α-helix length (Chang et al., 1978), we have obtained three best fitted results for three different numbers of α-helices and for a mean number N of residues per α-helix of 11 (Table 5).

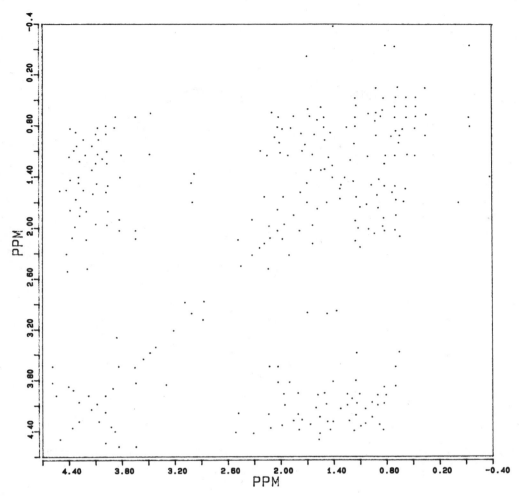

Figure 4.   (b) Consequent peakpicking.

Interestingly, the computerized determination (see Rauch et al., 1989) of the secondary structure of the H1 core from our modified GOR method is compatible with either three or four α-helices; three helices yield respectively 12, 13 and 8 residues (mean number N=11), 13% of β-sheet, 25% of random-coil and 23% of turn. The random coil and β-sheet regions together account for 38%, while the CD measurements at pH 7.7 yield 45% for the same regions.

The increased content of turns above pH 6 (Table 5) suggests a more compact folding of the protein.   At pH 2.6 and 1.9 no helices are detectable; the protein residues are nearly all in random-coil conformation. This result is also confirmed by 1D NMR spectrum of H1 core carried out at pH 1.9 and 23°C (not shown), which shows a more resolved spectrum and a complete disappearance of any resonance signal below 0.8 ppm - with respect to the NMR spectra at pH 4.1 at 23°C shown in Figure 8a. Polypeptides in random-coil conformation do not display chemical shifts below 0.8 ppm.

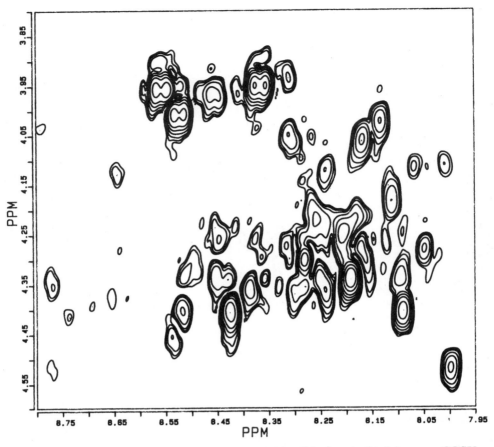

Figure 5.   Through-bond α-CH--N-H connectivities ("fingerprint") of the same COSY
spectrum of Fig. 2.

Table 5.   Secondary structure of histone H1 core from CD measurements.

	3 HELICES N=11				4 HELICES N=9				5 HELICES N=7			
	A	B	T	C	A	B	T	C	A	B	T	C
PH 7.7 :	38	--	17	45	37	--	20	3	35	--	24	41
PH 6.0 :	37	--	--	63	40	--	--	60	39	--	4	57
PH 4.4 :	40	--	--	60	41	--	3	56	40	--	8	52
PH 2.6 :	--	--	6	94	--	--	6	94	--	--	6	94
PH 1.9 :	--	--	2	98	--	--	2	98	--	--	2	98
error range:	±4	--	±8	±12	±4	--	±8	±12	±4	--	±8	±12

The numbers are the percentage of α-helix (A), β-sheet (B), turn (T) and random-coil (C). These percentages
were obtained from the best fitting of the experimental CD spectra with the standard protein reference spectra
(see methods). The error range is also expressed in terms of percentage of the total H1 globular region primary
sequence.

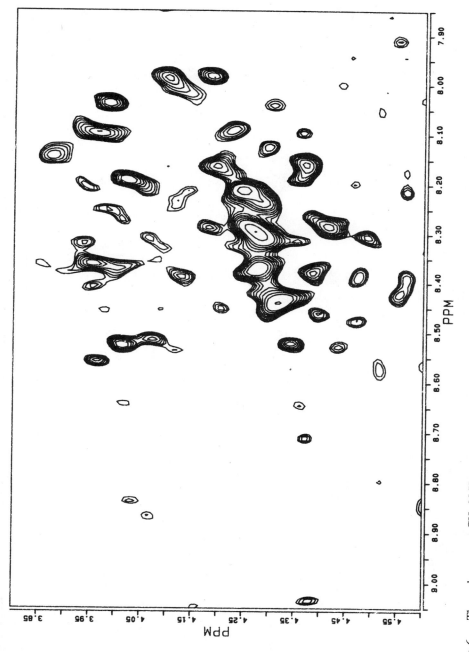

Figure 6.   Through-space α-CH−N-H connectivities of the H1 histone pure-absorption NOESY spectrum with a mixing time of 300 msec and processed as described in the text.

233

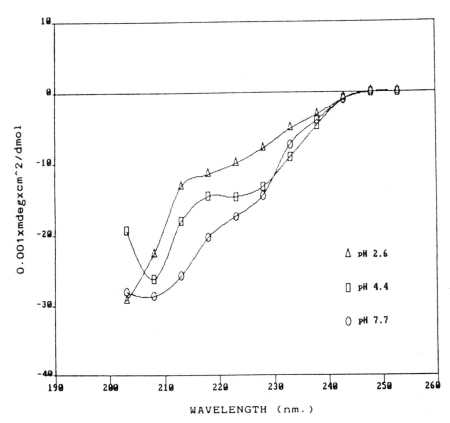

Figure 7. Circular Dichroism spectra of 3×10^{-6} M solutions of histone H1 core in 10 mM buffers with different pH values and 0.2 M NaF. The measurements were carried out at 25°C in a 1-mm path length cell. The errors were estimated of about 30% below 210 nm, of about 20% between 210 and 230 nm and of about 10% above 230 nm. These profiles were corrected for CD inactive sample artifacts.

Thermal denaturation studies of the H1 carried out by Differential Scanning Calorimetry appear relatively difficult due to the low heat exchange (work still in progress). For this reason the change in conformation following heat denaturation of the core of the H1 protein alone was also monitored by means of 1D-NMR spectroscopy (Figure 8). The NMR spectrum at 23°C is typical of a globular protein (Figure 8 bottom). In particular, the region of aromatic residues resonances (6.5 - 7.5 ppm) shows a dispersion of the chemical shifts of tyrosine-72 and phenylalanine-106 revealing that those amino acids are not in contact with solvent. The region of methyl resonances (below 1.5 ppm) shows five distinct chemical shifts at 0.7, 0.56, 0.46, 0.34 and -0.15 ppm which are never found in a random-coil polypeptide. These results are in agreement with those coming from CD measurements taken in similar pH and ionic strength, where samples from H1 core were found to have significant α-helix conformation. At 40°C (Figure 8 top) the NMR spectrum shows many changes indicating a certain loss of structure. Although the observed displacement of the aromatic resonances is small, a dispersion of the chemical shifts of tyrosine and phenylalanine is still present, suggesting that a sizeable structure is maintained even at 40°C. This is also indicated by the presence of only two methyl chemical shifts out of the five observed at 23°C, which are displaced to 0.63 and 0.53 ppm respectively. Other important changes from 23 to 40°C are shown in the α-carbon proton-resonance region

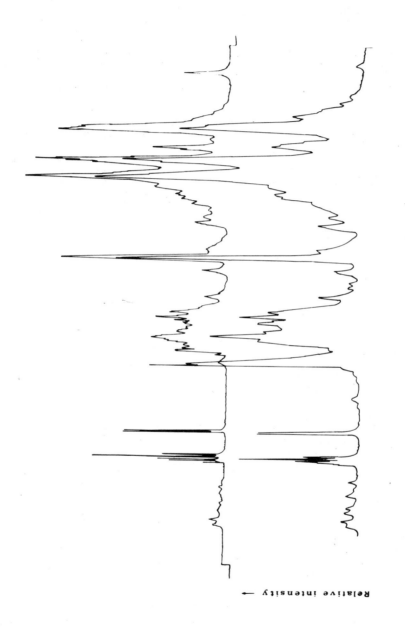

Figure 8. 1D-FT NMR spectra at 500 MHz of the histone H1 core, 5 mg/ml in D₂O, 0.5 M K₂SO₄ at pH 4.1: a) at 23°C (bottom); b) at 40°C (top).

(3.8 - 4.5 ppm) where a rearrangement of the chemical shifts is observed from the structured to the denatured state, even though resonance peaks cannot be easily identified in such a crowded pattern. A further indication of structure loss is the displacement of a methyl resonance of threonine (at 1.2 ppm) towards that of valine, leucine and isoleucine when temperature was increased to 40°C. When temperature was decreased from 40°C to 30°C and back to 23°C, NMR spectra (not shown) indicated that a good renaturation of the protein, although not complete, had taken place.

Analysis of the 2DFT NMR by PEPTO points to the identification of up to 68 spin systems (Table 1), 40 of them sequentially assigned (Table 2), yielding an initial list of the pairs of hydrogen atoms connected by strong, medium and weak NOEs (Table 3). In the H1 histone core in H_2O we observe a larger number of NOESY cross-peaks (Figure 6) than COSY cross-peaks (Figure 5), because of the presence of sequential and long-range NOEs. Several pairs of them are close together and can be unambiguously assigned at the same pairs of protons. Most COSY cross-peaks have indeed a NOESY counterpart, as suggested by the degree of correspondence among the two fingerprints.

In Table 3 a list of the observed interatomic short, medium and long-range NOEs is shown; from it we can delineate (within the core) two zones showing a significant amount of sequential and middle-range restraints that are pointing towards the presence of two α-helices: residues 50-57 and residues 76-83. The observed NOEs between the residues 43-50, 45-48, 46-47 indicate a parallel β-sheet. Two turns can also be seen at the two ends of the globular region (residues 40-43 and 115-118); they are characterized by d(i,i+1) and d(i,i+2) α-amide connectivities.

We show a possible description of the secondary structure in Table 6; after LEU83 there are not enough interresidual NOEs to assign the secondary structure elements unambiguously. Nevertheless, between residues 89-92 and 104-108 we can see d(i,i+3) α–β connectivities, indicators of α-helices.

Table 6. Secondary structure of the histone H1 core from 2DFT-NMR.

```
35              40              45              50              55
Lys-Ala-Ser-Gly-Pro-Pro-Val-Ser-Glu-Leu-Ile-Thr-Lys-Ala-Val-Ala-Ala-Ser-Lys-Glu-Arg-
 C   C   C   C   C   T   T   T   T   B   B   B   B   B   B   A   A   A   A   A   A
                60              65              70              75
Ser-Gly-Val-Ser-Leu-Ala-Ala-Leu-Lys-Lys-Ala-Leu-Ala-Ala-Ala-Gly-Tyr-Asp-Val-Glu-Lys-
 A   A   C   C   C   C   C   C   C   C   C   C   C   C   C   C   C   C   C   A
            80              85              90              95
Asn-Asn-Ser-Arg-Ile-Lys-Leu-Gly-Leu-Lys-Ser-Leu-Val-Ser-Lys-Gly-Thr-Leu-Val-Glu-Thr-
 A   A   A   A   A   A   A   C   C   C   C   C   A   A   A   A   A   C   C   C   C
         100             105             110             115
Lys-Gly-Thr-Gly-Ala-Ser-Gly-Ser-Phe-Lys-Leu-Asn-Lys-Lys-Ala-Ala-Ser-Gly-Glu-Ala-Lys-
 C   T   T   C   C   C   C   A   A   A   A   A   C   C   C   C   C   C   T   T   T
         120
Pro-Lys.
 T   C
```

Comparison between the different amounts of secondary structure calculated from NMR and CD approaches. The CD results shown here are reported from the 4 helices model at pH 4.4 in Table 5.

	NMR	CD
α-helix :	26 res (30%)	35 res (41%)
β-sheet :	6 res (7%)	--
turn :	10 res (12%)	2 res (3%)
random coil :	44 res (51%)	39 res (56%)

A = α-helix ; B = β-strand ; T = turn ; C = random coil.

Effect of H1 Removal on the Nucleosome Relation Parameters

Earlier nuclear magnetic resonance studies (Hogan and Jardetzky, 1980) have shown that internal structure of B-DNA experiences large fluctuations in nucleotide conformation which are strongly affected by the binding of ethidium bromide. Similarly, ^{31}P (Figure 9) and ^1H (not shown) NMR spectra of trinucleosomes suggest that as a result of chromosomal histone proteins binding nucleosomal DNA internal motions are greatly hindered. Indeed, we show that the selective removal of chromosomal proteins by progressively increasing the concentration up to 2 M NaCl, progressively increases the internal motions up to that of free DNA (Nicolini et al., 1989).

The H1 histone extraction at 0.6 M NaCl, as compared to the selective extraction of HMG non histone (0.35 M NaCl) and H2A-H2B histone (1.2 M NaCl), displays the largest variation of ^{31}P line width (Table 7). This is compatible with earlier suggestions on its role in the sealing of nucleosomes. While all histone proteins are known to be bound to the nucleosomal DNA, HMG non-histone proteins are possibly loosely bound to the DNA strand interconnecting single nucleosome within the trinucleosome solution analyzed here.

Table 7. Trinucleosome ^{31}P. Spectra half width - half height.

10 mM NaCl	525 Hz
0.35 M NaCl	465 Hz
0.6 M NaCl	404 Hz
1.2 M NaCl	404 Hz
2.0 M NaCl	323 Hz

DISCUSSION

The critical role of H1 core in maintaining the DNA internal motion within the nucleosome (Figure 10) is confirmed in the ^{31}P (Figure 9) and ^1H NMR study of trinucleosomes reported here. We have thereby a clear function: to modulate the sealing of nucleosome and through it of chromosome structure and expression by engineering H1 histone core.

In other words, H1 constitutes a good model system for studying structure-function relationships in mammalian nucleosome, and the knowledge of its native atomic structure and dynamics in solution is a prerequisite toward this end.

From the experimental CD, NMR, calorimetric and theoretical studies reported here, the histone H1 core appears to be consistently structured. This despite the fact that in the 2DNMR experiments the need to operate at relatively low pH (to reduce proton exchange in water) has probably caused the loss of some structure; alternatively at pH 4.1 a greater number of peaks could present a broadening of the lines due to a rapid exchange between different conformations. The consequent overlap of many peaks also hinders the complete assignments of the resonances (Table 2). However, observation of the short and medium range NOEs (see Table 4) allows us to say that a significant part of the protein is still structured and to obtain a possible secondary structure in agreement with the CD and statistical results (Table 5). The amount of α-helical secondary structure estimated from NMR data is identical to the earlier CD estimate (30%) of the same H1 core (Chang et al., 1978), but slightly less than that generated by the CD measurements reported here which have been corrected for inactive sample artefacts. The assignments and NOE data are poorer in the second half of the primary sequence, and the last two of the four

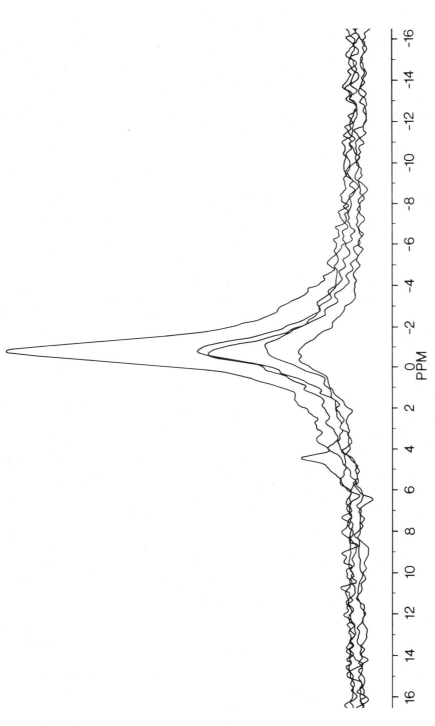

Figure 9. Effect of increasing NaCl concentration on the ^{31}P NMR spectrum of the trinucleosome solution. Samples are in 30 mM Tris, 3 mM Na_2EDTA, pH 7.2 and D_2O. Fourier transform NMR spectra were accumulated on an AM-500 Bruker spectrometer. Spectra were collected with non selective proton decoupling, acquisition time 2 sec., with a delay between pulses of 30 sec. (2000 transients). Chemical shifts are referenced to an external standard (85% H_3PO_4–15%D_2O), temperature was controlled at +23°C. DNA concentration was 4.0×10^{-3} M (basepairs/liter). From top to bottom: 2.0 M NaCl, 1.2 M NaCl, 0.6 M NaCl, 0.35 M NaCl, 10 mM Tris pH 7.

238

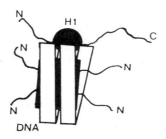

Figure 10. Present consensus on the structure of the nucleosomes, consisting of DNA wrapped around the "core" H2A-H2B, H3-H4 and sealed by H1.

helices proposed could be bigger. On the basis of long range NOE data (see Table 3), we can at present propose the 3D folding shown in Figure 11 for the H1 histone globular region.

One problem remains in the sequential 2DFT NMR resonance assignment of the H1, namely the little chemical-shift dispersion typically associated with the low number of aromatic residues (only 2). For instance, in a recent work (Zarbock et al., 1987) the avian H5 cross-peaks appear better resolved, probably because this protein (very similar to H1 from a functional point of view) has 7 aromatic rings that spread out the proton resonances. Therefore, we are running HOHAHA 2D NMR experiments on the H1 at different mixing times (i.e. 15 and 60 msec.). Furthermore, the H1 core is known to have several Glycines and Alanines whose resonances are very overlapped. To bypass this problem we are acquiring a TQF COSY for the suppression of AX and A_3X spin-systems, hoping to achieve a drastic simplification of the spectra.

Our major problem in this preliminary analysis, however, is due to the limiting assumption being made that the primary sequence of our H1 core preparation from calf liver is homogeneous and equal to the calf thymus H2 histone core amino acid sequence; the presence of a sizeable amount of H1$_0$, known to have significant differences in amino acid sequence even in the core

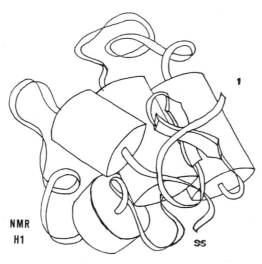

Figure 11. 3D folding of the H1 histone globular region as follows from 2D-NMR. See text for details.

from the rest of H1 subfractions, could explain our present difficulties. Work is in progress to overcome all above limitations.

ACKNOWLEDGMENTS

The research described in this article has been supported by research grants 001588 from NATO Scientific Affair Division and 88.00780.44 from Consiglio Nazionale delle Ricerche. The use of the NMR equipment was kindly provided by the Stanford Magnetic Resonance Laboratory.

REFERENCES

Barbero, J.L., Franco, L., Montero, F., and Moran, F., 1980, Structural studies on Histone H1, Circular Dichroism and Difference Spectroscopy of the Histones H1 and their trypsin-resistant cores from calf thymus and from the fruit fly *Ceratilis Capitata, Biochemistry* 19:4080.

Bradbury, E.M. and Baldwin, J.P., 1986, Neutron scatter and diffraction techniques applied to nucleosome and chromatin structure, *in:* "Nobel Symposium on Biosciences at the Physical Science Frontiers", Nicolini, C. ed. Humana Press, Clifton NJ., pp. 35-66.

Catasti, P., Carrara, E. and Nicolini, C., 1989, PEPTO: an expert system for automatic peak assignment of two-dimensional nuclear magnetic resonance spectra of proteins, *J. Comp. Chem.,* in press.

Chang, C.T., Wu, C.-S.C., & Yang, J.T., 1978, Circular Dichroism analysis of protein conformation: inclusion of the beta-turns, *Anal. Biochem.* 91:12.

Chapman, G.E., Hartman, P.G. and Bradbury, E.M., 1975, Studies and the roles and mode of operation in the very-lysine-rich Histone H1 in eukaryote chromatin. The isolation of the globular and non-globular regions of the Histone H1 molecule, *Eur.J.Biochem.* 61:69.

Chen, Y.-H., Yang, J.T. & Chau, K.H. ,1974, Determination of the helix and beta-form of proteins in acqueous solution by Circular Dichroism, *Biochemistry* 13:3350.

Dolby, T., Belmont, A., Borun, T. and Nicolini, C., 1981, DNA replication, chromatin structures, and histone phosphorylation altered by theophylline in synchronized HeLa S3 cells, *J. Cell Biol.* 89:78.

Hartman, P., Chapman, G., Moss, T., and Bradbury, E.M., 1977, Studies on the roles and mode of operation in the very-lysine-rich Histone H1 in eukaryote chromatin. The three structural regions of the Histone H1 molecule, *Eur. J. Biochem.* 71:45.

Hogan, M.E. and Jardetzky, O., 1980, Effect of ethidium bromide on deoxyribonucleic acid internal motion , *Biochemistry* 19:2079.

Jeener, J., Meier, B.H., Bachmann, P. and Ernst , R. R., 1979, Investigation of exchange process by two-dimensional NMR spectroscopy, *J. Chem. Phys.* 71:4546.

Marion D. and Wüthrich, K., 1983, Application of phase sensitive two-dimensional Correlated Spectroscopy (COSY) for measurements of 1H-1H spin-spin coupling constants in proteins, *Biochem Biophys. Res. Comm.* 113:967.

Nicolini, C., 1983, Chromatin structure from nuclei to genes, *Anticancer Research* 3:63.

Nicolini, C., 1986, "Biophysics and Cancer," Plenum Press, NY, 1-462.

Nicolini, C., Catasti P., Nizzari, M., Carrara E. and Szilágyi, L., 1989, H1 histone core solution structure by 2DFT NMR at high resolution, submitted to *Biophys. J.*

Nicolini, E.M., Szilágyi, L. and Yau, P., 1989, Effect of selective removal of chromosomal proteins on the DNA internal motion within nucleosomes: role of H1 histone, submitted to *Eur. J. Biophys.*

Pepe, M., Rauch, G., Catasti, P., Nizzari, M. and Nicolini, C., 1989, Histone H1 characterization by Differential Scanning Calorimetry, Circular Dichroism, 1DFT NMR and statistical algorithms, submitted for publication to *Biochemistry*.

Rance, M., Sorensen, D., Wagner, G. and Ernst, R. ,1984, Inclosed spectral resolution in COSY 1H-NMR spectra of proteins via double quantum filter, *Biochem. Biophys. Res. Comun.* 117:479.

Rauch, G., Catasti P., Nizzari, M., Pepe M., Panfoli, I. and Nicolini, C., 1989, A biophysical approach to the determination of the secondary structure of the histone H1 globular region, submitted to the *Int. J. Macr. Biol.*

Robson, B. and Garnier, J., 1986, "Introduction to Protein and Protein Engineering," Elsevier Science Publishers.

Sanders, C., 1977, A method for the fractionation of the High-mobility-group Non-histone chromosomal proteins, *Biochem. Biophys. Res. Comm.* 78:1034.

Wüthrich K., 1986, "NMR of Proteins and Nucleic Acids," J. Wiley and sons, New York.

Yang, J.T., Wu, C.-S.C., & Martinez, H.M., 1986, Calculation of protein conformation from Circular Dichroism, *Methods in Enzymol.* 130:208.

Yau, P., Thorne, A.W., Imai, B.S., Matthews, H.R. and Bradbury, E.M., 1982, Thermal denaturation studies of acetylated nucleosome and oligonucleotides, *Eur. J. Biochem.* 129:281.

Zarbock, J., Clore, G.M. and Gronenborn, A.M., 1986, Nuclear magnetic resonance study of the globular domain of chicken histone H5: Resonance assignment and secondary structure, *Proc. Nat. Acad. Sci. USA* 83:7628.

^{1}H NMR STUDIES OF GENETIC VARIANTS AND POINT MUTANTS OF MYOGLOBIN:

Modulation of Distal Steric Tilt of Bound Cyanide Ligand

Gerd N. La Mar[1], S. Donald Emerson[1], Krishnakumar Rajarathnam[1], Liping P. Yu[1], Mark Chiu[2] and Stephen A. Sligar[2]

[1]Department of Chemistry, University of California, Davis, CA 95616 and
[2]Department of Biochemistry, University of Illinois, Urbana, IL 61801, USA

INTRODUCTION

The oxygen-binding sites in the hemoproteins myoglobin, Mb, and hemoglobin, Hb, are considered to have evolved to provide a preferential binding site for the bent geometry of a bound O_2 molecule over the linear and perpendicular geometry of a bound CO (Dickerson and Geis, 1983). The preferential affinity of reduced heme iron for CO over O_2 in model compounds is modulated in proteins by steric destabilization of the linear Fe-CO unit and stabilization of the bent Fe-OO unit by hydrogen bonding by distal residues (Springer et al., 1989). X-ray structures have demonstrated distorted Fe-CO units in heme proteins (Kuriyan et al., 1986), but the tilt is difficult to characterize quantitatively. While hydrogen-bonding can be inferred from X-ray data by the position of the relevant heteroatoms (Phillips, 1980), the direct observation of the proton requires neutron diffraction (Phillips and Schoenborn, 1981). Functional studies of natural genetic variants (i.e., elephant Mb with E7 His→Gln) (Romero-Herrera et al., 1981) and, more recently, the sperm whale point mutants expressed in *E. coli* using a synthesized gene (Springer et al., 1989), have generally confirmed the perturbations expected to follow from systematic variations of the E7 residue. However, the details of the stereochemistry in the heme pocket for these variants are not yet known.

^{1}H NMR is ideally suited for providing these structural details (Wüthrich, 1986). Such NMR structural studies can be pursued in either diamagnetic (i.e., MbCO or MbO_2) or paramagnetic (i.e., high-spin $(S = 5/2)$ $metMbH_2O$, low-spin $(S = 1/2)$ metMbCN, spin-equilibrium $(S = 1/2 - 5/2)$ $metMbN_3$, deoxy Mb $(S = 2)$) forms (Satterlee, 1986). The former derivatives can yield the complete three dimensional solution structure by ^{1}H NMR using a variety of modern two-dimensional NMR methods (Mabbutt and Wright, 1985; Dalvit and Wright, 1987). On the other hand, NMR spectra of diamagnetic derivatives yield little useful information on the electronic structure of the heme or the magnetic properties of the iron, and do not provide insight into steric effects on distal ligand bonding. In a paramagnetic derivative, such as low-spin iron(III) metMbCN, the highly anisotropic paramagnetic susceptibility tensor χ, which causes the dipolar shifts for non-coordinated residues, is oriented by the molecular interactions between heme and protein that perturb the idealized axial symmetry of a model heme compound where the major magnetic axis is directed precisely along the heme normal (Shulman et al., 1971; Satterlee, 1986). The determination of the magnetic axes in a protein complex

allows identification of the interactions that are the determinants of the orientation. In turn, once the magnetic axes are determined for one reference protein (wild type sperm whale Mb) and the orientation interpreted in terms of heme-protein interactions, comparison among a series of closely related natural genetic variants or synthetic point mutants should allow monitoring the changes in these symmetry-perturbing interactions.

In high spin ferric met Mb derivatives, the pattern of the dominant contact shifts for the heme directly provides information on whether the sixth position on the iron is occupied by a water molecule. The presence of such water molecules can be interpreted in terms of the importance of hydrogen bonding by distal residues (La Mar et al., 1988). The azido complex of metMb exhibits a spin equilibrium whose position provides a sensitive probe of the effective axial ligand field strength (La Mar et al., 1983).

Detailed structural studies of paramagnetic Mb derivatives have been hampered until recently because of the unavailability of the needed assignments. Some of the heme resonances were located by isotope labeling (Mayer et al., 1974), but they provided no information on the important amino acid residues. The nuclear Overhauser effect, NOE, was initially supposed to be ineffective for paramagnetic systems because the paramagnetic leakage rendered the NOE too small to detect. Steady-state NOE spectra, however, have been shown to be readily obtainable for a wide variety of paramagnetic proteins. In fact, the paramagnetism is advantageous in allowing NOE studies of much larger paramagnetic (up to 150 kDa have been observed) than diamagnetic proteins because of the undermining spin diffusion. The 2D methods, NOESY (2D NOE) and COSY (spin correlation), which are the cornerstones of diamagnetic protein assignments and structure determination (Wüthrich, 1986), were also thought to be inapplicable to paramagnetic species. The effective paramagnetic relaxation was considered to cause NOESY cross peaks to decay too rapidly, and interfere with detecting COSY peaks because of both rapid loss of coherence and line broadening. However, it turns out that both NOESY and COSY experiments are highly useful for assignment of low-spin ferric complexes of myoglobin as well as larger hemoproteins (Emerson and La Mar, 1989a).

We show here how modified distal residues lead to altered tilt of the bound cyanide ligand for the case of elephant metMbCN and sperm whale point mutant [E7 His→Val]metMbCN. In both cases, the distal CN⁻ is tilted much less than for wild-type, (WT), sperm whale metMbCN. However, for the elephant Mb, we find that the dramatically altered functional properties arise not so much from the replacement of E7 His by Gln, but from the concomitant insertion of a new amino acid (Phe CD4) into the heme pocket that is found well outside the pocket in all other Mbs (Emerson et al., 1989).

METHODS

Our reference protein is chosen as WT sperm whale myoglobin in the cyanide-ligated ferric state, for which there are available several excellent X-ray crystal structures on related derivatives such as MbO_2 and MbCO. Cyanide is isoelectronic and isostructural with CO in preferring a linear Fe-C-N unit perpendicular to the heme, and has the added advantage of also acting as a hydrogen-bond acceptor towards the ubiquitous E7 distal His (Lecomte and La Mar, 1987). The determination of the orientation of the magnetic axes requires the assignment of the ^1H NMR signals close (< 7.5 Å) to the heme iron, the X-ray coordinates of protons in some arbitrary iron-based coordinate system x',y',z' (chosen as the pseudo-symmetry axes which include the heme normal and pass through the heme meso-Hs; see Figure 1), the components of the diagonal magnetic susceptibility tensor, and the Eulerian rotation matrix, $R(\alpha,\beta,\gamma)$, that transforms x',y',z' into the magnetic coordinate system, x,y,z (in which χ is diagonal).

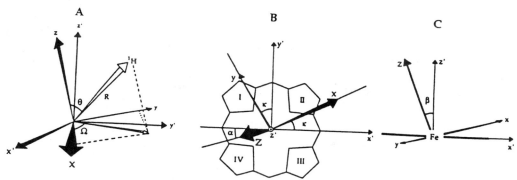

Figure 1.　(A) Molecular pseudo-symmetry coordinate system (x',y',z') refers to the iron-centered symmetry axes derived from the X-ray coordinates, and the magnetic coordinate system (x,y,z) refers to the axes in which the susceptibility tensor is diagonal. The Eulerian rotation matrix $R(\alpha,\beta,\gamma)$ converts x',y',z' to x,y,z. (B) and (C) are the face-on and edge-on view of the magnetic axes, x,y,z, for metMbCN that allow the best agreement between calculated and observed dipolar shifts. The angle β defines the tilt of the z axis from the heme normal, α defines the angle between the projection of the z axis on heme plane and x' axis, and the location of the magnetic rhombic axes are defined as $\kappa \sim \alpha + \gamma$ for small values of β.

Numerous assignments for metMbCN have been determined by a combination of isotope labeling of the heme and 1D NOEs. We have shown that the remainder of the assignment for amino acid residues in the heme pocket (Figure 2) can be derived from 2D NOESY and COSY maps, in spite of the protein being paramagnetic (Emerson and La Mar, 1989a). In the natural genetic variant, elephant Mb (E7 His → Gln) and the synthetic sperm whale Mb point mutant (E7 His → Val), we utilized 1D NOEs to assign the signals needed for our spectral interpretation. The rotation matrix, $R(\alpha,\beta,\gamma)$, was determined (Emerson and La Mar, 1989b) by carrying out a computer least-squares search for the values of the angles α,β,γ that minimize the error function, F/n, between the observed dipolar shifts, $\delta_{dip}(obs)$, and those calculated, $\delta'_{dip}(calc)$, in the crystal pseudo-symmetry coordinate system, x',y',z':

$$\frac{1}{n}F(\alpha,\beta,\gamma) = \sum_{\substack{n \\ \text{all non-coord. residues}}}[\delta_{dip}(obs) - \delta_{dip}(calc)]^2 , \qquad [1]$$

where:　　　$\delta_{dip}(calc) = \delta'_{dip}(calc)\ R(\alpha,\beta,\gamma)$ 　　　　　　　　　[2]

and　　$\delta'_{dip}(calc) = -\f(1,3N)\Delta\chi_{ax}(3\cos^2\theta' - 1)r^{-3} - \f(1,2N)\Delta\chi_{rh}\sin^2\theta'\cos2\phi'r^{-3},$ 　[3]

r,θ',ϕ' are the polar coordinates in the molecular coordinate system, x',y',z', and $\Delta\chi_{ax} = \chi_{zz} - 1/2(\chi_{xx} + \chi_{yy})$, and $\Delta\chi_{rh} = \chi_{xx} - \chi_{yy}$. The individual components of χ have been computed to yield $\Delta\chi_{ax} = 2198 \times 10^{-12}$ and, $\Delta\chi_{rh} = -573 \times 10^{-12}$ m^3/mol (Horrocks and Greenberg, 1973). The magnetic axes are considered determined when F/n is minimized and a plot of calculated versus observed dipolar shift for all signals yields a straight line with unit slope.

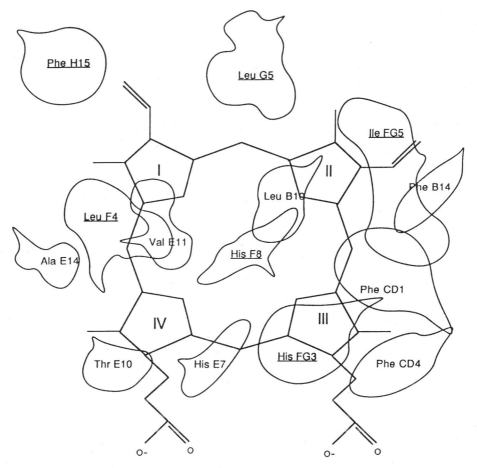

Figure 2. Relative positions of amino acid side chains which are in contact with the heme in sperm whale Mb. To differentiate the amino acids on the proximal and distal side, the former residues are underlined.

RESULTS AND DISCUSSION

The ^{1}H NMR spectra of WT sperm whale, elephant (E7 Gln), and sperm whale point mutant E7 His→Val metMbCN complexes are shown in Figure 3. 1D and 2D NOE assignments have been reported previously (Emerson and La Mar, 1989a; Emerson et al., 1989). The region of the 2D NOESY map of WT metMbCN illustrating the large number of detectable dipolar cross peaks between hyperfine-shifted peaks and the diamagnetic envelope is reproduced in Figure 4; selected cross peaks are identified in the caption. The observed shift, δ_{obs}, leads to the hyperfine shift, δ_{hf}, according to:

$$\delta(obs) = \delta_{hf} + \delta_{dia} , \qquad\qquad [4]$$

where δ_{dia} is the diamagnetic shift for the same residue in sperm whale MbCO (Dalvit & Wright, 1987). In general, the hyperfine shift contains both dipolar and contact contributions (Satterlee, 1986),

246

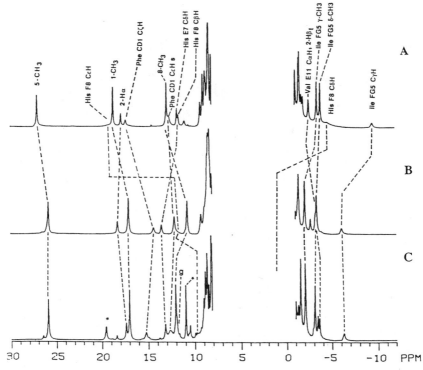

Figure 3.   500 MHz ^{1}H NMR spectra of the metMbCN complexes at 25 °C in ^{2}H$_2$O, pH = 8.6.
(A) Wild type sperm whale Mb; the assignments for the resolved resonances are
given (Emerson and La Mar, 1989a). (B) Sperm whale point mutant [E7 His →
Val]; the analogous assignments are connected with dotted lines. (C) Elephant Mb
[E7 Gln], assignments are indicated by dotted lines; the signals marked with an
asterisk arise from a residue in the heme pocket that is not present in sperm whale Mb
(Phe CD4). The signal marked 'g' arises from the E7 Gln.

$$\delta_{hf} = \delta_{con} + \delta_{dip} \ .$$
[5]

For non-coordinated residues, $\delta_{con} = 0$, so Eq. [4] directly yields $\delta_{dip}(obs)$ with available δ_{dia}.
The $\delta(obs)$ and resulting $\delta_{dip}(obs)$ for forty resonances have been assigned unambiguously and
the results for some of these protons are listed in Table 1 (Emerson and La Mar, 1989a). The 40
assigned peak shifts comprised the input data for our least-squares search for R(α,β,γ) using
Eqs. [1] - [3]. This search leads to a unique minimum with the magnetic axes x,y,z as depicted
in B and C of Figure 1 with $\alpha = 0°$, $\beta = -14.5°$, $\gamma = 35°$. The z axis is tilted ~ 15° (β) toward the
δ-meso-H, α is the angle between the projection of z on the heme plane and the x' axis and the
x,y axes make a projection on the heme plane that makes an angle $\kappa \sim \alpha + \gamma$ with the old x',y'
axes (Figure 1B). The resulting $\delta_{dip}(calc)$ data for some of the protons in this magnetic
coordinate system are listed in Table 1 and can be compared to the $\delta_{dip}(obs)$; a plot of $\delta_{dip}(obs)$
versus $\delta_{dip}(calc)$ is illustrated in Figure 5 which shows an excellent linear correlation with unit
slope and a mean standard deviation, $\bar{\sigma} = \sqrt{F/n} \leq 0.5$ ppm.

The determined magnetic axes reveal that the z-axis tilt coincides within experimental error
with the Fe-O vector determined for the tilted FeCO unit in the X-ray crystal structure of MbCO
(Kuriyan et al., 1986). Since FeCO and FeCN are isostructural, *we have proposed that the*

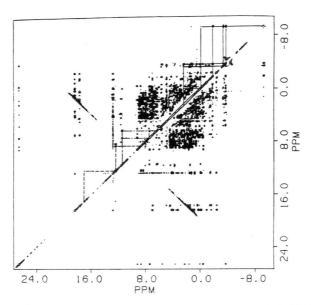

Figure 4. 360 MHz ^{1}H phase-sensitive NOESY spectrum of metMbCN in D_2O pH 8.6, at 30°C. The cross peaks for Phe CD1 (dashed line), His F8 (dash-dotted line), and Ile FG5 (solid line) are labelled. The mixing time was 50 msec. The residual water signal was suppressed by decouple power in the relaxation delay, evolution and mixing period. 256 blocks of 8192 data points were collected with alternate scans saved in adjacent blocks. The final data matrix was zero-filled to 2048 x 2048 complex points with a digital resolution of 7.6 Hz/point.

major molecular interaction that determines the orientation of the major magnetic axis is the tilt of the bound cyanide ligand by distal steric interactions (Emerson and La Mar, 1989b). The orientation of the in-plane axes, as reflected by the angle $\kappa = \alpha + \gamma$, coincides within experimental error, with the projection of the proximal His F8 imidazole projection on the heme plane. Thus the rhombic axes are determined by the axial iron-His F8 interaction, as previously proposed (Shulman et al., 1971). The conclusion on the location of the rhombic axes is the same as those proposed solely on the basis of the asymmetric π spin distribution of the heme, as reflected in the contact shift pattern of the heme methyls. The orbital hole determines both the rhombic magnetic axes as well as the unpaired spin distribution in the heme.

A noteworthy observation is that, while the heme hyperfine shifts in model compounds and metMbCN are very similar (La Mar et al., 1982), there is a strong discrepancy between the axially bound His F8 hyperfine shift pattern in the model and the protein (Emerson and La Mar, 1989a). While the two broad His F8 non-labile ring protons exhibit similar shifts in model and protein, the upfield peak arises from $C_\varepsilon H$ in models and $C_\delta H$ in the protein. Thus the hyperfine shift directions between $C_\varepsilon H$ and $C_\delta H$ of the axial imidazole are reversed between model and protein. This unprecedented discrepancy between model compound and protein hyperfine shift can be resolved upon considering the quantitative origin of the His F8 hyperfine shift. Since His F8 is coordinated to the iron, it exhibits both contact and dipolar shifts (Satterlee, 1986). The $\delta(obs)$ and $\delta_{hf}(obs)$ of His F8 are listed in Table 2. Using the magnetic coordinate system determined above, we determine $\delta_{dip}(calc)$, which with $\delta_{hf}(obs)$ and Eq.[5] yields δ_{con}. The resulting pattern of δ_{con} for the His F8 imidazole ring protons in metMbCN given in Table 2 is very similar to that found in model compounds, indicating that the axial iron-His bonding is relatively unperturbed in the protein and that the difference in hyperfine shift pattern must originate in δ_{dip} due to the rotated magnetic axes.

Table 1. Chemical Shift Data for Non-coordinated Heme Pocket Amino Acid Residues of Sperm Whale metMbCN[a]

Residue	Position	$\delta_{DSS}(obs)^{a,b}$	$\delta_{dia}{}^{c}$	$\delta_{dip}(obs)^{a,d}$	$\delta_{dip}(calc)^{a,e}$
Phe B14	$C_\delta Hs$	8.04	7.01	1.03	0.84
	$C_\epsilon Hs$	8.37	6.45	1.92	1.72
	$C_\zeta H$	8.40	5.22	3.18	3.40
Phe CD1	$C_\delta Hs$	8.70	7.29	1.41	1.57
	$C_\epsilon Hs$	12.58	6.08	6.50	7.64
	$C_\zeta H$	17.27	4.72	12.55	12.11
Phe CD4	$C_\delta Hs$	7.69	6.72	0.97	1.77
	$C_\epsilon Hs$	8.05	6.52	1.53	2.45
Thr E10	$C_\alpha H$	2.50	3.83	-1.33	-1.53
	$C_\beta H$	2.69	3.97	-1.28	-1.92
	$C_\gamma H_3$	-1.56	1.51	-3.07	-3.91
Ile FG5	$C_\alpha H$	2.35	4.29	-1.94	-2.10
	$C_\beta H$	-0.13	1.13	-1.26	-1.10
	$C_\gamma H$	-9.60	-0.28	-9.32	-9.57
	$C_\gamma H'$	-1.91	0.85	-2.76	-2.06
	$C_\gamma H_3$	-3.46	1.36	-4.82	-4.62
	$C_\delta H_3$	-3.83	1.47	-5.30	-5.32
His E7	$C_\beta H$	3.81	3.08	0.73	1.61
	$C_\beta H'$	4.41	2.72	1.69	2.96
	$C_\delta H$	11.71	4.97	6.74	10.31
	$N_\epsilon H$	23.67	3.88	19.79	25.59
	$C_\epsilon H$	f	7.17	--	-4.66

a   Shifts in ppm at 25 °C in ^2H_2O or ^1H_2O (for exchangeable protons) at pH 8.6.
b   Observed shift referred to 2,2-dimethyl-2-silapentane-5-sulfonate, DSS.
c   Diamagnetic shift obtained from MbCO data (Dalvit and Wright, 1987).
d   Observed dipolar shift obtained from Eqs. [4] and [5] with $\delta_{con} = 0$.
e   Dipolar shift calculated with Eqs. [2] and [3] with R(0°, -14.5°, 35°) using the MbCO crystal coordinates.
f   Not assigned to date.

The hypothesis that the altered His F8 hyperfine shift pattern results from z-axis (Fe-CN) tilting was tested by assuming δ_{con} in Table 2 is independent of the z-axis tilt angle β, and calculating δ_{dip} as a function of β while keeping $\kappa = \alpha + \gamma$ constant (Emerson and La Mar, 1989b). The results of such a calculation are shown in Figure 6; also included are the predicted $\delta_{dip}(calc)$ for Phe CD1 and Ile FG5. The His F8 shifts change with β most dramatically for the non-labile ring protons, and with slopes of opposite sign. As $\beta \to 0$, the $C_\epsilon H$ and $C_\delta H$ shift positions interchange, and are predicted to have values virtually identical to those observed in model compounds where the z axis is known to coincide with the heme normal. This observation of the dramatic change of the His F8 ring proton hyperfine shifts with z axis tilt indicates that the pattern of observed hyperfine shifts may serve as a probe of the orientation of

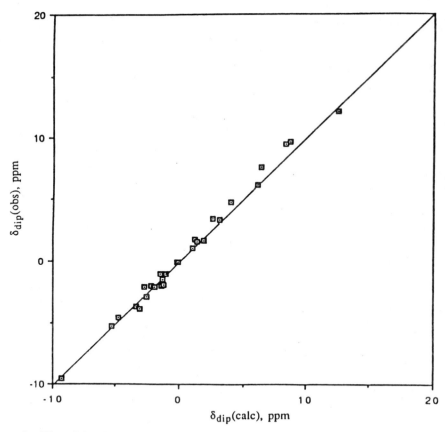

Figure 5. Plot of the observed dipolar shift versus the calculated dipolar shift. The calculated dipolar shift is based on equation [3] and the coordinate system described in Figure 1B and 1C with $\alpha = 0$, $\beta = -14.5°$ and $\gamma = 35°$. The unit slope confirms the quality of the fit.

the principal axes of the susceptibility tensor. The predicted changes for the two non-coordinated residues with the largest δ_{dip}, Ile FG5 and Phe CD1, would serve as auxiliary indicators of the magnetic axes. Inasmuch as the tilt of the z-axis from the heme normal coincides with the expected Fe-CN tilt in WT sperm whale metMbCN (Emerson and La Mar, 1989b), the His F8, in combination with Ile FG5 and Phe CD1 shifts, provide us with a valuable probe of variable cyanide tilt due to altered steric interaction in genetic variants and point mutants.

E7 His→Val Mb Mutant

The ^{1}H NMR trace of the met-cyano complex shown in Figure 3B is qualitatively very similar to that of WT metMbCN (Figure 3A). The resolved heme resonances, as well as those of Phe CD1, His F8 and Ile FG5, have been assigned by 1D NOEs, and are appropriately labeled in the Figure; the chemical shifts are listed in Table 3. It is observed that the four heme methyls move slightly closer, suggesting a negligible rotation of the in-plane rhombic magnetic axes with respect to WT protein. More importantly, the His F8 ring proton (either $C_\delta H$ or $C_\epsilon H$ as established by short $T_1 \sim 3$ ms) appears at 11.5 ppm: i.e., $C_\epsilon H$ has shifted dramatically upfield and $C_\delta H$ downfield compared to WT metMbCN. Also noted was the fact that both the Phe CD1 low-field dipolar shifts and the Ile FG5 upfield dipolar shifts are smaller. The g values

Table 2. Separation of Contact and Dipolar Shifts for Heme Methyls and His F8 Protons for Sperm Whale metMbCN

Resonances	δ_{DSS}(obs)[a,b]	δ_{dia}[a,c]	δ_{hf}[a,d]	δ_{dip}(calc)[a,e]	δ_{con}[a,f]
Heme					
1-CH₃	18.62	3.63	14.99	-3.41	18.40
3-CH₃	4.76	3.79	0.97	-5.70	6.67
5-CH₃	27.03	2.53	24.5	-3.32	27.82
8-CH₃	12.88	3.59	9.29	-6.51	15.80
His F8					
N_pH	13.2	7.15	6.05	5.98	0.07
$C_\alpha H$	7.51	2.90	4.61	5.35	-0.74
$C_\beta H$	6.43	1.72	4.71	7.85	-3.14
$C_\beta H'$	11.68	1.55	10.13	5.28	4.85
$C_\delta H$	-4.7	1.13	-5.83	0.39	-6.22
$N_\delta H$	20.11	9.36	10.75	16.32	-5.57
$C_\epsilon H$	19.2	1.66	17.54	33.32	-15.78

a   Shifts in ppm at 25 °C at pH 8.6 (upfield shifts are negative).
b   Observed shifts referenced to DSS.
c   Diamagnetic shift obtained from MbCO (Dalvit and Wright, 1987).
d   Hyperfine shift obtained via Eq. [4].
e   Dipolar shift calculated via Eqs. [2] and [3] and R(0°, -14.5°, 35°).
f   Contact shift calculated via Eq. [5].

are unchanged, indicating that the anisotropy is conserved and that the magnetic axes must be altered. The observed His F8, Phe CD1 and Ile FG5 shifts for [E7His→Val] metMbCN are semiquantitatively *predicted by the dipolar field of WT metMbCN if the z axis were tilted less (β ~ 7-8°; arrow B in Figure 6) from the heme normal than in the WT (β ~ 15°; arrow A in Figure 6).* Hence we conclude that the smaller E7 Val causes less steric strain on the bound CN⁻, and therefore causes less Fe-CN tilt than in WT (His E7). The ¹H shifts of the E7His→Gly mutant (not shown) suggest even less of a tilt, although assignments are incomplete at this time. Therefore the dramatic change in hyperfine shift patterns of the met-cyano complexes of sperm whale E7 point mutants appears to be directly related to the tilt of the major magnetic axis from the heme normal, and this z-axis directly monitors the steric tilt from the heme normal induced in the Fe-CN moiety.

It is also noted that, while the His E7 ring proton peak is missing in [E7 His→Val] metMbCN which is resolved in WT metMbCN (Figure 3A), as expected, no additional resolved peak is observed for the E7 Val (Figure 3B). This is consistent with the smaller dipolar shift expected for the smaller Val side chain and its resultant failure to be resolved from the diamagnetic envelope.

Elephant Mb (E7 His→Gln)

One-dimensional NOE spectra of this protein have led to the assignment of most of the heme and several important amino acid resonances resolved in the ¹H NMR spectrum shown in Figure 3C (Krishnamoorthi et al., 1984; Emerson et al., 1989). The observed shifts for the residues His F8, Ile FG5 and Phe CD1 are very similar to those of [E7 His→Val]metMbCN and are listed in Table 3. The broad peak at 9.5 ppm could not be unambiguously assigned by NOEs, but the large linewidth and short T_1 clearly identify it as either $C_\epsilon H$ or $C_\delta H$ of His F8. The

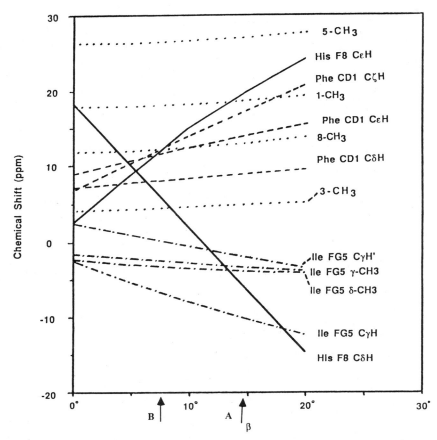

Figure 6. Plot of the predicted chemical shifts relative to DSS as a function of β. The angle β is the tilt of the magnetic coordinate system from the heme normal. The rhombic axes ($\kappa = \alpha + \gamma$), as well as the δ_{con} values from Table 2, were held constant. The optimum fit for WT sperm whale Mb was found to be at $\beta = -14.5°$ and is shown by arrow A, and the fit for the E7 His $\rightarrow$ Val and elephant Mb was found to be at $\beta = -7$ to $-8°$ as shown by arrow B.

pattern of the heme methyl shifts is again the same as that of WT sperm whale metMbCN, indicating that the in-plane magnetic axes are largely unchanged. The His F8, Ile FG5 and Phe CD1 shifts, however, are significantly altered in the same manner as for [E7 His→Val]metMbCN. Remarkably, the pattern in the His F8 ring hyperfine shifts, and Ile FG5 and Phe CD1 dipolar shifts are also reasonably well reproduced by our calculated hyperfine shift as a function of β, for the reduced value $\beta \sim 7°$ (indicated by arrow B in Figure 6).

We conclude that the z-axis of χ is tilted only 7° in elephant metMbCN and that this indicates that the bound CN is *tilted only half the amount in WT sperm whale metMbCN*. While E7 Gln is expected to offer less steric hindrance for a bound cyanide than E7 His, it is not clear that the E7 His $\rightarrow$ Gln is the only substitution taking place in the heme pocket. We find in elephant metMbCN hyperfine shifted resonances not only for all heme pocket residues in common with sperm whale Mb and [E7 His→Val]Mb, but in addition to the substituted Gln E7 peaks (labeled g in Figure 3C), we observe strongly shifted peaks that arise from *a residue not found in the heme pocket in sperm whale Mb* (Emerson et al., 1989). This residue appears to be an aromatic side chain suggestive of Phe CD4, and gives rise to the peaks labeled by

Table 3.   Observed Chemical Shift Data of Various Cyano-Met Mb Complexes[a,b]

Residue		WT (E7 His) Sperm Whale Mb	(E7 His → Val) Sperm Whale Mb	WT (E7 Gln) Elephant Mb
Phe B14	$C_\delta Hs$	8.04	7.70	c
	$C_\epsilon Hs$	8.37	7.90	c
	$C_\zeta H$	8.4	8.27	c
Phe CD1	$C_\delta Hs$	8.7	8.56	8.78
	$C_\epsilon Hs$	12.58	12.04	12.58
	$C_\zeta H$	17.27	14.22	15.16
Phe CD4	$C_\delta Hs$	7.69	c	8.12
	$C_\epsilon Hs$	8.05	7.57	10.91
	$C_\zeta H$	c	c	19.52
Thr E10	$C_\alpha H$	2.5	c	2.62
	$C_\beta H$	2.69	c	2.92
	$C_\gamma H_3$	-1.56	-1.37	-1.54
Ile FG5	$C_\alpha H$	2.35	2.94	2.97
	$C_\beta H$	-0.13	1.93	1.67
	$C_\gamma H$	-9.6	-6.09	-6.29
	$C_\gamma H'$	-1.91	-0.16	-0.43
	$C_\gamma H_3$	-3.46	-2.02	-2.03
	$C_\delta H_3$	-3.83	-3.31	-3.08
His F8	$N_p H$	13.2	13.85	13.38
	$C_\alpha H$	7.51	8.28	7.84
	$C_\beta H$	6.43	9.18	8.7
	$C_\beta H'$	11.08	13.4	13.08
	$C_\delta H$	-4.7	11.5[d]	9.5[d]
	$N_\delta H$	20.11	19.47	19.15
	$C_\epsilon H$	19.2	11.5[d]	9.5[d]

a   Shifts in ppm at 25 °C in D_2O at pH 8.6.
b   Observed shifts referred to DSS.
c   Not assigned to date.
d   The assignment is either the His F8 $C_\delta H$ or the $C_\epsilon H$ proton.

asterisks (*) in Figure 3C. Hence, while the functional and NMR spectral properties of elephant Mb are strongly perturbed in comparison to those of sperm whale Mb, the effect is possibly due more to the insertion of the new side chain than the replacement of His by Gln.

In order to establish whether the new amino acid is inserted into the heme pocket in physiologically relevant forms of elephant Mb, and to unequivocally identify this residue, we resort to the more conventional COSY maps of the aromatic region of *both the diamagnetic MbCO and MbO₂ complexes*. Of particular interest are the three known aromatic side chains in the distal heme pocket, Phe CD1, Phe CD4 and Phe B14 (see Figure 2). The aromatic ring spin systems of these three Phe residues are identified as shown in Figure 7. The assignments of these resonances are verified by relayed COSY, double quantum spectra, and NOESY maps (Yu et al., 1989). While the Phe CD1 and Phe B14 in elephant MbCO and MbO₂ have almost the same chemical shifts as in WT sperm whale MbCO (Mabbutt and Wright, 1985; Dalvit and

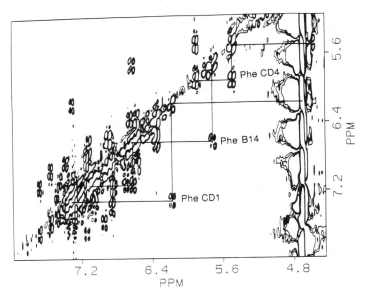

Figure 7.   Aromatic region of the phase-sensitive DQF-COSY spectrum of elephant MbCO in
D$_2$O, pH 8.7, at 30°C. The solid lines indicate the connectivities of the aromatic ring
spin systems of the three Phe residues (Phe CD1, Phe CD4, and Phe B14). The t_1
ridge at 4.71 ppm is due to residual water signal, which is only partially saturated.
136 blocks of 4096 data points were collected with alternate scans saved in adjacent
blocks. The final data matrix was zero-filled to 1024 x 1024 complex points with a
digital resolution of 5.1 Hz/point.

Wright, 1987), the resonances of Phe CD4 are found to be shifted more upfield by 2.0, 1.0, and
0.8 ppm for C$_\zeta$H, C$_\varepsilon$Hs, and C$_\delta$Hs, respectively, in elephant relative to WT sperm whale
MbCO. The further upfield shifts of Phe CD4 imply that this residue has moved much closer to
the heme center in elephant Mb, and experiences much larger heme ring current shifts. Based on
ring current calculations and NOE data, it is estimated that the C$_\zeta$H of Phe CD4 has moved
toward the iron by ~ 4 Å relative to that in WT sperm whale Mb. The observation of the
dramatic conformational change of the Phe CD4 in elephant Mb (E7 His → Gln) clearly indicates
that the simple interpretation of substitution of one amino acid residue (i.e., E7 His) by another
(i.e., Gln) is not valid, and that conformational changes of the substituted proteins have to be
considered for interpretation of their functional properties. Thus both the unexpectedly large
steric crowding of the heme pocket in elephant MbCO (Kerr et al., 1985), as well as the
increased redox potential (Bartnicki et al., 1983) and low rate of autoxidation (Romero-Herrera
et al., 1981), probably reflect more the added presence in the ligand binding pocket of the non-
polar Phe CD4 side chain than the substitution E7 His → Gln.

ACKNOWLEDGEMENTS

This work was supported by grants from the National Institutes of Health, HL16087, GM-
33775 and GM-31756.

REFERENCES

Antonini, E. and Brunori, M., 1971, "Hemoglobin and Myoglobin in their Reactions with Ligands", North Holland, Amsterdam.

Bartnicki, D.E., Mizukami, H. and Romero-Herrera, A.E., 1983, Interaction of ligands with the distal glutamine in elephant myoglobin, *J. Biol. Chem.* 258: 1599.

Dalvit, C. and Wright, P.E., 1987, Assignment of resonances in the ^{1}H-nuclear magnetic resonance spectrum of the carbon monoxide complex of sperm whale myoglobin by phase-sensitive two-dimensional techniques, *J. Mol. Biol.* 194: 313.

Dickerson, R.E. and Geis, I., 1983, "Hemoglobin: Structure, Function, Evolution and Pathology", Benjamin Cummings Publishing Company Inc., Menlo Park, CA.

Emerson, S.D. and La Mar, G.N., 1989a, Molecular and electronic structural characterization of met-cyano myoglobin in solution: resonance assignment of heme cavity residues by 2D NMR (submitted for publication).

Emerson, S.D. and La Mar, G.N., 1989b, Molecular and electronic structural characterization of cyano metmyoglobin in solution: the orientation of the magnetic susceptibility tensor as a probe of steric tilt of bound ligand (submitted for publication).

Emerson, S.D., La Mar G.N. and Krishnamoorthi, R., 1989, ^{1}H NMR resonance assignments and analysis of distal amino acid interactions in the metcyano complex of elephant myoglobin (manuscript in preparation).

Horrocks, Jr., W.D. and Greenberg, E.S., 1973, Evaluation of dipolar nuclear magnetic resonance shifts in low-spin hemin systems: ferricytochrome c and metmyoglobin cyanide, *Biochim. Biophys. Acta* 322: 38.

Kerr, E.A., Yu, N.-T., Bartnicki, D.E. and Mizukami, H., 1985, Resonance Raman studies of CO and O_2 binding to elephant myoglobin (distal His (E7) $\rightarrow$ Gln), *J. Biol. Chem.* 260: 8360.

Krishnamoorthi, R., La Mar, G.N., Mizukami, H. and Romero, A., 1984, A ^{1}H NMR comparison of the met-cyano complexes of elephant and sperm whale myoglobin, *J. Biol. Chem.* 259: 8826.

Kuriyan, J., Wilz, S., Karplus, M. and Petsko, G.A., 1986, X-ray structure and refinement of carbon-monoxy (Fe-II)-myoglobin at 1.5 Å resolution, *J. Mol. Biol.* 192: 133.

La Mar, G.N., Chatfield, M.J., Peyton, D.H., de Ropp, J.S., Smith, W.S., Krishnamoorthi, R., Satterlee, J.D. and Erman, J.E., 1988, Solvent isotope effects on NMR spectral parameters in high-spin ferric hemoproteins: an indirect probe for distal hydrogen bonding, *Biochim. Biophys. Acta* 956: 267.

La Mar, G.N., de Ropp, J.S., Chacko, V.P., Satterlee, J.D. and Erman, J.E., 1982, Axial histidyl imidazole non-exchangeable proton resonances as indicators of imidazole hydrogen bonding in ferric cyanide complexes of heme peroxidases, *Biochim. Biophys. Acta* 708: 317.

La Mar, G.N., Krishnamoorthi, R., Smith, K.M., Gersonde, K. and Sick, H., 1983, Proton nuclear magnetic resonance investigation of the conformation-dependent spin equilibrium in azide-ligated monomeric insect hemoglobins, *Biochemistry* 22: 6239.

Lecomte, J.T.J. and La Mar, G.N., 1986, ^{1}H NMR probe for hydrogen bonding of distal residues to bound ligands in heme proteins, *J. Am. Chem. Soc.* 109: 7219.

Mabbutt, B.C. and Wright, P.E., 1985, Assignment of heme and distal amino acid resonances in the ^{1}H NMR spectra of the carbon monoxide and oxygen complexes of sperm whale myoglobin, *Biochim, Biophys. Acta* 832: 175.

Mayer, A., Ogawa, S., Shulman, R.G., Yamane, T., Cavaleiro, J.A.S., Rocha Gonsalves, A.M. d'A., Kenner, G.W. and Smith, K.M., 1974, Assignment of the paramagnetically shifted heme methyl nuclear magnetic resonance peaks of cyanometmyoglobin by selective deuteration, *J. Mol. Biol.* 86: 749.

Phillips, S.E.V., 1980, Structure and refinement of oxymyoglobin at 1.6 Å resolution, *J. Mol. Biol.* 142: 531.

Phillips, S.E.V. and Schoenborn, B.P., 1981, Neutron diffraction reveals oxygen-histidine hydrogen bond in oxymyoglobin, *Nature* 292: 81.

Romero-Herrera, A.E., Goodman, M., Dene, H., Bartnicki, D.E. and Mizukami, H., 1981, An exceptional amino acid replacement on the distal side of the iron atom in proboscidean myoglobin, *J. Mol. Evol.* 17: 140.

Satterlee, J.D., 1986, Proton NMR studies of biological properties involving paramagnetic heme proteins, *in:* "Metal Ions in Biological Systems", Marcel Dekker, New York.

Shulman, R.G., Glarum, S.H. and Karplus, M., 1971, Electronic structure of cyanide complexes of hemes and heme proteins, *J. Mol. Biol.* 57: 93.

Springer, B.A., Egeberg, K.D., Sligar, S.G., Rohlfs, R.J., Mathews, A.J. and Olson, J.S., 1989, Discrimination between oxygen and carbon monoxide and inhibition of autoxidation by myoglobin, *J. Biol. Chem.* 264: 3057.

Wüthrich, K., 1986, "NMR of Proteins and Nucleic Acids", Wiley, New York.

Yu, L.P., La Mar, G.N. and Mizukami, H., 1989, ^{1}H NMR studies of the heme pocket structure of carbon monoxide and oxygen complexes of elephant myoglobin (manuscript in preparation).

SPECTROSCOPY OF MOLECULAR STRUCTURE AND DYNAMICS

Rudolf Rigler

Department of Medical Biophysics, Karolinska Institutet, Box 60400, S-104 01
Stockholm, Sweden

INTRODUCTION

Dynamic aspects of molecular structure at atomic resolution have become of increasing importance as experimental and theoretical methods have been improving for the analysis of molecular motions (Frauenfelder et al., 1979; Artemiuk et al., 1979; Karplus and McCammon, 1983). Simulating the motion of individual atoms from the knowledge of their position and interatomic forces (McCammon and Harvey, 1987; Brooks et al., 1988) has opened the way to predict molecular behavior and functional aspects of molecular assemblies such as complex enzymes and nucleic acids. An important part of this approach are techniques which can provide experimental data on structural variation and motions in a time range which is accessible to molecular dynamics simulation (10^{-12}-10^{-9}s).

Amongst various methods including inelastic neutron scattering, NMR and chemical relaxation methods, time resolved emission spectroscopy is of particular importance. (1) Its inherently high time resolution covering the ps and fs time domain overlaps the timescale for simulating structure and motion of molecular assemblies of molecular weights up to 10^4 kD (Rigler et al., 1984; Zinth et al., 1986). (2) The random motion of individual residues in specific positions can be investigated by photoselection in complex structures (Ehrenberg and Rigler, 1972; Rigler and Ehrenberg, 1976).

For peptides, proteins and enzymes, the aromatic amino acids phenylalanine, tyrosine and tryptophan can be used as specific reporter groups for side chain motion, interdomain flexibility, and overall motions. In addition, coenzymes such as NADH or riboflavin can provide additional information on dynamic aspects of enzyme structures. With regard to nucleic acids, nucleic acid bases with a significant fluorescent emission are only found in ribonucleic acids such as wybutine (Y-base) or 7 m guanosin in tRNA (Claesens and Rigler, 1986; Rigler and Claesens, 1986). The recent advances of nucleic acid synthesis provide means for incorporating highly fluorescent nucleic acid bases such as 2-aminopurine in synthetic deoxy- and ribonucleic acid chains at strategic positions (Rigler and Claesens, 1986; Nordlund et al., 1989). With the repertoire of gene reconstruction these probes can be introduced at various places in the gene.

ANALYSIS OF MOLECULAR CONFORMATIONS AND MOTIONS

Conformational States

From the analysis of the excited states of conformational probes, e.g., a tryptophan in a protein, information on conformational equilibria and their distribution can be obtained if the deexcitation rates k_d are very much larger than the rates of conformational transitions k_c

$$k_d = \frac{1}{\tau_d} \gg k_c \quad \text{where } \tau_d = \text{lifetime of the excited state.}$$

As example for this situation Tyrosine 13 in the calcium binding protein Calbindin D_{9k} (Figure 1) can be given which was analysed by pulsed fluorimetry (Rigler et al., 1989). Calbindin D_{9k} belongs to a superfamily of proteins binding Ca^{2+} in a cooperative manner. The existence of two excited states with lifetimes of 340 ps and 1.3 ns which are populated by 85% and 15%, respectively, were found for Tyr 13 (Figure 2a). A comparison of lifetimes of free tyrosine and tyrosine dipeptides (Ross et al., 1986) with those of Tyr 13 of Calbindin D_{9k} indicates that the 2 states in Tyr 13 are likely due to an interaction with the carboxyl group(s) of Glu 35:

Glu 35 = COO⁻ ⁺HO —⟨⟩— Tyr 13.

Comparison with molecular dynamic calculations by Alström et al. (1989) shows that the OH-group of Tyr 13 occurs during a 150 ps simulation to 80% in a position which would allow H-bonding to the carboxyl group of Glu 35. We are presently analyzing various point mutations of calbindin in order to find out whether they influence this conformational equilibrium.

Distribution of Conformational Substates (S)

After a short excitation pulse conformational substates exist in their excited state (S*) if their decay rate (k_d) is faster than the conformational exchange rate (k_c).

The observed fluorescence intensity decays with discrete exponentials:

$$I(t) = A_i \sum_i e^{-k_d^i t}$$

if the substates S_i^* populated by amplitude A_i are discrete

$$S_1^* \overset{k_c^*}{\leftrightarrow} S_2^* \leftrightarrow S_n^*$$

$$\downarrow k_d^1 \quad \downarrow k_d^2 \quad \downarrow k_d^n$$

$$S_1 \overset{k_c}{\leftrightarrow} S_2 \leftrightarrow S_n$$

or with

$$I(t) = \int f(k_d^i) e^{-k_d^i t} \, dk_d$$

if substates are distributed with a distribution function $f(k_d^i)$.

258

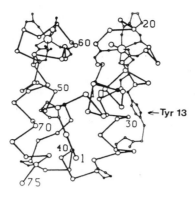

Figure 1. Structure of Calbindin D$_{9k}$ as represented by the backbone fold (after Szebenyi and Moffat, 1986).

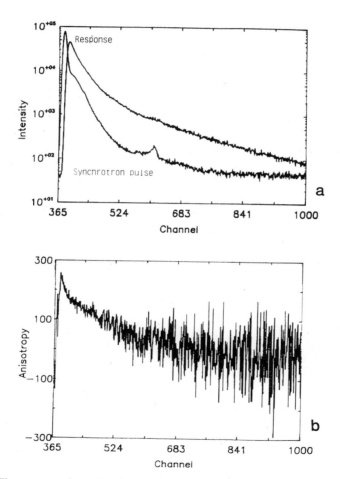

Figure 2. (a) Fluorescence intensity decay of wild type Calbindin and synchrotron excitation pulse at 280 nm, 1 channel = 21 ps. (b) Fluorescence anisotropy decay of wild type Calbindin at 25°C. 1 channel = 21 ps. Anisotropy scale x1000 (Rigler et al., 1989).

It has been argued whether side chains occur in conformations which are distributed around a mean value rather than in distinct conformations. We have been able to show that this is the case in the redox-enzyme Thioredoxin which contains tryptophans in position 28 and 31 (Merola et al., 1989). Comparison with mutant enzymes shows that both tryptophans occur in at least 3 local conformers, which have been related to rotoisomers of the indolring relative to the C_α-C_β bond (Petrich et al., 1983). Their distribution shows a strong temperature dependence (Figure 3). At low temperature very broad distributions are found that become very narrow at high temperatures, indicating the presence of many conformational substates at low temperature. Similar behavior was observed in other proteins by E. Gratton and coworkers (Gratton et al., 1987) and points to the general existence of conformation substates as proposed by H. Frauenfelder (Frauenfelder et al., 1979).

Side Chain Mobility in Proteins and Nucleic Acids

Analysis of crystallographic B-factors (Frauenfelder et al., 1979; Artemiuk et al., 1979) show that side chains as well as the peptide backbone atoms can undergo conformational transition leading to a static and/or dynamic equilibrium of substates, depending on the time interval of observation. The existence of stable and dynamic parts, which may have functional importance, can be analyzed in this way. Similarly, mean square deviations from the time average atomic position can be evaluated from molecular dynamics simulation of the time dependent motions of individual atoms.

It is therefore of interest to analyze side chain motion experimentally. The method that offers the highest time resolution and sensitivity rests on the observation of the decay of fluorescence anisotropy after excitation with a pulse of polarized light (Ehrenberg and Rigler, 1972; Rigler and Ehrenberg, 1976). We have by this method analyzed the motion of aromatic side chains in a variety of interesting enzymes and peptides and have compared it with results from molecular dynamics simulations. These data comprise the motion of singular tryptophan

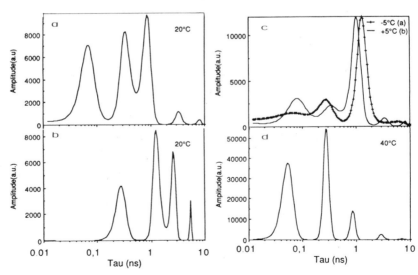

Figure 3. Distribution of conformational states of tryptophan residues of Thioredoxin in oxidized (a) and reduced (b) form at 20°C and for reduced Thioredoxin at -5°, +5°C (c) and 40°C (d) (Merola et al., 1989).

residues in the enzymes Ribonuclease T_1 (MacKerell et al., 1987; MacKerell et al., 1988) and thioredoxin (Merola et al., 1989) as well as in an alamethicin-like membrane spanning peptide (Vogel et al., 1988). In this case, it could be shown that the motion of tryptophan positioned at various places in the 21 amino acid long α-helical peptide

$$^{+}H_2(Ala\text{-}Aib\text{-}Ala\text{-}Aib\text{-}Ala)_x Trp(Ala\text{-}Aib\text{-}Ala\text{-}Aib\text{-}Ala)_{4-x}OMe \quad (x=0,1....4)$$

was dependent whether it was close to the surface of the lipid bilayer in which the peptide was immerged. Both amplitude and time constant of motion could be simulated by a molecular dynamic simulation. The use of tryptophan as a probe to sense conformational dynamics in enzymes exhibiting substrate specificity could be demonstrated in the case of Ribonuclease T_1 (MacKerell et al., 1987). Here the time constant of motion of Trp 59 depends on interaction between enzyme and substrate and senses the conformation of the binding site.

Rotational Relaxation

When excited with a pulse of polarized light, photoselection occurs, leading to an anisotropic distribution of a chromophor in its excited state. The randomization process due to Brownian motion can be followed by watching the decay of the fluorescence anisotropy as a function of time r(t). For the case of a rotor with a marker exhibiting internal motion that is much faster than the overall tumbling rate ($\phi_i^{-1} >> \phi_o^{-1}$)

$$r(t) = r_0 (1-S^2)e^{-t/\phi_i} + S^2 e^{-t/\phi_o}$$

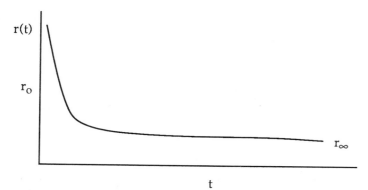

with r_0 = limiting anisotropy at t = 0
   r_∞ = anistotropy at $\phi_i < t < \phi_o$
   ϕ_i = rotational relaxation of internal motion
   ϕ_o = rotational relaxation time of overall motion

r_0 is related to the angle λ between absorption (μ_a) and excitation (μ_e) moment

$$r_0 = (3\cos^2-1)/5$$

The order parameter S is related to r_0 and r_∞ and describes the cone of wobbling with half angle θ (Lipari and Szabo, 1982)

$$S^2 = \frac{r_\infty}{r_0} = [\frac{1}{2} \cos\theta(1+\cos\theta)]^2$$

The anisotropy decay can be simulated by calculating the correlation function of the rotational fluctuation obtained from a molecular dynamic simulation run (Ichiye and Karplus, 1983)

$$r(t) = \frac{2}{5} < P_2[\hat{\mu}_a(0)\hat{\mu}_e(t)] >$$

where $\hat{\mu}_a(0)$ and $\hat{\mu}_e(t)$ are unit vectors along absorption and dipole moments.

Like tryptophan, tyrosine can be used as well as a dynamic probe. The Ca^{2+} binding protein Calbindin D_{9k} is an example. It could be shown that Tyr 13 exhibits a fast internal motion with a time constant of 490 ps and an appreciable amplitude ($\theta=46°$) (Figure 2b). Even in this case, agreement between experiment and simulation could be obtained (Ahlström et al., 1989).

Long Distance Geometry in Proteins and Peptides

The aromatic amino acid residues tryptophan, tyrosine and phenylalanine have overlapping emission and excitation spectra, and resonant energy transfer can occur between tyrosine and tryptophan, phenylalanine and tryptophan as well as phenylalanine and tyrosine. The three amino acids map in different regions in a 2 dimensional excitation-emission map as shown in the case of two proteins (Figure 4). Overlap of the energy surfaces causes resonant energy transfer.

The rate of fluorescence energy transfer (k_T) is proportional to the sixth power of interatomic distance (R_{DA}) as are also the population redistributions within a spin system that underlies the NOE effect:

$$k_T = k_0 \left(\frac{R_0}{R_{DA}}\right)^6.$$

R_0 which depends on spectroscopic properties of donor and acceptor molecules as well as on their mutual orientation (Förster, 1948) is defined as the distance where energy transfer has the same probability as direct emission ($k_T = k_0$).

A detailed treatment of the analysis of k_T as obtained from time resolved decay data in 1-, 2- and 3-dimensional space has been given by Yamazaki et al. (Tamai et al., 1985). The possibility of determining long range distances in proteins by site specific labelling has been demonstrated recently (Amir and Haas, 1987) and will provide important information in addition to a short-range geometry as determined by NOEs. This will be particularly important for proteins and peptides of nonglobular structure in solution.

The presence of aromatic residues provides suitable intrinsic markers between which distances can be determined. We are presently evaluating the end-to-end distance of the neuropeptide galanine, the solution structure of which has recently been determined by NMR (Wennerberg et al., 1989). In this sequence, a Trp and Tyr residue is positioned at the respective ends of the peptide chain.

For time resolved fluorescence spectroscopy, usually mode locked and synchronously pumped dye laser systems are used (Rigler et al., 1984). A beamline for time resolved spectroscopy has been established recently at the MAX-synchrotron at Lund (Sweden) providing a time resolution of better than 10 psec over a spectral range from the VUV to IR (Rigler et al., 1987). This facility has proven to be extremely useful in the study of aromatic amino acids.

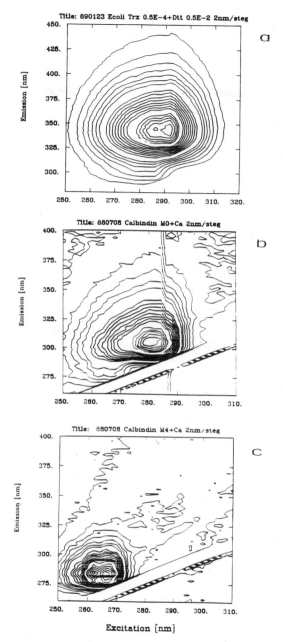

Figure 4. Excitation-emission spectra of (a) Thioredoxin -SH$_2$ (Trp 28,31), (b) Calbindin D$_{9k}$ wild type (Tyr 13), (c) Calbindin D$_{9k}$ mutant lacking Tyr 13 and exhibiting the emission of 5 phenylalanines which is quenched in the presence of Tyr 13.

Correlated Motions in Biomolecules

An important aspect of the dynamics of biopolymers is the coupling of motions occurring at locations in different parts of, e.g., a peptide or a nucleic acid chain. Little information about such couplings, which may provide a clue to important functional properties is available as yet. Allosteric as well as cooperative behavior may involve a linkage of structural dynamics.

In principle there are two ways of analyzing the coupling of events: (a) by applying a local perturbation in terms of site specific interaction, e.g., by a ligand or by optical resonance with a residue (photoselection) and watching the response of a site at a different location; (b) by observing the orientational fluctuation of residues at different positions in a structure, e.g., different spectroscopic markers at either end of a α-helix (Tyr and Trp).

From the intensity of fluctuation of markers moving in a beam of polarized light the rotational and translational motion can be analyzed by the autocorrelation function of the intensity fluctuation (Ehrenberg and Rigler, 1974; Rigler et al., 1979) of marker A or B.

$$G^A(t) = \; < I^A(t) \cdot I^A(0) >$$

$$G^B(t) \; 0 < I^B(t) \cdot I^B(0) >$$

Coupling between A and B can be analyzed from the observed cross correlation function

$$G^{AB}(t) = \; < I^A(t) \cdot I^B(0) >.$$

Significant improvements of the sensitivity and time range of correlation spectroscopy have been made recently (Kask et al., 1989; P. Kask and R. Rigler, unpublished observation) and allow the analysis of a few molecules in the observed volume element.

Comparable information can be obtained from the cross correlation of molecular trajectories as calculated from a molecular dynamic simulation.

Both spectroscopy as well as simulations need to be carried out in relevant structures in order to check the existence of long range dynamic coupling in biological structures (Careri and Wyman, 1984).

Intensity Fluctuation of Molecules

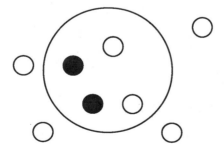

The intensity of molecules excited to fluorescence in a small volume element fluctuates due to translational and rotational motion. The autocorrelation function G(t) of the fluctuating intensity i(t)

$$G(t) = <i(t)i(0)> = \lim_{T \to \infty} \frac{1}{2T} \int_{-\infty}^{+\infty} <i(t+t_1)i(t_1)dt_1$$

is related to the rotational diffusion constant D_R and the translational diffusion constant D_T by

$$G(t) = <i>^2 \left[1 + \frac{1}{N} \left(\frac{4}{5} e^{-6D_R T} + \frac{1}{1+4Dt/\omega^2} - \frac{9}{5} e^{-t/\tau} \right) \right]$$

for the condition $\frac{1}{\tau} >> 6D_R = \frac{1}{\phi} >> \frac{4D_T}{\omega^2}$

N = number of molecules in the volume element
D_R = rotational diffusion constant
D_T = translational diffusion constant
ω = radius of volume element
τ = lifetime of excited state

ACKNOWLEDGEMENTS

These studies were supported by grants from the Swedish Natural Science Research Council and the K. & A. Wallenberg Foundation.

REFERENCES

Ahlström, P., Teleman, O., Kördel, J., Forsén, S. and Jönsson, B., 1989, A molecular dynamics simulation of Bovine Calbindin D_{9k}, *Biochemistry* 28:3205.

Amir, D. and Haas, E., 1987, Estimation of Intramolecular distance distribution in Bovine Pancreatic Trypsin Inhibitor by site-specific labeling and nonradiative excitation energy-transfer measurements, *Biochemistry* 26:2162.

Artemiuk, P.J., Blake, C.C.F., Grace, D.E.P., Oatley, S.J., Phillips, D.C. and Sternberg, M.J.E., 1979, Chrstallographic studies of the dynamic properties of lysozyme, *Nature* 280:563.

Brooks, Ch.L., Karplus, M. and Montgomery Pettitt, B., 1988, "Proteins," John Wiley & Sons, [n.p.].

Careri, G. and Wyman, J., 1984, Soliton-assisted unidirectional circulation in a biochemical cycle, *Proc. Natl. Acad. Sci USA* 81:4386.

Claesens, F. and Rigler, R., 1986, Conformational dynamics of the anticodon loop in yeast tRNA^{Phe} as sensed by the fluorescence of wybutine, *Eur. Biophys. J.* 13:831.

Ehrenberg, M. and Rigler, R., 1972, Polarized fluorescence and rotational Brownian motion, *Chem. Phys. Lett.* 14:539.

Ehrenberg, M. and Rigler, R., 1974, Rotational Brownian motion and fluorescence intensity fluctuations, *Chem. Phys.* 4:390.

Förster, Th.Z., 1948, Zwischenmolekulare Energiewanderung und Fluoreszenz, *Annalen der Physik* 2:437.

Frauenfelder, H., Petsko, G.A. and Tsernoglou, D., 1979, Temperature-dependent X-ray diffraction as a probe of protein structural dynamics, *Nature* 280:558.

Gratton, E., Alcala, J.R., Mariott, G. and Prendergast, F.J., 1987, Fluorescence lifetime distributions of single tryptophan proteins: a protein dynamics approach, *in* "Structure, Dynamics and Function of Biomolecules," Springer Series in Biophysics, A. Ehrenberg, R. Rigler, A. Gräslund and L. Nilsson, eds., Springer Verlag, [n.p.], 132.

Ichiye, T. and Karplus, M., 1983, Fluorescence depolarization of tryptophan residues in proteins: a molecular dynamics study, *Biochemistry* 22:2884.

Karplus, M. and McCammon, J.A., 1983, Dynamics of proteins: elements and function, *Ann. Rev. Biochem.* 52:263.

Kask, P., Piksarv, P., Pooga, M., Mets, U. and Lippmaa, E., 1989, Separation of the rotational contribution in fluorescence correlation experiments, *Biophys. J.* 55:213.

Lipari, G. and Szabo, A., 1982, Model-free approach to the interpretation of nuclear magnetic resonance relaxation in macromolecules. 1. Theory and range of validity. 2. Analysis of experimental results, *J. Am. Chem. Soc.* 104:4546.

MacKerell, A.D., Jr., Nilsson, L., Rigler, R. and Saenger, W., 1988, Molecular dynamics simulations of ribonuclease T1: analysis of the effect of solvent on the structure, fluctuations, and active site of the free enzyme, *Biochemistry* 27:4547.

MacKerell, A.D., Jr., Nilsson, L., Rigler, R., Heinemann, U. and Saenger, W., 1989, Molecular dynamics simulations of Ribonuclease T1: comparison of the free enzyme and the 2'GMP-enzyme conples, *Proteins. Structure, Function and Genetics,* in press.

MacKerell, A.D., Jr., Rigler, R., Nilsson, L., Hahn, U. and Saenger, W., 1987, Protein dynamics. A time-resolved fluorescence, energetic and molecular dynamics study of ribonuclease T1, *Biophys. Chem.* 26:247.

McCammon, J.A. and Harvey, S.C., 1987, "Dynamics of Proteins and Nucleic Acids," Cambridge University Press, Cambridge.

Merola, F., Rigler, R., Holmgren, A. and Brochon, J.Cl., 1989, Picosecond tryptophan fluorescence of Thioredoxin: evidence of discrete species in slow exchange, *Biochemistry* 28:3383.

Nordlund, T.M., Andersson, S., Nilsson, L., Rigler, R., Gräslund, G. and McLaughlin, L.W., 1989, Structure and dynamics of a fluorescent DNA oligomer containing the Eco RI recognition sequence: fluorescence, molecular dynamics, and NMR studies, *Biochemistry*, in press.

Petrich, J.W., Chang, D.B., McDonald, D.B. and Fleming, G.R., 1983, On the origin of nonexponential fluorescence decay in tryptophan and its derivatives, *J. Am. Chem. Soc.* 105:3824.

Rigler, R. and Claesens, F., 1986, Picosecond time domain spectroscopy of structure and dynamics in nucleic acids, *in* "Structure and Dynamics of RNA," P.H. Knippenberg and C.W. Hilbers, eds., Plenum Press, [n.p.], 45.

Rigler, R. and Ehrenberg, M., 1976, Fluorescence relaxation spectroscopy in the analysis of macromolecular structure and motion, *Quart. Rev. Biophys.* 13:831.

Rigler, R., Claesens, F. and Lomakka, G., 1984, Picosecond single photon fluorescence spectroscopy of nucleic acisds, *in* "Ultrafast Phenomena IV," D.H. Auston and K.B. Eisenthal, eds., Springer Verlag, [n.p.], 472.

Rigler, R., Grasselli, P. and Ehrenberg, M., 1979, Fluorescence correlation spectroscopy and application to the study of Brownian motion of biopolymers, *Physica Scripta* 19:486.

Rigler, R., Kristensen, O., Roslund, J., Thyberg, P., Oba, K. and Eriksson, M., 1987, Molecular structures and dynamics: beamline for time resolved spectroscopy at the MAX synchrotron in Lund, *Physica Scripta* T17:204.

Rigler, R., Roslund, J. and Forsén, S., 1989, Side chain mobility in Bovine Calbindin D_{9k}: rotational motion of Tyr 13, *Eur. Biochem. J.,* submitted.

Ross, J.B.A., Laus, W.R., Buker, A., Sutherland, J.C. and Wyssbrod, H.R., 1986, Time-resolved fluorescnece and ^{1}H NMR studies of tyrosyl residues in Oxytocin and small peptides: correlation of NMR-determined conformation of tyrosyl residues and fluorescence decay kinetics, *Biochemistry* 25:607.

Szebenyi, D.M.E. and Moffat, J., 1986, The refined structure of vitamin D-dependent calcium-binding protein from bovine intestine, *J. Biol. Chem.* 261:8761.

Tamai, N., Yamazaki, T. and Yamazaki, I., 1985, Diffusion effect on excitation energy transfer in solution: analysis by means of picosecond time-resolved fluorimeter, *Chem. Phys. Lett.* 120:24.

Vogel, G., Nilsson, L., Rigler, R., Voges, K.P. and Jung, G., 1988, Structural fluctuations of a helical polypeptide traversing a lipid bilayer, *Proc. Natl. Acad. Sci. USA* 85:5067.

Wennerberg, A.B.A., Cooke, R.M., Carlquist, M., Rigler, R., and Campbell, I.D., 1989, A ^{1}H-NMR study of the solution conformation of the neuropeptide Galanin, *FEBS Letts.,* submitted.

Zinth, W., Dobler, J. and Kaiser, W., 1986, Femtosecond spectroscopy of the primary events of bacterial photosynthesis, *in* "Ultrafast Phenomena V," G.R. Fleming and R.F. Siegman, eds., Springer Verlag, [n.p.], 379.

MOLECULAR DYNAMICS: Applications to Proteins

Martin Karplus

Department of Chemistry, Harvard University, Cambridge, MA 02138, USA

INTRODUCTION

Molecular dynamics of macromolecules of biological interest began in 1977 with the publication of a paper on the simulation of a small protein, the bovine pancreatic trypsin inhibitor (McCammon et al., 1977). Although the trypsin inhibitor is rather uninteresting from a dynamical viewpoint (its function is to bind to trypsin) experimental and theoretical studies of this model system - the "hydrogen atom" of protein dynamics - served to initiate explorations in this field.

The most important consequence of the first simulations of biomolecules was that they introduced a conceptual change (Karplus and McCammon, 1981; Brooks et al., 1988). Although to chemists and physicists it is self-evident that polymers like proteins and nucleic acids undergo significant fluctuations at room temperature, the classic view of such molecules in their native state had been static in character. This followed from the dominant role of high-resolution x-ray crystallography in providing structural information for these complex systems. The remarkable detail evident in crystal structures led to an image of biomolecules with every atom fixed in place. D.C. Phillips, who determined the first enzyme crystal structure, wrote, "the period 1965-75 may be described as the decade of the rigid macromolecule. Brass models of DNA and a variety of proteins dominated the scene and much of the thinking" (Phillips, 1981). Molecular dynamics simulations have been instrumental in changing the static view of the structure of biomolecules to a dynamic picture. It is now recognized that the atoms of which biopolymers are composed are in constant motion at ordinary temperatures. The x-ray structure of a protein provides approximate average atomic positions, but the atoms exhibit fluid-like motions of sizable amplitudes about these averages. Crystallographers have acceded to this viewpoint and have come so far as to sometimes emphasize the parts of a molecule they do not see in a crystal structure as evidence of motion or disorder (Marquart et al., 1980). Thus, the knowledge of protein dynamics subsumes the static picture in that use of the average positions still allows discussion of many aspects of biomolecule function in the language of structural chemistry. The recognition of the importance of fluctuations opens the way for more sophisticated and accurate interpretations.

Simulation studies on biomolecules have the possibility of providing the ultimate detail concerning motional phenomena (Brooks et al., 1988). The primary limitation of simulation methods is that they are approximate. It is here that experiment plays an essential role in validating the simulations; that is, comparisons with experimental data can serve to test the accuracy of the calculations and to provide criteria for improving the methodology. When experimental comparisons indicate that the simulations are meaningful, their capacity for providing detailed results often makes it possible to examine specific aspects of the atomic motions far more easily than by making measurements.

In what follows, a brief introduction to molecular dynamics will be given, followed by applications that illustrate its utility for increasing our understanding of proteins, including enzymes, and for interpreting experiments in a more effective way.

METHODOLOGY

To study theoretically the dynamics of a macromolecular system, it is essential to have a knowledge of the potential energy surface, which gives the energy of the system as a function of the atomic coordinates. The potential energy can be used directly to determine the relative energies of the different possible structures of the system; the relative populations of such structures under conditions of thermal equilibrium are given in terms of the potential energy by the Boltzmann distribution law (McQuarrie, 1976). The mechanical forces acting on the atoms of the systems are simply related to the first derivatives of the potential with respect to the atom positions. These forces can be used to calculate dynamical properties of the system, e.g., by solving Newton's equations of motion to determine how the atomic positions change with time (McQuarrie, 1976; Hansen and McDonald, 1976). From the second derivatives of the potential surface, the force constants for small displacements can be evaluated and these can be used to find the normal modes (Levy and Karplus, 1979); this serves as the basis for an alternative approach to the dynamics in the harmonic limit (Levy and Karplus, 1979; Brooks and Karplus, 1983)

Although quantum mechanical calculations can provide potential surfaces for small molecules, empirical energy functions of the molecular mechanics type are the only possible source of such information for proteins and the surrounding solvent. Since most of the motions that occur at ordinary temperatures leave the bond lengths and bond angles of the polypeptide chains near their equilibrium values, which appear not to vary significantly throughout the protein (e.g., the standard dimensions of the peptide group first proposed by Pauling [Pauling et al., 1951]), the energy function representation of the bonding can have an accuracy on the order of that achieved in the vibrational analysis of small molecules. Where globular proteins differ from small molecules is that the contacts among nonbonded atoms play an essential role in the potential energy of the folded or native structure. From the success of the pioneering conformational studies of Ramachadran and co-workers (Ramachadran et al., 1963) that made use of hardsphere nonbonded radii, it is likely that relatively simple functions (Lennard-Jones nonbonded potentials supplemented by electrostatic interactions) can adequately describe the interactions involved.

The energy functions used for proteins are generally composed of terms representing bonds, bond angles, torsional angles, van der Waals interactions and electrostatic interactions. The expression used in the program CHARMM (Brooks et al., 1983) has the form:

$$E(\bar{R}) = \frac{1}{2} \sum_{\text{bonds}} K_b(b-b_0)^2 + \frac{1}{2} \sum_{\text{bond angles}} K_\theta(\theta-\theta_0)^2 +$$

$$\frac{1}{2} \sum_{\text{torsional angles}} K_\phi[1+\cos(n\phi-\delta)] + \sum_{\text{nb pairs}} \left(\frac{A}{r^{12}} - \frac{C}{r^6} + \frac{q_1 q_2}{Dr}\right) \qquad [1]$$

$$r < 8\text{Å}$$

The energy is a function of the Cartesian coordinate set, $\bar{R}$, specifying the positions of all the atoms involved, but the calculation is carried out by first evaluating the internal coordinates for bonds (b), bond angles (θ), dihedral angles (ϕ), and interparticle distances (r) for any given geometry, $\bar{R}$, and using them to evaluate the contributions to Eq. [1], which depend on the bonding energy parameters K_b, K_θ, K_ϕ, Lennard-Jones parameters A and C, atomic charges q_i, dielectric constant D, and geometrical reference values b_0, θ_0, n, and δ. For most simulations, use has been made of a representation that replaces aliphatic groups (CH_3, CH_2, CH) by single extended atoms. Although the earliest studies employed the extended atom representation for all hydrogens, present calculations treat hydrogen-bonding hydrogens explicitly. In the most detailed simulations, every protein atom (including aliphatic hydrogens) and explicit solvent molecules (e.g., a three-site or five-site model for each water molecule) is included (Brooks et al., 1983).

Given a potential energy function, one may take any of a variety of approaches to study protein dynamics. The most detailed information is provided by molecular dynamics simulations, in which one uses a computer to solve the Newtonian equations of motion for the atoms of the protein and any surrounding solvent (McCammon et al., 1977; McCammon et al., 1979; van Gunsteren and Karplus, 1982). With currently available computers, it is possible to simulate the dynamics of small proteins for periods of up to a nanosecond. Such periods are long enough to characterize completely the librations of small groups in the protein and to determine the dominant contributions to the atomic fluctuations. To study slower and more complex processes in proteins, it is generally necessary to use other than the straightforward molecular dynamics simulation method. A variety of dynamical approaches, such as stochastic dynamics (Chandrasekhar, 1943), harmonic dynamics (Levy and Karplus, 1979; Brooks and Karplus, 1983), and activated dynamics (Northrup et al., 1982), can be introduced to study particular problems (Brooks et al., 1988).

Since molecular dynamics simulations have been used most widely for studying protein motions, we briefly describe the methodology. To begin a dynamical simulation, one must have an initial set of atomic coordinates and velocities. These are usually obtained from the x-ray coordinates of the protein by a preliminary calculation that serves to equilibrate the system (Brooks et al., 1983). The x-ray structure is first refined using an energy minimization algorithm to relieve local stresses due to non-bonded atomic overlaps, bond length distortions, etc. The protein atoms are then assigned velocities at random from a Maxwellian distribution corresponding to a low temperature, and a dynamical simulation is performed for a period of a few psec. The equilibration is continued by alternating new velocity assignments (chosen from Maxwellian distributions corresponding to successively increased temperatures) with intervals of dynamical relaxation. The temperature, T, for this microcanonical ensemble is measured in terms of the mean kinetic energy for the system composed of N atoms as

$$\frac{1}{2} \sum_{i=1}^{N} m_i \langle v_i^2 \rangle = \frac{3}{2} Nk_BT \qquad [2]$$

where m_i and $\langle v_i^2 \rangle$ are the mass and average velocity squared of the ith atom, and k_B is the Boltzmann constant. Any residual overall translational and rotational motion for an isolated protein can be removed to simplify analysis of the subsequent conformational fluctuations; in a solution simulation, the protein can diffuse through the solvent. The equilibration period is considered finished when no systematic changes in the temperature are evident over a time of about 10 psec (slow fluctuations could be confused with continued relaxation over shorter intervals). It is necessary also to check that the atomic momenta obey a Maxwellian distribution and that different regions of the protein have the same average temperature. The actual dynamical simulation, which provides coordinates and velocities for all the atoms as a function of time, is then performed by continuing to integrate the equations of motion for the desired time period. The available simulations for proteins range from 25 to 300 psec. Several different algorithms for integrating the equations of motion in Cartesian coordinates are being used in protein molecular dynamics calculations. Most common are the Gear predictor-corrector algorithm, familiar from small molecule trajectory calculations (McCammon et al., 1979) and the Verlet algorithm (Verlet, 1967), widely used in statistical mechanical simulations (van Gunsteren and Berendsen, 1977).

INTERNAL MOTIONS AND THE UNDERLYING POTENTIAL SURFACE

For native proteins with a well-defined average structure, two extreme models for the internal motions have been considered. In one, the fluctuations are assumed to occur within a single multidimensional well that is harmonic of quasiharmonic as a limiting case (Karplus and Kushick, 1981; Brooks and Karplus, 1983; Levitt et al., 1985). The other model assumes that there exist multiple minima or substates; the internal motions correspond to a superposition of oscillations within the wells and transitions among them (Austin et al., 1975; Frauenfelder et al., 1979; Levy et al., 1982; Swaminathan et al., 1982; Brooks and Karplus, 1983; Debrunner and Frauenfelder, 1982). Experimental data have been interpreted with both models, but it has proved difficult to distinguish between them (Agmon and Hopfield, 1983; Ansari et al., 1985).

To characterize the protein potential surface structurally and energetically (Elber and Karplus, 1987), we use a 300 ps molecular dynamics simulation of the protein myoglobin at 300°K; details of the simulation method have been presented (Levy et al., 1985). Myoglobin was chosen for study because it has been examined experimentally by a variety of methods and the two motional models have been applied to it (Austin et al., 1975; Frauenfelder et al., 1979; Levy et al., 1982; Agmon and Hopfield, 1983; Bialek and Goldstein, 1985). It is ideally suited for the present analysis, because its well defined secondary structure (a series of α-helices connected by loops) facilitates a detailed characterization of the dynamics.

The topography of the potential surface underlying the dynamics can be explored by finding the local energy minima associated with coordinate sets sequential in time (Stillinger and Weber, 1982; Stillinger and Weber, 1984). Thirty-one coordinate sets (one every 10 psec) were selected and their energy was minimized with a modified Newton-Raphson algorithm suitable for large molecules (Brooks et al., 1983). Since the coordinate sets all corresponded to different minima, structures separated by shorter time periods were examined to determine how long the trajectory remains in a given minimum. Seven additional coordinate sets (one every 0.05 psec) were chosen and their behavior on minimization was examined; if two coordinate sets converged, they corresponded to the same minimum; if they diverged, they corresponded to different minima. The measure for the distance between two structures is their rms coordinate difference after superposition.

Analysis of the short time dynamics demonstrates that convergence occurs for intervals up to 0.15 ± 0.05 ps. Thus, the 300 ps simulation samples on the order of 2000 different minima; this is a sizable number but it may nevertheless be small relative to the total (finite) number of minima available to such a complex system in the neighborhood of the native average structure (that is, conformations that are native-like and significantly populated at room temperature). The rms differences among the minimized structures reach a maximum value of approximately 2 Å at about 100 psec. Thus, the difference vector $(\underline{R}_K - \underline{R}_{K'})$, where $\underline{R}_K$ represents the coordinates of all the atoms in a native-like conformation K, is restricted to a volume bounded by a radius of 2 Å.

Comparison of the energies of the minimized structures shows that they vary over about 20°K (40 cal/mole) per degree of freedom. Since this difference in energy between the "inherent" structures (Stillinger and Weber, 1982; Stillinger and Weber, 1984) is small, they are significantly populated at room temperature. Further, the large number of such structures sampled by the room temperature simulation suggests that the effective barriers separating them are low and that the protein is undergoing frequent transitions from one structure to another. The fluctuations within a well can be described by a harmonic or quasiharmonic model while the transitions among the wells cannot. Estimates based on the time development of the rms atomic fluctuations for mainchain atoms at room temperature (Swaminathan et al., 1982) indicate that 20 to 30 percent of the rms fluctuations are contributed by oscillations within a well and 70 to 80 percent arise from transitions among wells; for sidechains the contribution from transitions among the multiple wells is expected to be larger. Since energy differences among some of the wells are small, molecules may be trapped in metastable states at low temperatures, in analogy to third law violations in crystals (e.g., crystals of CO) and models for the glassy state (Ziman, 1979; Stillinger and Weber, 1982; Stillinger and Weber, 1984; Toulouse, 1984; Ansari et al., 1985; Stein, 1985). A number of experiments suggest that the transition temperature for myoglobin is in the neighborhood of 200°K (Austin et al., 1975; Parak et al., 1982; Debrunner and Frauenfelder, 1982; Ansari et al., 1985). Because large scale, collective motions that involve the protein surface are important in the fluctuations (Swaminathan et al., 1982), it is likely that the observed transition is due to the freezing of the solvent matrix (Swaminathan et al., 1982; Parak et al., 1982).

Because the details of the native structure of a protein play an essential role in its function, it is important to determine the structural origins of the multiminimum surface obtained from the dynamics analysis. The general features of the structure (helices and turns) are preserved throughout the simulation and the differences in position are widely distributed. The motions are associated primarily with loop displacements or relative displacements of helices which individually behave as nearly rigid bodies. Rearrangements within individual loops are the elementary step in the transition from one minimum to another; they are coupled with associated helix displacements. Which loop or turn changes in a given time interval appears to be random. Specific loop motions may be initiated by sidechain transitions in the helix contacts, mainchain dihedral angle transitions of the loops themselves, or a combination of the two. As the time interval between two structures increases, more loop transitions have occurred. At room temperature, the transition probabilities are such that for an interval 100 psec or longer between two structures, some transitions will have taken place in all of the flexible loop regions. However, since the rms differences between structures continue to increase up to 200 psec, the configuration space available to the molecule includes a range of structures for the loop regions that are not completely sampled in a 100 psec.

To characterize the helix motions that are coupled with the loop rearrangements, the internal structural changes of the helices were separated from their relative motions. Individual helices and loops were superimposed and the rms differences for the mainchain calculated for the set of structures; the rms difference for the internal structure of the helices is generally less than 1 Å. Corresponding results for the loop regions show that they undergo much larger internal structural changes on the order of 2.5 Å.

In analyzing the relative motion of the helices, it is of particular interest to examine the behavior of helix pairs that are in van der Waals contact; these are helix pairs A-H, B-E, B-G, F-H and G-H for all of which at least three residues from each helix are interacting. Each helix was fitted to a straight line and the fluctuations of the distance between the helix centers of mass and the relative orientations of the lines were compared. The relative translations found in this case have rms values of 0.3 to 0.7 Å and the relative rotations have rms values of 1 to 14°; the maximum differences are 1.3 to 2.2 Å and 5 to 39°, respectively.

The dynamical results for the helix motions can be compared with structural data from two sources; the first is derived from proteins of a given sequence in different environments (e.g., two different crystal forms, deoxy and oxy hemoglobin [Chothia and Lesk, 1985]) and the second from homologous proteins with different sequences (e.g., the globins [Lesk and Chothia, 1980]). The maximum dynamical displacements are, in fact, larger than those observed in different x-ray structures of a given protein. The values are of the same order as the differences (2 to 3 Å, 15 to 30°; there are some larger changes) found in comparing a series of different globins with known crystal structures and sequence homology in the range 16 to 88 percent. Thus, the range of conformations sampled by a single myoglobin trajectory is similar to that found in the evolutionary variation among crystal structures of the globin series. This suggests a molecular plasticity which is likely to have played an important role in the evolution of protein sequences.

The comparison of the various globin structures (Lesk and Chothia, 1980) suggested that the range of helix packings is achieved primarily by changes in sidechain volumes resulting from amino acid substitutions. In the dynamics, it is the correlated motions of sidechains that are in contact, plus the rearrangements of loops, that make possible the observed helix fluctuations. Different positions within wells and transitions between wells for sidechains (e.g., ±60°, 180° for χ_1) are involved. This is in accord with the results of high-resolution x-ray studies that show significant disorder in sidechain orientations (Smith et al., 1986; Kuriyan et al., 1987). Further, correlated dihedral angle changes differentiate the various minima. Since more than one set of sidechain orientations is consistent with a given set of helix positions, the known globin crystal structures probably represent only a small subset of the possible local minima.

Myoglobin at normal room temperatures samples a very large number of different minima that arise from the inhomogeneity of the system. This is expected to have important consequences for the interpretation of myoglobin function and, more generally, for the functions of other proteins, including enzymes. There are solid-like microdomains (the helices), whose mainchain structure is relatively rigid, and liquid-like regions (the loops and the sidechain clusters at interhelix contacts) that readjust as the helices move from one minimum to another. Since the minima have similar energies, myoglobin is expected to be glass-like at low temperatures. Freezing in of the liquid-like regions could result in a transition to the glassy state (Stein, 1985).

ATOMIC FLUCTUATION AND X-RAY DIFFRACTION

Since atomic fluctuations are the basis of protein dynamics, it is important to have experimental tests of the accuracy of the simulation results concerning them. For the magnitudes of the motions, the most detailed data are provided, in principle, by an analysis of the Debye-Waller or temperature factors obtained in crystallographic refinements of x-ray structures. Averages over the fluctuations can be obtained also from incoherent neutron scattering (Doster et al., 1989).

It is well known from small-molecule crystallography that the effect of thermal motion must be included in the interpretation of the x-ray data to obtain accurate structural results. Detailed models have been introduced to take account of anisotropic and anharmonic motions of the atoms and these molecules have been applied to high-resolution data for small molecules (Zucker and Schultz, 1982). In protein crystallography, the limited data available relative to the large number of parameters that have to be determined, have made it necessary in most cases to assume that the atomic motions are isotropic and harmonic. Then the structure factor, $F(Q)$, which is related to the measured intensity by $I(Q) = |F(Q)|^2$, is given by

$$F(\underset{\sim}{Q}) = \sum_{j=1}^{N} f_j(\underset{\sim}{Q})\ e^{i\underset{\sim}{Q}\cdot<\underset{\sim}{r}_j>}\ e^{W_j(\underset{\sim}{Q})} \qquad [3]$$

where Q is the scattering vector, $<r_j>$ is the average position of atom j with atomic scattering factor $f_j(Q)$ and the sum is over the N atoms in the asymmetric unit of the crystal. The Debye-Waller factor, $W_j(Q)$, is defined by

$$W_j(\underset{\sim}{Q}) = -\tfrac{8}{3}\ \pi^2<\Delta r_j^2>s^2 = -B_j s^2 \qquad [4]$$

where $s = |Q|/4\pi$. The quantity B_j is usually referred to as the temperature factor, which is directly related to the mean-square atomic fluctuations in the isotropic harmonic model. More generally, if the motion is harmonic but anisotropic, a set of six parameters

$$B_j^{xx} = <\Delta x_j^2>,\ B_j^{xy} = <\Delta x_j \Delta y_j>,... B_j^{z} = <\Delta z_j^2> \qquad [5]$$

is required to fully characterize the atomic motion. Although in the earlier x-ray studies of proteins, the significance of the temperature factors was ignored (presumably because the data were not at a sufficient level of resolution and accuracy), more recently attempts have been made to relate the observed temperature factors to the atomic motions (Frauenfelder et al., 1979; Artymiuk et al., 1979).

In principle, the temperature factors provide a very detailed measure of these motions because information is available for the mean-square fluctuation of every heavy atom. In practice, there are two types of difficulties in relating the B factors obtained from protein refinements to the atomic motions. The first is that, in addition to thermal fluctuations, any static (lattice) disorder in the crystal contributes to the B factors; i.e., since a crystal is made up of many unit cells, different molecular geometries in the various cells have the same effect on the average electron density, and therefore the B factor, as atomic motions. For the iron atom of myoglobin there has been an experimental attempt to determine the disorder contribution (Hartmann et al., 1982). Since the Mossbauer effect is not altered by static disorder (i.e., each nucleus absorbs independently), but does depend on atomic motions, comparisons of Mossbauer and x-ray data have been used to estimate a disorder contribution for the iron atom; the value obtained is

$$<\Delta r_{Fe}^2> = 0.08\ \text{Å}^2$$

Although the value is only approximate, it nevertheless indicates that the observed B factors (e.g., on the order of 0.44 Å^2 for backbone atoms and 0.50 Å^2 for sidechain atoms) are dominated by the motional contribution. Most experimental B factor values are compared directly with the molecular dynamics results (i.e., neglecting the disorder contribution) or are rescaled by a constant amount (e.g., by setting the smallest observed B factor to zero) on the assumption that the disorder contribution is the same for all atoms (Petsko and Ringe, 1984). The second difficulty is that, since simulations have shown that the atomic fluctuations are highly anisotropic and, in some cases, anharmonic, there may be significant errors in the refinement due to the assumption of isotropic and harmonic motion. A direct experimental estimate of the errors is difficult because sufficient data are not yet available for protein crystals, although incoherent neutron scattering can provide information independent of static disorder

(Doster et al., 1989). Moreover, any data set includes other errors which would obscure the analysis. As an alternative to an experimental analysis of the errors in the refinement of proteins, a purely theoretical approach can be used (Kuriyan et al., 1986). The basic idea is to generate x-ray data from a molecular dynamics simulation of a protein and to use these data in a standard refinement procedure. The error in the analysis can then be determined by comparing the refined x-ray structure and temperature factors with the average structure and the mean-square fluctuations from the simulation. Such a comparison, in which no real experimental results are used, avoids problems due to inaccuracies in the measured data (exact calculated intensities are used), to crystal disorder (there is none in the model), and to approximations in the simulation (the simulation is exact for this case). The only question about such a comparison is whether the atomic motions found in the simulation are a meaningful representation of those occurring in proteins. As has been shown (Petsko and Ringe, 1984; Karplus and McCammon, 1983), molecular dynamics provides a reasonable picture of the motions in spite of errors in the potentials, the neglect of the crystal environment, and the finite-time classical trajectories used to obtain the results. However, these inaccuracies do not affect the exactitude of the computer "experiment" for testing the refinement procedure that is described below.

In this study (Kuriyan et al., 1986), a 25-psec molecular dynamics trajectory for myoglobin was used (Levy et al., 1985). The average structure and the mean-square fluctuations from that structure were calculated directly from the trajectory. To obtain the average electron density, appropriate atomic electron distributions were assigned to the individual atoms and the results for each coordinate set were averaged over the trajectory. Given the symmetry, unit cell dimensions, and position of the myoglobin molecule in the unit cell, average structure factors, $<F(Q)>$, and intensities, $I(Q)=|<F(Q)>|^2$ were calculated from the Fourier transform of the average electron density, $<\tilde{\rho}(r)>$, as a function of position r in the unit cell. Data were generated at 1.4 Å resolution, since this is comparable to the resolution of the best x-ray data currently available for proteins the size of myoglobin (Kuriyan et al., 1987). The resulting intensities at Bragg reciprocal lattice points were used as input data for the widely applied crystallographic program, PROLSQ (Konnert and Hendrickson, 1980). The time-averaged atomic positions obtained from the simulation and a uniform temperature factor provide the initial model for refinement. The positions and an isotropic, harmonic temperature factor for each atom were then refined iteratively against the computer-generated intensities in the standard way. Differences between the refined results for average atomic positions and their mean-square fluctuations and those obtained from the molecular dynamics trajectory are due to errors introduced by the refinement procedure.

The overall rms error in atomic positions ranged from 0.24 Å to 0.29 Å for slightly different restrained and unrestrained refinement procedures (Kuriyan et al., 1986). The errors in backbone positions (0.10 - 0.20 Å) are generally less than those for sidechain atoms (0.28 - 0.33 Å); the largest positional errors are on the order of 0.6 Å. The backbone errors, although small, are comparable to the rms deviation of 0.21 Å between the positions of the backbone atoms in the refined experimental structures of oxymyoglobin and carboxymyoglobin (Kuriyan et al., 1987; Phillips, 1980). Further, the positional errors are not uniform over the whole structure. There is a strong correlation between the positional error and the magnitude of the mean-square fluctuation for an atom, with certain regions of the protein, such as loops and external sidechains, having the largest errors.

The refined mean-square fluctuations are systematically smaller than the fluctuations calculated directly from the simulation. The magnitudes and variation of temperature factors along the backbone are relatively well reproduced, but the refined sidechain fluctuations are almost always significantly smaller than the actual values. The average backbone B factors from different refinements are in the range 11.3 to 11.7 Å^2, as compared with the exact value of

12.4 Å^2; for the sidechains, the refinements yield 16.5 to 17.6 Å^2, relative to the exact value of 26.8 Å^2. Regions of the protein that have high mobility have large errors in temperature factors as well as in positions. Examination of all atoms shows that fluctuations greater than about 0.75 Å^2 (B = 20 Å^2) are almost always underestimated by the refinement. Moreover, while actual mean-square atomic fluctuations have values as large as 5 Å^2, the x-ray refinement leads to an effective upper limit of about 2 Å^2. This arises from the fact that most of the atoms with large fluctuations have multiple conformations and that the refinement procedure picks out one of them.

To do refinements that take some account of anisotropic motions for all but the smallest proteins, it has been necessary to introduce assumptions concerning the nature of the anisotropy. One possibility is to assume anisotropic rigid body motions for sidechains such as tryptophan and phenylalanine (Artymiuk et al., 1979; Glover et al., 1983). An alternative is to introduce a "dictionary" in which the orientation of the anisotropy tensor is related to the stereochemistry around each atom (Konnert and Hendrickson, 1980); this reduces the six independent parameters of the anisotropic temperature factor tensor B_j to three parameters per atom. An analysis of a simulation for BPTI (Yu et al., 1985) has shown that the actual anisotropies in the atomic motions are generally not simply related to the local stereochemistry; an exception is the mainchain carbonyl oxygen, which has its largest motion perpendicular to the C=O bond. Thus, use of stereochemical assumptions in the refinement can yield incorrectly oriented anisotropy tensors and significantly reduced values for the anisotropies. The large-scale motions of atoms are collective and sidechains tend to move as a unit so that the directions of largest motion are not related to the local bond direction and have similar orientations in the different atoms forming a group that is undergoing correlated motions. Consequently, it is necessary to use the full anisotropy tensor to obtain meaningful results. This is possible with proteins that are particularly well ordered, so that the diffraction data extend to better than 1 Å resolution.

X-RAY REFINEMENT BY SIMULATED ANNEALING

Crystallographic structure determinations by x-ray or neutron diffraction generally proceed in two stages. First, the phases of the measured reflections are estimated and a low- to medium-resolution model of the protein is constructed and, second, more precise information about the structure is obtained by refining the parameters of the molecular model against the crystallographic data (Wyckoff et al., 1985). The refinement is performed by minimizing the crystallographic R factor, which is defined as the difference between the observed ($|F_{obs}(h,k,l)|$) and calculated ($|F_{calc}(h,k,l)|$) structure factor amplitudes,

$$R = \sum_{h,k,l} ||F_{obs}(h,k,l)| - |F_{calc}(h,k,l)|| \ / \ \sum_{h,k,l} |F_{obs}(h,k,l)| \qquad [6]$$

where h,k,l are the reciprocal lattice points of the crystal.

Conventional refinement involves a series of steps, each consisting of a few cycles of least-squares refinement with stereochemical and internal packing constraints or restraints (Sussman et al., 1977; Jack and Levitt, 1978; Konnert and Hendrickson, 1980; Moss and Morffew, 1982) that are followed by manual rebuilding of the model structure by use of interactive computer graphics. Finally, solvent molecules are included and alternative conformations for some protein atoms may be introduced. The standard refinement procedure is time consuming, because the limited radius of convergence of least-squares algorithms (approximately 1 Å) necessitates the periodic examination of electron density maps computed with various combinations of F_{obs} and F_{calc} as amplitudes, and with phases calculated from the model structure. Also, the least-squares refinement process is easily trapped in a local minimum so that human intervention is necessary.

Simulated annealing (Kirkpatrick et al., 1983), which makes use of Monte Carlo or molecular dynamics (Brünger et al., 1987b) simulations to explore the conformational space of the molecule can help to overcome the local-minimum problem. This has been demonstrated in the application of molecular dynamics to structure determination with nuclear magnetic resonance (NMR) data. In contrast to the NMR application (Brünger et al., 1986), the initial model for crystallographic refinement cannot be arbitrary. It has to be relatively close to the correct geometry to provide an adequate approximation to the phases of the structure factors.

To employ molecular dynamics in crystallographic refinement, an effective potential

$$E_{sf} = S \sum_{h,k,l} [|F_{obs}(h,k,l)| - |F_{calc}(h,k,l)|]^2 \qquad [7]$$

was added to the empirical energy potential given in Eq. [1]. The effective potential E_{sf} describes the differences between the observed structure factor amplitudes and those calculated from the atomic model; it is identical to the function used in standard least-squares refinement methods (Jack and Levitt, 1978). The scale factor S was chosen to make the gradient of E_{sf} comparable in magnitude to the gradient of the empirical energy potential of a molecular dynamics simulation with S set to zero.

As in the case of the NMR analysis, simulated annealing refinement was also tested on crambin, for which high-resolution x-ray diffraction data and a refined structure, determined by resolved anomalous phasing and conventional least-squares refinement with model-building, are available (Hendrickson and Teeter, 1981). The initial structure for the MD-refinement was obtained from the NMR structure determination; the orientation and position of the NMR-derived crambin molecule in the unit cell was determined by molecular replacement (Brünger et al., 1987a). The root-mean-square (rms) differences for residue positions of this initial structure and the final manually refined structure (Hendrickson and Teeter, 1981) are as large as 3.5 Å, with particularly large differences for residues 34 to 40; the R factor of the initial structure was 0.56 at 2 Å resolution. MD-refinement at 3000 K starting with 4 Å resolution data for 2.5 ps, extending to 3 Å resolution for 2.5 ps, and finally to 2 Å resolution for 5 ps, followed by several cycles of minimization, reduces the atomic rms deviations to 0.34 and 0.56 Å for the backbone and sidechain atoms, respectively. During the MD-refinement, some atoms in residues 35 to 40 moved by more than 3 Å. The essential point is that the refinement of the crambin structure was achieved starting from the initial NMR-structure without human intervention. The R factor (0.294) of the MD-refined structure is somewhat higher than the R factor (0.258) of the manually refined structure without solvent and with constant temperature factors; minor model-building would correct this difference. Other annealing protocols using higher temperatures (e.g., 7000 to 9000°K) yield structures that are still closer to the manually refined structure. The refinement required approximately one hour of central processing unit (CPU) time on a CRAY-1; structure factor calculations accounted for about half this time. The latter portion of the calculations has been considerably reduced in time by use of Fast Fourier transform (FFT) methods (Brünger, 1989).

As a control, the initial NMR-derived structure was refined without rebuilding by a restrained least-squares method (Konnert and Hendrickson, 1980), starting at 4 Å resolution and then increasing the resolution to 3 Å, and finally to 2 Å. The R factor dropped to 0.381, but the very bad stereochemistry and large deviation from the manually refined structure indicate that this structure has not converged to the correct result; residues 34 to 40 have not moved and substantial model-building would be required to correct the structure. Thus, restrained least-squares refinement in the absence of model-building did not produce the large conformational changes that occurred in MD-refinement by simulated annealing. With a version of the molecular dynamics program CHARMM optimized for x-ray refinement (the program X-PLOR [Brünger et al., 1989]), many applications of simulated annealing have been made and shown to be of considerable utility in decreasing the human effort involved (e.g., Navia et al., 1989).

USE OF NUCLEAR MAGNETIC RESONANCE DATA FOR DYNAMICS AND STRUCTURE

Nuclear magnetic resonance (NMR) is an experimental technique that has played an essential role in the analysis of the internal motions of proteins (Campbell et al., 1978; Gurd and Rothgeb, 1979; Dobson and Karplus, 1986). Like x-ray diffraction, it can provide information about individual atoms; unlike x-ray diffraction, NMR is sensitive not only to magnitude but also to the time scales of the motions. Most nuclear relaxation processes are dependent on atomic motions on the nanosecond to picosecond time scale. Although molecular tumbling is generally the dominant relaxation mechanism for proteins in solution, internal motions contribute as well; for solids, the internal motions are of primary importance. In addition, NMR parameters, such as nuclear spin-spin coupling constants and chemical shifts, depend on the protein environment. In many cases different local conformations exist but the interconversion is rapid on the NMR time scale, here on the order of milliseconds, so that average values are observed. When the interconversion time is on the order of the NMR time scale or slower, the transition rates can be studied by NMR; an example is provided by the reorientation of aromatic rings (Campbell et al., 1976; Brooks et al., 1988).

In addition to supplying data on the dynamics of proteins, NMR can also be used to obtain structural information. With recent advances in techniques it is now possible to obtain a large number of approximate interproton distances for proteins by the use of nuclear Overhauser effect measurements (Noggle and Schirmer, 1971). If the protein is relatively small and has a well resolved spectrum, a large portion of the protons can be assigned and several hundred distances for these protons can be determined by the use of two-dimensional NMR techniques (Wagner and Wüthrich, 1982). Clearly, these distances can serve to provide structural information for proteins, analogous to their earlier use for organic molecules (Noggles and Schirmer, 1971; Honig et al., 1971). Of great interest is the demonstration that enough distance information can be measured to determine the high resolution structure of a protein in solution. In the last few years it has been shown how such NMR structures can serve to supplement results from x-ray crystallography, particularly for proteins that are difficult to crystallize (Wüthrich, 1989).

In what follows we consider two questions related to structure determination. The first concerns the effect of motional averaging on the accuracy of the apparent distances obtained from the NOE studies and the second, the use of molecular dynamics simulated annealing to obtain structural results from the NOE data.

For spin-lattice relaxation, such as observed in nuclear Overhauser effect measurements, it is possible to express the behavior of the magnetization of the nuclei being studied by the equation (Olejniczak et al., 1984; Solomon, 1955)

$$\frac{d(I_z(t)-I_0)_i}{dt} = -\rho_i(I_z(t)-I_0)_i - \sum_{i \neq j} \sigma_{ij}(I_z(t)-I_0)_j \qquad [8]$$

where $I_z(t)$ and I_{0i} are the z components of the magnetization of nucleus i at time t and at equilibrium, ρ_i is the direct relaxation rate of nucleus i, and σ_{ij} is the cross relaxation rate between nuclei i and j. The quantities ρ_i and σ_{ij} can be expressed in terms of spectral densities

$$\rho_i = \frac{6\pi}{5} \gamma_i^2 \gamma_j^2 h^2 \sum_{i \neq j} [1/3 J_{ij}(\omega_i - \omega_j) + J_{ij}(\omega_i) + 2J_{ij}(\omega_i + \omega_j)]$$

[9]

$$\sigma_{ij} = \frac{6\pi}{5} \gamma_i^2 \gamma_j^2 h^2 [2J_{ij}(\omega_i + \omega_j) - 1/3 J_{ij}(\omega_i - \omega_j)]$$

where ω_i is the resonance frequency of nucleus i. The spectra density functions can be obtained from the correlation functions for the relative motions of the nuclei with spins i and j (Olejniczak et al., 1984; Levy et al., 1981),

$$J_{ij}^n(\omega) = \int_0^\infty \frac{<Y_n^2(\theta_{lab}(t)\phi_{lab}(t))Y_n^{2*}(\theta_{lab}(0)\phi_{lab}(0))>}{r_{ij}^3(0)r_{ij}^3(t)} \cos(\omega t) dt$$

[10]

where $Y_n^2(\theta(t)\phi(t))$ are second-order spherical harmonics and the angular brackets represent an ensemble average which is approximated by an integral over the molecular dynamics trajectory. The quantities $\theta_{lab}(t)$ and $\phi_{lab}(t)$ are the polar angles at time t of the internuclear vector between protons i and j with respect to the external magnetic field and r_{ij} is the interproton distance. In the simplest case of a rigid molecule undergoing isotropic tumbling with a correlation time τ_o this reduces to the familiar expression

$$J_{ij}(\omega) = \frac{1}{4\pi_{ij}^0} \left(\frac{\tau_o}{1+(\omega\tau_o)^2} \right)$$

[11]

The nuclear Overhauser effect corresponds to the selective enhancement of a given resonance by the irradiation of another resonance in a dipolar coupled spin system. Of particular interest for obtaining motional and distance information are measurements that provide time-dependent NOEs from which the cross relaxation rates ϕ_{ij} (see Eq. [9]) can be determined directly or indirectly by solving a set of coupled equations (Eqs. [8] and [9]). Motions on the picosecond timescale are expected to introduce averaging effects that decrease the cross-relaxation rates by a scale factor relative to the rigid model. A lysozyme molecular dynamics simulation (Ichiye et al., 1986) has been used to calculate dipole vector correction functions (Olejniczak et al., 1984) for proton pairs that have been studied experimentally (Olejniczak et al., 1981; Poulsen et al., 1980). Four proton pairs on three sidechains (Trp 28, Ile 98, and Met 105) with very different motional properties were examined. Trp 28 is quite rigid, Ile 98 has significant fluctuations, and Met 105 is particularly mobile in that it jumps among different side-chain conformations during the simulation. The rank order of the scale factors (order parameters) is the same in the theoretical and experimental results. However, although the results for the Trp 28 protons agree with the measurements to within the experimental error, for both Ile 98 and Met 105 the motional averaging found from the NOE's is significantly greater than the calculated value. This suggests that these residues are undergoing rare fluctuations involving transitions that are not adequately sampled by the simulation.

If nuclear Overhauser effects are measured between pairs of protons whose distance is not fixed by the structure of a residue, the strong distance-dependence of the cross-relaxation rates ($1/r^6$) can be used to obtain estimates of the interproton distances (Poulsen et al., 1980; Olejniczak et al., 1981; Wagner and Wüthrich, 1982; Clore et al., 1985). The simplest application of this approach is to assume that proteins are rigid and tumble isotropically. The lysozyme molecular dynamics simulation was used to determine whether picosecond fluctuations are likely to introduce important errors into such an analysis (Olejniczak et al., 1984). The results show that the presence of the motions will cause a general decrease in most NOE effects observed in a protein. However, because the distance depends on the sixth root of the observed NOE, motional errors of a factor of two in the latter lead to only a 12% uncertainty in the

280

distance. Thus, the decrease is usually too small to produce a significant change in the distance estimated from the measured NOE value. This is consistent with the excellent correlation found between experimental NOE values and those calculated using distances from a crystal structure (Poulsen et al., 1980). Specific NOEs can, however, be altered by the internal motions to such a degree that the effective distances obtained are considerably different from those predicted for a static structure. Such possibilities must, therefore, be considered in any structure determination based on NOE data. This is true particularly for cases involving averaging over large fluctuations, such as may occur for external sidechains and mobile loop regions.

Because of the inverse sixth power of the NOE distance dependence, experimental data so far are limited to protons that are separated by less than 5 Å. Thus, the information required for a direct protein structure determination is not available. To overcome this limitation it is possible to introduce additional information provided by empirical energy functions (Brooks et al., 1983). One way of proceeding is to do molecular dynamics simulated annealing with the approximate interproton distances introduced as restraints in the form of skewed biharmonic potentials (Clore et al., 1985; Brünger et al., 1986); the force constants can be chosen to correspond to the experimental uncertainty in the distance.

A model study of the small protein crambin, which is composed of 46 residues, was made with realistic NOE restraints (Brünger et al., 1986). Two hundred forty approximate interproton distances less than 4 Å were used, including 184 short-range distances (i.e., those connecting protons in two residues that are less than 5 residues apart in the sequence) and 56 long-range distances. The molecular dynamics simulations converged to the known crambin structure (Hendrickson and Teeter, 1981) from different initial extended structures. The average structure obtained from the simulations with a series of different protocols had rms deviations of 1.3 Å for the backbone atoms, and 1.9 Å for the sidechain atoms. Individual converged simulations had rms deviations in the range 1.5 to 2.1 Å and 2.1 to 2.8 Å for the backbone and sidechain atoms, respectively. Further, it was shown that a dynamics structure with significantly large deviations (5.7 Å) could be characterized as incorrect, independent of a knowledge of the crystal structure because of its higher energy and the fact that the NOE restraints were not satisfied within the limits of error. The incorrect structure resulted when all NOE restraints were introduced simultaneously, rather than allowing the dynamics to proceed first in the presence of only the short-range restraints followed by introduction of the long-range restraints. Also of interest is the fact that although crambin has three disulfide bridges it was not necessary to introduce information concerning them to obtain an accurate structure.

The folding process as simulated by the restrained dynamics is very rapid. At the end of the first 2 ps the secondary structure is essentially established while the molecule is still in an extended conformation. Some tertiary folding occurs even in the absence of long-range restraints. When they are introduced, it takes about 5 ps to obtain a tertiary structure that is approximately correct and another 6 ps to introduce the small adjustments required to converge to the final structure.

It is of interest to consider the relation between the results obtained in the restrained dynamics simulation and actual protein folding. That correctly folded structures are achieved only when the secondary structural elements are at least partly formed before the tertiary restraints are introduced is suggestive of the diffusion-collision model of protein folding (Bashford et al., 1984). Clearly, the specific pathway has no physical meaning since it is dominated by the NOE restraints. Also, the time scale of the simulated folding process is 12 orders of magnitude faster than experimental estimates. About 6 to 9 orders of magnitude of the rate increase are due to the fact that the secondary structure is stable once it is formed, in contrast to real protein folding where the secondary structural elements spend only a small fraction of time in the native conformation until coalescence has occurred. The remainder of the artificial rate increase presumably arises from the fact that the protein follows a single fairly direct path to the folded state in the presence of the NOE restraints, instead of having to go through a complex search process.

Many applications of NMR data to structure determinations have been made. Both distance geometry methods and molecular dynamics have been employed for reducing the data (Wüthrich, 1989; Clore and Gronenborn, 1989).

STRUCTURAL ROLE OF ACTIVE-SITE WATERS IN RIBONUCLEASE A

To achieve a realistic treatment of solvent-accessible active sites, a molecular dynamics simulation method, called the stochastic boundary method, has been implemented (Brooks and Karplus, 1983; Brünger et al., 1984; Brooks et al., 1985). It makes possible the simulation of a localized region, approximately spherical in shape, that is composed of the active site with or without ligands, the essential portions of the protein in the neighborhood of the active site, and the surrounding sovlent. The approach provides a simple and convenient method for reducing the total number of atoms included in the simulation, while avoiding spurious edge effects.

The stochastic boundary method for solvated proteins starts with a known x-ray structure; for the present problem the refined high-resolution (1.5 to 2 Å) x-ray structures provided by Petsko and coworkers were used (G. Petsko, private communication). The region of interest (here the active site of ribonuclease A) was defined by choosing a reference point (which was taken at the position of the phosphorous atom in the CpA inhibitor complex) and constructing a sphere of 12 Å radius around this point. Space within the sphere not occupied by crystallographically determined atoms was filled by water molecules, introduced from an equilibrated sample of liquid water. The 12-Å sphere was further subdivided into a reaction region (10 Å radius) treated by full molecular dynamics and a buffer region (the volume between 10 and 12 Å) treated by Langevin dynamics, in which Newton's equations of motion for the nonhydrogen atoms are augmented by a frictional term and a random-force term; these additional terms approximate the effects of the neglected parts of the system and permit energy transfer in and out of the reaction region. Water molecules diffuse freely between the reaction and buffer regions, but are prevented from escaping by an average boundary force (Brünger et al., 1984). The protein atoms in the buffer region are constrained by harmonic force derived from crystallographic temperature factors (Brooks et al., 1985). The forces on the atoms and their dynamics were calculated with the CHARMM program (Brooks et al., 1983); the water molecules were represented by the ST2 model (Stillinger and Rahman, 1974).

One of the striking aspects of the active site of ribonuclease is the presence of a large number of positively charged groups, some of which may be involved in guiding and/or binding the substrate (Matthew and Richards, 1982). The simulation demonstrated that these residues are stabilized in the absence of ligands by well-defined water networks. A particular example includes Lys-7, Lys-41, Lys-66, Arg-39 and the doubly protonated His-119. Bridging waters, some of which are organized into trigonal bipyramidal structures, were found to stabilize the otherwise very unfavorable configuration of near-neighbor positive groups because the interaction energy between water and the charged $C-NH_n^+$ (N = 1, 2, or 3) moieties is very large; e.g., at a donor-acceptor distance of 2.8 Å, the $C-NH_3^+-H_2O$ energy is -19 kcal/mole with the empirical potential used for the simulation (Brooks et al., 1983), in approximate agreement with accurate quantum-mechanical calculations (Desmeules and Allen, 1980) and gas-phase ion-molecule data (Kebarle, 1977). The average stabilization energy of the charged groups (Lys-7, Lys-41, Lys-66, Arg-39, and His-119) and the 106 water molecules included in the simulation is -376.6 kcal/mole. This energy is calculated as the difference between the simulated system and a system composed of separate protein and bulk water. Unfavorable protein-protein charged-group interactions are balanced by favorable water-protein and water-water interactions. The average energy per molecule of pure water from an equivalent stochastic boundary simulation (Brünger et al., 1984) was -9.0 kcal/mole, whereas that of the waters included in the active-site simulation was -10.2 kcal/mole; in the latter a large contribution to the energy came from the interactions between the water molecules and the protein atoms. It is such energy differences that are essential to a correct evaluation of binding equilibria and the changes introduced by site-specific mutagenesis (Fersht et al., 1985).

During the simulation, the water molecules involved in the charged-group interactions oscillated around their average positions, generally without performing exchange. On a longer time scale, it is expected that the waters would exchange and that the sidechains would undergo larger-scale displacements. This is in accord with the disorder found in the x-ray results for lysine and arginine residues (e.g., Lys-41 and Arg-39) (Gilbert et al., to be published; Wlodawer, 1985), a fact that makes difficult a crystallographic determination of the water structure in this case. It is also of interest that Lys-7 and Lys-41 have an average separation of only 4 Å in the simulation, less than that found in the x-ray structure. That this like-charged pair can exist in such a configuration is corroborated by experiments that have shown that the two lysines can be cross-linked (Marfey et al., 1965); the structure of this compound has been reported recently (Weber et al., 1985) and is similar to that found in the native protein.

In addition to the role of water in stabilizing the charged groups that span the active site and participate in catalysis, water molecules make hydrogen bonds to protein polar groups that become involved in ligand binding. A particularly clear example is provided by the adenine-binding site in the CpA simulation. The NH_2 group of adenine acted as a donor, making hydrogen bonds to the carbonyl of Asn-67, and the ring N^{1A} of adenine acted as an acceptor for a hydrogen bond from the amide group of Glu-69. Corresponding hydrogen bonds were present in the free ribonuclease simulation, with appropriately bound water molecules replacing the substrate. These waters and those that interact with the pyrimidine-site residues Thr-45 and Ser-123 help to preserve the protein structure in the optimal arrangement for binding. Similar substrate "mimicry" has been observed in x-ray structures of lysozyme (Blake et al., 1983) and penicillopepsin (James and Sielecki, 1983), but has not yet been seen in ribonuclease.

A COOPERATIVITY MUTANT IN HEMOGLOBIN

Hemoglobin has long been a subject of experimental and theoretical studies because it is the classic example of cooperativity in biological systems. Since the determination of the x-ray structure of the unliganded (deoxy) and liganded (oxy) tetramers by Perutz and coworkers (Perutz, 1970; Fermi and Perutz, 1981), attention has been focused on the atomic details of the cooperative mechanism. In particular, structural and thermodynamic measurements on native, mutant, and modified hemoglobins have been utilized in attempts to isolate the essential amino acids and to determine their contributions to cooperativity. It has been suggested on the basis of such studies (Pettigrew et al., 1982; Perutz, 1970) and theoretical analyses (Gelin et al., 1983) that the interactions between the C helix of one chain and the FG corner region of another ($C\alpha_1$-FGB_2) play an important role in the coupling of relative stabilities of the quaternary structures of the tetramer to the tertiary changes induced in the allosteric core by ligand binding to individual subunits. Asp $\beta99$ (G1) is one of the residues that have been studied in great detail. A series of naturally occurring mutants all have significantly reduced cooperativity and increased oxygen affinity relative to normal hemoglobin (Dickerson and Geis, 1983; Bunn and Forget, 1980). From a comparison of the deoxy and oxy normal hemoglobin tetramer structures, it has been suggested that the essential role of Asp β99 (G1) is to stabilize the deoxy tetramer by making hydrogen bonds to Tyr α42 (C7) and to Asn α97 (G4), which are absent in the oxy tetramer (Morimoto et al., 1971). It is now possible to supplement such observational conclusions by free energy simulations. We here employ the simulation method to show that the observed free energy changes result from interactions of Asp β99 (G1) with several amino acids and with the solvent. Although both Tyr α42 (C7) and Asn α97 (G4) are found to play a significant role, other interactions are shown to be of equal or greater importance. In what follows, we focus on the mutant Asn β99 (G1) to Ala (Hb Radcliffe) (Weatherall et al., 1977) which is of the "deletion" type (Fersht, 1987) and is therefore expected to lead to only localized structural changes (Shih et al., 1985) that are simplest to interpret.

The free energy difference ΔG between two states A and B (here A corresponds to normal hemoglobin and B to a mutant hemoglobin) is obtained by thermodynamic integration with the formula (Kirkwood, 1935; Kirkwood, 1942; Fleischman et al., 1990)

$$\Delta G = \int_0^1 <\Delta V>_\lambda \, d\lambda \qquad [12]$$

where $\Delta V = V_B - V_A$ and λ is a parameter, such that $V_\lambda = (1-\lambda)V_A + \lambda V_B$; the quantities V_A and V_B are empirical energy functions describing, respectively, the normal and the mutant hemoglobin molecule system. The essential part of the calculation is the evaluation of the thermodynamic average $<\Delta V>_\lambda$, where the subscript λ implies that average is over the hybrid system described by V_λ. For calculating the integral (Eq. [12]), a series of λ values are used (Fleischman et al., 1990; Brooks et al., 1983). A stochastic boundary simulation, which followed the procedure described previously (Brünger et al., 1985), was employed to obtain the required averages.

To determine the effect of the mutation Asp β99 (G1) → Ala on cooperativity, the free energy change of the deoxy and oxy tetramers resulting from this mutation has been calculated by Eq. [12] (Gao et al., 1989) (see Table 1). Both the deoxy and the oxy tetramer are destabilized by the mutation (66 and 60.5 kcal/mole, respectively, per interface), but it is the deoxy tetramer which is *more* destabilized, leading to the reduced cooperativity and increased ligand affinity. The differences between them (5.5 kcal/mole) can be compared with the measurements of Ackers et al. (3.4 kcal/mole [private communication]). The experimental and theoretical results have the same sign and are of the same order, suggesting that the simulation may be meaningfully analyzed to obtain insight into the interactions that contribute to the free energy differences.

To analyze the results, we make use of the fact that due to the linear form of Eq. [12], the free energy can be decomposed into the contribution from interactions between the mutated residue and any other residues or water molecules; in all cases, we consider the change in the free energy induced by the mutation. The change in the solvent interactions, which are essentially electrostatic, is more destabilizing for the oxy than for the deoxy tetramer. This is in accord with the x-ray structures since in the oxy tetramer the Asp sidechain is more exposed than in the deoxy tetramer. With respect to the protein interactions, the mutation stabilizes the oxy tetramer and destabilizes the deoxy tetramer. There are both inter- and intrasubunit contributions. As to the intersubunit terms, Tyr α42 (C7) does indeed stabilize the deoxy form in accord with the analysis of Morimoto et al. (1971). By contrast, Asn α97 (G4) favors the oxy form by a relatively small amount. The interaction with Asp α94 (G1) is unfavorable in both the deoxy and oxy form; i.e., the free energy of interaction between Asp α94 and Asp β99 is destabilizing in both tetramers, so that the replacement by the nonpolar Ala stabilizes the deoxy tetramer. Also of interest are the contributions that arise from within the β2 subunit, which are by definition the result of tertiary structural changes that accompany the quaternary transition.

Table 1. Free Energy for the Mutation Asp G1(99)β → Ala[a]

Contribution	ΔG(deoxy)	ΔG(oxy)	ΔΔG(oxy-deoxy)
Solvent	46.0	68.5	22.5
Protein[b]	20.0	-8.0	-28.0
Asp G1(99)β_2[c]	-8.8	-11.0	-2.2
Inter (α_1)	<u>2.8</u>	<u>-24.4</u>	<u>-27.2</u>
Tyr C7(42)	8.4	-4.3	-12.7
Asp G1(94)	-22.0	-44.4	-22.4
Val G3(96)	1.6	7.1	5.5
Asn G4(97)	9.7	13.0	3.3
Intra (β_2)	<u>26.1</u>	<u>27.4</u>	<u>1.3</u>
His FG4(97)	-2.1	-3.3	-1.2
Pro G2(100)	8.2	5.4	-2.8
Glu G3(101)	-11.2	6.9	18.1
Asn G4(102)	14.3	10.1	-4.2
TOTAL	<u>66.0</u>	<u>60.5</u>	<u>-5.5</u>

[a] All values in kcal/mole are given for one $\alpha_1\beta_2$ interface; a term in ΔG with a <u>positive</u> sign corresponds to the fact that the given contribution <u>destabilizes</u> the mutant (Ala) relative to the wild type (Asp). When the effect of the Asp residue by itself is discussed in the text, <u>stabilizing</u> contributions have a <u>positive</u> sign.
[b] Only the residues which contribute more than 1.5 kcal/mole to both the deoxy and oxy forms are listed.
[c] Internal energy contribution.

All the residues involved are close to the mutated residue Asp β99 and the largest contribution involves Glu β101. Apparently, the Asp β99 / Glu β101 interaction is stabilizing in the oxy tetramer and destabilizing in the deoxy tetramer.

It is evident that the free energy simulations provide new insights into the nature of the interactions in proteins and the possible consequences of mutations. Although free energy simulations are a recent development in molecular dynamics, so that their reliability is not fully established, it is likely that even if the quantitative values obtained here are not correct, the qualitative insights are still of interest. It is clear that a relatively small overall change in free energy may involve contributions from several large terms. Also, the balance between protein-protein and protein-solvent interactions plays an essential role. Finally, the intrasubunit contribution to cooperativity has not been considered previously.

CONCLUSION

Molecular dynamics is now playing an important role in the study of the properties of macromolecules of biological interest. It is also being used effectively in the analysis of experimental data and, in particular, has been shown to provide a new approach to structure determination by NMR and x-ray crystallography. Because molecular dynamics simulations are relatively new they have so far been employed primarily by theoreticians. It is to be hoped that experimentalists, as well, will begin to use molecular dynamics as a research tool for obtaining a deeper understanding of the biomolecules with which they work.

ACKNOWLEDGEMENTS

I wish to thank my collaborators who have contributed to the specific studies described here. They include: C.L. Brooks III, A.T. Brünger, G.M. Clore, C.M. Dobson, R. Elber, J. Gao, A.M. Gronenborn, T. Ichiye, K. Kuczera, J. Kuriyan, R.M. Levy, E.T. Olejniczak, G.A. Petsko, and B. Tidor. The work was supported in part by grants from the National Science Foundation and the National Institutes of Health. The present text is essentially that published previously in other reviews.

REFERENCES

Agmon, M. and Hopfield, J.J., 1983, CO binding to heme proteins: A model for barrier height distributions and slow conformational changes, *J. Chem. Phys.* 79:2042.

Ansari, A., Berendzen, J., Bowne, S.F., Frauenfelder, H., Iben, I.E.T., Sauke, T.B., Shyamsunder, E. and Young, R.D., 1985, Protein states and proteinquakes, *Proc. Natl. Acad. Sci. USA* 82:5000.

Artymiuk, P.J., Blake, C.C.F., Grace, D.E.P., Oatley, S.J., Phillips, D.C. and Sternberg, M.J.E., 1979, Crystallographic studies of the dynamic properties of lysozyme, *Nature* 280:563.

Austin, R.H., Beeson, K.W., Eisenstein, L., Frauenfelder, H. and Gunsalus, I.C., 1975, Dynamics of ligand binding to myoglobin, *Biochemistry* 14:5355.

Bashford, D., Weaver, D.L. and Karplus, M., 1984, Diffusion-collision model for the folding kinetics of λ-repressor operator-binding domain, J. Biomol. Struct. Dyns. 1:1243.

Bialek, W. and Goldstein, R.F., 1985, Do vibrational spectroscopies uniquely describe protein dynamics? The case for myoglobin, *Biophys. J.* 8:1027.

Blake, C.C.F., Pulford, W.C.A. and Artymiuk, P.J., 1983, X-ray studies of water in crystals of lysozyme, *J. Mol. Biol.* 167:693.

Brooks, B.R., Bruccoleri, R.E., Olafson, B.D., States, D.J., Swaminathan, S. and Karplus, M., 1983, CHARMM: a program for macromolecular energy, minimization, and dynamics calculations, *J. Comp. Chem.* 4:187.

Brooks, B.R. and Karplus, M., 1983, Harmonic dynamics of proteins: normal modes and fluctuations in bovine pancreatic trypsin inhibitor, *Proc. Natl. Acad. Sci. USA* 80:6571.

Brooks III, C.L., Brünger, A.T. and Karplus, M., 1985, Active site dynamics in protein molecules: a stochastic boundary molecular-dynamics approach, *Biopolymers* 24:843.

Brooks III, C.L., Karplus, M. and Pettitt, B.M., 1988, Proteins: a theoretical perspective of dynamics, structure and thermodynamics, *Adv. Chem. Phys.* LXXI, John Wiley & Sons, New York.

Brünger, A.T., 1989, A memory-efficient fast Fourier transformation algorithm for crystallographic refinement on supercomputers, *Acta Cryst.* A45:42.

Brünger, A.T., Brooks III, C.L. and Karplus, M., 1984, Stochastic boundary conditions for molecular dynamics simulations of ST2 water, *Chem. Phys. Letters* 105:495.

Brünger, A.T., Brooks III, C.L. and Karplus, M., 1985, Active site dynamics of ribonuclease, *Proc. Natl. Acad. Sci. USA* 82:8458.

Brünger, A.T., Campbell, R.L., Clore, G.M., Gronenborn, A.M., Karplus, M., Petsko, G.A. and Teeter, M.M., 1987a, Solution of a protein crystal structure with a model obtained from NMR interproton distance restraints, *Science* 235:1049.

Brünger, A.T., Clore, G.M., Gronenborn, A.M. and Karplus, M., 1986, Three-dimensional structure of proteins determined by molecular dynamics with interproton distance restraints: application to crambin, *Proc. Natl. Acad. Sci. USA* 83:3801.

Brünger, A.T., Karplus, M., and Petsko, G.A., 1989, Crystallographic refinement by simulated annealing: application to crambin, *Acta Cryst.* BA45:50.

Brünger, A.T., Kuriyan, J. and Karplus, M., 1987b, Crystallographic R factor refinement by molecular dynamics, *Science* 235:458.

Bunn, H.F. and Forget, B.G., 1980, "Hemoglobin: molecular, genetic clinical aspects," Saunders, New York.

Campbell, I.D., Dobson, C.M., Moore, G.R., Perkins, S.J. and Williams, R.J.P., 1976, Temperature dependent molecular motion of a tyrosine residue of ferrocytochrome c, *FEBS Lett.* 70:96.

Campbell, I.D., Dobson, C.M. and Williams, R.J.P., 1978, Structures and energetics of proteins and their active sites, *Adv. Chem. Phys.* 39:55.

Chandrasekhar, S., 1943, Stochastic problems in physics and astronomy, *Rev. Mod. Phys.* 15:1.

Chothia, C. and Lesk, A.M., 1985, Helix movements in proteins, *TIBS* 10:116.

Clore, G.M. and Gronenborn, A.M., 1989, Determination of three-dimensional structures of proteins and nucleic acids in solution by nuclear magnetic resonance spectroscopy, *CRC Crit. Rev. Biochem.*, in press.

Clore, G.M., Gronenborn, A.M., Brünger, A.T. and Karplus, M., 1985, Solution conformation of a heptadecapeptide comprising the DNA binding helix F of the cyclic AMP receptor protein of *Escherichia coli*: combined use of ^{1}H nuclear magnetic resonance and restrained molecular dynamics, *J. Mol. Biol.* 186:435.

Debrunner, P.G. and Frauenfelder, H., 1982, Dynamics of proteins, *Ann. Rev. Phys. Chem.* 33:283.

Desmeules, P.J. and Allen, L.C., 1980, Strong, positive-ion hydrogen bonds: the binary complexes formed from NH_3, OH_2, FH, PH_3, SH_2, and ClH, *J. Chem. Phys.* 72:4731.

Dickerson, R.E. and Geis, I., 1983, "Hemoglobin: structure, function, evolution, and pathology," Benjamin/Cummings, Menlo Park.

Dobson, C.M. and Karplus, M., 1986, Internal motion of proteins: nuclear magnetic resonance measurements and dynamic simulations, *in* "Methods in Enzymology," 131, C.H.W. Hirs and S.N. Timasheff, eds., Academic Press, Inc., New York.

Doster, W., Cusack, S. and Petry, W., 1989, Dynamical transition of myoglobin revealed by inelastic neutron scattering, *Nature* 337:734.

Elber, R. and Karplus, M., 1987, Multiple conformational states of proteins: a molecular dynamics analysis of myoglobin, *Science* 235:318.

Fermi, G. and Perutz, M.F., 1981, "Haemoglobin and myoglobin," Atlas of Molecular Structures in Biology: 2, Clarendon, Oxford.

Fersht, A.R., 1987, The hydrogen bond in molecular recognition, *Trends Biochem. Sci.* 12:301.

Fleischman, S.H., Tidor, B., Brooks III, C.L. and Karplus, M., Free energy simulation methodology, *J. Comp. Chem.*, to be published.

Frauenfelder, H., Petsko, G.A. and Tsernoglou, D., 1979, Temperature-dependent x-ray diffraction as a probe of protein structural dynamics, *Nature* 280:558.

Gao, J., Kuczera, K., Tidor, B. and Karplus, M., 1989, Hidden thermodynamic analysis mutant proteins: A molecular dynamics analysis, *Science* 244:1069.

Gelin, B.R., Lee, A.W.-M. and Karplus, M., 1983, Hemoglobin tertiary structural change on ligand binding: its role in the co-operative mechanism, *J. Mol. Biol.* 171:489.

Glover, I., Haneef, I., Pitts, J., Wood, S., Moss, D., Tickle, I. and Blundell, T., 1983, Conformational flexibility in a small globular hormone: x-ray analysis of avian pancreatic polypeptide at 0.98-Å resolution, *Biopolymers* 22:293.

Gurd, F.R.N. and Rothgeb, J.M., 1979, Motions in proteins, *Adv. Prot. Chem.* 33:73.

Hansen, J.P. and McDonald, I.R., 1976, "Theory of Simple Liquids," Academic Press, New York.

Hartmann, H., Parak, F., Steigemann, W., Petsko, G.A., Ringe Ponzi, D. and Frauenfelder, H., 1982, Conformational substates in a protein: structure and dynamics of metmyoglobin at 80°K, *Proc. Natl. Acad. Sci. USA* 79:4967.

Hendrickson, W.A. and Teeter, M.M., 1981, Structure of the hydrophobic protein crambin determined directly from the anomalous scattering of sulphur, *Nature* 290:107.

Honig, B., Hudson, B., Sykes, B.D. and Karplus, M., 1971, Ring orientation in β-ionone and retinals, *Proc. Natl. Acad. Sci. USA* 68:1289.

Ichiye, T., Olafson, B.D., Swaminathan, S. and Karplus, M., 1986, Structure and internal mobility of proteins: a molecular dynamics study of hen egg white lysozyme, *Biopolymers* 25:1909.

Jack, A. and Levitt, M., 1978, Refinement of large structures by simultaneous minimization of energy and R factor, *Acta Cryst.* A34:931.

James, M.N.G. and Sielecki, A.R., 1983, Structure and refinement of penicillopepsin at 1.8 Å resolution, *J. Mol. Biol.* 163:299.

Karplus, M. and Kushick, J.N., 1981, Method for estimating the configurational entropy of macromolecules, *Macromolecules* 14:325.

Karplus, M. and McCammon, J.A., 1981, The internal dynamics of globular proteins, *CRC Crit. Rev. Biochem.* 9:293.

Karplus, M. and McCammon, J.A., 1983, Dynamics of proteins: elements and function, *Ann. Rev. Biochem.* 53:263.

Kebarle, P., 1977, Ion thermochemistry and solvation from gas phase ion equilibria, *Ann. Rev. Phys. Chem.* 28:445.

Kirkpatrick, S., Gelatt, Jr., C.D. and Vecchi, M.P., 1983, Optimization by simulated annealing, *Science* 220:671.

Kirkwood, J.G. and Boggs, E.M., 1942, The radial distribution function in liquids, *J. Chem. Phys.* 10:394.

Kirkwood, J.G., 1935, Statistical mechanics of fluid mixtures, *J. Chem. Phys.* 3:300.

Konnert, J.H. and Hendrickson, W.A., 1980, A restrained-parameter thermal-factor refinement procedure, *Acta Cryst.* A36:344.

Kuriyan, J., Karplus, M. and Petsko, G.A., 1987, Estimation of uncertainties in x-ray refinement results by use of perturbed structures, *Proteins* 2:1.

Kuriyan, J., Petsko, G.A., Levy, R.M. and Karplus, M., 1986, Effect of anisotropy and anharmonicity on protein crystallographic refinement: an evaluation by molecular dynamics, *J. Mol. Biol.* 190:227.

Lesk, A.M. and Chothia, C., 1980, How different amino acid sequences determine similar protein structures: the structure and evolutionary dynamics of the globins, *J. Mol. Biol.* 136:225.

Levitt, M., Sander, C. and Stern, P.S., 1985, Protein normal-mode dynamics: trypsin inhibitor, crambin, ribonuclease and lysozyme, *J. Mol. Biol.* 181:423.

Levy, R.M. and Karplus, M., 1979, Vibrational approach to the dynamics of an α-helix, *Biopolymers* 18:2465.

Levy, R.M., Karplus, M. and Wolynes, P.G., 1981, NMR relaxation parameters in molecules with internal motion: exact Langevin trajectory results compared with simplified relaxation models, *J. Am. Chem. Soc.* 103:5998.

Levy, R.M., Perahia, D. and Karplus, M., 1982, Molecular dynamics of an α-helical polypeptide: temperature dependence and deviation from harmonic behavior, *Proc. Natl. Acad. Sci. USA* 79:1346.

Levy, R.M., Sheridan, R.P., Keepers, J.W., Dubey, G.S., Swaminathan, S. and Karplus, M., 1985, Molecular dynamics of myoglobin at 298° K: results from a 300 ps computer simulation, *Biophys. J.* 48:509.

Marfey, P.S., Uziel, M. and Little, J., 1965, Reaction of bovine pancreatic ribonuclease A with 1,5-difluoro-2,4-dinitrobenzene, *J. Biol. Chem.* 240:3270.

Marquart, M., Deisenhofer, D., Huber, R. and Palm, W., 1980, Crystallographic refinement and atomic models of the intact immunoglobulin molecule Kol and its antigen-binding fragment at 3.0Å and 1.9Å resolution, *J. Mol. Biol.* 141:369.

Matthew, J.B. and Richards, F.M., 1982, Anion binding and pH-dependent electrostatic effects in ribonuclease, *Biochemistry* 21:4989.

McCammon, J.A., Gelin, B.R. and Karplus, M., 1977, Dynamics of folded proteins, *Nature* 267:585.

McCammon, J.A., Wolynes, P.G. and Karplus, M., 1979, Picosecond dynamics of tyrosine side chains in proteins, *Biochemistry* 18:927.

McQuarrie, D.A., 1976, "Statistical Mechanics," Harper & Row, New York.

Morimoto, H., Lehmann, H. and Perutz, M.F., 1971, Molecular pathology of human haemoglobin: stereochemical interpretation of abnormal oxygen affinities, *Nature* 232:408.

Moss, D.S. and Morffew, A.J., 1982, Restrain: a restrained least squares refinement program for use in protein crystallography, *Comput. & Chem.* 6:1.

Navia, M.A., Fitzgerald, P.M.D., McKeever, B.M., Leu, C.-T., Heimbach, J.C., Herber, W.K., Sigal, I.S., Darke, P.L. and Springer, J.P., 1989, Three-dimensional structure of aspartyl protease from human immunodeficiency virus HIV-1, *Nature* 337:615.

Noggle, J.H. and Schirmer, R.E., 1971, "The Nuclear Overhauser Effect," Academic Press, New York.

Northrup, S.H., Pear, M.R., Lee, C.-Y., McCammon, J.A. and Karplus, M., 1982, Dynamical theory of activated processes in globular proteins, *Proc. Natl. Acad. Sci. USA* 79:4035.

Olejniczak, E.T., Dobson, C.M., Levy, R.M. and Karplus, M., 1984, Motional averaging of proton nuclear Overhauser effects in proteins. Predictions from a molecular dynamics simulation of lysozyme, *J. Am. Chem. Soc.* 106:1923.

Olejniczak, E.T., Poulsen, F.M. and Dobson, C.M., 1981, Proton nuclear Overhauser effects and protein dynamics, *J. Am. Chem. Soc.* 103:6574.

Parak, F., Knapp, E.W. and Kucheida, D., 1982, Protein dynamics. Mössbauer spectroscopy on deoxymyoglobin crystals, *J. Mol. Biol.* 161:177.

Pauling, L., Corey, R.B. and Branson, H.R., 1951, The structure of proteins: two hydrogen-bonded helical configurations of the polypeptide chain, *Proc. Natl. Acad. Sci. USA* 37:205.

Perutz, M.F., 1970, Haem-Haem interaction and the problem of allostery, *Nature* 228:726.

Petsko, G.A. and Ringe, D., 1984, Fluctuations in protein structure from x-ray diffraction, *Ann. Rev. Biophys. Bioeng.* 13:331.

Pettigrew, D.W., Romeo, P.H., Tsapis, A., Thillet, J., Smith, M.L., Turner, B.W. and Ackers, G.K., 1982, Probing the energetics of proteins through structural perturbation: sites of regulatory energy in human hemoglobin, *Proc. Natl. Acad. Sci. USA* 79:1849.

Phillips, D.C., 1981, Closing Remarks, *in* "Biomolecular Stereodynamics," R.H. Sarma, ed., Adenine, New York.

Poulsen, F.M., Hoch, J.C. and Dobson, C.M., 1980, A structural study of the hydrophobic box region of lysozyme in solution using nuclear Overhauser effects, *Biochemistry* 19:2597.

Ramachandran, G.N., Ramakrishnan, C. and Sasisekharan, V., 1963, Stereochemistry of polypeptide chain configurations, *J. Mol. Biol.* 7:95.

Shih, H.H.-L., Brady, J. and Karplus, M., 1985, *Proc. Natl. Acad. Sci. USA* 82:1697.

Smith, J.L., Hendrickson, W.A., Honzatko, R.B. and Sheriff, S., 1986, Structural heterogeneity in protein crystals, *Biochemistry* 25:5018.

Solomon, I., 1955, Relaxation processes in a system of two-spins, *Phys. Rev.* 99:559.

Stein, D.L., 1985, A model of protein conformational substates, *Proc. Natl. Acad. Sci. USA* 82:3670.

Stillinger, F.H. and Rahman, A., 1974, Improved simulation of liquid water by molecular dynamics, *J. Chem. Phys.* 60:1545.

Stillinger, F.H. and Weber, T.A., 1982, Hidden structure in liquids, *Phys. Rev.* A25:978.

Stillinger, F.H. and Weber, T.A., 1984, Packing structures and transitions in liquids and solids, *Science* 225:983.

Sussman, J.L., Holbrook, S.R., Church, G.M., Kim, S.-H., 1977, A structure-factor least-squares refinement procedure for macromolecular structures using constrained and restrained parameters, *Acta Cryst.* A33:800.

Swaminathan, S., Ichiye, T., van Gunsteren, W.F. and Karplus, M., 1982, Time dependence of atomic fluctuations in proteins: analysis of local and collective motions in bovine pancreatic trypsin inhibitor, *Biochemistry* 21:5230.

Toulouse, G., 1984, Progrés récent dans la physique des systèmes désordonnés, *Helv. Phys. Acta* 57:459.

van Gunsteren, W.F. and Berendsen, M.J.C., 1977, Algorithms for macromolecular dynamics and constraint dynamics, *Mol. Phys.* 34:1311.

van Gunsteren, W.F. and Karplus, M., 1982, Effect of constraints on the dynamics of macromolecules, *Macromolecules* 15:1528.

Verlet, L., 1967, Computer "experiments" on classical fluids, I. Thermodynamical properties of Lennard-Jones molecules, *Phys. Rev.* 159:98.

Wagner, G. and Wüthrich, K., 1982, Amide proton exchange and surface conformation of the basic pancreatic trypsin inhibitor in solution, *J. Mol. Biol.* 160:343.

Weatherall, D.J., Clegg, J.B., Callender, S.T., Wells, R.M.G., Gale, R.E., Huehns, E.R., Perutz, M.F., Viggiano, G. and Ho, C., 1977, Haemoglobin Radcliffe ($\alpha2\beta2$ 99 (G1)Ala): a high oxygen-affinity variant causing familial polycythaemia, *British J. Haemotology* 35:177.

Weber, P.C., Salemme, F.R., Lin, S.H., Konishi, Y. and Scheraga, H.A., 1985, Preliminary crystallographic data for cross-linked (lysine[7]-lysine[41])-ribonuclease A, *J. Mol. Biol.* 181:453.

Wlodawer, A., 1985, Structure of bovine pancreatic ribonuclease by x-ray and neutron diffraction, *in* "Biological Macromolecules and Assemblies: Vol. 2, Nucleic Acids and Interactive Proteins," F.A. Jurnak and A. McPherson, eds., Wiley, New York.

Wüthrich, K., 1989, The development of nuclear magnetic resonance spectroscopy as a technique for protein structure determination, *Acc. Chem. Res.* 22:36.

Wyckoff, H.W., Hirs, C.H.W. and Timasheff, S.N., eds., 1985, "Diffraction Methods for Biological Macromolecules," Part B, *Methods Enzymology* 115.

Yu, H.-A., Karplus, M. and Hendrickson, W.A., 1985, Restraints in temperature-factor refinement for macromolecules: an evaluation by molecular dynamics, *Acta Cryst.* B41:191.

Ziman, J.M., 1979, "Models of Disorder: The Theoretical Physics of Homogeneously Disordered Systems," Cambridge University Press, New York.

PROTEIN ENGINEERING AND BIOPHYSICAL STUDIES OF METAL BINDING PROTEINS

Sture Forsén

Department of Physical Chemistry 2, Chemical Center, Lund University, POB 124, 221 00 Lund, Sweden

INTRODUCTION

As will be amply illustrated during this Advanced Summer School, basic studies of enzyme mechanisms, protein stability, folding-unfolding pathways, protein-protein and protein-nucleic acid interactions have been greatly facilitated through the introduction of recombinant DNA methods and site-specific mutations. Mammalian proteins, which are often difficult to isolate and purify from natural sources, may - provided the gene can be cloned or synthesized - be expressed in a microorganism in substantial quantities. That aspect alone may be crucial for characterization of a protein with biophysical techniques that often require respectable amounts of purified material - NMR spectroscopy is a case in point. Expression of mammalian proteins in a rapidly growing microorganism also makes possible the application of isotope labelling (^{2}H, ^{13}C, ^{15}N, etc.). This is a technique of increasing importance in the assignment of NMR signals of larger proteins for which spectral crowding and overlap is an experimental obstacle. By means of spectral editing through isotope labelling, it may eventually be possible to study in detail selected areas in large proteins - for example individual domains or the active regions of enzymes (Redfield, 1989) - without the need for a complete assignment of the whole NMR spectrum.

In addition, the recombinant proteins can be "engineered," specific amino acids can be replaced by others or deleted, additional amino acids may be inserted in the sequence, S-S bonds may be introduced, whole blocks of the protein may be removed, N-terminals and C-terminals may be moved to new positions, etc. The possibilities opened up by modern molecular genetic techniques in protein studies are virtually limitless and our creative fantasy has a great play.

In this chapter, we shall mainly limit the discussion to one aspect of protein structure/ function studies: molecular properties of importance for the binding of metal ions in metalloproteins. As a particular example we will take a calcium binding protein, calbindin D_{9k}, which belongs to the calmodulin superfamily of intracellular regulatory proteins. Although our studies of this protein have been going on for several years, we are still far from understanding all aspects of its Ca^{2+}-binding properties or its stability. Nor do we know today to what extent our results on calbindin D_{9k} may be carried over to other metal ion binding proteins.

Since the strategies we have employed and many of our results are either already published, or soon to appear in print, the present account is somewhat condensed and personal. Only certain aspects are highlighted - but efforts have been made to make this contribution reasonably readable and self-contained.

CALBINDIN D_{9k} AND OTHER INTRACELLULAR Ca^{2+}-BINDING PROTEINS

It has been established for some time that Ca^{2+} ions play a very important role in the regulation of a wide variety of cellular activities (Rasmussen, 1968). To make a long and interesting story short - and drastically simplified - the mode of action of Ca^{2+} ions goes roughly as follows. In a resting eucaryotic cell - we may think of a mammalian cell - the cytosolic Ca^{2+} concentration is generally low. So low, in fact - about 100 - 200 nM - it was not until recently that it could be measured with some confidence. As a consequence of, say, hormonal action at a plasma membrane receptor, the intracellular Ca^{2+} concentration may be transiently increased several orders of magnitude. This in turn has the result that proteins with Ca^{2+}-binding sites characterized by binding constants of the order of 10^6 - 10^7 M^{-1} now will have their sites largely occupied by Ca^{2+} ions. In a particular family of homologous intracellular proteins, Ca^{2+} binding is accompanied by conformational changes that will influence their ability to interact with enzymes or other functional proteins. The biological activity of these target proteins, and consequently that of the whole cell, will then become altered. Through energy requiring transport proteins, Ca^{2+} levels will, after a short while, be restored back to normal and the regulatory Ca^{2+}-binding proteins will release the Ca^{2+} ions and switch back to their "Ca^{2+}-free" ("apo") conformations. One of the first proteins in this family of proteins to be discovered and characterized was troponin C ($M_r \approx 18,000$; 4 Ca^{2+} sites) that is found associated with skeletal muscle. Here Ca^{2+} binding, following a neural stimulus, triggers a contractile cycle. Whereas troponin C is expressed only in specialized muscle cells, a related protein, calmodulin ($M_r \approx 16,700$; 4 Ca^{2+} sites) appears ubiquitous and is found in virtually all eucaryotic cells. X-ray diffraction studies show that in the crystal both skeletal muscle troponin C and calmodulin have an unusual, dumbbell-like structure with two globular domains - each with two Ca^{2+} sites.

Needless to say both calmodulin (CaM) and troponin C (TnC) are of utmost biological interest, but they are at the same time almost too complicated for detailed structure-function studies (Forsén et al., 1986). For one thing, it is difficult to measure four binding constants accurately or to disentangle the kinetic scheme for the dissociation of four Ca^{2+} ions. Nevertheless, their genes have been cloned and the proteins have been expressed in *E. coli*. Site-directed mutagenesis has been employed in order to identify which surface amino acids of CaM and TnC are involved in interactions with their target proteins. Attempts have been made to establish whether or not the dumbbell-like shape also occurs in solution.

As a first step towards comprehensive studies of structure-function relations in the calmodulin superfamily of intracellular regulatory proteins, we have chosen to concentrate on bovine calbindin D_{9k} ($M_r \approx 8,500$; 75 amino acids). This protein has a crystal structure that strongly resembles that of the globular domains of CaM and TnC (Szebeni and Moffat, 1986). The schematic structure of calbindin D_{9k} is shown in Figure 1. The protein has two Ca^{2+} sites - one of which (site II) has the same helix-loop-helix fold and amino acid distribution as in the archetypal "EF-hand" sites of parvalbumin, CaM and TnC, whereas the other site (site I) constitutes a variant hand with two extra amino acids in the loop, one of which is a proline. A comparison of the amino acid sequences of the Ca^{2+}-binding loops of calbindin with that of some other related Ca^{2+}-binding proteins is shown in Table 1.

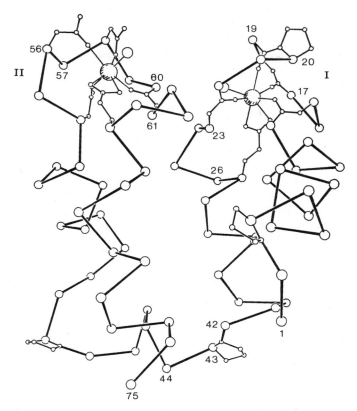

Figure 1. Schematic structure of bovine calbindin D_{9k} in the Ca^{2+}-loaded form (adapted after Szebenyi and Moffat, 1986).

Before we say a few words about gene synthesis and expression of wild-type and mutant calbindin D_{9k}, let us briefly summarize some, but by no means all, specific questions concerning the Ca^{2+}-binding proteins of the calmodulin superfamily:

* Why do EF-hand Ca^{2+} sites mainly appear in pairs?

* Can we rationalize the four to five orders of magnitude variation in the Ca^{2+} affinity in this class of proteins in terms of interactions between the ion(s) and amino acid residues of the protein?

* What is the molecular basis of the cooperative Ca^{2+} binding that has been inferred in many of the CaM superfamily proteins?

* What structural rearrangements and changes in dynamic properties of the protein accompany binding, and release, of Ca^{2+} ions?

* What interactions govern the stability of the proteins in solution?

It should be stated right away that we have as yet only very tentative and vague answers to some of the above queries.

Table 1. A comparison of the calcium-binding regions of several calcium binding proteins. Abbreviations: PA = parvalbumin; sTnC = skeletal muscle troponin C; cTnC = cardiac muscle troponin C; CaM = calmodulin. The single-letter amino acid code is used. X,Y,Z,-Y,-X and -Z are Ca^{2+}-ligand coordinates as specified by Kretsinger (Kretsinger, 1980).

Protein		X	Y	Z	-Y	-X	-Z
PA	CD	I-D-Q-D-K-S-G-F-I-E-E-D-E-L					
	EF	G-D-S-D-G-D-G-K-I-G-V-D-E-F					
STNC	I	F-D-A-D-G-G-G-D-I-S-V-K-E-L					
	II	V-D-E-D-G-S-G-T-I-D-F-E-E-F-					
	III	F-D-R-N-A-D-G-Y-I-D-A-E-E-L					
	IV	G-D-K-D-N-D-G-R-I-D-F-D-E-F-					
CTNC	I	F-V G-A-E-D-G-C-I-S-T-K-E-L					
		$\backslash /$					
		L					
CaM	I	F-D-K-D-G-N-G-T-I-T-T-K-E-L					
	II	V-D-A-D-G-N-G-T-I-D-F-P-E-F					
	III	F-D-K-D-G-N-G-Y-I-S-A-A-E-L					
	IV	A-D-I-D-G-D-G-E-V-N-Y-E-E-F					
Calbindin D_{9k}	I	^{13}Y A-K-E-G-D-P Q-L-S-K-E-E-L^{28}					
		$\backslash /$ $\qquad$ $\backslash /$					
		A $\qquad$ N					
	II	^{53}L-D-K-N-G-D-G-E-V-S-F-E-E-F^{66}					

GENE SYNTHESIS AND EXPRESSION, PURIFICATION OF MUTANT PROTEINS

Gene Synthesis and Expression

In order to produce bovine calbindin D_{9k} in *E. coli* a gene was designed encoding the known amino acid sequence of the protein. Codons observed to be preferentially utilized in highly expressed *E. coli* genes were mainly used, but a few moderately utilized codons were also used when restriction enzyme sites were to be introduced or avoided. To obtain independence from the existence of terminators in different expression vectors, a DNA sequence corresponding to the terminator of transcription of the *rrnc* gene was introduced behind the coding sequence. The gene was constructed from 34 overlapping oligonucleotides synthesized by the very rapid segmented microscale method with cellulose discs as solid support (Matthes et al., 1987). The oligonucleotides were assembled into five DNA segments in different bacteriophage M13 derivatives (Mathes et al., 1987) using the "shotgun ligation" technique (Grundström et al., 1985). In this method, correctly ligated oligonucleotides are obtained through direct biological selection instead of using gel purification. The method is very convenient for construction of large DNA segments from oligonucleotides. The DNA segments constructed were delimited by restriction enzyme sites placed at convenient distances in the gene, and were cloned into the corresponding sites of different M13 bacteriophages. Progeny phages were isolated from all five ligations and the nucleotide sequences of the DNA segments were verified by dideoxy sequencing. The complete gene was then assembled from the DNA segments, and a phage with the desired structure was isolated. The assembly of the gene from "DNA building blocks" bordered by restriction enzyme sites facilitates the construction of genes encoding calbindin molecules with amino acid substitutions, because only the DNA segment containing the substitutions desired has to be reassembled from oligonucleotides for each mutation. In order to further facilitate the construction of mutations in the N-terminal part of the protein, containing calcium binding site I, a truncated version of the gene was constructed (Linse et al., 1989). In this gene a 101 base pair segment corresponding to the first 33 amino acids in the proteins was replaced by a short polylinker sequence (Figure 2). The restriction enzymes sites bordering this polylinker were subsequently used for reassembling the complete gene from the original set of oligonucleotides with replacements of modified oligonucleotides containing the codons to be changed.

Two different expression plasmides were developed to overproduce bovine calbindin D_{9k} in E. coli. Both carry a strong inducible hybrid promoter, the "tac" promoter, in front of a polylinker containing several restriction enzyme sites, and a ribosome binding site with high homology to the 3' end of 16S ribosomal RNA, and with the most commonly found distance between this homology and the initiation codon. The first expression plasmid used, pICBI, (Brodin et al., 1986) is a vector based upon the high copy number plasmid pBR 322. In order to keep the tac promoter repressed on such a multicopy plasmid in the absence of the inducer IPTG, the lac repressor has to be overproduced.

Therefore, a 1.2 kb DNA fragment carrying $lacI^Q$, a lac repressor overproducing gene, was cloned into the plasmid. The second expression vector, pRCB1, is a derivative of the plasmid pBEU50 (Uhlin et al., 1983) where the EcoRI-HpaI restriction fragment containing a part of the tet gene has been replaced by a DNA segment from PICB1 containing the tac promoter, the synthetic calbindin gene and the $lacI^Q$ gene. This plasmid is temperature sensitive in the control of its copy number. At 30°C the vector is present in a moderate number of copies per cell, but when the temperature is raised above 35°C the replication control is lost and the number of copies per cell increases continuously, thereby increasing the gene dosage of the calbindin gene. Cell growth and protein synthesis continue normally for many generations at the higher temperature. By the use of this so called runaway vector, it has been possible to obtain an about five fold higher level of overproduction of the mutated calbindins than with the pICBI type of plasmids previously reported. This vector is also very useful for isotope labelling of recombinant calbindin D_{9k} - the labelled compounds are added concomitant with a raise in temperature (Brodin et al., 1989).

Stability problems have been encountered with foreign proteins in E. coli. A protease encoded from lon gene has been shown to be responsible for a considerable part of this protein breakdown. For this reason, expression of calbindin from pICBI was studied both in lon^+ and in lon^- E. coli strains by SDS polyacrylamide gel electrophoresis of total protein extracts. The level of expression was found to be approximately three times higher from a lon^- than from a lon^+ strain. The higher level of expression corresponds to one to two percent by weight of total E. coli protein (Brodin et al., 1986). We have found that the runaway expression system based upon vector pRCB1 is not compatible with the lon^- strain. However, this was not a problem because the high gene dosage that could be achieved by using the runaway vector pRCB1 leads to a high expression of calbindin in lon^+ E. coli reaching five to ten weight per cent of total protein. Considering that calbindin has a more than five times lower molecular weight than an average E. coli protein, this represents in molar terms a very high level of expression.

Protein Purification

The purification of proteins in considerable quantities for biophysical measurements is always an important task, and proteins expressed in E. coli are no exception. In this subsection we will briefly discuss some of the problems we encountered during the initial stages of our

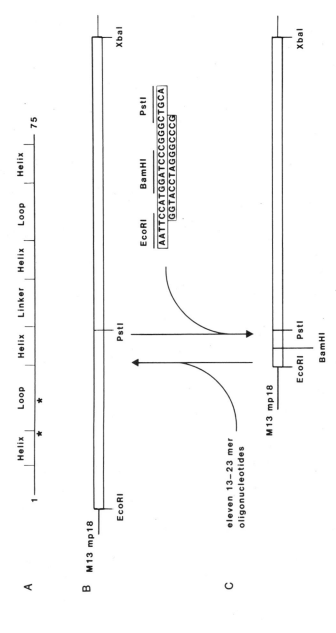

Figure 2. Construction of synthetic genes encoding bovine calbindin D_{9K} with amino acid substitutions and deletions. (A) shows the different structural parts of the calbindin molecule. The N-terminal amino acid is marked 1, and the C-terminal amino acid is marked 75. The asterisks show the positions in the protein where amino acid changes were introduced. (B) shows the entire calbindin gene, as previously constructed from oligonucleotides and inserted into bacteriophage M13mp18 (Brodin et al., 1986). The EcoRI-PstI DNA segment containing the N-terminal part of the gene was replaced by the shown polylinker sequence to obtain the truncated gene displayed in (C). Full-length calbindin genes with nucleotide substitutions and/or deletions were subsequently assembled by ligation of sets of eleven oligonucleotides, 13–23 nucleotides long, corresponding to both DNA strands, between the EcoRI and PstI sites bordering the polylinker.

work and the ways we solved them. It is our experience from talking to colleagues at meetings that similar problems have been encountered by others in studies of genetically engineered proteins, but few written accounts have been published.

The wild-type protein, "M0," and our initial set of mutant proteins, "M1, M2, M3 and M4," were purified as described in detail elsewhere (Brodin et al., 1986; Linse et al., 1989). The same methods have been employed also for the more recently expressed mutants. The steps consisted of sonication of the cell and fractionation on DEAE ion exchange and Sephadex G-50 columns. An urea step was found necessary to remove a low molecular weight contaminant that did not reveal itself on SDS-polyacrylamide or agarose gels but was observed in UV- and ^{1}H NMR spectra. We have not identified the impurity with certainty, but it appears to be a proteoglycan fragment from the *E. coli* cell wall (Brodin et al., 1986). It could not be removed from the protein using dialysis under normal conditions. However, urea treatment (8M urea) prior to the Sephadex G-50 fractionation was effective as was a heat treatment at 95°C. We have now discarded the urea treatment since it was shown to result in deamidation (of Asn-56) and now only use a rapid heat treatment (cf. Chazin et al., 1989).

The purity of the mutant proteins was checked by SDS-polyacrylamide and agarose gel electrophoresis in the presence of either 2mM EDTA or 2mM Ca^{2+}. This combination of gels is quite useful. Whereas all mutant proteins have the same mobility on the SDS gels, minor changes in charge show up on agarose gels. During the last year we have had access to a Pharmacia "Phast System" with thin gel electrofocusing. This technique has proved to be of great value in assessing the homogeniety of our protein preparations and has in some cases forced us to repurify samples regarded as pure on the basis of SDS and agarose gel electrophoresis. This is particularly important since deamidation ($R-CONH_2 \rightarrow R-COOH$), of Asn in particular, may occur during protein purification, and aging and deamidation of Asn-56 does occur in the case of calbindin (Chazin et al., 1989). ^{1}H NMR spectra were always obtained on the protein preparations during the final stages of purification. Not only did ^{1}H NMR permit the detection of contaminants due to the expression system, but this method also allowed facile detection of EDTA used in the ion exchange steps. The binding of EDTA to calbindin is strong and it is difficult to remove by dialysis using normal procedures. The binding is found to be increased in the mutants where a carboxylic acid has been replaced with an amide group.

Calbindin D_{9k} is not unique in its interaction with EDTA. In our experience, many proteins will bind EDTA very strongly and this chelator is not easily removed. EDTA contamination in metalloprotein preparations appears in fact to be more common than is generally appreciated, and it may have been the cause of diverging results on metal binding properties from different laboratories and on different preparations in the same laboaratory. A case in point is the metalloenzyme alkaline phosphatase. For several years the metal-ion binding properties of this enzyme were in dispute, with widely different stoichiometries and affinities reported by different research groups. It was finally convincingly shown, through the use of ^{14}C labelled EDTA, that the enzyme had an extraordinary affinity for this chelator and that rather drastic measures had to be taken to remove it from allegedly metal-free preparations of the protein (Csopak et al., 1972). Once such measures were taken, consistent metal binding results were obtained.

BIOPHYSICAL STUDIES

Under this heading we will first discuss results from studies on a select number of mutants aimed at probing the importance of charged amino acids at the surface of bovine calbindin for the protein's Ca^{2+} ion binding ability and its stability towards unfolding. Then we will briefly describe how we have established that bovine calbindin D_{9k} exists in solution as an equilibrium mixture of two isoforms due to *cis-trans* isomerization around the Gly-42 - Pro-43 amide bond - and that this conformation change is accomodated essentially locally in the loop region where Pro-43 is located.

Surface Charges and Ca^{2+} Affinity

Inspection of the X-ray structure of bovine calbindin D_{9k} reveals that it has a markedly asymmetrical charge distribution. There is an excess of negatively charged amino acids at and around the region of the Ca^{2+} binding sites, while positive charges tend to concentrate on the opposite side of the protein. What role does this charge asymmetry play for the functional properties of the protein? What is the effect on Ca^{2+} affinity and stability?

The cluster of negative surface residues around the two sites of calbindin, with some relevant distances indicated is shown in Figure 3. The beauty of site-specific mutagenesis is that negative charges can be switched off - with a minimal change in side-chain structure and volume by substituting Asn for Asp and Gln for Glu. In the first series of experiments, three of the negative surface residues, Glu-17, Asp-19 and Glu-26, were neutralized. Note that these residues are not directly liganded to the Ca^{2+} ion but some 6 to 17 Å away. First, a set of three mutants was prepared in which only one of these amino acids was neutralized. Then three mutants in which the amino acids were pairwise neutralized. Finally, a mutant protein was prepared in which all three surface charges were neutralized - in all seven mutants. The macroscopic (or "stoichiometric") Ca^{2+}-binding constants, K_1 and K_2, of the wild type and seven mutants were accurately determined using a competition method involving fluorescent Ca^{2+} chelators with Ca^{2+} affinities similar to that of the proteins (Quin 2, 5,5'-Br$_2$-BAPTA). The results are summarized in Table 2. Neutralization of the negative surface residues gradually reduces the Ca^{2+} affinity. The trend is perhaps best seen in the total free-energy change, ΔG_{tot}, for the binding of two Ca^{2+} ions: $\Delta G_{tot} = -RT \ln (K_1 \cdot K_2)$. On the average, each of the three

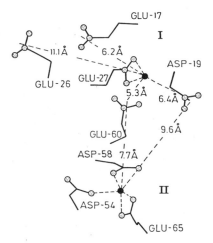

Figure 3. The cluster of negatively charged amino acid side chains around the two Ca^{2+} binding sites of calbindin D_{9k}. The view is from the top of Figure 1. According to X-ray diffraction studies, only the carboxylate groups of Glu-27, Asp-54, Asp-58 and Glu-65 are part of the inner ligand sphere of sites I and II, respectively. Distances between the Ca^{2+} ions and the non-liganded carboxylate groups (midpoint between the oxygens) are indicated.

Table 2. Macroscopic binding constants (K_1 and K_2), free energy of binding of two Ca^{2+} ions ($\Delta G_{tot} = -RT \ln (K_1 \cdot K_2)$) at low ionic strength and lower limit of the free energy of interaction between the two Ca^{2+} sites ($-\Delta\Delta G_{\eta=1} = RT \ln (4K_2/K_1)$) at low ionic strength and at [KCl] = 0.1 M.

Calbindin D_{9k} variant	$\log_{10}K_1$	$\log_{10}K_2$	ΔG_{tot} (kJ/mol) [a]	$-\Delta\Delta G_{\eta=1}$ (kJ/mol)	
				[KCl] $\approx$ 0	[KCl] = 0.1M
wild-type	8.3	8.6	-97	6.9±2.0	5.2±1.0
E17Q	7.4	8.1	-89	7.5±3.0	4.0±1.5
D19N	7.6	8.0	-89	6.1±1.0	2.1±0.5
E26Q	7.8	8.4	-93	11.8±2.0	7.0±1.5
E17Q; D19N	7.4	7.1	-83	1.7±0.3	-1.5±0.3
E17Q; E26Q	6.8	7.8	-83	9.0±2.0	3.8±0.5
D19N; E26Q	6.9	7.6	-83	7.2±1.5	2.6±0.5
E17Q; D19N; E26Q	6.8	6.7	-77	2.8±0.5	-0.5±0.2

[a] Standard errors approximately ± 0.5 kJ/mol.

surface charges contributes -7 kJ/mol to ΔG_{tot}, the effect being additive within experimental errors. This finding immediately tells us that attempts to rationalize the Ca^{2+} affinity of EF-hand proteins exclusively in terms of the nature and spatial arrangement of groups directly involved as ligands are likely to fail because of the strong influence of surface charges. This finding is not totally surprising: electrostatic effects are, after all, long-range and decay as $(distance)^{-1}$.

The influence of the surface charges on Ca^{2+} binding is reduced but not eliminated by ionic screening even at physiological ionic strength. We have recently finished an investigation of the effect of added KCl on the Ca^{2+} binding constants of wild-type calbindin and the seven mutants of Table 2. The results are presented in Figure 4. For clarity the error bars have been left out. The values ΔG_{tot} are generally accurate to ± 0.5 kJ/mole and we may thus compare the ionic strength effects on the individual mutants. We note that the contributions to ΔG_{tot} from the individual negative side-chains of Glu-17, Asp-19 and Glu-26 are not equally large. Glu-26 contributes only about half as much to ΔG_{tot} as does Glu-17 or Asp-19 - a difference in line with the relative distances of these surface groups with respect to the Ca^{2+} ions (cf. Figure 2). It is also evident that the difference in Ca^{2+} affinity among the various mutants is most pronounced at low ionic strengths. Extrapolation of the ΔG_{tot} vs. [KCl] curves furthermore indicates that ΔG_{tot} of the wild type is reduced by approximately 30% at very high ionic strengths. This might be entirely due to screening of electrostatic interactions but we cannot at present rule out an effect due to competitive binding interactions between K^+ and Ca^{2+}. Studies of the Ca^{2+} affinity in the presence of NaCl are in progress and it is also quite possible that direct NMR studies of $^{39}K^+$ and $^{23}Na^+$ can resolve the ambiguity.

In proteins with two or more metal binding sites, it is of interest to know not only the total free energy of ion binding, ΔG_{tot}, but also the free energy of interaction between the sites, $\Delta\Delta G$. In the case of calbindin D_{9k}, positive cooperativity ($\Delta\Delta G<0$) is observed for Ca^{2+} binding in the wild-type and all mutant calbindins listed in Table 2. A limiting value of $-\Delta\Delta G$, $-\Delta\Delta G_{\eta=1}$, can be obtained solely from the macroscopic binding constants K_1 and K_2, as (Linse et al., 1989; Linse et al., 1988):

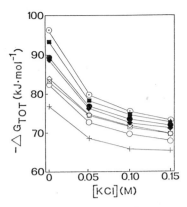

Figure 4. Apparent free energy of binding of two Ca^{2+} ions, ΔG_{tot}, as a function of KCl concentration for: (⊙) the wild-type protein, (●) E17Q, (◆) D19N, (■) E26Q, (○) (E17Q, D19N), (□) (E17Q, E26Q), (◇) (D19N, E26Q), (+) (E17Q, D19N, E26Q). ΔG_{tot} is highly reproducible between individual titrations of the same protein at the same value of [KCl]. The uncertainties in ΔG_{tot} are ±0.5 kJ/mol.

$$-\Delta\Delta G_{\eta=1} = RT \ln \left(\frac{4K_2}{K_1}\right) \qquad [1]$$

calculated values of $-\Delta\Delta G_{\eta=1}$ at low ionic strength ([KCl] ≈ 0.003M) and at [KCl] = 0.1 M are given in the two last columns of Table 2. Several results emerge. First, we note that cooperativity in Ca^{2+} binding is observed both at high and low ionic strength - although it possibly becomes reduced with increasing KCl concentration. The observation of cooperative ion binding goes against a simplistic picture of binding involving no conformational changes which would predict anticooperativity: the binding of the first Ca^{2+} ion reduces the negative charges of the binding region and therefore the second Ca^{2+} ion should bind more weakly. Second, we find that removal of surface charges may result in increased cooperativity. It appears, furthermore, that the negative side chain of Asp-19 contributes to positive cooperativity, whereas that of Glu-26 seems to counteract it.

The intramolecular interactions that are responsible for the cooperative Ca^{2+} binding in calbindin D_{9k} are yet to be identified. We are very likely dealing with a delicate balance of several effects. A rearrangement of the spatial distribution of charges at the binding site region - in particular at the loops - are, however, likely to accompany binding of the Ca^{2+} ions. Somewhat simplistically, we may imagine that binding of the first Ca^{2+} ion will bring Asp-19 closer to the second site while Glu-26 moves in the opposite way.

Surface Charges and Rates of Ca^{2+} Dissociation and Association

Neutralization of surface charges some 6 to 16 Å away from the Ca^{2+} sites results in a markedly reduced Ca^{2+} affinity, as we have seen in the preceeding subsection. Is this effect due to an increased rate of Ca^{2+} dissociation, or to a reduced rate of association, or both? These questions may be answered through kinetic measurements - we have used a combination of stopped-flow studies and ^{43}Ca NMR. In the stopped-flow experiments, the fluorescent chelator Quin 2 was used to monitor the release of Ca^{2+} from the initially Ca^{2+}-loaded proteins in a manner described elsewhere (Bayley et al., 1984; Martin et al., 1985). Reaction rates and amplitudes obtained at 20°C, under conditions essentially identical to those used in the determination of binding constants, are summarized in Table 3. In the ^{43}Ca NMR experiments, the line shape of the ^{43}Ca signal (~90% isotope enriched ^{43}Ca was used) was recorded on protein solutions with a slight excess of $^{43}Ca^{2+}$. ^{43}Ca NMR signals from "free" $^{43}Ca^{2+}$ and from the two Ca^{2+} binding sites could be resolved at room temperature (Vogel et al., 1985). Rates of Ca^{2+} exchange are obtained from the variation in line-shape with temperature as described in detail elsewhere (Drakenberg et al., 1983). Results from the ^{43}Ca NMR studies - limited to a selected set of mutants - are given in the last column of Table 3.

Table 3. Ca^{2+} dissociation rates (K) obtained from stopped flow and ^{43}Ca NMR experiments on genetically modified calbindins D_{9k} at 20°C. Experimental conditions = 20 mM Pipes/KOH, pH 7.0.

Calbindin D_{9k} variant	Stopped-flow exp.		^{43}Ca NMR exp
	k_{slow} (s^{-1})	k_{fast} (s^{-1})	k_{fast} (s^{-1})
wild-type	3.5^a		
Glu-17→Gln	4.3	-	-
Asp-19→Asn	8.3	21.5	19^b
Glu-26→Gln	2.4	12.4	-
Glu-17→Gln; Asp-19→Asn	6.3	15.3	-
Glu-26→Gln; Glu-17→Gln	3.9	18.3	-
Glu-26→Gln; Asp-19→Asn	4.8^a	-	-
Glu-26→Gln; Glu-17→Gln; Asp-19→Asn	3.4	≥ 400	370^b

[a] The amplitude corresponds to the dissociation of 2 Ca2+ ions with the same effective rate.
[b] The NMR data are obtained assuming that exchange from only one site contributes to the line broadening. If both sites exchange with equal rates, the rate constants calculated from the experiments will be reduced by approximately 50%.

A detailed analysis of the kinetic parameters of wild-type calbindin and mutants may be made going out from the scheme presented in Figure 5. Even without an extensive analysis we may, however, note that the nearly 100-fold reduction in Ca^{2+} affinity (both in K_1 and K_2), experimentally observed as the three surface charges are neutralized is only reflected in a corresponding increase in Ca^{2+} dissociation rate from one of the sites. Even with one or two surface charges neutralized, the Ca^{2+} dissociation rate for the fastest of the two steps seen is only a factor of 3 to 5 higher than that observed for the 2 Ca^{2+}-step in the wild-type. We may conclude, therefore, that the observed reduction in Ca^{2+} affinity is primarily due to a reduced rate of Ca^{2+} association.

There is some ambiguity in the calculation of on-rates and off-rates in the scheme of Figure 5, since we do not know which of the pathways of Ca^{2+} dissociation is the preferred one. If the upper pathway is assumed, and if we identify the slowest observed off-rate (k_{slow}) with k^I_{off} then $k^I_{on} = k_{slow} \cdot K_1$. But since the maximum value of K_I equals K_1 (i.e., the first macroscopic Ca^{2+} binding constant), the maximum value of k^I_{on} is equal to $k_{slow} \cdot K_1$.

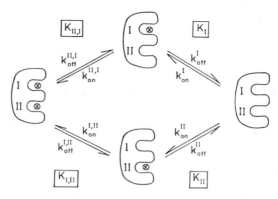

Figure 5. Schematic diagram showing the different modes of Ca^{2+} association and dissociation in bovine calbindin D_{9k}. The relation between the individual rate constants and the different site binding constants ($K = k_{on}/k_{off}$) is indicated. It should be noted that the stoichiometric binding constants, K_1 and K_2, are related to the site binding constants through the relations: $K_1 = K_I + K_{II}$ and $K_2 = K_{I,II} \cdot K_{II}/K_1$.

Conversely, if the lower pathway is assumed we may conclude that $k^{II}_{on} = k_{slow} \cdot K_1$. Using these equations, we calculate the following values for the maximum association rate constant for binding either to site I or site II under conditions when the second site is unoccupied:

wild type	:	$8 \cdot 10^8 \ M^{-1}s^{-1}$
E17Q	:	$1.7 \cdot 10^8 \ M^{-1}s^{-1}$
D19N	:	$3.3 \cdot 10^8 \ M^{-1}s^{-1}$
E17Q;D19N	:	$1.3 \cdot 10^8 \ M^{-1}s^{-1}$
E17Q:E26Q	:	$1.5 \cdot 10^7 \ M^{-1}s^{-1}$
D19N;E26Q	:	$6.9 \cdot 10^7 \ M^{-1}s^{-1}$
E17Q;D19N;E26Q	:	$2.1 \cdot 10^7 \ M^{-1}s^{-1}$

It appears thus that the charge neutralization has resulted in a significant reduction in the rate of Ca^{2+} association, at least for the first Ca^{2+} ion to bind to the protein.

Surface Charges and Thermochemical Properties

In the previous subsections we have seen that: (1) surface charges on calbindin D_{9k}, 5 to 16 Å away from the Ca^{2+} binding sites, have a marked effect on the Ca^{2+} affinity and, (2) the reduced Ca^{2+} affinity observed when these surface charges are neutralized is largely due to a reduced rate of Ca^{2+} association. To further investigate the role of the surface charges for the properties of calbindin, we could also try to separate the free energy changes of Ca^{2+} binding, ΔG_{tot} into entropic ($T\Delta S_{tot}$) and enthalpic (ΔH_{tot}) contributions. Such a division may be accomplished through calorimetric measurements of the enthalpy for the Ca^{2+} binding process:

$$2(Ca^{2+})_{aq} + \text{apo-calbindin } D_{9K} \xrightarrow{\Delta H_{tot}} (Ca^{2+})_2 \text{ calbindin } D_{9K} + aq \qquad [2]$$

where we have put a subscript on Ca^{2+} to indicate that water of hydration is liberated in the reaction. Needless to say, water of hydration of groups on the protein will also become reorganized.

We have recently completed a series of microcalorimetric measurements of the enthalpy change in Eq. [2] using an LKB Batch Microcalorimeter Model LKB 10700-1 fitted with an LKB titration assembly (Chen and Wadsö, 1982). The data have been analyzed using a modified version of the program KALORI, which takes into account that the binding process is a reversible reaction (Karlsson and Kullberg, 1976). The enthalpy change is for all seven mutants nearly linearly dependent on the total amount of Ca^{2+} added up to a Ca^{2+}-to-protein ratio of about 2. This is in line with our previous results showing positive cooperativity of Ca^{2+} binding. From the values of ΔH_{tot} and the known value of ΔG_{tot} (= - RT ln ($K_1 \cdot K_2$)) values of $T\Delta S_{tot}$ are obtained. The resulting data are summarized in the diagram in Figure 6. Only average values for ΔH_{tot} and $T\Delta S_{tot}$ for the mutants with one or two charges neutralized, respectively, are indicated in this diagram - the spread of values within each charge group is not large and we will ignore this in the present discussion.

As is evident from Figure 6, the changes in free energy of Ca^{2+} binding, ΔG_{tot}, as successive surface charges are neutralized is *not* due to changes in enthalpy of binding. ΔH_{tot} is small throughout the series and stays approximately constant at about -15 to -17 kJ/mol. The changes in ΔG_{tot} are almost entirely due to changes in the entropy term, $T\Delta S_{tot}$! An interpretation of this somewhat surprising result in molecular terms is certainly not straightforward. What appears likely is, however, that differential solvation plays a very important role. A possible interpretation of the thermochemical results would be that the more surface charges that are neutralized the less becomes the number of water molecules liberated when the two Ca^{2+} ions are bonded. This model implies that water molecules in the vicinity of a charged surface group are restricted in their mobility not only by the local electric field, but also by fields from charged groups 5 to 16 Å away.

NMR EVIDENCE FOR THE EXISTENCE OF STRUCTURAL HETEROGENEITY IN CALBINDIN D9K: *CIS/TRANS* ISOMERISM AT Pro-43

We now turn our attention to yet another aspect of calbindin D9k - an unexpected discovery that the protein in solution exists as an equilibrium between two isoforms.

All ^{1}H NMR spectra of calbindin show that approximately half of the proton resonances are resolved into major and minor resonances at a ratio of 3:1. Figure 7 shows a small region of the backbone fingerprint in a two-dimensional (2D) scalar correlated spectroscopy (COSY) spectrum

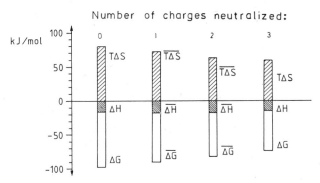

Figure 6. Schematic illustration of the thermodynamic parameters characterizing the binding of Ca^{2+} ions to wild-type and the mutant calbindins in which non-liganded carboxylate groups are neutralized. In the case where one or two charges have been neutralized, the parameters shown are average values for three different mutants (cf. text).

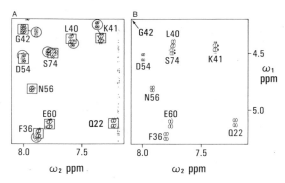

Figure 7. Expanded COSY spectrum of A) native calbindin D_{9k} and B) the Pro-43→Gly mutant. Major resonances are boxed and, when distinguishable, minor resonances are circled (Chazin et al., 1989).

of calbindin D_{9k}. The resolution of major and minor resonances is indicated and it is clear that the chemical shift difference between the two forms is variable. NMR studies at elevated temperatures clearly showed that the two forms are in equilibrium and the interconversion rate is estimated to 10^{-1} - 10^{-2} s^{-1} at room temperature (Chazin et al., 1989). To gain further insight into the origin of this conformational heterogeneity, the complete sequence specific assignments are needed (Kördel et al., 1989; Kördel et al., submitted). In summary, these were acquired using a strategy integrating relayed coherent transfer and multiple quantum techniques (Chazin et al., 1988) to identify the spin system of each amino acid. These amino acid spin systems were thereafter assigned to their location in the sequence by use of the sequential assignment procedure (Wüthrich, 1986). For a general review, see Drakenberg et al., 1989.

Figure 8A depicts the chemical shift differences between the backbone protons of the two forms as a function of sequence. It appears that the global structure is essentially identical for the two forms and the structural differences, as deduced from chemical shifts, are primarily localized to the linker region between the two domains of the protein (cf. Figure 1). The exchange rate and the sequential location of the perturbation indicate that *cis/trans* isomerism at Pro-43 is the cause of the multiple conformations. This conclusion is strengthened by the fact that in the ^{1}H NMR spectra of the mutant protein Pro-43→Gly only one set of resonances appear (cf. Figure 7B).

As Figure 9 shows, it is possible to distinguish *cis* from *trans* Gly-Pro peptide bonds from the through-space proton pair connectivities. While the proline C$^\delta$ protons, but not the C$^\alpha$ proton, are within 5 Å of the glycine C$^\alpha$ protons in the *trans* form (Figure 9B), the opposite is true for the *cis* form where only the C$^\alpha$ protons of glycine and that of proline are within 5 Å of each other (Figure 6B). Using 2D nuclear Overhauser spectroscopy (NOESY) experiments adjusted to detect only proton pairs closer than 5 Å, the major form of calbindin could be shown to correspond to *trans*-Pro-43 calbindin and the minor form to the *cis*-Pro-43 isoform (Chazin et al., 1989). This conformational heterogeneity is also present in porcine calbindin (Drakenberg et al., 1989), as well as in deamidation products of calbindin (Chazin et al., 1989). The structural heterogeneity can be eradicated by substituting Pro-43 with a glycine residue. As can be seen in Figure 8B, 2D ^{1}H NMR analysis of this mutant demonstrates that the global conformation is very similar to that of the two wild-type forms, while there is no conformational heterogeneity as confirmed in Figure 7B.

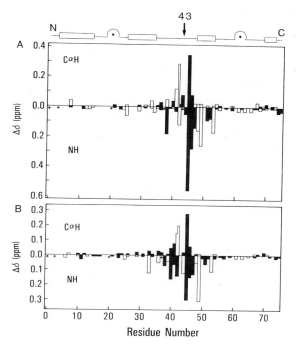

Figure 8. Chemical shift differences as a function of sequence for the backbone protons. A) (*trans*-Pro-43) - (*cis*-Pro-43 calbindin). B) (*trans*-Pro-43) - (Pro-43→Gly calbindin). Filled bars represent a negative chemical shift difference and open bars a positive. The (helix-loop-helix)$_2$ secondary structure of calbindin as a function of the sequence is outlined at the top of the figure.

CONCLUDING REMARKS

At this point it is certainly appropriate to return to the short list of questions raised above and try to summarize what we have learned so far about the molecular characteristics of bovine calbindin D_{9k}.

We have first of all established that Ca^{2+} binding in calbindin is a cooperative process. This finding goes against a simplistic model of Ca^{2+} binding to a number of negative surface charges in a structurally rigid protein - a model that would predict anticooperativity. Thus Ca^{2+} binding must be accompanied by structural changes in the binding region, although it is still too early to detail these changes. Cooperativity may be quantified as the free energy of interaction between the two Ca^{2+} sites, $\Delta\Delta G$. We have been able to put lower limits on $-\Delta\Delta G$ in wild-type and mutant calbindin D_{9k} and we have established that (1) cooperativity becomes more pronounced when certain negative surface residues on the protein are neutralized and, (2) the cooperativity is not much dependent on ionic strength. It clearly appears, then, that pairs of EF-hands are functional units that allow for cooperative Ca^{2+} binding. Studies of other proteins in the calmodulin superfamily indicate that this result has a general validity. Cooperative Ca^{2+} binding in intracellular regulatory proteins has direct implications for the cellular response to changes in Ca^{2+} concentration levels.

If we now turn to our second question, we have seen that, in order to rationalize differences in Ca^{2+} affinity between different proteins, we must not only look at the groups directly liganded to the ions but also look at the effect of charged amino acids located considerable distances

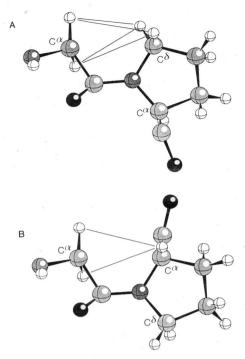

Figure 9. A Gly-Pro dipeptide in A) *trans* conformation and B) in *cis* conformation. Nuclear Overhauser enhancements possible to observe in NOESY experiments are indicated with solid lines.

away. This result very likely has validity also for a number of other proteins to which metal ions or other ionic species are bonded.

The third and fourth questions have hardly been addressed in the studies discussed here. In order to gain information on these points, one would first of all like to compare the three-dimensional structures of calbindin D_{9k} in both the Ca^{2+}-free and the Ca^{2+}-loaded forms. As yet only the crystal structure of the latter is known. 2D ^1H NMR studies of the solution structures of the two forms are in progress, however, and results should be arriving shortly.

It would be pretentious to say that we have been able to give an answer to the fifth question. It will take many more years of biophysical studies of proteins before we can hope to understand protein stability. We have, however, in the case of calbindin D_{9k} made the surprising discovery that this protein in solution exists as an equilibrium mixture of two isoforms due to *cis* ⇌ *trans* isomerisation around a proline amide bond. Furthermore, it has been possible to establish that the conformation change accompanying the isomerisation is essentially accomodated within a stretch of 8 - 10 amino acids in, or immediately adjacent to, the loop in which the proline (Pro-43) is located. These findings may not be unique to calbindin D_{9k} but occur in other proteins. Further studies on this point are in progress at Lund or elsewhere.

ACKNOWLEDGEMENTS

The present chapter is a brief summary of work done by a large team of coworkers - whose names appear in the appropriate references. This author is indebted to all of them for their generous support, but would like to express his warmest appreciation for the encouragement and help extended by two persons in particular: CarlJohan Kördel and Sara Linse.

The work has been supported by a grant from the Swedish Natural Science Research Council.

REFERENCES

Bayley, P.M., Ahlström, P., Martin, S.R., Forsén, S., 1984, The kinetics of calcium binding to calmodulin: Quin 2 and ANS stopped-flow fluorescence studies, *Biochem. Biophys. Res. Comm.* 120:185.

Billeter, M., Braun, W. and Wüthrich, K., 1982, Sequential assignments in protein ^{1}H NMR spectra, *J. Mol. Biol.* 155:321.

Brodin, P., Drakenberg, T., Thulin, E., Forsén, S. and Grundström, T., 1988/89, Selective proton labelling of amino acids in deuterated bovine calbindin D_{9k}. A way to simplify ^{1}H NMR spectra, *Protein Eng.*, 2:353.

Brodin, P., Grundström, T., Hofmann, T., Drakenberg, T., Thulin, E. and Forsén, S., 1986, Expression of bovine intestinal calcium binding protein from a synthetic gene in *E. coli* and characterization of the product, *Biochemistry* 25:5371.

Chazin, W.J., Kördel, J., Drakenberg, T., Thulin, E., Hofmann, T. and Forsén, S., 1989, Identification of an iso-aspartyl linkage formed upon deamidation of bovine calbindin D_{9k} and structural characterization by 2D ^{1}H NMR, *Biochemistry*, in press.

Chazin, W.J., Kördel, J., Thulin, E., Drakenberg, T., Brodin, P., Grundström, T. and Forsén, S., 1989, Proline isomerism leads to multiple folded conformations of calbindin D_{9k}: direct evidence from two-dimensional ^{1}H NMR, *Proc. Natl. Acad. Sci. USA* 86:2195.

Chazin, W.J., Rance, M. and Wright, P, 1988, Complete assignment of the ^{1}H NMR spectrum of French bean plastocyanin. Application of an integrated approach to spin system identification in proteins, *J. Mol. Biol.* 202:603.

Chen, A.T. and Wadsö, I., 1982, Simultaneous determination of ΔG, ΔH and ΔS by an automatic microcalorimetric titration technique. Application to protein ligand binding, *J. Biochem. and Biophys. Meth.* 6:307.

Csopak, H., Falk, K.E. and Szajn, H., 1972, Effect of EDTA on *E. coli* alkaline phosphatase, *Biochem Biophys. Acta* 258: 466.

Drakenberg, T., Forsén, S. and Lilja, H., 1983, ^{43}Ca NMR studies of calcium binding to proteins. Interpretation of experimental data by bandshape analysis, *J. Magn. Res.* 53:412.

Drakenberg, T., Hofmann, T. and Chazin, W.J., 1989, ^{1}H NMR studies of porcine calbindin D_{9k} in colution: sequential resonance assignments, secondary structure and global fold, *Biochemistry*, in press.

Forsén, S., Vogel, H.J. and Drakenberg, T., 1986, Biophysical studies of calmodulin, *in* "Calcium and Cell Function," Vol. VI, W.Y. Cheung, ed., Academic Press, New York, 113.

Grundström, T., Zenke, W.Z., Wintzerith, M., Mathes, H.W.D., Staub, A. and Chambon, P., 1985, Oligonucleotide-directed mutagenesis by microscale "shot-gun" gene synthesis, *Nucleic Acids Res.* 13:3305.

Karlsson, R. and Kullberg, L., 1976, A computer method for simultaneous calculation of equilibrium constants and enthalpy changes from calorimetric data, *Chem. Scr.* 9:54.

Kördel, J., Forsén, S. and W.J. Chazin, 1989, ^{1}H NMR sequential assignments, secondary structure and global fold in solution of the major (*trans*-Pro43) form of bovine calbindin D_{9k}, *Biochemistry* 28:7065.

Kördel, J., Forsén, S., Drakenberg T. and Chazin, W.J., Structure differences due to *cis-trans* isomerization in calbindin D_{9k}: NMR studies of the *cis*-Pro43 isoform and the Pro43 Gly mutant, submitted.

Kretsinger, R.H., 1980, Structure and evolution of calcium-modulated proteins, *CRC Crit. Rev. Biochem.* 8:119.

Linse, S., Brodin, P., Drakenberg, T., Thulin, E., Sellers, P., Elmdén, K., Grundström, T. and Forsén, S., 1987, Structure-function relationships in EF-hand Ca^{2+}-binding proteins. Protein engineering and biophysical studies of calbindin D_{9k}, *Biochemistry*, 26:6723.

Linse, S., Brodin, P., Johansson, C., Thulin, E., Grundström, T. and Forsén, S., 1988, The role of protein surface charges in ion binding, *Nature*, 335:651.

Martin, S.R., Andersson-Teleman, A., Bayley, P.M., Drakenberg, T. and Forsén, S., 1985, Kinetics of calcium dissociation from calmodulin and its tryptic fragments. A stopped-flow fluorescence study using Quin 2 reveals a two-domain structure, *Eur. J. Biochem.* 151:543.

Matthes, H.W.D, Staub, S. and Chambon, P., 1987, The segmented paper method: DNA synthesis and mutagenesis by rapid microscale "shotgun ligation synthesis", *in* "Methods in Enzymology," R. Wu, ed., Academic Press, New York.

Rasmussen, H., 1986, The calcium messenger system, *N. Engl. J. Med.* 314:1064 & 1164.

Redfield, A.G., 1989, "Isotope-filtered Proton NMR of Macromolecules," Abstract 33, Annual Meeting Am. Biophys. Soc., *Biophys. J.* 55:225a.

Szebenyi, D.M.F. and Moffat, K., 1986, The refined structure of vitamin D-dependent calcium-binding protein from bovine intestine. Molecular details, ion-binding, and implications for the structure of other calcuim-bindng proteins, *J. Biol. Chem.* 261:8761.

Uhlin, B.E., Schweikart, V. and Clark, A.J., 1983, New runaway-replication-plasmid cloning vectors and suppression of runaway replication by novobicin, *Gene* 22:255.

Vogel, H.J., Drakenberg, T., Forsén, S., O'Neil, J.D.J. and and Hofmann, T., 1985, Structural differences in the two calcium binding sites of the porcine intestinal calcium binding protein: a multinuclear NMR study, *Biochemistry* 24:3870.

Weber, G., 1975, Energetics of ligand binding to proteins, *Adv. Prot. Chem.* 29:1.

Wüthrich, K., 1986, "NMR of Proteins and Nucleic Acids," J. Wiley and Sons, Inc., New York.

GROWTH HORMONES: EXPRESSION, STRUCTURE AND PROTEIN ENGINEERING

K. G. Skryabin[1], P. M. Rubtsov[1], V. G. Gorbulev[1], A. A. Schulga[1], A. Sh. Parsadanian[1], M. P. Kirpichnikov[1], A. A. Bayev[1], A. G. Pavlovskii[2], S. N. Borisova[2], B. K. Vainstein[2] and A. A. Bulatov[3]

[1]Engelhardt Institute of Molecular Biology, Academy of Sciences of the USSR
[2]Institute of Crystallography, and [3]Institute of Experimental Endocrinology and Hormone Chemistry, Academy of Medical Sciences of the USSR, Moscow

INTRODUCTION

The growth hormones (GHs) are polypeptide hormones with molecular weight of 22 kD consisting of one chain. GHs are synthesized in an anterior lobe of hypophysis as a precursors with molecular weight of 25 kD. These hormones are necessary for the normal growth and development of organism and play an essential role in regulation of anabolic processes. The characteristic features of GHs are species specificity and multiple biological activity. Cloning and sequencing of DNA complementary to the rat GH messenger RNA in 1977 represented one of the first mammalian cDNA to be analyzed (Seeburg et al., 1977). Since then a variety of cDNAs corresponding to members of GH family from different vertebrates have been isolated and analyzed. These cDNAs have been used to produce the hormones in bacteria and eucaryotic cells by recombinant DNA techniques.

This paper summarizes the results of our work on cloning of cDNAs coding for GHs from different species and synthesis in bacteria the natural and mutant variants of GHs, three-dimensional structure of hGH and structure-functional relations in GHs.

CLONING AND ANALYSIS OF cDNAs AND EXPRESSION OF NATURAL GHs IN *ESCHERICHIA COLI*

Initially DNAs complementary to messenger RNAs from human pituitary adenoma were cloned and plasmids containing full-length cDNA encoding human GH precursor were isolated as it was described previously by us (Rubtsov et al., 1985). The insert from one of such plasmids phGH18 was completely sequenced. The primary structure of this insert is presented on Figure 1.

It contains 5'-nontranslated sequence (29 bp), sequences coding for signal peptide and mature GH (573 bp), 3'-nontranslated region (108 bp) and polyA-track (20 bp). This sequence is identical to that published earlier by Martial et al. (1979).

The cloned human GH cDNA was used as a probe for the screening of cDNA library prepared for mRNAs isolated from bovine pituitary and pbGH18 plasmid carrying full-length

309

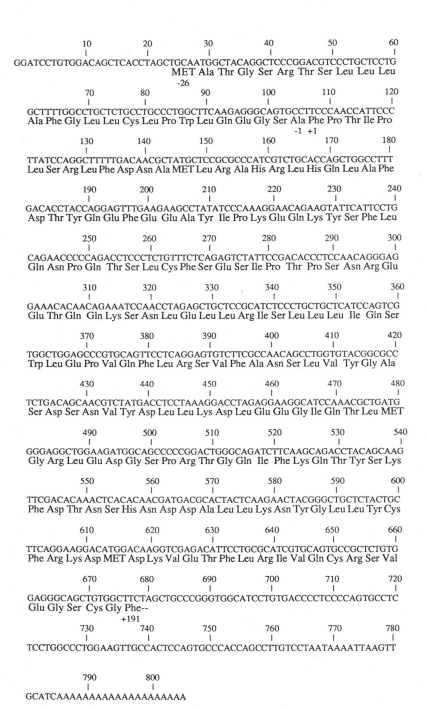

```
            10        20        30        40        50        60
            |         |         |         |         |         |
GGATCCTGTGGACAGCTCACCTAGCTGCAATGGCTACAGGCTCCCGGACGTCCCTGCTCCTG
                          MET Ala Thr Gly Ser Arg Thr Ser Leu Leu Leu
                          -26
            70        80        90       100       110       120
            |         |         |         |         |         |
GCTTTTGGCCTGCTCTGCCTGCCCTGGCTTCAAGAGGGCAGTGCCTTCCCAACCATTCCC
Ala Phe Gly Leu Leu Cys Leu Pro Trp Leu Gln Glu Gly Ser Ala Phe Pro Thr Ile Pro
                                                          -1 +1
           130       140       150       160       170       180
            |         |         |         |         |         |
TTATCCAGGCTTTTTGACAACGCTATGCTCCGCGCCCATCGTCTGCACCAGCTGGCCTTT
Leu Ser Arg Leu Phe Asp Asn Ala MET Leu Arg Ala His Arg Leu His Gln Leu Ala Phe

           190       200       210       220       230       240
            |         |         |         |         |         |
GACACCTACCAGGAGTTTGAAGAAGCCTATATCCCAAAGGAACAGAAGTATTCATTCCTG
Asp Thr Tyr Gln Glu Phe Glu Glu Ala Tyr Ile Pro Lys Glu Gln Lys Tyr Ser Phe Leu

           250       260       270       280       290       300
            |         |         |         |         |         |
CAGAACCCCCAGACCTCCCTCTGTTTCTCAGAGTCTATTCCGACACCCTCCAACAGGGAG
Gln Asn Pro Gln Thr Ser Leu Cys Phe Ser Glu Ser Ile Pro Thr Pro Ser Asn Arg Glu

           310       320       330       340       350       360
            |         |         |         |         |         |
GAAACACAACAGAAATCCAACCTAGAGCTGCTCCGCATCTCCCTGCTGCTCATCCAGTCG
Glu Thr Gln Gln Lys Ser Asn Leu Glu Leu Leu Arg Ile Ser Leu Leu Leu Ile Gln Ser

           370       380       390       400       410       420
            |         |         |         |         |         |
TGGCTGGAGCCCGTGCAGTTCCTCAGGAGTGTCTTCGCCAACAGCCTGGTGTACGGCGCC
Trp Leu Glu Pro Val Gln Phe Leu Arg Ser Val Phe Ala Asn Ser Leu Val Tyr Gly Ala

           430       440       450       460       470       480
            |         |         |         |         |         |
TCTGACAGCAACGTCTATGACCTCCTAAAGGACCTAGAGGAAGGCATCCAAACGCTGATG
Ser Asp Ser Asn Val Tyr Asp Leu Leu Lys Asp Leu Glu Glu Gly Ile Gln Thr Leu MET

           490       500       510       520       530       540
            |         |         |         |         |         |
GGGAGGCTGGAAGATGGCAGCCCCCGGACTGGGCAGATCTTCAAGCAGACCTACAGCAAG
Gly Arg Leu Glu Asp Gly Ser Pro Arg Thr Gly Gln Ile Phe Lys Gln Thr Tyr Ser Lys

           550       560       570       580       590       600
            |         |         |         |         |         |
TTCGACACAAACTCACACAACGATGACGCACTACTCAAGAACTACGGGCTGCTCTACTGC
Phe Asp Thr Asn Ser His Asn Asp Asp Ala Leu Leu Lys Asn Tyr Gly Leu Leu Tyr Cys

           610       620       630       640       650       660
            |         |         |         |         |         |
TTCAGGAAGGACATGGACAAGGTCGAGACATTCCTGCGCATCGTGCAGTGCCGCTCTGTG
Phe Arg Lys Asp MET Asp Lys Val Glu Thr Phe Leu Arg Ile Val Gln Cys Arg Ser Val

           670       680       690       700       710       720
            |         |         |         |         |         |
GAGGGCAGCTGTGGCTTCTAGCTGCCCGGGTGGCATCCTGTGACCCCTCCCCAGTGCCTC
Glu Gly Ser Cys Gly Phe--
                +191
           730       740       750       760       770       780
            |         |         |         |         |         |
TCCTGGCCCTGGAAGTTGCCACTCCAGTGCCCACCAGCCTTGTCCTAATAAAATTAAGTT

           790       800
            |         |
GCATCAAAAAAAAAAAAAAAAAAAAAA
```

Figure 1.   Nucleotide sequence of human preGH cDNA (phGH18) and deduced amino acid sequence of protein.

bovine GH precursor cDNA was isolated. The insert in pbGH18 was 830 bp long and contained 33 bp of 5'-nontranslated region, 651 bp encoding GH precursor, 104 bp of 3'-nontranslated region and polyA-tail (Figure 2).

The comparison of pbGH18 insert with the sequence of bovine pre GH cDNA reported by Miller et al. (1980) for pBP348 revealed several differences. Five of them were found in the coding region of GH. All these substitutions were silent: His[22]-CAT(instead of CAC), Arg[34]-CGC(CGT), Ser[55]-TCT(TCC), Phe[92]-TTC(TTT) and Leu[121]-CTG(TTG). However all these codons were coincided with those identified in bovine GH cDNA by other groups (Woychik et al., 1982; Seeburg et al., 1983).

By the screening of the porcine pituitary cDNA library two clones hybridizing with bovine GH cDNA were identified. The plasmids from these clones pPGH1 and pPGH2 contained the inserts with the length of 500 and 760 bp (Zvirblis et al.,1988). The insert from pPGH1 was found to contain the partial cDNA of porcine GH starting from the codon for the 66th amino acid of mature hormone. Another plasmid pPGH2 bore the entire sequence of mature GH and codons for last 9 amino acids of signal peptide. However the cDNA in this clone was severe rearranged probably during the cloning procedure (Zvirblis et al., 1988). Despite of it we were able to reconstitute the proper cDNA sequence starting from the position of -6 of signal peptide codons by ligation of subfragments isolated from pPGH1 and pPGH2 (Zvirblis et al., 1988). This reconstituted sequence is presented in Figure 3.

The comparison of nucleotide sequence of inserts from pPGH1 and pPGH2 with the data published by other groups for cDNA (Seeburg et al., 1983; Movva and Schulz, 1984) and genomic gene (Vize and Wells, 1987) revealed a number of differences summarized in Table 1.

Table 1. The comparison of porcine GH cDNA sequences.

Position in the cDNA	Porcine GH cDNA variants cloned by:				
	A	B	C	This work	
				pPGH1	pPGH2
coding region:					
-2	GGA(Gly)	GGC(Gly)	GGC(Gly)	-	GGA(Gly)
+25	GCC(Ala)	GCT(Ala)	GCT(Ala)	-	GCC(Ala)
+48	CAG(Gln)	CAA(Gln)	CAG(Gln)	-	CAG(Gln)
+56	ACC(Thr)	ACG(Thr)	ACC(Thr)	-	ACC(Thr)
+100	CTG(Leu)	TTG(Leu)	CTG(Leu)	CTG(Leu)	CTG(Leu)
+127	GAA(Glu)	GAG(Glu)	GAG(Glu)	GAG(Glu)	GAA(Glu)
+134	GGA(Gly)	GGC(Gly)	GGA(Gly)	GGA(Gly)	GGA(Gly)
+136	ATC(Ile)	ATC(Ile)	ATC(Ile)	ACC(Thr)	ATC(Ile)
+190	TTC(Phe)	TCC(Ser)	TTC(Phe)	TTC(Phe)	TTC(Phe)
3'-noncoding region:					
44-51	GGTGCCCT	GACCT	GACCCT	GACCCT	GACCCT
83-87	TCCCT	TCCT	TCCT	TCCT	TCCT
94-97	ACCA	ACA	ACCA	ACCA	ACCA
105-107	TCG	TCG	TCG	TTG	TTG

A - Seeburg et al., 1983
B - Movva and Schulz, 1984
C - Vise and Wells, 1987

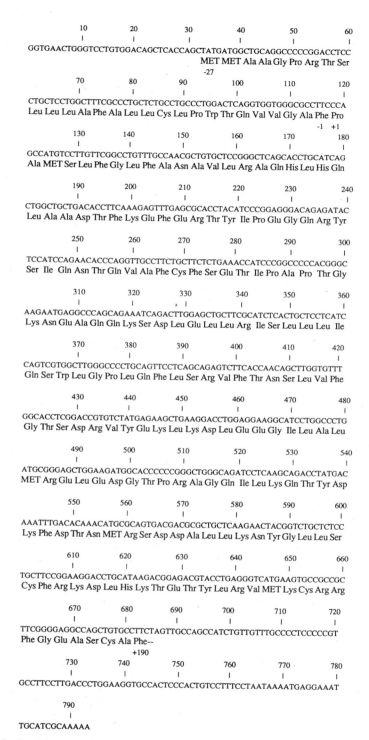

```
          10        20        30        40        50        60
          |         |         |         |         |         |
GGTGAACTGGGTCCTGTGGACAGCTCACCAGCTATGATGGCTGCAGGCCCCCGGACCTCC
                                       MET MET Ala Ala Gly Pro Arg Thr Ser
                                               -27

          70        80        90       100       110       120
          |         |         |         |         |         |
CTGCTCCTGGCTTTCGCCCTGCTCTGCCTGCCCTGGACTCAGGTGGTGGGCGCCTTCCCA
Leu Leu Leu Ala Phe Ala Leu Leu Cys Leu Pro Trp Thr Gln Val Val Gly Ala Phe Pro
                                                              -1 +1

         130       140       150       160       170       180
          |         |         |         |         |         |
GCCATGTCCTTGTTCGGCCTGTTTGCCAACGCTGTGCTCCGGGCTCAGCACCTGCATCAG
Ala MET Ser Leu Phe Gly Leu Phe Ala Asn Ala Val Leu Arg Ala Gln His Leu His Gln

         190       200       210       220       230       240
          |         |         |         |         |         |
CTGGCTGCTGACACCTTCAAAGAGTTTGAGCGCACCTACATCCCGGAGGGACAGAGATAC
Leu Ala Ala Asp Thr Phe Lys Glu Phe Glu Arg Thr Tyr Ile Pro Glu Gly Gln Arg Tyr

         250       260       270       280       290       300
          |         |         |         |         |         |
TCCATCCAGAACACCCAGGTTGCCTTCTGCTTCTCTGAAACCATCCCGGCCCCCACGGGC
Ser Ile Gln Asn Thr Gln Val Ala Phe Cys Phe Ser Glu Thr Ile Pro Ala Pro Thr Gly

         310       320       330       340       350       360
          |         |         |         |         |         |
AAGAATGAGGCCCAGCAGAAATCAGACTTGGAGCTGCTTCGCATCTCACTGCTCCTCATC
Lys Asn Glu Ala Gln Gln Lys Ser Asp Leu Glu Leu Leu Arg Ile Ser Leu Leu Leu Ile

         370       380       390       400       410       420
          |         |         |         |         |         |
CAGTCGTGGCTTGGGCCCCTGCAGTTCCTCAGCAGAGTCTTCACCAACAGCTTGGTGTTT
Gln Ser Trp Leu Gly Pro Leu Gln Phe Leu Ser Arg Val Phe Thr Asn Ser Leu Val Phe

         430       440       450       460       470       480
          |         |         |         |         |         |
GGCACCTCGGACCGTGTCTATGAGAAGCTGAAGGACCTGGAGGAAGGCATCCTGGCCCTG
Gly Thr Ser Asp Arg Val Tyr Glu Lys Leu Lys Asp Leu Glu Glu Gly Ile Leu Ala Leu

         490       500       510       520       530       540
          |         |         |         |         |         |
ATGCGGGAGCTGGAAGATGGCACCCCCCGGGCTGGGCAGATCCTCAAGCAGACCTATGAC
MET Arg Glu Leu Glu Asp Gly Thr Pro Arg Ala Gly Gln Ile Leu Lys Gln Thr Tyr Asp

         550       560       570       580       590       600
          |         |         |         |         |         |
AAATTTGACACAAACATGCGCAGTGACGACGCGCTGCTCAAGAACTACGGTCTGCTCTCC
Lys Phe Asp Thr Asn MET Arg Ser Asp Asp Ala Leu Leu Lys Asn Tyr Gly Leu Leu Ser

         610       620       630       640       650       660
          |         |         |         |         |         |
TGCTTCCGGAAGGACCTGCATAAGACGGAGACGTACCTGAGGGTCATGAAGTGCCGCCGC
Cys Phe Arg Lys Asp Leu His Lys Thr Glu Thr Tyr Leu Arg Val MET Lys Cys Arg Arg

         670       680       690       700       710       720
          |         |         |         |         |         |
TTCGGGGAGGCCAGCTGTGCCTTCTAGTTGCCAGCCATCTGTTGTTTGCCCCTCCCCCGT
Phe Gly Glu Ala Ser Cys Ala Phe--
                +190

         730       740       750       760       770       780
          |         |         |         |         |         |
GCCTTCCTTGACCCTGGAAGGTGCCACTCCCACTGTCCTTTCCTAATAAAATGAGGAAAT

         790
          |
TGCATCGCAAAAA
```

Figure 2. Nucleotide sequence of bovine preGH cDNA (pbGH18) and deduced amino acid sequence of protein.

312

```
          10        20        30        40        50        60
           |         |         |         |         |         |
ACTCAGGAGGTGGGAGCCTTCCCAGCCATGCCCTTGTCCAGCCTATTTGCCAACGCCGTG
Thr Gln Glu Val Gly Ala Phe Pro Ala MET Pro Leu Ser Ser Leu Phe Ala Asn Ala Val
 -6                    -1  +1
          70        80        90       100       110       120
           |         |         |         |         |         |
CTCCGGGCCCAGCACCTGCACCAACTGGCTGCCGACACCTACAAGGAGTTTGAGCGCGCC
Leu Arg Ala Gln His Leu His Gln Leu Ala Ala Asp Thr Tyr Lys Glu Phe Glu Arg Ala

         130       140       150       160       170       180
           |         |         |         |         |         |
TACATCCCGGAGGGACAGAGGTACTCCATCCAGAACGCCCAGGCTGCCTTCTGCTTCTCG
Tyr Ile  Pro Glu Gly Gln Arg Tyr Ser Ile Gln Asn Ala Gln Ala Ala Phe Cys Phe Ser

         190       200       210       220       230       240
           |         |         |         |         |         |
GAGACCATCCCGGCCCCCACGGGCAAGGACGAGGCCCAGCAGAGATCGGACGTGGAGCTG
Glu Thr Ile Pro Ala Pro Thr Gly Lys Asp Glu Ala Gln Gln Arg Ser Asp Val Glu Leu

         250       260       270       280       290       300
           |         |         |         |         |         |
CTGCGCTTCTCGCTGCTGCTCATCCAGTCGTGGCTCGGGCCCGTGCAGTTCCTCAGCAGG
Leu Arg Phe Ser Leu Leu Leu Ile Gln Ser Trp Leu Gly Pro Val Gln Phe Leu Ser Arg

         310       320       330       340       350       360
           |         |         |         |         |         |
GTCTTCACCAACAGCCTGGTGTTTGGCACCTCAGACCGCGTCTACGAGAAGCTGAAGGAC
Val Phe Thr Asn Ser Leu Val Phe Gly Thr Ser Asp Arg Val Tyr Glu Lys Leu Lys Asp

         370       380       390       400       410       420
           |         |         |         |         |         |
CTGGAGGAGGGCATCCAGGCCCTGATGCGGGAGCTGGAGGATGGCAGCCCCCGGGCAGGA
Leu Glu Glu Gly Ile Gln Ala Leu MET Arg Glu Leu Glu Asp Gly Ser Pro Arg Ala Gly

         430       440       450       460       470       480
           |         |         |         |         |         |
CAGACCCTCAAGCAAACCTACGACAAATTTGACACAAACTTGCGCAGTGATGACCGCCTG
Gln Thr Leu Lys Gln Thr Tyr Asp Lys Phe Asp Thr Asn Leu Arg Ser Asp Asp Arg Leu

         490       500       510       520       530       540
           |         |         |         |         |         |
CTTAAGAACTACGGGCTGCTCTCCTGCTTCAAGAAGGACCTGCACAAGGCTGAGACATAC
Leu Lys Asn Tyr Gly Leu Leu Ser Cys Phe Lys Lys Asp Leu His Lys Ala Glu Thr Tyr

    550   560   570   580   590   600
      |     |     |     |     |     |
CTGCGGGTCATGAAGTGTCGCCGCTTCGTGGAGAGCAGCTGTGCCTTCTAGTTGCTGGGC
Leu Arg Val MET Lys Cys Arg Arg Phe Val Glu Ser Ser Cys Ala Phe--
                                             +190
         610       620       630       640       650       660
           |         |         |         |         |         |
ATCTCTGTTGCCCCTCCCCAGTACCTCCCCTGACCCTGGAAAGTGCCACCCCAATGCCTG

         670       680       690       700       710
           |         |         |         |         |
CTGTCCTTTCCTAATAAAACCAGGTTGCATTGTAAAAAAAAAAAAAAAAAAA
```

Figure 3.   Nucleotide sequence of reconstituted porcine GH cDNA (pPGH1 and pPGH2) and deduced amino acid sequence of protein.

```
          10        20        30        40        50        60
          |         |         |         |         |         |
GTTCAAGCAACACCTGAGCAACTCTCCCGGCAGGAATGGCTCCAGGCTCGTGGTTTTCTCCT
                                   MET Ala Pro Gly Ser Trp Phe Ser Pro
                                   -26

          70        80        90        100       110       120
          |         |         |         |         |         |
CTCCTCATCGCTGTGGTCACGCTGGGACTGCCGCAGGAAGCTGCTGCCACCTTCCCTGCC
Leu Leu Ile Ala Val Val Thr Leu Gly Leu Pro Gln Glu Ala Ala  Ala Thr Phe Pro Ala
                                                         -1  +1

          130       140       150       160       170       180
          |         |         |         |         |         |
ATGCCCCTCTCCAACCTGTTTGCCAACGCTGTGCTGAGGGCTCAGCACCTCCACCTCCTG
MET Pro Leu Ser Asn Leu Phe Ala Asn Ala Val Leu Arg Ala Gln His Leu His Leu Leu

          190       200       210       220       230       240
          |         |         |         |         |         |
GCTGCCGAGACATATAAAGAGTTCGAACGCACCTATATTCCGGAGGACCAGAGGTACACC
Ala Ala Glu Thr Tyr Lys Glu Phe Glu Arg Thr Tyr Ile Pro Glu Asp Gln Arg Tyr Thr

          250       260       270       280       290       300
          |         |         |         |         |         |
AACAAAAACTCCCAGGCTGCGTTTTGTTACTCAGAAACCATCCCAGCTCCCACGGGGAAG
Asn Lys Asn Ser Gln Ala Ala Phe Cys Tyr Ser Glu Thr Ile Pro Ala Pro Thr Gly Lys

          310       320       330       340       350       360
          |         |         |         |         |         |
GATGACGCCCAGCAGAAGTCAGACATGGAGCTGCTTCGGTTTTCACTGGTTCTCATCCAG
Asp Asp Ala Gln Gln Lys Ser Asp MET Glu Leu Leu Arg Phe Ser Leu Val Leu Ile Gln

          370       380       390       400       410       420
          |         |         |         |         |         |
TCCTGGCTGACTCCCGTGCAATACCTAAGCAAGGTGTTCACGAACAACTTGGTTTTTGGC
Ser Trp Leu Thr Pro Val Gln Tyr Leu Ser Lys Val Phe Thr Asn Asn Leu Val Phe Gly

          430       440       450       460       470       480
          |         |         |         |         |         |
ACCTCAGACAGAGTGTTTGAGAAACTAAAGGACCTGGAAGAAGGGATCCAAGCCCTGATG
Thr Ser Asp Arg Val Phe Glu Lys Leu Lys Asp Leu Glu Glu Gly Ile Gln Ala Leu MET

          490       500       510       520       530       540
          |         |         |         |         |         |
AGCGAGCTGGAGGACCGCAGCCCGCGGGGCCCCGCAGCTCCTCAGACCCACCTACGACAAG
Ser Glu Leu Glu Asp Arg Ser Pro Arg Gly Pro Gln Leu Leu Arg Pro Thr Tyr Asp Lys

          550       560       570       580       590       600
          |         |         |         |         |         |
TTCGACATCCACCTGCGCAACGAGGACGCCCTGCTGAAGAACTACGGCCTGCTGTCCTGC
Phe Asp Ile His Leu Arg Asn Glu Asp Ala Leu Leu Lys Asn Tyr Gly Leu Leu Ser Cys

          610       620       630       640       650       660
          |         |         |         |         |         |
TTCAAGAAGGATCTGCACAAGGTGGAGACCTACCTGAAGGTGATGAAGTGCCGGCGCTTC
Phe Lys Lys Asp Leu His Lys Val Glu Thr Tyr Leu Lys Val MET Lys Cys Arg Arg Phe

          670       680       690       700       710       720
          |         |         |         |         |         |
GGAGAGAGCAACTGCACCATCTGAAGGCCCCGTGCCTGCGCCATGGCTGATGGCCCTGTC
Gly Glu Ser Asn Cys Thr Ile--
                +190
          730       740       750       760       770       780
          |         |         |         |         |         |
CCCCCCCCCCCCCCTTCCTCCCCGTCACCAAAAACACGAGGAATAAACCCCACAGCGCCAA

          790
          |
AAAAAAAAAAAAAA
```

Figure 4.   Nucleotide sequence of chicken GH precursor cDNA and deduced amino acid sequence of protein.

The screening of the chicken pituitary cDNA library by the hybridization with bovine GH cDNA probe revealed that approximately 2.5% of clones contained the GH-specific sequences. The plasmids from 2 clones (pcGH1 and pcGH2) with 800 bp long inserts were analyzed (Zvirblis et al.,1987). The insert from pcGH1 was completely sequenced (Figure 4). As it is shown on Figure 4, the cDNA in pcGH1 is 795 bp long and included 5'-nontranslated region (35 bp), GH precursor coding sequence (648 bp), 3'-nontranslated region (96 bp) and polyA-track (16 bp). The insert from pcGH2 contains 4 additional nucleotides (AAAC) at 5'-end. By the comparison of chicken GH cDNA cloned by us with that published earlier (Souza et al., 1984) several differences including the additional unique HinfI-site (GACTC) at the position corresponding to codons for Leu86, Thr87 and Pro88 (CTGACTCCC instead of CTCACCCCC) were found.

The cloned human, bovine and porcine GH cDNAs described above were used for construction of genes for expression of these hormones in bacteria. To design the genes to be expressed in *E. coli* we applied different strategies shown in Figure 5. For creation of gene for human GH expression 5'-nontranslated and signal peptide coding region of cDNA were deleted and synthetic adaptor containing ATG-codon was inserted in the front of the sequence coding for mature GH. The bovine and porcine GH genes were assembled by the joining of corresponding cDNA subfragments lacking 5'-terminal part with the synthetic fragments encoding N-terminal part of hormones. Thus the quasisynthetic bovine and porcine GH genes were obtained. The ovine GH gene was created by means of oligonucleotide-directed mutagenesis of bovine GH gene.

All genes constructed as described were inserted into expression vectors containing *trp*-promoter. The special work on optimization of expression was done, e.g. a number of variants of plasmids with different distance between SD-sequence and ATG-codon and different sequence in this area were created. Thus *E. coli* strains producing human (Rubtsov et al., 1984), bovine, ovine and porcine GH were obtained. The level of synthesis of desired proteins could be evaluated as 15-25% from the total cell protein, as it is shown for bovine GH synthesis (Figure 6).

THREE-DIMENSIONAL STRUCTURE OF HUMAN GROWTH HORMONE

Despite of great success in investigation of the human and animal GHs due to the application of recombinant DNA technique the molecular mechanism of its action are poorly

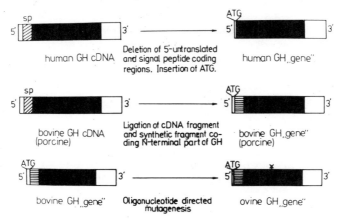

Figure 5. The scheme of construction of GH genes designed for expression in *E. coli*.

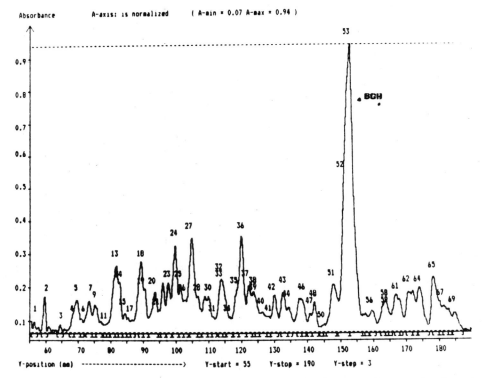

Figure 6.   Densitogram of SDS-polyacrylamide gel used for the separation proteins from the bovine GH producing strain.

understood. The lack of knowledge about the three dimensional structure of these hormones was one of the obstacle in the study of structure-functional interrelationship of different parts of GH molecule. The successful synthesis of human and animal GHs in bacteria opens the possibility of producing the homogeneous hormones in an active form in the quantities sufficient for physico-chemical studies. As a result of this breakthrough the crystals of porcine GH (Abdel-Meguid et al.,1986) and human GH (Jones et al., 1987; Borisova et al., 1988) suitable for X-ray diffraction study were grown (Figure 7). The tertiary structures of porcine GH at 2.8 Å resolution (Abdel-Meguid et al., 1987) and of human GH at 3 Å resolution (Pavlovsky et al., 1989) were determined.

Crystals of HGH belong to space group p4$_3$2$_1$2, with unit-cell parameters: a=b=80.6 Å , c=61.7 Å .

Search for heavy-atom derivatives was done by soaking experiments and by iodination of the protein. Two derivatives, with K$_2$PtCl$_4$ and HgBr$_2$, as well as the iodinated derivative produced marked changes in X-ray intensities without changes in unit-cell parameters.

The structure was solved using three heavy-atom derivatives and solvent flattering procedure at resolution 4 Å and then extended to 3 Å resolution by model building on the basis of a partially interpreted electron density map (Figure 8).

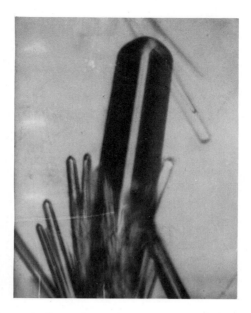

Figure 7.   Crystals of genetically engineered variant of human growth hormone. The largest
crystals measure about 0.3mm x 0.3mm x 0.5mm.

The HGH molecule has a four - helical motif. The position of each helix in the amino acid
sequence was established by trial and error fitting bulky side chains in regions of the electron
density map near each helix:

Helix A (residues 6 - 32)

Helix B (residues 75 - 96, bent in the middle at the position of Pro 89)

Helix C (residues 106 - 128)

Helix D (residues 155 - 181).

Irregular parts between helices were well traced for a short connection between the helices B
and C and a longer one between the helices C and D. The region between the helices A and B
(residues 34 - 74) was poorly seen and was reconstructed by a procedure suggested by Jones
and Thirup (1986) which is based on a simulation of short segments of polypeptide chain by
standard pieces selected from the Bank of known protein structures.

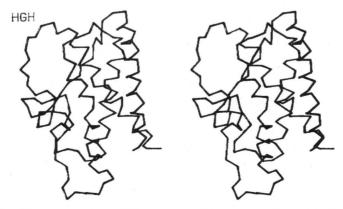

Figure 8.   Graphic representation of human growth hormone polypeptide chain extended to
3 Å resolution.

The course of polypeptide chain was traced throughout of about 93% of its length. For about 50% of residues position of a side chain was also found. One of the two S - S binds between residues 53 and 165 is also well seen. At present four residues from the N-terminal and ten residues from the C-terminal have no features on the electron density map.

THE CONSTRUCTION OF GENES AND EXPRESSION OF HUMAN GH ANALOGS AND DERIVATIVES IN E. COLI

The knowledge of the tertiary structure of GHs and extensive sequencing and homology comparison of GHs open good prospects for the investigation of GHs by the protein engineering methods. We have started the work on the synthesis of some analogs and derivatives of human GH.

There are the total of the 37 amino acids residues which are the very conservative among the hormones of the somatotropin family. It was shown by Watahiki et al. (1989) by means of the plural alignment of the sequences of the 18 various representatives of the growth hormone family (the growth hormones, placental lactogens, prolactins and proliferins of various species of animals). These amino acids form the five clusters GD1-GD5 along the polypeptide chains of the hormones.

The chromosomal gene of the somatotropin has five exons. The exon 2 codes amino acids 1-31 and therefore completely the helix A. There is a very conservative segment, called GD1, in the helix from 6th to 31th aa.

The exon 3 codes a very long string of amino acids with irregular structure, which connects the helices A and B of the hormone. This loop is arranged on the outside of the molecule. It contains also the very conservative region from 53th to 68th aa, which is called GD2.

The exon 4 codes the stretch of amino acids from 72th to 126th aa, which constitute the two helices B and C and the short loop (97- 105aa) between them. There are two the very conservative segments in this region-GD3 (from 75th to 93th aa) and GD4 (from 114th to 130th aa).

The exon 5 is the longest. It codes the amino acids from 131th to 191th of the hormone and contains completely the helix D and the irregular segment which connects the helices C and D. There the highest conservative region is situated, which is called GD5(162-191 aa).

Based on the study of the growth hormone family, Nicoll et al. (1986) have concluded that some residues in the conservative region GD1 are specifically connected with the growth-promoting activity of the somatotropins. A variant form of human GH (hGH-V) has approximately 50% of the activity of normal hGH (Pavlakis et al., 1981). From the 37 aa conserved among the GHs, only one -His21 is replaced by Tyr in hGH-V (Watahiki et al., 1989). Moreover it is known that the human placental lactogen shows very high structural homology to human GH. However it exhibits low growth-promoting activity (Niall et al., 1971; Nicoll et al., 1986; Shine et al., 1977). Human placental lactogen differentiates from the human GH only in two of 37 conserved positions - 16th and 20th. It allows consideration of the residues - Arg16, Leu20 and His21 as important ones for the developing of growth-promoting activity of the somatotropins (Watahiki et al., 1989).

Another potential region which possibly takes part in the forming of growth-promoting determinant, is a fragment 77-107 aa. Bulatov et al., 1980 and Bulatov et al., 1983 examined

318

this fragment very extensively. The fragment, which encompasses completely the conservative region GD3 was purified from the tryptic digestion of the sperm whale somatotropin. It was determined that the fragment 77-107 aa, making up only 16% of the amino acid chain of the somatotropin, is capable of reproducing the effects of the hormone, associated with the stimulation of the growth. Like the somatotropin, the fragment 77-107 aa had the considerable residual growth-promoting activity in the "tibia-test" and increased the level of secretion of somatomedins in the blood on the hypophysectomized rats. It also stimulated the biosynthesis of DNA in the culture of fibroblasts (Table 2).

Table 2. Comparison of the biological activity spectrum of somatotropin and its fragment 77-107

Biological effects	Somatotropin	Fragment
Growth-promoting activity:		
1. "tibia-test"	+	+
2. stimulation of secretion of somatomedins *in vivo*	+	+
3. stimulation of biosynthesis of DNA in the culture of fibroblasts	+	+
Fast metabolic effect:		
1. increasing of the uptake of glucose by intercostal muscle *in vitro*	+	-
2. stimulation of the synthesis of glycogen in intercostal muscle *in vitro*	+	-
3. stimulation of lipolyse *in vitro*	+	-

Some of the segments in the hormone may play an important role of the stabilizing factors in the forming of growth-promoting determinant. From this point of view the fragment 140-191aa is very attractive. It was shown in the works Li and Graf (1974) and Graf et al. (1976) that the fragment 1-134 of the ovine somatotropin in the pure form did not exhibit any growth-promoting activity in the rat "tibia-test". However the reconstitution of the molecule of the growth hormone by means of non-covalent interaction of segments 1- 134 aa and 140-191 aa gave the restoration of full biological activity of the native hormone in rat "tibia-test". So the segment 140-191 may interact with the fragments, forming the growth-promoting determinant, and stabilize it.

From the work of Niall et al., 1971 it is known that the human somatotropin has four regions of internal homology. These are segments 15-22 aa, 92-109 aa, 127-146 aa and 162-179 aa. All the discussed above segments 8-31(GD1), 77-107 aa and 140-191aa are localized within the fragments with internal homology, and at the same time the segments 8-31 aa, 77-107 aa, and 127-146 aa are exposed on the surface of the molecule. Does the fragment 127-146 also exhibit the growth-promoting activity? Or is this fragment the important one in the stabilizing of the growth-promoting determinant?

The segment 129-154 aa is very susceptible to the action of the various proteinases. The functional meaning of such property of the fragment is unknown. It was speculated that the cutting in this place leads to the bioactivation of growth-promoting activity of GH (Mittra, 1984) or conversely it starts up the process of biodegradation of the hormone. Thus the fragment may play the role of the peptide trigger-"turning on" or "turning off" the activity of the hormone. The data obtaining in the work Lewis et al., 1981 suggests that the process of such switching can be regulated by the deamidation of the amino acids residues- Asp152 and Glu137 in the hormone.

The deamidation can provide the different character of the hydrolysis by the proteinases and therefore the different hydrolysis peptides products.

The region 32-75 aa in the growth hormone is peculiar. This fragment is considered to be introduced in the structure of the hormones after consecutive duplications of the primordial ancestor gene of the somatotropins. Thus it is some "foreign body" in the growth hormone molecules. Feasibly the fragment brought itself the functionally important regions with the early insulin-like and diabetagenic activities.

It was shown by Yudaev et al., 1983 that fragment 31-44 aa of hGH which was isolated or chemically synthesized have the insulin-like activity. From 5 to 10% of the somatotropin are synthesizing in the pituitary gland in the form of 20 kD-variant of GH (Lewis et al., 1978). The 20 kD hGH is completely homologous to the major 22 kD form of GH, excluding the segment 32-46 aa (Lewis et al., 1980; Chapman et al., 1981). Lack of the fragment leads to the 80% loss of the early insulin-like activity (Kostyo et al., 1985). Moreover it is known that fragment 32-4 6 aa of the teleost GHs (Tilapia) has no homology with the analogous fragments of somatotropins of the other species of animals. This fact is in a good agreement with the result that teleost GH has very low insulin-like activity (Cameron et al., 1985; Watahiki et al., 1989). Therefore the fragment responsible for the early insulin-like activity locates at the position 32-46 aa.

The fragment 44-77 aa may exhibit the diabetagenic activity (Lostroch and Krahl, 1976; Lostroch and Krahl, 1978).

Several types of modified products could be produced in bacteria or other expression systems to search for the HGs domains functions. First of all the human GH can be used for construction of some fusion polypeptides.

The second type of the products we are trying to synthesize in *E. coli* are the deleted forms of human GH. It was shown that 20 kD variant is a product of an alternative splicing of human GH pre-mRNA (De Noto et al., 1981). Physico-chemical studies of 20 kD variant suggest that its conformation is similar to that of 22 kD human GH. It was also reported that 20 kD form has growth- promoting activity in the hypophysectomized rats (Lewis et al., 1979) and lactogenic activity in the pigeon crop sack assay (Lewis et al., 1978) equal to those of the 22 kD variant. It means that at least some of the regions responsible for the specific biological activities have the same conformation in both forms of the human GH. Recently Lecomte et al. (1987) have cloned three distinct human GH cDNAs encoding respectively 22 kD, 20 kD and yet unknown 17.5 kD variant which lacks the sequence corresponding to whole exon 3 (residues 32-71). The possibility of synthesis of 17.5 kD human GH variant *in vivo* and its potential biological activities remain to be investigated. As the first step in the study of the structure and function of both deleted forms of the human GH we have constructed the genes encoding 20 and 17.5 kD variants and expressed them in *E. coli*. The genes with deletions of 45 and 120 bp were inserted into the expression vectors and 20 and 17.5 kD variants were synthesized in *E. coli* (Figure 9). We obtained that the monoclonal antibodies, made against 22 kD hGH, react also with 20 kD hGH. At the same time 17.5 kD hGH variant is not capable to interact with these monoclonals. Now the work on purification of these deleted forms of human GH is in progress in our laboratory.

The third type of the products we are trying to synthesize and study are the human GH "muteins" - analogs of GH which contain mutations of predetermined amino acids. It is well documented that GHs are the polyfunctional hormones. Considerable experimental evidences suggest that the multiple biological actions are produced by the different parts of the hormone molecule (Paladini et al., 1979; Bulatov et al., 1983). As we discussed above many enzymes such as plasmin, thrombin, trypsin, bacterial proteases cleave the molecule within the region 129-154 at one or more sites. Thrombin cleaves specifically only one peptide bond between ARG[134] and THR[135] (Graf et al., 1976). Two fragments produced remain be connected by disulfide bridge. Plasmin cleaves preferentially two bonds: ARG[134]-THR[135] and LYS[140]-GLN[141] (Li and Graf, 1974). In this case the hexapeptide is cleaved out and the fragments 1-134 and 141-191 are held together by S-S-bridge. We believe that construction and study of GH "muteins" resistant to specific proteolysis would help to answer some functional questions.

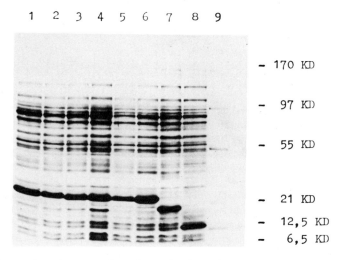

Figure 9. Electrophoretic separation of proteins from *E. coli* strains producing: 1 - wild type human GH, 2-6 - different mutated human GH variants, 7 - 20 kD human GH, 8-17.5 kD human GH, 9 - molecular weights markers.

By the use of the oligonucleotide-directed mutagenesis we have changed the codons in the region 134-140 of human GH. Mutated genes were inserted into the vector with *trp* promoter. The level of expression of human GH variants was approximately the same as for wild type protein (Figure 9). All mutant proteins obtained interact with polyclonal and monoclonal antibodies with the same efficiency as wild type GH. It allowed us to utilize for purification of mutant variants the affinity chromatography on the monoclonal antibodies coupled to Sepharose. For the testing of proteolytic resistance we used the analogs of human GH *in vivo* labeled by S[35]-methionine. It was shown that purified analogs of human GH were resistant to the cleavage by proteases trombin or plasmin *in vitro*. The study of biological activities of these GH variants is now in progress in our laboratory.

Authors thank Drs. B.K. Chernov, Yu.B. Golova and G.E. Pozmogova for the providing of synthetic oligonucleotides widely used throughout this study and Drs. G.A. Popkova, P.G. Sveshnikov, I.A. Prudovsky, K.I. Plishkina and R. Fidler for the providing of poly- and monoclonal antibodies to the GHs and S.T. Sadiev, P.S. Sverdlova, V.V. Chupeeva for the help in some of experiments.

REFERENCES

Abdel-Meguid, S.S., Smith, W.W., Violand, B.N. and Bentle, L.A.J., 1986, Crystallization of mathionyl porcine somatotropin, a genetically engineered variant of porcine growth hormone, *Mol. Biol.* 192:159.

Abdel-Meguid, S.S., Shien, H.-S., Smith, W.W., 1987, Three dimensional structure of a genetically engineered variant of porcine growth hormone, *Proc. Natl. Acad.Sci.USA* 84:6434.

Borisova, S.N., Pavlovsky, A.G., Naktinis, V.J., Janulaitis, E.-A.A., Rubtsov, P.M., Skryabin, K.G., Bayev, A.A. and Vainstein, B.K., 1988, Crystallization and preliminary X-ray analysis of human somatotropin sunthesized by the method of genetic engineering, *Doklady Akad. Nauk SSSR* 301:474.

Bulatov, A.A., Osipova, T.A. and Pankov, Yu.A., 1980, Hydrophobic peptide from tryptic digestion of sperm whale growth hormone possessing growth-promiting activity, *Problems of Endokrinology (USSR)* 6:54.

Bulatov, A.A., Osipova, T.A., Terehov, S.M., Sazina, E.T. and Pankov,Yu.A., 1983, New evidence on the biological activbity of pituitary somatotropin gragment 77-107, *Biokimia (USSR)* 48:1305.

Cameron, C.M., Kostyo, J.L. and Papkiff, H., 1985, Mammalian growth hormones have diabetogenic and insulin-like activity, *Endocrinology* 116:1501.

Chapman, G.E., Rogers, K.M., Brittain,T., Bradshaw, R.A., Bates, O.J., Turner, C., Cary, P.D. and Crane-Robinson, C., 1981, The 20,000-molecular weight variant of human growth homorne. Preparation and some physical and chemical properties, *J. Biol. Chem.* 256:2395.

De Noto, F.M., Moore, D.D. and Goodman, H.M., 1981, Human growth hormone DNA sequence and mRNA structure: possible alternative splicing, *Nucleic Acids Res.* 9:3719.

Graf, L., Barat, E., Borvendeg, J., Hermann, I. and Patthy, A., 1976, Action of trombin on ovine, bovine and human pituitary growth hormones, *Eur. J. Biochem.* 64:333.

Jones, N.D., De Honiesto, J., Tackitt, P.M. and Becker, G.W., 1987, Crystallization of authentic recombinant human growth hormone, *BIOTECHNOLOGY* 5:499.

Kostyo, J.L., Cameron, C.M., Olson, K.C., Jones, A.J.S. and Pai, R.-C., 1985, Biosynthetic 20-kinodalton methionyl-human growth hormone has diabetogenic and insulin-like activities, *fProc. Natl., Acad. Sci. USA* 82:4250.

Lecomte, C.M., Renard, A. and Martial, J.A., 1987, A new natural hGH variant-17.5 ks-produced by alternative splicing. An additional consensus sequence which might play a role in branchpoint selection, *Nucleic Acids Res.* 15:6331.

Lewis, U.J. Dunn, J.T., Bonewald, L.F., Seavey, B.L. and VanderLaan, W.P., 1978, A naturally occuring structural variant of human growth hormone, *J. Biol. Chem.* 253:2679.

Lewis, U.J., Singh, R.N.P. and Tutwiler, G.F., 1979, Hyperglycaemic activity of the 20,000D variant of human growth hormone, *Endocr. Res. Commun.* 91:778.

Lewis, U.J., Bonewald, L.F. and Lewis, L.J., 1980, The 20,000-dalton variant of human growth hormone: location of the amino acid deletions, *Biochem. Biophys. Res. Commun.* 92:511.

Lewis, U.J., Singh, R.N.P., Bonewald, F. and Seavey, B.K., 1981, Altered proteolytic cleavage of human growth hormone as a result of deamidation, *J. Biochem.* 256:11645.

Li, C.H. and Graf, L., 1974, Human pituitary growth hormone: isolation and properties of two biologically active fragments from plasmin digest, *Proc. Natl. Acad.Sci.USA* 71:1197.

Lostroch, A.J. and Krahl, M.E., 1976, Diabetogenic peptide from human growth hormone: partical purificaiton from peptic digest and long-term action in ob/ob mice, *Proc. Natl. Acad. Sci.USA* 73:4706.

Lostroch, A.J. and Krahl, M.E., 1978, Synthetic fragment of human growth hormone with the hyperglycaemic properties: residues 44-77, *Diabetes* 27:597.

Martial, J.A., Hallewell, R.A., Baxter, J.D. and Goodman, H.M., 1979, Human growth hormone: complementary DNA cloning and expression in bacteria, *Science* 205:602.

Miller, W.L., Martial, J.A. and Baxter, J.D., 1980, Molecular cloning of DNA complementary to bovine growth hormone mRNA, *J. Biol. Chem.* 255:7521.

Mittra, I., 1984, Somatomedins and proteolytic bioactivation of prolactin and growth hormone. Minireviews, *Cell* 38:347.

Movva, N. and Schulz, M.-F., 1984, Patent EPO 104920 A1.

Niall, H.D., Hogan, M. L., Sauer, R., Rosenblum, I.Y. and Greenwood, F.C., 1971, Sequences of pituitary and placental lactogenic and growth hormones: evolution from a primordial peptide by gene reduplicaiton, *Proc. Natl. Acad. Sci. USA* 68:866.

Nicoll, C.S., Mayer, G.L. and Russel, S.M., 1986, Structural prpperties of prolactins and growth hormones which can be related to their biological properties, *Endocr. Rev.* 7:169.

Paladini, A.C., Pena, C. and Retegui, L.A., 1979, The intriguing nature of the multiple actions of GH, *TIBS* 4:256.

Pavlakis, G.N., Hizuka, N., Gorden, P., Seeburg, P.H. and Hammer, D.H., 1981, Expression of two human growth hormone genes in monkey cells infected by simian virus 40 recombinants, *Proc. Natl. Acad. Sci. USA* 78:7398.

Pavlovsky, A.G., Borisova, S.N., Strokopytov, B.V., Vagin, A.A., Vainstein, B.K., Alkimavichius, G.A., Naktinis, V.I., Janulaitis, E.-A.A., Rubtsov, P.M., Skryabin, K.G. and Bayev, A.A., 1989, Three-dimensional structure of human somatotropin at 3Å resolution, *Doklady Akad. Nauk SSSR* 305:861.

Rubtsov, P.M., Parsadanian, A.Sh., Sverdlova, P.S., Chupeeva, V.V., Lashas, L.V., Skryabin, K.G. and Bayev, A.A., 1984, Characterization of human somatotropin synthesized in bacteria by the genetic engineering methods, *Doklady Akad. Nauk SSSR* 276:762.

Rubtsov, P.M., Chernov, B.K., Gorbulev, V.G., Parsadanian, .SH., Sverdlova, P.S., Chupeeva, V.V., Golova, Yu.B., Batchikova, N.V., Zbirblis, G.S., Skryabin, K.G. and Bayev, A.A., 1985, Genetic engineering of peptide hormones, *Molekulyarnaya Biologiya (USSR)* 19:267.

Seeburg, P.H., Shine, J., Martial, J.A. and goodman, H.M., 1977, Nucleotide sequence and amplification in bacteria of structural gene for rat growth hormone, *Nature* 270: 486.

Seeburg, P.H., Sias, S., Adelman, J., De Boer, H.A., Hayflick, J., Jhurani, P., Goeddel, D.V. and Heyneker, H.L., 1983, Efficient bacterial expression of bovine and porcine growth hormone, *DNA* 2:37.

Shine, J., Seeburg, P.H., Martial, J.A., Baxter, J.D. and Goodman, H.M., 1977, Nucleotide sequence and smplification in bacteria of structural gene for growth hormone, *Nature* 270:494.

Souza, L.M., Boone, T.C., Murdock, S., Langley, K., Wypych, J., Fenton, D., Johnson, S., Lai, P.H., Everett, R., Hsu, R.-Y. and Bosselman, B., 1984, The application of recombinant DNA technologies to studies on chicken growth hormone, *J. Exp. Zoology* 232:465.

Vize, P.D. and Wells, J.R.E., 1987, Isolation and characterization of the porcine growth hormone gene, *Gene* 55:339.

Watahiki, M., Yamamoto, M., Yamakawa, M., Tanaka, M. and Nakashima, K. 1989, Consedrved and unique amino acid residues in the domains of the growth hormones. Flounder growth hormone deduced from the cDNA sequence has the minimal size in the growth hormone prolactin gene family, *J. Biol. Chem.* 264:312.

Woychik, R.P., Camper, S.A., Lyons, R.H., Horowitz, S., Goodwin, E.C. and Rottman, F.M., 1982, Cloning and nucleotide sequencing of the bovine growth hormone gene, *Nucleic Acids Res.* 10:7197.

Yudaev, N.A., Pankov, Yu.A., Keda, Yu.M., Sazina, E.T., Osipova, T.A., Shwachkin, Yu.P. and Ryabtsev, M.N., 1983, The effect of synthetic fragment 31-44 of human growth hormone on glucose uptake by isolated adipose tissue, *Biochem. Biophys. Res. Commun.* 110:866.

Zvirblis, G.S., Gorbulev, V.G., Rubtsov, P.M., Chernov, B.K., Golova, Yu.B., Pozmogova, G.E., Skryabin, K.G. and Bayev, A.A., 1988, Genetic engineering of peptide hormones. III. The cloning of the porcine growth hormone cDNA and the construction of the gene suitable for the hormone expression in bactreria, *Molekulyarnaya Biologiya (USSR)* 22:145.

Zvirblis, G.S., Gorbulev, V.G., Rubtsov, P.M., Karapetyan, R.V., Zuravlev, I.V., Fisinin, V.I., Skryabin, K.G. and Bayev, A.A., 1987, Genetic engineering of peptide hormones. I. Cloning and primary structure of chicken growth hormone cDNA, *Molekulyarnaya Biologiya (USSR)* 21:1620.

CLONING, SEQUENCING AND EXPRESSION OF A NEW β-GALACTOSIDASE FROM THE EXTREME THERMOPHILIC *SULFOLOBUS SOLFATARICUS*

Mosè Rossi[1,2], Maria Vittoria Cubellis[2], Carla Rozzo[1], Marco Moracci[1] and Rocco Rella[1]

[1]Istituto di Biochimica delle Proteine ed Enzimologia, Via Toiano, 6 - 80072 Arco Felice, Naples, Italy & [2]Dipartimento di Chimica Organica e Biologica Università di Napoli, Via Mezzocannone, 16 - 80134 Naples, Italy

INTRODUCTION

The number of isolated microorganisms surviving and growing in extreme environmental conditions (60-110°C) is rapidly increasing (Stetter, 1986). Comparison of the biology of these thermophilic microorganisms and their successful strategy of living at such high temperatures requires an understanding of both their metabolism and the relationship between the structure and function of their cellular components. In particular, interest in the enzymes from these organisms is growing because their peculiar properties render them a considerable biotechnological potential and industrial tool. In fact, whereas conventional enzymes are irreversibly inactivated by heat, these enzymes, in addition to their thermophilic and thermostable characteristics, show an enhanced activity, in the presence of the common protein denaturants and organic solvents as well as proteolytic enzymes.

One of the most studied microorganisms of this type is *Sulfolobus solfataricus (S. solfataricus)*, a thermoacidophilic microorganism belonging to the recently classified independent archaeobacteria kingdom (Woese and Wolfe, 1985). In contrast to the abundance of data on archaeobacteria stable RNAs, only a few sequences of protein coding genes have been studied (Reiter et al., 1987; Lechner and Boeck, 1987; Souillard et al., 1988; Itoh, 1988; Fabry et al., 1989). Some of these have been cloned due to their ability to direct protein synthesis and complement mutations in *E. coli* (Beckler and Reeve, 1986), thus implying that these genes utilize the universal genetic code and suggesting that they do not contain introns. Yet, the occurrence of genes containing introns and/or flanking regions not accidentally recognized by the eubacterial transcriptional and translational machinery could have been underestimated. Hence, different approaches have to be adopted to sample more representative specimens of protein-encoding genes.

Of great interest is the need to more fully characterize the expression of genes codifying for enzymes from extreme thermophilic bacteria in typical mesophilic hosts, such as *E. coli* . This could clarify the regulation mechanisms of these genes and explore the possibility of their expression in heterologous hosts by their own regulation sequences that are recognized by other expression systems such as that of *E. coli*.

From an application viewpoint, it may be possible to produce proteins that are heat-stable and resistant to common protein denaturing agents in well known mesophilic hosts and using well characterized fermentative methodologies. In addition, enzymes isolated from extreme thermophilic bacteria are natural examples of thermostable, thermophilic and solvent-resistant proteins which can be used as models for designing and producing, by protein engineering, modified structures having the desired properties.

This contribution describes the cloning, sequencing and expression in *E. coli* of the gene coding for a new β–galactosidase from the extreme archaeobacterium *S. solfataricus*.

Purification and Properties of β–galactosidase

β–galactosidase was purified to homogeneity from cell extracts of *S. solfataricus* strain MT-4, grown at 87°C and ph 3.0 (Pisani et al., manuscript submitted). The results of gel filtration, glycerol gradient centrifugation, treatment with cross-linking reagents, amino acid composition and SDS-PAGE experiments demonstrated that this enzyme was a tetramer (Mr 240,000 ± 8,000) composed of four similar or identical subunits (Mr 60,000 ± 2,000). β–galactosidase from *E. coli* and other thermophilic and mesophilic bacterial sources differ from the *S. solfataricus* enzyme by their higher molecular weight despite their tetrameric structure.

The thermophilicity of *S. solfataricus* β–galactosidase was striking even when compared to other enzymes from thermophiles. The activity temperature relationship showed a maximum at 95°C for β–ONPG hydrolysis; from 35 to 95°C the Arrhenius plot was not linear, having a breakpoint around 65°C. The thermal stability of β–galactosidase from *S. solfataricus* was notably greater than that reported from the galactosidase from either *E. coli* or *T. aquaticus* (Ulrich et al., 1972); even at 75°C its activity was enhanced by salts and protein denaturants.

Polyclonal rabbit antibodies were prepared against the enzyme and did not cross-react with *E. coli* β–galactosidase. The -NH$_2$ terminal sequence (Gas-Phase Applied Biosystems Sequencer) and -COOH terminal residues (carboxypeptidase) were determined to enable identification, sequencing, cloning and expression of the gene.

Construction of DNA Libraries and Isolation of β-galactosidase Clones

A genomic library from *S. solfataricus* MT-4 was constructed by inserting DNA partially digested with DNAse into vector λgt11 (Cubellis et al., submitted). DNA fragments of about 1000 bps were isolated by agarose gel electrophoresis, treated with EcoRI methylase and T4 DNA polymerase and ligated to EcoRI linkers. After digestion with EcoRI endonuclease, the *S. solfataricus* DNA was ligated to λgt11 dephosphorilated arms and packaged *in vitro*.

The λgt11 genomic library was screened with a mixture of ^{32}P and labeled probes of 53mer oligonucleotides synthesized on the basis of the aminoterminal sequence of the protein. The oligodeoxynucleotides used as hybridization probes or sequencing primers were synthesized by the Applied Biosystems 381A DNA Synthetizer or directly purchased from Applied Biosystems. The oligonucleotides used were as follows:

5' GGG TCT TCG TG(C/G) CC(G/T) CCC ATT TGG TGT TCG
AA(G/T) CC(G/T) GCT TCG TGC CA 3'

Four clones out of 24,000 were found to hybridize to the ^{32}P-labeled oligonucleotides. Their EcoR1 inserts were isolated and subcloned into vector pEMBL18. Restriction analysis of the inserts revealed that they overlapped and covered a total region of 1,800 bps centred around the target of the oligonucleotide probe (Figure 1). All the clones contained an identical nucleotide sequence coding for the previously determined 33 aminoterminal residues of β–galactosidase and pH1 was chosen for further investigation.

Furthermore, the λgt11 library was screened with anti-β–galactosidase antibodies: of 14 positive clones over 40,000 plaques only three hybridized to the insert of pH1 and the expression was found to be independent of the presence of *lac* inducer IPTG. This result suggests that the *S. solfataricus* β–galactosidase gene can be expressed in *E. coli* exploiting its own 5' flanking regions.

A second genomic library was constructed by ligating *S. solfataricus* DNA partially digested with MboI and phosphatase treated with suitably prepared arms of the EMBL3 phage; the fragments were longer than 9 kb. The methodologies were essentially those described by Maniatis et al. (1982) or recommended by the manufacturers. This library was screened using the pH1 insert as probe. One positive clone (λC1) was analyzed by restriction mapping and a 2.9 kb XbaI fragment hybridizing to pH1 was subcloned into pEMBL18 generating the plasmid pD22 (Figure 1).

The β–galactosidase Gene

The nucleotide sequence of pD22 includes a long open reading frame (ORF) from residue 206 to 1696 (Figure 2). This ORF was searched for the first possible initiation site considering the triplets ATG, GTG and TTG, which are the most frequently encountered in archaeobacteria (Souillard and Sibold, 1986; Bokranz and Klein, 1987). Starting from ATG at position 229, an amino acid sequence is encoded, identical to that obtained from the purified β–galactosidase, suggesting that the first residue of the mature protein is also the first translated amino acid. The

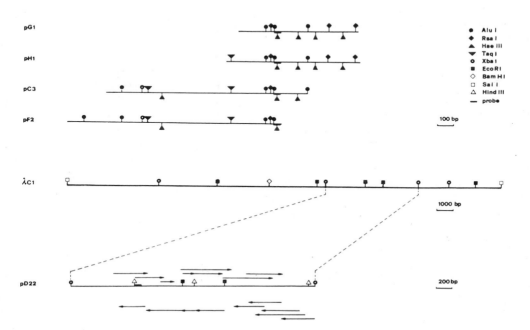

Figure 1. Cloning strategy and restriction analysis of the archaebacterial β-galactosidase gene.

```
          10        20        30        40        50        60        70        80        90
AAGGAGAAACTTGGCAGTTTATAACTTGACAGTAGGTTGTGGAGTGACTGGATCCAATACTAGGAGGAGTAGCATATAATTACGTTAC

         100       110       120       130       140       150       160       170       180
ACAATTTTATAACCCAATATATTCAATAGACCTTATGCTTATCCTATCCTCTATTCTAAGATTCTCGGTATCTCCCCTATTCTTGACCAT

         190       200       210       220       230       240       250       260       270
AAAAGATACTCGCTCAAAGCTTAAATAATATTAATCATAAATAAAGTCATGTACTCATTTCCAAATAGCTTTAGGTTTGGTTGGTCCCAG
                                              MetTyrSerPheProAsnSerPheArgPheGlyTrpSerGln

         280       290       300       310       320       330       340       350       360
GCCGGATTTCAATCAGAAATGGGAACACCAGGGTCAGAAGATCCAAATACTGACTGGTATAAATGGGTTCATGATCCAGAAAACATGGCA
AlaGlyPheGlnSerGluMetGlyThrProGlySerGluAspProAsnThrAspTrpTyrLysTrpValHisAspProGluAsnMetAla

         370       380       390       400       410       420       430       440       450
GCGGGATTAGTAAGTGGAGATCTACCAGAAAATGGGCCAGGCTACTGGGGAAACTATAAGACATTTCACGATAATGCACAAAAAATGGGA
AlaGlyLeuValSerGlyAspLeuProGluAsnGlyProGlyTyrTrpGlyAsnTyrLysThrPheHisAspAsnAlaGlnLysMetGly

         460       470       480       490       500       510       520       530       540
TTAAAAATAGCTAGACTAAATGTGGAATGGTCTAGGATATTTCCTAATCCATTACCAAGGCCACAAAACTTTGATGAATCAAAACAAGAT
LeuLysIleAlaArgLeuAsnValGluTrpSerArgIlePheProAsnProLeuProArgProGlnAsnPheAspGluSerLysGlnAsp

         550       560       570       580       590       600       610       620       630
GTGACAGAGGTTGAGATAAACGAAAACGAGTTAAAGAGACTTGACGAGTACGCTAATAAAGACGCATTAAACCATTACAGGGAAATATTC
ValThrGluValGluIleAsnGluAsnGluLeuLysArgLeuAspGluTyrAlaAsnLysAspAlaLeuAsnHisTyrArgGluIlePhe

         640       650       660       670       680       690       700       710       720
AAGGATCTTAAAAGTAGAGGACTTTACTTTATACTAAACATGTATCATTGGCCATTACCTCTATGGTTACACGACCCAATAAGAGTAAGA
LysAspLeuLysSerArgGlyLeuTyrPheIleLeuAsnMetTyrHisTrpProLeuProLeuTrpLeuHisAspProIleArgValArg

         730       740       750       760       770       780       790       800       810
AGAGGAGATTTTACTGGACCAAGTGGTTGGCTAAGTACTAGAACAGTTTACGAATTCGGCTAGATTCTCAGCTTATATAGCTTGGAAATTC
ArgGlyAspPheThrGlyProSerGlyTrpLeuSerThrArgThrValTyrGluPheGlyAlaArgPheSerAlaTyrIleAlaTrpLysPhe

         820       830       840       850       860       870       880       890       900
GATGATCTAGTGGATGAGTACTCAACAATGAATGAACCTAACGTTGTTGGAGGTTTAGGATACGTTGGTGTTAAGTCCGGTTTTCCCCCA
AspAspLeuValAspGluTyrSerThrMetAsnGluProAsnValValGlyGlyLeuGlyTyrValGlyValLysSerGlyPheProPro

         910       920       930       940       950       960       970       980       990
GGATACCTAAGCTTTGAACTTTCCCGTAGGCATATGTATAAACATCATTCAAGCTCACGCAAGAGCGTATGATGGGATAAAGAGTGTTTCT
GlyTyrLeuSerPheGluLeuSerArgArgHisMetTyrAsnIleIleLeuGlnAlaHisAlaArgAlaTyrAspGlyIleLysSerValSer

        1000      1010      1020      1030      1040      1050      1060      1070      1080
AAAAAACCAGTTGGAATTATTTACGCTAATAGCTCATTCCAGCCGTTAACGGATAAAGATATGGAAGCGGTAGAGATGGCTGAAAATGAT
LysLysProValGlyIleIleTyrAlaAsnSerSerPheGlnProLeuThrAspLysAspMetGluAlaValGluMetAlaGluAsnAsp

        1090      1100      1110      1120      1130      1140      1150      1160      1170
AATAGATGGTGGTTCTTTGATGCTATAATAAGAGGTGAGATCACCAGAGGAAACGAGAAGATTGTAAGAGATGACCTAAAGGGTAGATTG
AsnArgTrpTrpPhePheAspAlaIleIleArgGlyGluIleThrArgGlyAsnGluLysIleValArgAspAspLeuLysGlyArgLeu

        1180      1190      1200      1210      1220      1230      1240      1250      1260
GATTGGATTGGAGTTAATTATTACACTAGGACTGTTGTGAAGAGGACTGAAAAGGGATACGTTAGCTTAGGAGGTTACGGTCACGGATGT
AspTrpIleGlyValAsnTyrTyrThrArgThrValValLysArgThrGluLysGlyTyrValSerLeuGlyGlyTyrGlyHisGlyCys

        1270      1280      1290      1300      1310      1320      1330      1340      1350
GAGAGGAATTCTGTAAGTTTAGCGGGATTACCAACCAGCGACTTCGGCTGGGAGTTCTTCCCAGAAGGTTTTATATGACGTTTTGACGAAA
GluArgAsnSerValSerLeuAlaGlyLeuProThrSerAspPheGlyTrpGluPhePheProGluGlyLeuTyrAspValLeuThrLys

        1360      1370      1380      1390      1400      1410      1420      1430      1440
TACTGGAATAGATATCATCTCTATATGTACGTTACTGAAAATGGTGATGCCGATTATCAAAGGCCCTATTATTTAGTATCT
TyrTrpAsnArgTyrHisLeuTyrMetTyrValThrGluAsnGlyIleAlaAspAspAlaAspTyrGlnArgProTyrTyrLeuValSer

        1450      1460      1470      1480      1490      1500      1510      1520      1530
CACGTTTATCAAGTTCATAGAGCAATAAATAGTGGTGCAGATGTTAGAGGGTATTTACATTGGTCTCTAGCTGATAATTACGAATGGGCT
HisValTyrGlnValHisArgAlaIleAsnSerGlyAlaAspValArgGlyTyrLeuHisTrpSerLeuAlaAspAsnTyrGluTrpAla

        1540      1550      1560      1570      1580      1590      1600      1610      1620
TCAGGATTCTCTATGAGGTTTGGTCTGTTAAAGGTCGATTACAACACTAAGAGACTATACTGGAGACCCTCAGCACTAGTATATAGGGAA
SerGlyPheSerMetArgPheGlyLeuLeuLysValAspTyrAsnThrLysArgLeuTyrTrpArgProSerAlaLeuValTyrArgGlu

        1630      1640      1650      1660      1670      1680      1690      1700      1710
ATCGCCACAAATGGCGCAATAACTGATGAAATAGAGCACTTAAATAGCGTACCTCCAGTAAAGCCATTAAGGCACTAAACTTTCTCAAGT
IleAlaThrAsnGlyAlaIleThrAspGluIleGluHisLeuAsnSerValProProValLysProLeuArgHis

        1720      1730      1740      1750      1760      1770      1780      1790      1800
CTCACTATACCAAATGAGTTTTCTTTTAATCTTATTCTAATCTCATTTTCATTAGATTGCAATACTTTCATACCTTCTATATTATTTATT

        1810      1820      1830      1840      1850      1860      1870      1880      1890
TTGTACCTTTTGGGATCTACACTTAATGTTAGCCTAATTGGAAAGTCATTTAGATTTAATACTGTTACCAGTCCATCCCTTTTAATTATT

        1900      1910      1920      1930      1940      1950      1960      1970      1980
AATGAAAATAAGAAGGGATAAGTAGCGATAGCCCTTATTCCGATATGGTCTCCAACAATATCCCTTATTATCTGCCTTGCAACACTAGGG

        1990      2000      2010      2020      2030      2040      2050      2060      2070
TAGAACTCTGAAATCAGATATGGTAGGTAAGTTGTAAGTGATAGGACGTAAACTTTAGAGTTAGAGTAAGTGTTCTGAAAGACTACTGGG

        2080      2090      2100      2110      2120      2130      2140      2150      2160
TGCAATTCGACACCGTTATAGGCGTAAAGGATTGGCGTAGCTCCGTTTAATGAAAATATAGGTCCTACAGGGAAATTGGCTTGCCTCTTG

        2170      2180      2190      2200      2210      2220      2230      2240      2250
TAATATGACCAATAGAACGTTTTCCCATCCCTGGTTAACGCATTGACACTAACACTATCGTAAATCAAGTTACCGACACCAAGAATTTTC

        2260      2270      2280      2290      2300      2310      2320      2330      2340
AGTGCAGTATCCCCAAGACTTCAATAAGCTTTTTAGCTGCACTTGCTGTAAACATTAAGTTAACTCCCCTATTAAGTAAATCCACAATA

TCTAGA
```

Figure 2. Sequence of XbaI DNA fragment containing the β-galactosidase gene.

328

ORF between position 230 and 1696 encodes for a protein of 489 amino acids with a predicted molecular weight of 56,650 Da, which is in agreement with the amino acid composition and molecular weight determined directly from the purified protein. In addition, the amino acids immediately preceding the termination codon correspond to the carboxy-terminal residues of the purified protein determined by carboxypeptidase digestion. We are confident that the entire S. solfataricus β–galactosidase gene is contained in the pD22 plasmid.

These data strongly suggest the absence of large introns in the β–galactosidase gene even if they do not exclude the existence of short intervening sequences. No archaeobacterial protein gene sequenced so far appears to contain introns which viceversa are widely distributed among the genes for stable RNAs (Wich et al., 1987).

Sequence analysis of the β–galactosidase gene indicates a codon usage not observed in the gene encoding structural proteins of the Sulfolobus virus, such as particle SSV1 described by Reiter et al. (1897) and appears to be more similar to that of eukaryote than to that of eubacterium organisms.

Comparing the sequence around ATG at position 230 with 3' end and 16S rRNA from S. solfataricus, two ribosome binding sites can be recognized (Figure 2). The first one overlaps the ATG start site and the second is located ten nucleotides upstream, in agreement with the data of Reiter et al. (1987). Upstream to the β–galactosidase gene two regions are found resembling the consensus sequence drawn for archaeobacterial promoters. The first is constituted by a box A at position 19 and a box B at position 48 while the second is constituted by a box A at position 97 and a box B at position 126. No other region compatible with the consensus sequence of archaeobacterial promoters is found in the β–galactosidase gene or in its 3' flanking regions.

A TTTTCTTTT sequence is encountered 30 bp downstream the stop codon of β–galactosidase and at least three short inverted repeats partly comprising this polypyrimidine stretch are found. At the moment archaeobacterial terminators are poorly characterized but the sequence TTTTTYT is frequently localized at the end of genes (Larsen et al., 1986; Reiter et al., 1988).

Expression of the β–galactosidase Gene in *E. Coli*

In order to verify the possibility of expressing this gene in *E. coli*, the 3.0 kb XbaI DNA fragment was subcloned into the XbaI site of plasmid pEMBL18, the same used for the sequence analysis, in the two possible orientations that produce two different vectors called pD22 and pD23. These plasmids were used to transform β-gal⁻ *E. coli* strain JM109 and in the overnight culture of these *E. coli* lysates a β–galactosidase activity was found using an enzymatic assay at high temperature. This activity was not detected in cultures of *E. coli* JM109 lac⁻ or transformed with pEMBL18, carrying the *E. coli* β-gal gene.

Remarkably, the activity was found regardless of the DNA fragment orientation in the vector, suggesting that the expression of this archaeobacterial gene was due to its own regulative structure recognized by the *E. coli* expression machinery.

Isolation of the Expressed Enzyme and Comparison with the Native

A β–galactosidase activity was isolated from JM109 *E. coli* strain containing the plasmid pEMBL18 carrying in a 3.0 kb XbaI DNA fragment the *S. solfataricus* β–galactosidase gene. A partial purification (10-20 fold) was achieved by heating the *E. coli* centrifuge homogenate for 30 min. at 75°C discarding the denatured proteins and precipitating the enzyme at pH 5.0. The partial characterization of the β–galactosidase reported in Table 1 indicates that the expressed enzyme has properties similar to the native enzyme.

Table 1. Comparison of the Native and Expressed Enzyme

	Native Enzyme	Expressed Enzyme
Km (β-ONPG)	0.23 mM	1.0 mM
Optimal pH	6.5	6.5
Optimal temperature activity	>95°C	>95°C
Isoelectric point	4.5	4.5
Residual activity after 2 hr. at 75°C	100%	100%

Work is in progress to purify this enzyme to homogeneity and fully characterize it.

ACKNOWLEDGMENTS

This work was partially supported by the EEC Biotechnology Action Programme Contract No. 0052-I and by the Progetto Finalizzato Biotecnologie CNR, Italy.

REFERENCES

Beckler, G.S., and Reeve, J.N., 1986, Conservation of primary structure in the hisI gene of the archaebacterium, *Methanococcus vanielii*, the subacterium *Escherichia coli*, and the eucaryote *Saccharomyces cerevisiae*, *Mol. Gen. Genet.* 204: 133.

Bokranz, M. and Klein, A., 1987, Nucleotide sequence of the methyl coenzyme M reductase gene cluster from *Methanosarcina barkeri*, *Nucl. Acid. Res.* 15: 4350.

Cubellis, M.V., Rozzo, C., Montecucchi, P. and Rossi, M. (manuscript in preparation).

Fabry, S., Lang, J., Niermann, T., Vingron, M. and Hensel, R., 1989, Nucleotide sequence of the glyceraldehyde-3-phosphate dehydrogenase gene from the mesophilic methanogenic archaebacteria *Methanobacterium bryantii* and *Methanobacterium formicicum*, *Eur. J. Biochem.* 179: 405.

Itoh, T., 1988, Complete nucleotide sequence of the ribosomal "A" protein operon from the archaebacterium *Halobacterium halobium*, *Eur. J. Biochem.* 176: 297.

Larsen, N., Leffer, H., Kjems, J. and Garret, R.A., 1986, Evolutionary divergence between the ribosomal RNA operons of *Halococcus morrhuae* and *Desofurococcus mobilis*, *System Appl. Microbiol.* 7: 49.

Lechner, K. and Boeck, A., 1987, Cloning and nucleotide sequence of the gene for an archaebacterial protein synthesis elongation factor Tu, *Mol. Gen. Genet.* 208: 523.

Maniatis, T., Fritsch, E.F. and Sambrook, J., 1982, Molecular Cloning, *in:* "A Laboratory Manual," Cold Spring Harbor Laboratory, Cold Spring Harbor, New York.

Pisani, F.M., Rella, R., Rozzo, C., Raia, C.A., Nucci, R., Gambacorta, A., De Rosa, M. and Rossi, M., 1989, Thermostable β-galactosidase from the archaebacterium *Sofolobus solfataricus*. Purification and properties, *Eur. J. Biochem.* (in press).

Reiter, W.D., Palm, P., Henschen, A., Lottspeich, F., Zillig, W. and Grampp, B., 1987, Identification and characterization of the genes creating three structural proteins of the *Sulfolobus* virus-like particle SSV 1, *Mol. Gen. Genet.* 206: 144.

Reiter, W.D., Palm, P. and Zillig, W., 1988, Transcription termination in the archaebacterium *Sulfolobus*: signal structure and linkage to transcription initiation, *Nucl. Acids Res.* 16: 2245.

Souillard, N. and Sibold, L., 1986, Primary structure and expression of a gene homologous to nifH (Nitrogenase Fe protein) from the archaebacterium *Methanococcus voltae*, *Mol. Gen. Genet.* 203: 21.

Souillard, N., Magot, M., Possot, O. and Sibold, L., 1988, Nucleotide sequence of regions homologous to nifH (nitrogenase Fe protein) from the nitrogen-fixing archaebacteria *Methanococcus thermolithotrophicus* and *Methanobacterium ivanovii*, *J. Mol. Evol.* 27: 65.

Stetter, K.O., 1986, *in:* "Thermophiles: general molecular and applied microbiology," T.D. Brock and J.K. Zeikus, eds., Wiley.

Ulrich, J.T., McFeters, G.A. and Temple, K.L., 1972, Induction and characterization of β-galactosidase in an extreme thermophile, *J. Bacteriol.* 110: 691.

Wich, G., Leinfelder, W. and Boeck, A., 1987, Genes for stable RNA in the extreme thermophile *Thermoproteus tenax*: introns and transcription signals, *EMBO J.* 6: 523.

Woese, C.R. and Wolfe, R.S., 1985, *in:* "The Bacteria," C.R. Woese and R.S. Wolfe, eds., 8: 561, Academic Press, New York.

SITE-DIRECTED MUTAGENESIS AND THE MECHANISM OF FLAVOPROTEIN DISULPHIDE OXIDOREDUCTASES

Richard N. Perham, Alan Berry, Nigel S. Scrutton and Mahendra P. Deonarain

Department of Biochemistry, University of Cambridge, Tennis Court Road
Cambridge CB2 1QW, UK

INTRODUCTION

Glutathione plays a critical role in the maintenance of reduced thiol groups in the cell and is of particular importance in the biosynthesis of DNA [for a review, see Holmgren, 1985]. Glutathione itself is maintained in a reduced form at the expense of NADPH by the action of the enzyme glutathione reductase (EC 1.6.4.2):

$$GSSG + NADPH + H^+ = 2GSH + NADP^+$$

Glutathione reductase, trypanothione reductase, dihydrolipoamide dehydrogenase, mercuric reductase and thioredoxin reductase are all members of a growing family of enzymes, the flavoprotein disulphide oxidoreductases.

Trypanothione reductase is an analogue of glutathione reductase restricted to trypanosomatids (Shames et al., 1986; Krauth-Siegel et al., 1987). It catalyses the NADPH-dependent reduction of trypanothione (bis-γ glutathionylspermidine), which replaces glutathione in these organisms. Dihydrolipoamide dehydrogenase plays an essential part in the mechanism of the 2-oxo acid dehydrogenase multi-enzyme complexes (Reed, 1974; Perham et al., 1987) in which it acts to oxidise the dihydrolipoyl groups of the lipoate acyltransferase components in an NAD^+-dependent reaction:

$$Lip(SH)_2 + NAD^+ = LipS_2 + NADH + H^+$$

Mercuric reductase in bacteria is part of a plasmid-encoded system for the detoxification of mercuric ions, catalysing the following reaction (Fox and Walsh, 1982):

$$Hg^{2+} + NADPH = Hg^0 + NADP^+ + H^+$$

Thioredoxin reductase catalyses the reduction of the small, heat-stable protein thioredoxin, which is a cofactor in the reduction of ribonucleoside to deoxyribonucleoside diphosphates by ribonucleotide reductase (Holmgren, 1985):

$$Thioredoxin - S_2 + NADPH + H^+ = Thioredoxin - (SH)_2 + NADP^+$$

All these enzymes are dimers with an M_r of about 105,000 and each possesses a redox-active disulphide bridge which is essential to the catalytic mechanism [reviewed by Williams (1976)]. Considerable homology, consistent with their evolution from a common ancestor, exists between the enzymes glutathione reductase, dihydrolipoamide dehydrogenase and mercuric reductase, as shown by studies of the amino acid sequences around their redox-active disulphides (Perham et al., 1978; Williams et al, 1982; Packman and Perham, 1982; Krauth-Siegel et al, 1982; Fox and Walsh, 1983). Thioredoxin reductase, however, is sufficiently different to suggest that it has probably arisen independently, a possible example of convergent evolution (Perham et al., 1978).

Various structural genes for members of this family, including the *lpd* gene of *Escherichia coli* encoding dihydrolipoamide dehydrogenase (Stephens et al, 1983), the *merA* gene of the transposon Tn501 from *Pseudomonas aeruginosa*, encoding mercuric reductase (Brown et al., 1983), the *gor* gene of *E.coli* encoding glutathione reductase (Greer and Perham, 1986), the gene encoding trypanothione reductase from *Trypanosoma congolense* (Shames et al., 1988), and the *trxB* gene of *E.coli* encoding thioredoxin reductase (Russel and Model., 1988), have been cloned and their nucleotide sequences determined. This has enabled the complete amino acid sequences of these enzymes to be inferred and their structures compared, confirming that thioredoxin reductase is at best only distantly related to the other enzymes in the family. However, the best structural information for any of these enzymes is that available for human glutathione reductase. The amino acid sequence is known (Krauth-Siegel et al., 1982) and the X-ray crystallographic structure (Thieme et al., 1981) has recently been refined to 0.154nm resolution (Karplus and Schulz, 1987). These studies have led to a detailed appreciation of the reaction mechanism for glutathione reductase (Pai and Schulz, 1983; Karplus et al., 1989).

The homology between the primary structures of these enzymes is indicative of closely similar three-dimensional structures. This view has been borne out, for example, by a direct attempt to fit the amino acid sequence of the *E.coli* dihydrolipoamide dehydrogenase to the three-dimensional structure of human glutathione reductase (Rice et al., 1984). The recent solution at low resolution of the crystal structure for dihydrolipoamide dehydrogenase from *Azotobacter vinelandii* has fully substantiated this conclusion (Schierbeek et al., 1989).

The cloning and sequence analysis of the *E.coli gor* gene and the strong homology between the human and *E.coli* glutathione reductases (Greer and Perham, 1986) has thus made it possible to learn more about the reaction mechanism of this enzyme and by inference of other flavoprotein disulphide oxidoreductases, by the methods of site-directed mutagenesis and protein engineering.

OVER-EXPRESSION OF THE *E. COLI GOR* GENE

In any study of this kind, an important pre-requisite is over-expression of the target gene. A little time ago, we described an expression system for the *E. coli gor* gene based on the plasmid pKGR, in which the *gor* gene was placed under the control of the powerful *tac* promoter, inducible with isopropyl β-D-thiogalactoside (IPTG). This system over-expresses the protein about 200 times above the level obtained from the chromosomal *gor* gene (Scrutton et al., 1987). We have recently improved on this level of expression by deleting the intervening sequence (approximately 700bp, including a poly T stretch) that lay between the *tac* promoter and the ribosome-binding site of the *gor* gene, to generate the new plasmid, pKGR4 (Figure 1).

When transformed into *E. coli* TG1 *lac*Iq cells, plasmid pKGR4 was found to express glutathione reductase activity at high levels, even without IPTG induction (Deonarain et al.,

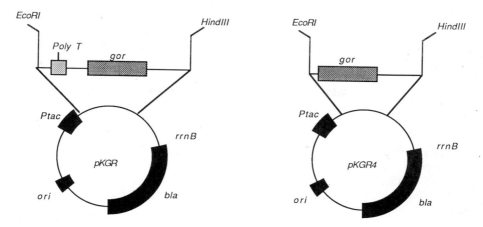

Figure 1. Expression plasmids used to over-express the *gor* gene in *E. coli*.

1989). The overproduction of glutathione reductase was accompanied by a marked increase in the yellow appearance of the cell-free extract, implying that flavin biosynthesis is also stimulated in cells carrying plasmid pKGR4. This could be due to sequestration of FAD by the high levels of apoenzyme generated in these cells, which in turn might stimulate the biosynthesis of flavins. SDS/polyacrylamide gel electrophoresis of samples of the cell-free extract revealed that it contained large amounts of a protein with an electrophoretic mobility identical to that of purified glutathione reductase (Figure 2). However, the specific activity of glutathione reductase in the cell-free extract was no higher than that found in cell-free extracts of cells expressing the *gor*

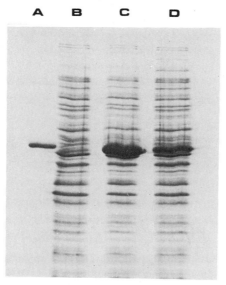

Figure 2. Expression of the *gor* gene in *E. coli* strain TG1. Samples of cell extracts of *E. coli* strain TG1 were prepared and analysed by means of SDS/polyacrylamide gel electrophoresis. Protein bands were visualized by staining with Coomassie Brilliant Blue R-250. Track A, purified wild-type glutathione reductase. Track B, cell extract of *E. coli* strain TG1. Track C, cell extract of *E. coli* strain TG1 transformed with plasmid pKGR4. Track D, cell extract of *E. coli* strain TG1 transformed with plasmid pKGR4 induced with 2mM IPTG.

gene from plasmid pKGR (Scrutton et al., 1987), indicating that large amounts of either apoenzyme or inactive holoenzyme were present in pKGR4-transformed cells. This could be reconstituted to form fully active enzyme during its purification by adding excess FAD to the cell-free extract before an ammonium sulphate fractionation. On the basis of these reconstituted activities and the SDS/polyacrylamide gel electrophoresis, it was estimated that non-induced cells carrying plasmid pKGR4 were expressing glutathione reductase at levels approx. 40000 times above that of untransformed *E. coli* (Deonarain et al., 1989).

ENGINEERING AN INTERSUBUNIT DISULPHIDE BRIDGE

There are two notable differences in the sequence alignments of the human enzyme and its *E. coli* counterpart: the former possesses an additional N-terminal segment of about 18 amino acid residues, which appears to be flexible in the crystal structure (Thieme et al., 1981; Karplus and Schulz, 1987), and a single intersubunit disulphide bridge between Cys-90 and Cys-90', both of which features are absent from the *E. coli* enzyme (Greer and Perham, 1986).

We have begun an investigation into subunit assembly in the dimer by studying the effects of replacing Thr-75 in the *E. coli* enzyme (the position equivalent to Cys-90 in the human enzyme) with a cysteine residue, thereby conferring the potential to form an intersubunit disulphide bridge. The folding of protein monomers and their assembly into active enzyme multimers is a fundamental part of protein biosynthesis. However, comparatively little is known of the detailed mechanisms involved in such macromolecular assembly (Shaw, 1987; Jaenicke, 1987). A closely related problem is that of the role of disulphide bridges in the stabilization of protein conformation (Creighton, 1988). Previous attempts at engineering disulphide bridges into proteins have mainly been confined to intrachain bridges and have met with varying degrees of success (Perry and Wetzel, 1986; Wells and Powers, 1986; Wetzel et al., 1988). The only reported example of an engineered disulphide bridge between subunits is for the N-terminal domain of the dimeric λ-repressor (Sauer et al., 1986).

The directed mutation of Thr-75 to a cysteine residue in *E. coli* glutathione reductase was found to generate an enzyme that contains an intersubunit disulphide bridge, one which was formed without the need for added oxidizing agents (Scrutton et al., 1988). We cannot say, any more than can be said of the human enzyme, whether this disulphide link exists in the dimer *in vivo* or whether it was formed by oxidation of the pair of juxtaposed thiol groups during the aerobic preparation of the enzyme. However, the bridge has no discernible effect on the catalytic activity of the enzyme *in vitro* and, as far as we know, is the first intersubunit disulphide bridge to be introduced into an enzyme with full retention of biological activity.

A study of the thermal stabilities of the mutant and wild-type proteins failed to detect any increase in stability of the T75C mutant over the wild-type enzyme. This is not as surprising as it may at first appear, given the nature of the dimer interface of glutathione reductase (Scrutton et al., 1988). In the human enzyme the interface interactions lie in two clearly defined areas of protein-protein contact (Karplus and Schulz, 1987). The "upper" area between the two interface domains contributes a much greater proportion of the binding energy and is the most ordered part of the protein. In contrast, the "lower" area contains the only portion of polypeptide chain (other than the N-terminal segment) where the main chain geometry is not clearly defined in the electron density map. The "upper" area is therefore much more important for dimer interaction than the "lower" one (Karplus and Schulz, 1987). The same is likely to be true for the interface in the *E. coli* enzyme, since the enzymes show 92% sequence identity in the "upper" interface area compared with 52% identity overall. The insertion of the Cys-75/Cys-75' disulphide bridge in the T75C mutant therefore places a covalent link across the "lower" interface area; any increase in stability afforded by this interaction is most probably masked by the stronger interactions across the "upper" contact area.

The ability to insert an inter-subunit disulphide bridge so readily into the *E. coli* glutathione reductase offers compelling evidence for the predicted close similarity between its structure and that of the human enzyme. Further, it lends confidence to the use of the crystal structure of the human enzyme in site-directed mutagenesis experiments on the *E. coli* enzyme.

PROTECTION OF REDUCED FLAVIN AND AN APPARENT SWITCH IN KINETIC MECHANISM

The active site of human glutathione reductase is shown in schematic form in Figure 3. It lies in the cleft between subunits of the dimer. When comparison is made with the amino acid sequence of the *E. coli* enzyme (Greer and Perham, 1986), the conservation of amino acid residues important in the catalytic mechanism (Pai and Schulz, 1983) is almost perfect. Only one residue is changed: His-219 in the human enzyme is replaced by a lysine (Lys-199) residue

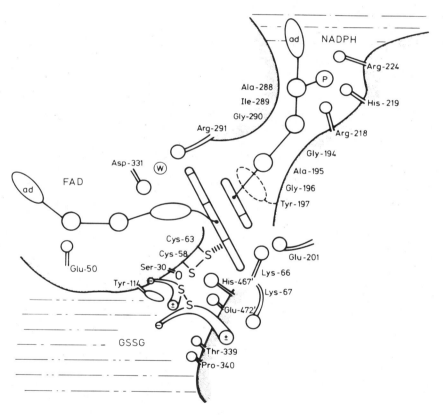

Figure 3.   Schematic diagram illustrating the active site of human glutathione reductase. The two substrate-binding sites are shown separated by the isoalloxazine ring of the enzyme-bound FAD. The amino acids thought to be involved in binding and catalysis are also shown. GSSG is bound in the active site across the two subunits and the position of the Cys-63/FAD charge-transfer interaction is shown. [Reproduced with permission from Pai & Schulz (1983)]

(the difference in numbering is a result of the omission from the *E. coli* enzyme of the first 17 amino acids of the human enzyme). To test the proposed role of some of these residues in the mechanism, we have designed and made several site-directed mutations in *E. coli* glutathione reductase.

Tyrosine-177 and the Kinetic Mechanism of Glutathione Reductase

In the human apoenzyme the bulky side chain of Tyr-197 lies in the NADPH-binding pocket and shields the isoalloxazine ring of the flavin from solvent. On binding NADPH, the tyrosine side-chain moves to allow the nicotinamide ring of the NADPH to come close to the flavin for electron transfer (Pai and Schulz, 1983), and it has been postulated (Rice et al., 1984) that this residue may act as a 'lid' in the NADPH-binding pocket to prevent adventitious loss of electrons from the reduced flavin moiety. Three mutations in position 177 have been made to date, one (Y177F) of tyrosine to phenylalanine (to preserve the aromatic ring), one (Y177S) of tyrosine to serine (to preserve the hydroxyl group), and one (Y177G) of tyrosine to glycine (to remove the side-chain altogether).

The wild-type and mutant enzymes were purified from extracts of an *E. coli* strain bearing a deletion of the chromosomal *gor* gene (strain SG5) over-expressing (about 200-fold) the wild-type or mutated *gor* gene from plasmid pKGR (Berry et al., 1989). The mutant Y177F was found to be almost as active as wild-type enzyme, whereas mutant Y177S showed only about 25% and mutant Y177G only about 3% of the wild-type activity (Table 1). However, despite the substantial fall in catalytic activity, all three mutants were clearly active.

During the catalytic cycle of glutathione reductase, the enzyme is reduced by NADPH to generate a two-electron reduced form of the enzyme, thought to be a charge-transfer complex. Any adventitious oxidation of the reduced enzyme (conveniently generated with sodium borohydride) can be monitored by the disappearance of the charge-transfer band at 540nm (Williams, 1976). Under anaerobic conditions, the borohydride-reduced wild-type and mutant (Y177F, Y177S, Y177G) enzymes showed no 'apparent oxidase' activity. Under aerobic conditions in air-saturated buffer, the half-life of the reduced flavin in the wild-type enzyme

Table 1. Specific Catalytic Activities and Kinetic Parameters of Wild-type and Mutant Forms of *E. coli* Glutathione Reductase

	Wild-type	Y177F	Y177S	Y177G
SPECIFIC ACTIVITY				
(U/mg)[a]	252	246	72	7
FORWARD REACTION				
Mechanism	Ping-Pong	Ping-Pong	Sequential	Sequential
K_m NADPH (μM	38 ± 4	24 ± 5	30 ± 8	18 ± 9
K_m GSSG (μM)	97 ± 12	53 ± 9	2 ± 1.5	5 ± 2.5
k_{cat} (min^{-1})	36000 ± 2600	31000 ± 3900	8200 ± 1400	280 ± 70
REVERSE REACTION				
Mechanism	Ter-Bi Sequential	Ter-Bi Sequential	Ter-Bi Sequential	Not determined
K_m NADP$^+$ (μM)	120 ± 20	70 ± 13	80 ± 9	
K_m GSH (μM)	1300 ± 1000	1200 ± 300	700 ± 100	
k_{cat} (min^{-1})	3000 ± 250	1400 ± 70	100 ± 3	

[a] Enzyme specific activities were measured at saturating concentrations of all substrates.

was about 60 min., whereas the half-lives in the mutant forms of the enzyme were all 20-25 min. The only comparison we have is with $FADH_2$, which free in aqueous solution has a half-life of less than 1 second. This would be consistent with a significant protection of the reduced flavin in the enzyme. Given that large differences in the size and nature of the amino acid side-chains could be engineered at position 177 without much change in the oxidase rates, it appears that protection of the reduced flavin in glutathione reductase is largely due to burial of the isoalloxazine ring within the protein (Thieme et al., 1981) and not to the particular properties of Tyr-177.

The kinetic mechanism envisaged for human glutathione reductase is Ping-Pong, addition of NADPH leading to formation of an EH_2 intermediate, which is subsequently reoxidized by GSSG with formation of GSH (Williams, 1976; Pai and Schulz, 1983). The results of a detailed study of the kinetics of the wild-type E. coli enzyme and of the mutant enzymes (Berry et al., 1989) are also shown in Table 1. The wild-type E. coli enzyme displayed Ping-Pong kinetics, consistent with the mechanism postulated for the human enzyme (Williams, 1976; Pai and Schulz, 1983), and the kinetic mechanism for mutant Y177F was also found to be predominantly Ping-Pong. However, the kinetic plots for the mutants Y177S and Y177G indicated an Ordered Sequential mechanism (Figure 4). Moreover, the mutations Y177S and Y177G were found to have caused a substantial fall in the value of K_m for GSSG (Table 1). This was also unexpected since GSSG binds at a separate site on the enzyme, remote from the point of mutation in the NADPH-binding pocket (Thieme et al., 1981; Pai and Schulz, 1983). No change was detected in the kinetic mechanism of the reverse reaction (Table 1). It is likely that the effects of the mutations are due to subtle changes in and around the active site rather than to major structural changes in the enzyme.

These unusual results are most readily explained by supposing that E. coli glutathione reductase actually follows a hybrid kinetic mechanism (Figure 5), such as that postulated for the yeast enzyme (Mannervik, 1973). The mutations Y177S and Y177G appear to divert flux from the Ping-Pong loop, which dominates in wild-type enzyme, to the Ordered Sequential loop. This can be accounted for in two possible ways, both of which involve a change in the partitioning of the intermediate E-NADPH. First, if the mutation slowed or blocked the step from E-NADPH to EH_2 (Figure 5), it would tend to force the intermediate E-NADPH more through the Ordered Sequential pathway. This pathway might coincidentally have a lower K_m for GSSG, thereby accounting for the change observed in this kinetic parameter (Table 1). Alternatively, if the mutation caused the observed lowering of the K_m for GSSG in mutants Y177S and Y177G, it would be equivalent to raising the concentration of GSSG, and high concentrations of GSSG favour the Ordered Sequential pathway in a hybrid Ping-Pong/Ordered Sequential mechanism (Figure 5). The latter explanation might require some action at a distance in the protein, given that the binding site for glutathione is about 1.8nm from the NADPH-binding pocket (Thieme et al., 1981; Pai and Schulz, 1983). However, a proper understanding of the molecular basis of the apparent change in kinetic mechanism must await a crystallographic analysis of the structures of the mutant enzymes.

CHANGING PUTATIVE PROTON DONORS

An important feature of the catalytic mechanism (Figure 6) of glutathione reductase is the need for a proton donor/acceptor in the glutathione-binding site. In the related enzyme dihydrolipoamide dehydrogenase, this role has been assigned to a histidine residue (Matthews et al., 1977), and X-ray crystallographic work has shown that His-467 is suitably positioned in the glutathione-binding pocket of human glutathione reductase to act in this capacity (Pai and Schulz, 1983; Karplus et al., 1989). We have tested this prediction by systematically removing

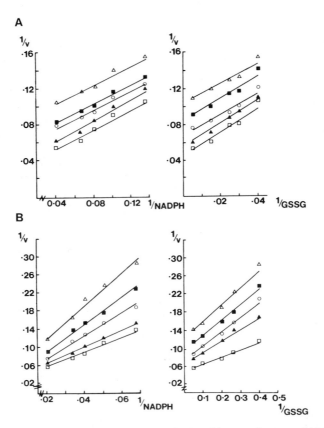

Figure 4.   Kinetic plots for mutant forms of *E. coli* glutathione reductase. (A) Mutant Y177F: left-hand panel, GSSG concentrations are (top to bottom) 25, 33.3, 40, 70, and 250μM; right-hand panel, NADPH concentrations are (top to bottom) 7.5, 10, 12.5, 15, and 25μM. (B) Mutant Y177S: left-hand panel, GSSG concentrations are (top to bottom) 2.5, 3.75, 5, 10, and 18.75μM; right-hand panel, NADPH concentrations are (top to bottom) 7.5, 10, 12.5, 15, and 25μM. The wild-type enzyme behaved similarly to mutant Y177F and mutant Y177G behaved similarly to mutant Y177S, although the inferred kinetic parameters were different (see Table 1).

the possible proton donors/acceptors in the glutathione-binding pocket of *E. coli* glutathione reductase, notably His-439 and Tyr-99, which are the highly conserved counterparts of His-467 and Tyr-114 in the human enzyme (Berry et al., 1989; Deonarain et al., 1989).

Replacement of Histidine-439

In the first experiment, His-439 was replaced by a glutamine residue. The first surprise was that the mutant (H439Q) enzyme retained approximately 1% of the catalytic activity of the wild-type enzyme (Berry et al., 1989). In human glutathione reductase, a solvent water molecule, which is ideally placed to serve as proton donor, is hydrogen-bonded to His-467 (Karplus and Schulz, 1987). The residual activity in the H439Q mutant of the *E. coli* enzyme might therefore be due to the H-bonding capacity of the glutamine residue mimicking that of a histidine residue. To test this possibility, we removed the potential for H-bonding at position 439 by replacing His-439 with an alanine residue. Mutant H439A was purified to homogeneity and found to possess about 0.3% of the specific activity of the wild-type enzyme (Table 2),

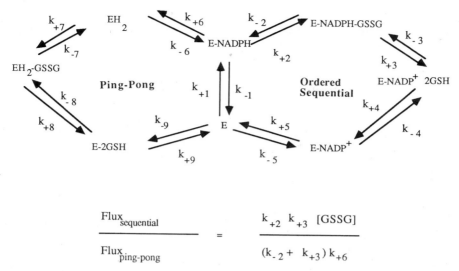

$$\frac{\text{Flux}_{\text{sequential}}}{\text{Flux}_{\text{ping-pong}}} = \frac{k_{+2}\, k_{+3}\, [\text{GSSG}]}{(k_{-2} + k_{+3})\, k_{+6}}$$

Figure 5. Kinetic mechanism for glutathione reductase. The hybrid Ping-Pong/Bi-Bi Ordered Sequential mechanism is that proposed for the yeast enzyme (Mannervik, 1973).

comparable with that of the H439Q mutant (Deonarain et al., 1989). Thus, a protonable side chain at position 439 is not absolutely essential for activity. Moreover, the small residual catalytic activity in these mutants (H439A and H439Q) cannot be due to any ability of the side-chain at this position to hydrogen-bond to a solvent water molecule, allowing it to serve as proton donor . It must be, therefore, that the enzyme has recruited an alternative proton donor in order to turn over, albeit inefficiently.

The second surprise came from a detailed kinetic analysis of mutants H439Q and H439A (Berry et al., 1989; Deonarain et al., 1989). Both mutants were found to exhibit a much-diminished Michaelis constant for NADPH (Table 2). However, although the overall activity of the enzyme was severely impaired, the catalytic competence of the NADPH-binding site was unaltered, as judged by the ability of the mutant enzymes (H439Q and H439A) to catalyse the NADPH/thio-NADP^{+} transhydrogenase activity as efficiently as the wild-type enzyme (Table 2).

Thus, in a result that echoes the findings with the Tyr-177 mutants described above, mutations in the GSSG-binding pocket are manifesting themselves in substantial changes in the K_m value for NADPH bound in a separate and distant pocket. Once again a crystallographic analysis of the mutants is needed to provide a structural explanation for this observation.

Replacement of Tyrosine-99

In human glutathione reductase, binding of GSSG is accompanied by the movement of the phenol ring of Tyr-114 by about 0.1nm to lie between the two glycyl moieties of the glutathione. The tyrosine hydroxyl group is in van der Waals contact with both sulphurs of GSSG (Pai and Schulz, 1983; Karplus et al., 1989). In this position it could conceivably donate a proton to glutathione in the reaction mechanism, although, given the relatively high pK$_a$ of unperturbed phenolic hydroxyl groups, such a proposal is possible but not probable. We assessed its contribution to the mechanism by replacing the corresponding residue (Tyr-99) in E. coli glutathione reductase with phenylalanine.

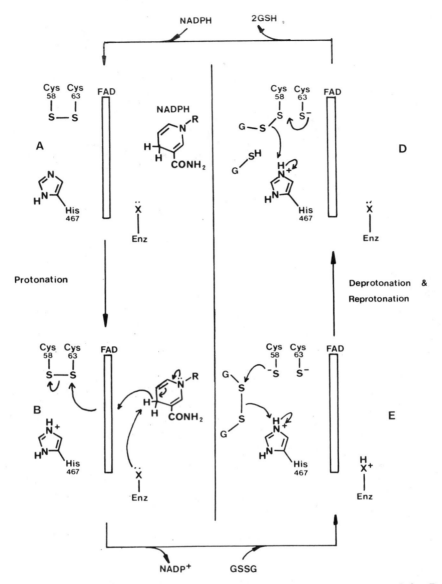

Figure 6. The postulated reaction mechanism for human glutathione reductase [after Pai & Schulz (1983) and Wong et al. (1988)]. The scheme should not be taken to imply that electron transfers are concerted. The base X has tentatively been identified as Lys-66 (Pai & Schulz, 1983).

The Y99F mutant form of *E. coli* glutathione reductase had a specific activity similar to that of the wild-type enzyme, it followed Ping-Pong kinetics, and the K_m values for GSSG and NADPH were similar to those of the wild-type enzyme (Table 2). Thus, although the phenolic hydroxyl group of Tyr-99 may be intimately associated with the bound GSSG, it does not appear to participate substantially in the binding of this substrate, nor is it an essential contributor to the mechanism (Deonarain et al., 1989).

In a further experiment, we united the H439Q and Y99F mutations into a single gene to encode a double mutant protein and examined its catalytic activity. The Y99FH439Q double

Table 2. Specific Catalytic Activities and Kinetic Parameters of Wild-Type and Mutant Forms of *E. coli* Glutathione Reductase

| Enzyme | Specific Activities (U/mg)[a] | | Kinetic Parameters for Reduction of GSSG | | |
	NADPH-dependent reduction of GSSG	Transhydrogenase activity	K_m [GSSG] (μM)	K_m [NADPH] (μM)	k_{cat} (min^{-1})
Wild-type	252	1.2	97 ± 12	38 ± 4	36000 ± 2600
H439Q[b]	3.0	1.2	310 ± 30	<2	140 ± 10
H439A[b]	0.95	3.2	219 ± 34	<2	43 ± 4
Y99F	210	1.0	81 ± 17	53 ± 11	36000 ± 5300
Y99FH439Q[b]	1.1	0.9	66 ± 6	<2	64 ± 3

[a] Enzyme specific activities were measured at saturating concentrations of all substrates. The transhydrogenase activity was measured at 30°C by the thio-NADP$^+$-dependent oxidation of NADPH, with both coenzymes at an initial concentration of 100μM.

[b] The true value of K_m for NADPH in this mutant could not be measured since discrimination in rate could not be achieved even at a concentration of NADPH as low as 2μM.

mutant enzyme was purified in a similar manner to the H439Q single mutant (Berry et al., 1989). In its kinetics, it behaved almost identically to the H439Q mutant. Thus, it retained about 1% of the wild-type specific catalytic activity, exhibited a much-diminished Michaelis constant for NADPH compared with wild-type enzyme, and had an unaffected transhydrogenase activity (Table 2). These results prove conclusively that Tyr-99 is not acting as a surrogate proton donor/acceptor during the catalytic cycle of the H439Q mutant of glutathione reductase and render it unlikely that Tyr-99 plays any role in proton transfer in the wild-type enzyme. It is conceivable that some as yet unidentified protein side-chain acts in this capacity but it could well be that in the mutants lacking His-439 the proton is simply acquired from the aqueous solution bathing this somewhat open part of the active site.

CONCLUSIONS

The flavoprotein disulphide oxidoreductases have an unusual and interesting mechanism. The two active sites lie between the subunits in the dimer of glutathione reductase (Thieme et al., 1981; Pai and Schulz, 1983) and an active monomer cannot therefore be envisaged. The experiments we have described on the successful engineering of a disulphide bridge across the dimer interface represent the first step of a systematic and, we hope, illuminating study of the pathway of assembly of the active enzyme dimer.

The limutations we have described of key residues in the active site, of Tyr-177 in the NADPH-binding site and of His-439 and Tyr-99 in the GSSG-binding site, also have thrown up some interesting and novel results. Of particular interest are the way the kinetic mechanism of the forward reaction changes in response to mutations of Tyr-177, and the accompanying substantial fall in the K_m value for GSSG which is bound at a physically distinct site. Similarly, mutations of His-439, the putative proton donor in the GSSG-binding site, are accompanied by a substantial fall in the K_m value for NADPH.

The advantage that accrues to the enzyme from placing a histidine residue at position 439, an approximately 100-fold increase in k_{cat} compared with the H439Q or H439A mutants, is very clear. Thus, it is probable that, in the wild-type enzyme, the imidazole side-chain of His-439 does act as proton donor/acceptor in the way originally envisaged (Williams, 1976; Matthews et al., 1977; Pai and Schulz, 1983). But again, it is of interest to note the way in which the enzyme is able to recruit an alternative proton donor if the glutathione-binding pocket is deprived of this particular protonatable residue.

X-ray crystallographic analysis of the mutant forms of the enzyme will be needed to settle the molecular basis of these unexpected effects. Even at this stage, the results serve warning of the need to undertake a rigorous analysis of the effects of mutation. Our knowledge of protein structure is still too insecure to predict the outcome of mutations with total confidence.

ACKNOWLEDGEMENTS

This work was supported by the Science and Engineering Research Council and its Cambridge Centre for Molecular Recognition, the Royal Society and the Royal Commission for the Exhibition of 1851. A.B. is a Royal Society 1983 University Research Fellow. N.S.S. is a Research Fellow of the Royal Commission for the Exhibition of 1851. M.P.D. was supported by a Research Studentship from the SERC and a Benefactors' Research Scholarship from St. John's College, Cambridge.

REFERENCES

Berry, A., Scrutton, N.S. and Perham, R.N., 1989, Switching kinetic mechanism and putative proton donor by directed mutagenesis of glutathione reductase, *Biochemistry* 28:1264.

Brown, N. L., Ford, S. J., Pridmore, D. and Fritzinger, D. C., 1983, Nucleotide sequence of a gene from the *Pseudomonas* transposon Tn501 encoding mercuric reductase, *Biochemistry* 22:4089.

Creighton, T.E.,1988, Disulphide bonds and protein stability, *BioEssays* 8:57.

Deonarain, M.P., Berry, A., Scrutton, N.S. and Perham, R.N., 1989, Alternative proton donors/acceptors in the catalytic mechanism of the glutathione reductase of *Escherichia coli*: the role of His-439 and Tyr-99, *Biochemistry*, in press.

Fox, B. and Walsh, C.T., 1982, Mercuric reductase. Purification and characterization of a transposon-encoded flavoprotein containing an oxidation-reduction0active disulphide, *J. Biol Chem.* 257:2498.

Fox, B. and Walsh, C.T., 1983, Mercuric reductase: Homology to glutathione reductase and lipoamide dehydrogenase. Iodoacetamide alkylation and sequence of the active site peptide, *Biochemistry* 22:4082.

Greer, S. and Perham, R.N., 1986, Glutathione reductase from *Escherichia coli:* cloning and sequence analysis of the gene and relationship to other flavoprotein disulphide oxidoreductases, *Biochemistry* 25:2736 .

Holmgren, A., 1985, Thioredoxin, *Annu. Rev. Biochem.* 54:237.

Jaenicke, R., 1987, Folding and association of proteins, *Progress in Biophys. and Mol. Biol.* 49:117.

Karplus, P.A. and Schulz, G.E., 1987, Refined structure of glutathione reductase as 1.54 Å resolution, *J. Mol. Biol.* 195:701.

Karplus, P.A., Pai, E.F. and Schulz, G.E., 1989, A crystallographic study of the glytathione binding site of glytathione reductase at 0.3nm resolution, *Eur. J. Biochem.* 178:693.

Krauth-Siegel, R.L., Blatterspiel, R., Saleh, M., Schulz, G.E., Schirmer, R.H. and Untucht-Grau, R., 1982, Glutathione reductase from human erythrocytes. The sequences of the NADPH domain and of the interface domain, *Eur. J. Biochem.* 121:259.

Krauth-Siegel, R.L., Enders, B., Henderson, G.B., Fairlamb, A.H. and Schirmer, H.R., 1987, Trypanothione reductase from *Trypanosoma cruzi*. Purification and characterization of the crystalline enzyme, *Eur. J. Biochem.* 164:123.

Mannervik, B., 1973, A branching mechanism of glutathione reductase, *Biochem. Biophys. Res. Commun.* 53:1151.

Matthews, R. G., Ballou, D. P., Thorpe, C. and Williams, C. H., Jr., 1977, Ion pair formation in pig heart lipoamide dehydrogenase. Rationalization of pH profiles for reactivity of oxidized enzyme with dihydrolipoamide and 2-electron-reduced enzyme with lipoamide and iodoacetamide, *J. Biol. Chem.* 252:3199.

Packman, L.C. and Perham, R.N., 1982, An amino acid sequence in the active site of lipoamide dehydrogenase from *Bacillus stearothermophilus*, *FEBS Lett.* 139:155.

Pai, E.F. and Schulz, G.E., 1983, The catalytic mechanism of glytathione reductase as derived from X-ray diffraction analyses of reaction intermediates, *J. Mol. Biol.* 258:1751.

Perham, R.N., Harrison, R.A. and Brown, J.P., 1978, The lipoamide dehydrogenase component of the 2-oxo acid dehydrogenase multienzyme complexes of *Escherichia coli*, *Biochem. Soc. Trans.* 6:47.

Perham, R.N., Packman, L.C. and Radford, S.E., 1987, 2-Oxo acid dehydrogenase multienzyme complexes: in the beginning and halfway there, *in:* "Kreb's citric acid cycle - half a centry and still turning," J. Kay and P.D.J. Weitzman, eds., *Biochem. Soc. Symp.* 54:67.

Perry, L.J., and Wetzel, R., 1986, Unpaired cysteine-54 interferes with the ability of an engineered disulfide to stabilize T4 lysozyme, *Biochemistry* 25:733.

Reed, L.J., 1974, Multienzyme complexes, *Acc. Chem. Res.* 7:40.

Rice, D.W., Schulz, G.E. and Guest, J.R., 1984, Structural relationship between glutathione reductase and lipoamide dehydrogenase, *J. Mol. Biol.* 174:483.

Russel, M. and Model, P., 1988, Sequence of thioredoxin reductase from *Escherichia coli.* Relationship to other flavoprotein disulphide oxidoreductases, *J. Biol. Chem.* 263:9015.

Sauer, R.T., Hehir, K., Stearman, R.S., Weiz, M.A., Jeitler-Nilsson, A., Suchanek, E.G., and Pabo, C.O., 1986, An engineered intersubunit disulfide enhances the stability and DNA binding of the N-terminal domain of λ-repressor, *Biochemistry* 25:5992.

Schierbeek, A.J., Swarte, M.B.A., Dijksta, B.W., Vriend, G., Read, R.J., Hol, W.G.J., Drenth, J. and Betzel, C., 1989, X-ray structure of lipoamide dehydrogenase from *Azotobacter vinelandii* determined by a combination of molecular and isomorphous replacement techniques, *J. Mol. Biol.* 206:365.

Scrutton, N.S., Berry, A. and Perham, R.N., 1987, Purification and characterization of glutathione reductase encoded by a cloned and over-expressed gene in *Escherichia coli*, *Biochem. J.* 245:875.

Scrutton, N.S., Berry, A. and Perham, R.N., 1988, Engineering of an intersubunit disulphide bridge in glutathione reductase from *Escherichia coli*, *FEBS Lett.* 24:46.

Shames, S.L., Fairlamb, A.H., Cerami, A. and Walsh, C.T., 1986, Purification and characterization of trypanothione reductase from *Crithidia fasciculata*, a newly discovered member of the family of disulphide-containing flavoprotein reductases, *Biochemistry* 25:3519.

Shames, S.L., Kimmel, B.E., Peoples, O.P., Agabian, N. and Walsh, C.T., 1988, Trypanothione reductase of *Trypanosoma congolense:* Gene isolation, primary sequence determination, and comparison to glutathione reductase, *Biochemistry* 27:5014.

Shaw, W.V., 1987, Protein Engineering. The design, synthesis and characterization of fictitious proteins, *Biochem. J.* 246:1.

Stephens, P.E., Lewis, H.M., Darlison, M.G. and Guest, J.R., 1983, Nucleotide sequence of the lipoamide dehydrogenase gene of *Escherichia coli* K12, *Eur. J. Biochem.* 135:519.

Thieme, R., Pai, E.F., Schirmer, R.H. and Schulz, G.E., 1981, Three-dimensional structure of glutathione reductase at 2 Å resolution *J. Mol. Biol.* 151:763.

Wells, J.A. and Powers, D.B., 1986, *In vivo* formation and stability of engineered disulfide bonds in subtilisin, *J. Biol. Chem.* 261:6564.

Wetzel, R., Perry, L.J., Baase, W.A. and Becktel, W.J., 1988, Disulfide bonds and thermal stability in T4 lysozyme, *Proc. Natl. Acad. Sci. U.S.A.* 85:401.

Williams, C. H., Jr., 1976, Flavin containing dehydrogenases, *in:* "The Enzymes" 3rd edn., P.D. Boyer, ed., Academic Press, New York. 13:89.

Williams, C.H., Jr., Arscott, L.D. and Schulz, G.E., 1982, Amino acid sequence homology between pig heart lipoamide dehydrogenase and human erythrocyte glutathione reductase, *Proc. Natl. Acad. Sci. U.S.A.* 79:2199.

Wong, K.K., Vanoni, M.A. and Blanchard, J.S., 1988, Glutathione reductase: Solvent equilibrium and kinetic isotope effects, *Biochemistry* 27:7091.

RESONANCE RAMAN AND SITE-DIRECTED MUTAGENESIS STUDIES OF

MYOGLOBIN DYNAMICS

Paul M. Champion

Department of Physics, Northeastern University, Boston, MA 02115, USA

INTRODUCTION

The protein structure-function relationships involved in the binding of ligands to heme proteins have been the focus of a wide variety of physical, biological and chemical investigations during the last several decades. A well studied process from an experimental point of view is the geminate recombination of CO to myoglobin:

$$Mb \cdot CO \underset{\{k\}}{\overset{\gamma}{\rightleftarrows}} Mb + CO \qquad [1]$$

Equation 1 denotes the photolysis of carbon monoxy myoglobin (MB $\cdot$ CO) by light (γ) and the subsequent rebinding ($\{k\}$). The curly brackets around the (k) indicate that a single rate is not sufficient to describe the rebinding and that a distribution in rates is necessary to explain the non-exponential kinetics observed at low temperature (Austin et al., 1975). We believe that the distribution in rates is due to protein structural fluctuations that are frozen in below T_f (Srajer et al., 1988). Above T_f the protein fluctuations are rapid compared to kinetic time scales and a single exponential rate, averaged over the fluctuations, is observed.

In this paper we discuss two separate types of structural fluctuations and their effect on the rebinding rates. The first type involves a nearly continuous distribution of protein "microstates" which distributes the iron-porphyrin out-of-plane equilibrium position in deoxy Mb. This type of finely grained distribution affects the proximal terms in the expression for the barrier height and leads to the stretched exponential rebinding behavior at low temperature. A similar type of fine grained proximal distribution is needed to account for the optical absorption and Raman excitation profile of deoxy Mb (Srajer et al., 1986). The second type of structural fluctuation is much more discrete, or coarsely grained, and corresponds to distinct protein "macrostates" that we associate with an "open" or "closed" distal pocket. These structures affect the distal terms in the expression for the barrier height.

In Figure 1 we sketch how the probability distributions involving the generalized protein coordinates might look within a simple model. Here, the various protein conformational states that affect the iron-porphyrin out-of-plane displacement are denoted as x_p, and are taken to be finely grained so that a continuous Gaussian distribution can approximate $P(x_p)$. The width of the distribution is dependent upon temperature and a restoring force constant (f) that maintains

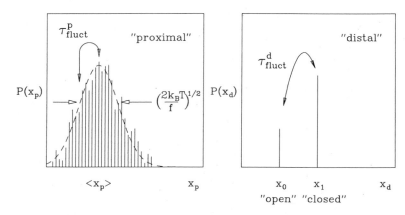

Figure 1. The probability distributions representing the generalized protein coordinates x_p and x_d. The proximal distribution is fine grained and can be approximated by a Gaussian. The important distal conformations appear to be much more discrete and the finely grained fluctuations, which are surely present, can be neglected to first order. The time scales τ_p and τ_d are probably somewhat different with $\tau_d > \tau_p$.

the "most probable" conformation, $<x_p>$. At temperatures above freezing (T_f) the fluctuation time, τ_{fluct}, is rapid compared to the rebinding time scales so that an average rebinding rate is observed. Below T_f a distribution of x_p is frozen into the ensemble and τ_{fluct} is long compared to the geminate rebinding. The distal pocket fluctuations can be treated in much the same way except that the relevant states, observed using IR and Raman spectroscopy, appear to be much more discrete. In Figure 1 we have depicted a very simple model that characterizes the distal pocket as "open" (when HisE7 swings out towards the solvent) or "closed" (when HisE7 hinders the CO ligand).

PROXIMAL EFFECTS

In Figure 2 we show a schematic diagram of the iron-porphyrin system. We believe that the work needed to bring the iron-porphyrin system into a planar configuration contributes important terms to the rebinding barrier height. We write these proximal terms as:

$$H_p = \frac{1}{2} K a^2 \tag{2}$$

where a is the iron-porphyrin displacement. The distribution in a tracks with the protein coordinate, x_p, and is taken as a Gaussian. This leads directly to the *non-Gaussian* distribution in H needed to explain the rebinding kinetics at low temperature (Srajer et al., 1988). In contrast to the model of Agmon and Hopfield (1983), a formal average over the distribution leads to the observed high temperature rebinding rate. The H are the rebinding barrier heights in the Arrhenius expression for the rates:

$$k(H) = k_0 e^{-H/k_B T} \tag{3}$$

where

$$H = H_p + H_D. \tag{4}$$

348

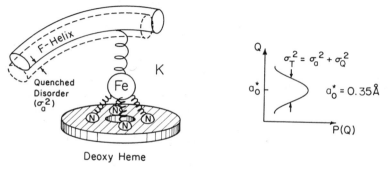

Deoxy Heme

Figure 2. A schematic diagram showing the source of the "proximal pocket work," H_p. The constant K represents all linear restoring forces involved in displacing the iron porphyrin system toward the planar transition state. For photolyzed Mb at low temperature, we take the mean out of plan displacement to be $a_0^* = 0.35\text{Å}$.

Figure 3 and Table 1 demonstrate the kinetics experiment (Austin et al., 1975) and the parameters needed to fit the data (Srajer et al., 1988). Typical barrier height distribution functions are shown in Figure 4 for a variety of systems. The breadth of these curves is a universal feature due primarily to the proximal effects discussed above. The shifts along the axis, due to distal pocket alterations and/or pH, are carried by the distal pocket term H_D.

DISTAL EFFECTS

The distal effects are more easily probed by experiments involving site-directed mutants of Mb (Morikis et al., 1989). We have used resonance Raman spectroscopy to demonstrate that when the distal histidine is replaced by other amino acids, the sub-populations of heme-CO states become independent of pH (see Table 2 and Figure 5). In the native material a strong pH dependence is observed (Figures 5, 6). This indicates that histidine E7 is the titratable group in native MbCO. Moreover, the pH dependence of the population dynamics is found to be inconsistent with a simple two-state Henderson-Hasselbalch analysis. Instead, we suggest a four-state model involving the coupling of histidine protonation and conformational change

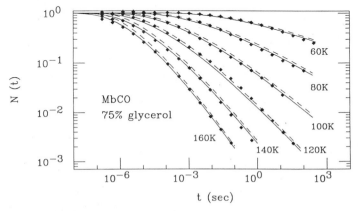

Figure 3. A fit of the MbCO rebinding kinetics using the method of Srajer et al., 1988. The dashed lines represent a closed form approximation for N(t), the amount of unbound Mb at time T after photolysis. The solid line is a computer integration of the exact equations. The parameters are listed in Table 1.

Table 1. Parameters for MB· CO rebinding[a].

Parameter	Rebinding Fit	Independent Values	Independent Technique
a_0^*	0.35 Å	0.45 Å	x-ray diffraction
		0.35 Å	EXAFS
σ_a	0.11 Å	0.3 Å[b]	x-ray T=200 K
		0.24 Å	Mössbauer T=200 K
		0.25 Å	Soret Absorption
		0.07 Å	Dynamics calculation
K	13.8 N/m	5.1 N/m	Soret Absorption
		$\geq$2.1 N/m[c]	Mössbauer
		88 N/m	Raman $\tilde{\nu}_{Fe-N_{His}}$=220 cm^{-1}
k_0	2.8×10^9 s^{-1}	-	-
H$_D$	7.0 kJ/mole	6.9 kJ/mole	High temperature rebinding limit $< H > = 12.2$ kJ/mole

(a) Useful conversions: N/m=10^{-2} mdyne/Å=1.44 kcal/moleÅ^2,
kcal/mole=4.18 kJ/mole=350 cm^{-1}.

(b) The values of σ_{Fe} obtained from x-ray and Mössbauer spectroscopy are expected to be somewhat larger than σ_a. This is primarily because they are absolute rather than relative measures of iron disorder.

(c) One can not simply use the slope of σ_{Fe}^2 vs. T in the linear region ($\sigma_{Fe}^2 \cong k_B T/K_{Fe}$) to extract the low frequency force constant, since the coupling constants that relate the cartesian and normal coordinates are also needed. When the zero-point contributions to σ_{Fe}^2 are analyzed, the coupling constant for the low frequency mode(s) can be determined to be on the order of 0.1 (P. Debrunner, private communication). This sets a lower limit of K$\geq$2 N/m from Mössbauer spectroscopy.

(Scheme 1). Within this model, the pK of the distal histidine is found to be 6.0 in the "open" configuration and 3.8 in the "closed" configuration. This corresponds to a 3kcal/mole destabilization of the positively charged distal histidine within the hydrophobic distal pocket and suggests how protonation can lead to a larger population of the "open" configuration. Clearly, other residues must also begin to protonate below pH 4.0 and the four state model of Scheme 1 is an over simplification in this region.

Scheme 1.

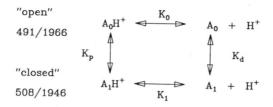

"open"
491/1966

"closed"
508/1946

At pH 7 approximately 3% of the population remains in the open conformation. This is likely to be important in accounting for ligand access to the heme. In the "closed" conformation, determined from the X-ray structure, there is no way for the CO to approach the heme. Thus, the dynamics of these conformational interconversions could play an important role in ligand binding and release under physiological conditions. We have also observed that ionic strength affects the relative populations of the "open" and "closed" forms. This suggest the possibility of an ion specific regulatory control mechanism in the Mb system.

Finally, we note that the separate subpopulations of protein macrostates, observed spectroscopically, can be associated with the different barrier height distributions as shown in Figure 4. In the open form favored at low pH, or in the mutants, the absence of His E7 reduces the value of H_D and shifts the distributions to lower energy. Recent spectroscopically resolved kinetics measurements (Doster et al., 1982; Ansari et al., 1987) support this point of view. The explicit interactions responsible for the variation of the spectroscopic frequencies and the parameter H_D have yet to be determined, but a direct histidine-CO interation cannot be responsible for the closed A_1 state, since it is clearly observed in the glycine mutant, probably due to a partially collapsed pocket. Ligand orientation and pocket polarization properties are two of the ideas that have been suggested.

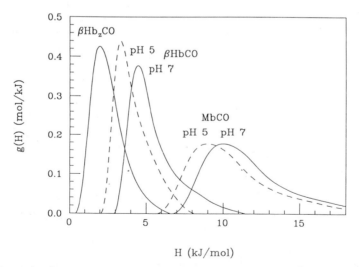

Figure 4. The activation enthalpy probability distributions, g(H), for heme protein recombination. The distributions for Mb, the β chains of Hb and Hb Zurich are shown to depend on pH and distal pocket structure. Within the model, H_D shifts the distributions along the energy axis without altering the shape.

Table 2. Raman frequencies and relative intensities of MbCO[a].

sample	vibration	A_0	A_1	A_3	I_{A_0}/I_{A_1}	I_{A_3}/I_{A_1}
wild type pH 7	ν_{Fe-CO}	491	508	518	0.08±0.02[b]	0.03±0.02
native pH 7	"	491	508	518	0.05±0.02	0.02±0.02
native pH 3.9[c]	"	491	508	518	0.96±0.10	0.04±0.02
crystal	"	491	508	518	0.89±0.10	0.01±0.01
GLY pH 9.5	"	492	506	†	2.41±0.30	-
GLY pH 7	"	492	506	†	2.54±0.30	-
GLY pH 5.5	"	492	506	†	2.67±0.30	-
GLY pH 4.1	"	492	506	†	3.34±0.30	-
MET pH 8.5	"	495	506	†	3.3-9.4[d]	-
MET pH 7	"	495	506	†	3.5-9.4	-
MET pH 6.1	"	495	506	†	3.8-9.7	-
wild type pH 7	ν_{C-O}	†	1946	1932	-	0.35±0.10
native pH 7	"	†	1946	1932	-	0.26±0.10
native pH 3.9	"	1966	1946	1932	0.92±0.20	0.15±0.15
GLY pH 7	"	1965	1944	†	2.81±0.60	-
MET pH 7	"	1964	1947	†	5.9-13.7	-

[a]Frequencies and areas are obtained from a non-linear least square fit using Lorentzian lineshapes. All frequencies in cm^{-1}. Excitation wavelength is 420 nm and sample is held at room temperature in a spinning cell with $\geq$ 90 % in MbCO bound state.

[b]The quoted uncertainties reflect fitting error and estimated experimental error which can be large for small peaks.

[c]Acetate buffer was used and later found to slightly affect the observed ratios when compared to phosphate and citrate/ phosphate buffers.

[d]The quoted range in the peak areas of the methionine sample reflects fits with linewidths of 9.0-8.5 cm^{-1} for A_0 and 9.0-3.5 cm^{-1} for A_1. The large uncertainty in the width of the A_1 mode is due to the low resolution. In the native material the widths for all peaks were fixed to ca. 9.0 cm^{-1} while the widths for the glycine mutant were found to be 9.5 cm^{-1} for A_0 and 6.2 cm^{-1} for A_1.

† Not measured within experimental signal to noise.

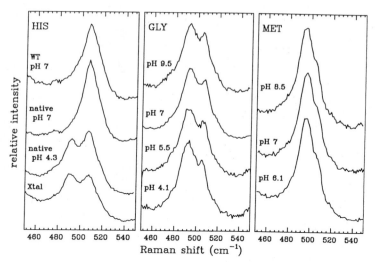

Figure 5. High resolution resonance Raman scans of the $\nu_{Fe\text{-}CO}$ region. Notice the pH
dependence of the $\nu_{Fe\text{-}CO}$ mode in the native material and the similarity of the low pH
spectrum to that of the single crystal. The mutants E7Gly and E7Met show a pH
independent line shape.

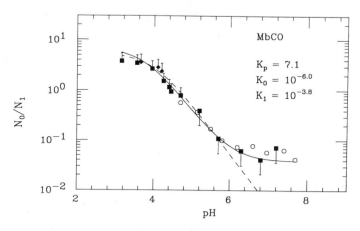

Figure 6. The population ration N_0/N_1 in native MbCO is plotted logarithmically as a function of
pH. The solid line is the result of using the four state model of Scheme 1 and the
parameters listed. A standard two-state analysis leads to a straight line on the
logarithmic scale.

REFERENCES

Agmon, N. and Hopfield, J., 1983, CO binding to heme proteins: A model for barrier feight distributions and slow conformational changes, *J. Chem. Phys.* 79:2042.

Ansari, A., Berendzen, J., Braunstein, D., Cowen, B., Frauenfelder, H., Hong, H., Iben, I., Johnson, J., Ormas, P., Sauke, T., Scholl, R., Schulte, A., Steinbach, P., Vittiton, J. and Young, R., 1987, Rebinding and relaxation in the myoglobin pocket, *Biophys. Chem.* 26:337.

Austin, R., Beeson, K., Eisenstein, L., Frauenfelder, H. and Gunsalus, I., 1975, Dynamics of ligand bonding to myoglobin, *Biochemistry* 14:5355.

Doster, W., Beece, D., Bowne, S., Ditorio, E., Eisenstein, L., Frauenfelder, H., Reinisch, L., Shyamsunder, E., Winterhalter, K. and Yue, K., 1982, Control and pH dependence of ligand binding to heme proteins, *Biochemistry* 21:4831.

Morikis, D., Champion, P.M., Springer, B. and Sligar, S., 1989, Resonance Raman investigations of site-directed mutants of myoglobin, *Biochemistry* 28:4791.

Srajer, V., Reinisch, L. and Champion, P.M., 1988, Protein fluctuations, distributed coupling and the binding of ligands to heme proteins, *J. Am. Chem. Soc.* 110:6656.

Srajer, V., Schomacker, K. and Champion, P.M., 1986, Spectral broadening in biomolecules, *Phys. Rev. Lett.* 57:1267.

COMPARISON OF THE SECONDARY STRUCTURES OF HUMAN CLASS I AND CLASS II MHC ANTIGENS BY FTIR AND CD SPECTROSCOPY

Joan C. Gorga[1], Aichun Dong[2], Mark C. Manning[2], Robert W. Woody[2], Winslow S. Caughey[2] and Jack L. Strominger[1]

[1]Department of Biochemistry and Molecular Biology, Harvard University Cambridge, MA 02138 & [2]Department of Biochemistry, Colorado State University, Fort Collins, CO 80523, USA

INTRODUCTION

Considerable evidence exists that the structures of class I and class II histocompatibility antigens are similar. Most of the evidence (summarized in Kappes and Strominger, 1988 and Brown et al., 1988) is based on sequence homologies and similarities in domain structure at both the protein and DNA levels. In addition, some T cells that are specific for either class I or class II molecules use the same receptor (Rupp et al., 1985; Marrack and Kappler, 1986). However, the secondary structures of purified class I and class II antigens have not been directly compared.

In the crystal structure of the papain-solubilized class I antigen HLA-A2 (A2$_{pap}$) approximately 42% of the amino acid residues form antiparallel β-pleated sheets (Bjorkman et al., 1987a). The membrane-proximal domains, $\alpha3$ and β_2-microglobulin (β_2m), are β-sandwich structures similar to the structure described for immunoglobulin constant regions. The membrane-distal domains, $\alpha1$ and $\alpha2$, which do not show sequence homology to immunoglobulin constant regions, consist of an antiparallel β-pleated sheet under a long α-helical region. Approximately 20% of the amino acid residues compose this set of α-helices, which form the sides of a peptide-binding groove at the membrane-distal surface of the molecule (Bjorkman et al., 1987b).

The two membrane-proximal domains, $\alpha2$ and $\beta2$, of class II antigens also show strong sequence homology to immunoglobulin constant regions, while the membrane-distal domains, $\alpha1$ and $\beta1$, do not (Kappes and Strominger, 1988). A model for the membrane-distal domains of class II antigens based on the crystal structure of A2$_{pap}$ has been proposed (Brown et al., 1988). In this manuscript, Fourier transform infrared spectroscopy (FTIR) and circular dichroism (CD) are used to compare the secondary structure of the papain-solubilized class II antigen DR1 (DR1$_{pap}$) with the secondary structures of the papain-solubilized class I antigens HLA-A2 (A2$_{pap}$) and HLA-B7 (B7$_{pap}$), and with those of purified human β_2m, bovine immunoglobulin G (IgG), and lysozyme, in order to address the question of whether a class II antigen has amounts of α-helix and β-sheet consistent with a class II model based on the class I crystal structure.

MATERIALS AND METHODS

Protein sources. Purified human urinary β_2m was obtained from A.R. Sanderson (Seriological Reagents Limited, England); bovine IgG and lysozyme were purchased from Sigma; papain-solubilized HLA-A2 and HLA-B7 were purified as described (Turner et al., 1975; Parham et al., 1977); DR1 was purified by immunoaffinity chromatography and then solubilized by digestion with papain (Gorga et al., 1987).

FTIR spectroscopy. Spectra for solutions of A2$_{pap}$, B7$_{pap}$, DR1$_{pap}$, β_2m, IgG, and lysozyme at 20°C were measured in cells with CaF$_2$ windows (Gorga et al., 1985) and a pathlength of 6 μm using a Perkin Elmer Model 1800 FTIR spectrophotometer at 2 cm^{-1} resolution in the single beam mode. 1000 scans were averaged for data recorded at 1 cm^{-1} intervals from 4000 to 1000 cm^{-1}. Background spectra were recorded under identical conditions with only buffer in the cell. The protein spectrum was obtained by digital subtraction of the spectrum of the medium from the spectrum observed for the protein solution. Prior to any further manipulation, the data were smoothed with a nine-point Savitsky-Golay function (Savitsky and Golay, 1964) to remove the possible white noise. The Perkin-Elmer Enhance function, which is analogous to the method developed by Kauppinen, et al. (1981), with the parameters set to a half-bandwidth of 16 cm^{-1} and a K value of 2.3 was used to achieve spectral enhancement; second derivative spectra were obtained with Savitsky-Golay derivative function software for a five data point window. The method used to determine protein secondary structure from second derivative Amide I spectra will be described in detail elsewhere (A. Dong, P. Huang, and W.S. Caughey, manuscript in preparation).

CD Spectroscopy. CD spectra for A2$_{pap}$, B7$_{pap}$, DR1$_{pap}$, and β_2m were obtained with each protein at approximately 600 μg/ml in 5 mM sodium phosphate, pH 7.75 or 5 mM Tris, pH 7.75. Protein concentration was determined by amino acid analysis of the samples used for CD spectroscopy. The spectra were obtained on a JASCO J41C circular dichrograph using a silica cell of 0.1 or 0.5 mm pathlength with a spectral band width of 1 nm and a time constant of 4 sec. The results of 4-5 scans were averaged. The instrument was calibrated (Chen and Yang, 1977) with (+)-10-camphorsulfonic acid.

RESULTS

FTIR spectroscopy has been shown to be a sensitive indicator of protein secondary structure (Krimm and Bandekar, 1986). Because of the strong absorption of water in the infrared, FTIR of proteins has normally been done in D$_2$O. Recent technical advances in FTIR allow infrared spectroscopy of proteins in aqueous solution (Alvarez et al., 1987), and therefore allow direct comparison with the results of CD studies. With the use of very short pathlengths (in the range of 6-10 μm) it is possible to measure the IR spectra of proteins in aqueous solution with excellent signal-to-noise ratios. In addition, FTIR microscopy can be used to take IR spectra of very small samples (Kwiatkoski and Reffner, 1987), and of single cells (Dong et al., 1988) or crystals.

The Amide I region of protein IR spectra contains strong absorption bands due almost entirely to the C-O stretch vibration of the peptide linkages that constitute the backbone structure (Susi, 1972). FTIR spectral enhancement (Surewicz and Mantsch, 1988) and second derivative analysis has been shown to be a reliable indicator of the component peak positions (Susi and Byler, 1986). The peak intensities can be used for a quantitative measure of the relative amount of each component. The Amide I regions of FTIR spectra of lysozyme, IgG, and β_2m in aqueous solution, along with FTIR spectral enhancement and second derivative analysis (Figure 1), revealed bands that could be assigned to α-helix, β-sheet, turns, or unordered chain

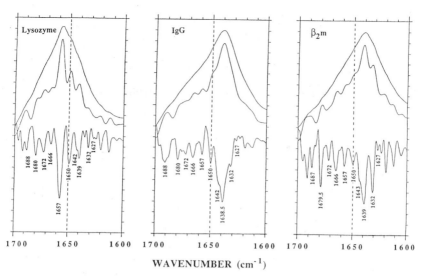

WAVENUMBER (cm⁻¹)

Figure 1. Infrared spectra of lysozyme, immunoglobulin G (IgG), and β2-microglobulin (β2m) in the Amide I region. In each, the upper curve shows the difference spectrum, i.e., the observed spectrum of the protein solution minus the spectrum of the medium in which the protein was dissolved, the middle curve shows the FTIR enhanced spectrum, and the lower curve shows the second derivative of the FTIR enhanced spectrum. The spectra were obtained with the proteins at concentrations of between 10 and 15 mg/ml in 5 mM sodium phosphate, pH 7.4.

(A. Dong, P. Huang, and W. S. Caughey, manuscript in preparation). Amide I band frequencies are sensitive to secondary structure as demonstrated in the lysozyme and IgG spectra (Figure 1). The lysozyme spectrum consists of a major absorption centered at 1657 cm⁻¹ due to α-helix, a set of peaks between 1627 and 1642 cm⁻¹ due to β-sheet, a set of peaks between 1666 and 1688 cm⁻¹ due to turns, and a peak at 1650 cm⁻¹ due to unordered structures. This secondary structure assignment is consistent with the crystal structures of lysozyme (Levitt and Greer, 1977; Provencher and Glockner, 1981). In contrast, the IgG FTIR spectrum shows only a small amount of α-helix (at 1657 cm⁻¹) and a large contribution from β-sheet (the second derivative bands at 1642, 1638.5, 1632, and 1627 cm⁻¹) with some unordered structure present (at 1650 cm⁻¹). These IgG assignments are also consistent with X-ray crystal (Amzel and Poljak, 1979) and CD (Sears and Beychok, 1973) data (see also Table 1).

The β2m Amide I spectrum is strikingly dissimilar to the lysozyme spectrum and is nearly identical to the IgG spectrum. As expected, and in agreement with X-ray (Bjorkman et al., 1987a; Becker and Reeke, 1985) and CD data (Karlsson, 1974; Isenman et al., 1975; Lancet at al., 1979; Trägårdh et al., 1979, and Figure 3), β2m is rich in β-sheet with some unordered structure and very little α-helix.

The Amide I spectra of the class I antigens A2pap and B7pap and the class II antigen DR1pap are shown in Figure 2. All three proteins show peaks ascribable to α-helix (at 1657.5 cm⁻¹) and β-sheet (at 1642-3, 1638.5, 1632, and 1627 cm⁻¹), as well as peaks ascribable to unordered chain and turns. The peaks at 1657.5 cm⁻¹, due to α-helix, are considerably stronger than those due to α-helix at 1657 cm⁻¹ in the IgG and β2m spectra. The characters of the β-sheet bands differ among the three antigens, with those for A2pap and DR1pap very similar to each other, but

Table 1. Secondary structures of human class I and class II MHC antigens as measured by IR and CD spectra and X-ray methods

| Protein | Buffer | Secondary Structure (%) | | | | | Methods* |
		α-helix	β-sheet	Turns	Random	Other++	
A2pap		17	41	28	14		IR
	phosphate	8	74	--	--	13	CD
	Tris	13	77	--	--	10	CD
		20	42	--	--		X-ray[1]
DR1pap		10	53	24	13		IR
	phosphate	17	42	--	--	41	CD
	Tris	23	38	--	--	39	CD
B7pap		8	48	32	12		IR
	phosphate	20	29	--	--	51	CD
	Tris	20	39	--	--	41	CD
β2m		6	52	33	9		IR
	phosphate	0	59	--	--	41	CD
		0	48	--	--		X-ray[1]
IgG		3	64	28	5		IR
		3	67	18	12		X-ray[2]
Lysozyme		40	19	27	14		IR
		45	19	23	13		X-ray[2]
		41	16	23	20		X-ray[3]

*CD spectra were analyzed by the method of Provencher and Glockner (1981). CD data in the range of 240-190 nm were used in the analyses for A2pap and DR1pap, and 240-195 nm for B7pap and β2m.

++Turns + Random coil not calculated separately by CD

[1]Bjorkman et al, 1987a

[2]Levitt and Greer, 1977

[3]Provencher and Glockner, 1981

different from the B7pap bands; however, the relative proportions of α-helix and β-sheet are quite similar for all three proteins. The relative proportions of secondary structures were quantitatively estimated from the areas under the peaks in the second derivative spectra (Table 1).

The secondary structures of the MHC antigens were also examined by circular dichroism spectroscopy. CD is a well-established technique for examining the secondary structure of proteins (Johnson, Jr., 1988). Estimates of α-helix content are generally quite reliable, whereas β-sheet contents are subject to greater uncertainty (Yang et al., 1986). The CD spectrum of β2m (Figure 3) was in good agreement with previously reported studies (Karlsson, 1974; Isenman et al., 1975; Lancet et al., 1979; Trägårdh et al., 1979), and was consistent with substantial amounts of β-sheet and little or no α-helix (Yang et al., 1986). Quantitative analysis of the CD spectra (Table 1) gave 0% α-helix and 59% β-sheet, in reasonable agreement with the X-ray data.

Figure 3 also shows the CD spectra for two class I antigens, A2pap and B7pap, dissolved in phosphate buffer. The amplitude of the 217 nm band for B7pap agreed well with that reported previously (Lancet at al., 1979), and was somewhat higher than that reported for a mixture of

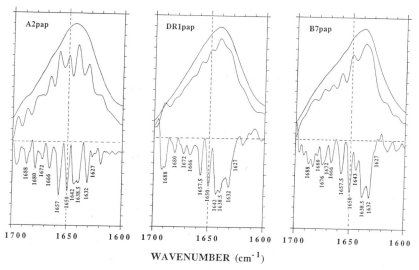

WAVENUMBER (cm⁻¹)

Figure 2. Infrared spectra of papain-solubilized HLA-A2 (A2$_{pap}$), HLA-B7 (B7$_{pap}$), and DR1 (DR1$_{pap}$)in the Amide 1 region, as described in Figure 1.

class I antigens (Trägårdh et al., 1979). For both proteins, the position of the negative band at 217 nm is characteristic of proteins in which β-sheet is the predominant secondary structure, but the higher intensity and greater breadth of this band, as well as the position of the shorter wavelength positive band, indicate the presence of some α-helix. Quantitative analysis bears this out (Table 1). For A2$_{pap}$, helix contents of 8-13% were obtained, depending on the buffer used, while the β-sheet content was estimated at ca. 75%. Comparison with the X-ray structure

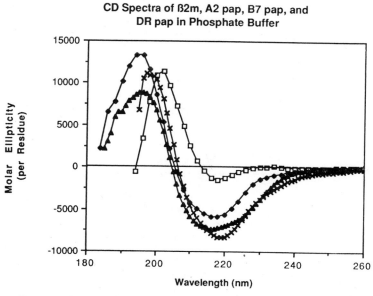

Figure 3. CD spectra of β2m (□), A2$_{pap}$ (◆), B7$_{pap}$ (X), and DR1$_{pap}$ (▲) in 5 mM sodium phosphate, pH 7.75.

359

(Bjorkman et al., 1987a) indicates that the α-helix content was underestimated to some extent, whereas the β-sheet content was greatly overestimated. A different method of analysis (Bolotina et al., 1981) gave results in much better agreement with the X-ray structure (β-sheet content of ca. 40%), but the fit to the CD spectrum was very poor. Comparison with the FTIR results showed that this discrepancy cannot be due to a genuine crystal-solution difference.

The CD of B7$_{pap}$ gave 20% α-helix and 29-39% β-sheet. Contrary to the case of A2$_{pap}$, the CD estimate of α-helix in B7$_{pap}$ was higher than that from FTIR, while that for the β-sheet was lower. However, the discrepancy for the β-sheet was much less than in the case of A2$_{pap}$.

Relative to the class I proteins, the class II antigen DR1$_{pap}$ displayed a broader minimum near 217 nm with an unresolved shoulder near 222 nm (Figure 3). This pattern was indicative of increased α-helix content. Quantitative analysis gave an α-helix content of 17-23% and a β-sheet content of 38-42%. As in the case of B7$_{pap}$, CD gave a higher β-helix content and a lower β-sheet content than FTIR.

In all three of the MHC proteins studied, the CD spectrum was observed to be more intense in Tris buffer than in phosphate buffer, as shown for the case of A2$_{pap}$ in Figure 4. The corresponding differences in secondary structure content (Table 1) are within the error limits of the method, so one cannot conclude that Tris increases helix content in A2$_{pap}$ or DR1$_{pap}$. Nevertheless, the observed CD differences suggest minor buffer-induced conformational changes.

DISCUSSION

Analysis of the secondary structures of proteins with low amounts of α-helix, such as proteins of the immunoglobulin superfamily, has in the past been difficult (Johnson, Jr., 1988).

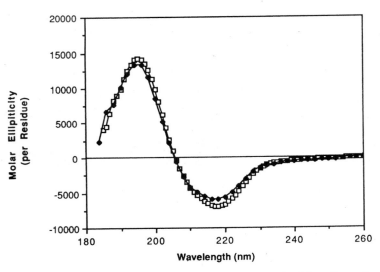

A2 pap CD Spectra in Tris,Phosphate Buffers

Figure 4. CD spectra of A2$_{pap}$ in 5 mM Tris, pH 7.75 (□) and 5 mM sodium phosphate, pH 7.75 (◆).

Indeed, it was easy to assume that the small amount of α-helix detected in B7$_{pap}$ by CD (Lancet et al., 1979) was insignificant, especially with spectra that extended to only 200 nm. Improvements in methods of CD analysis now permit more accurate determination of small amounts of α-helix. However, larger uncertainties remain in the determination of β-sheet content. With the recent advances in FTIR spectroscopy of proteins in aqueous solution, it has become possible to assign peaks in the Amide I region to different forms of secondary structure, and to quantitate the results (A. Dong, P. Huang, and W. S. Caughey, manuscript in preparation). Thus, it is possible to use both FTIR spectroscopy and CD to determine the secondary structures of proteins in solution.

Several factors limit the accuracy with which FTIR and CD can determine protein secondary structure. Both methods assume that elements of secondary structure contribute independently to the observed spectroscopic property. It is also assumed that variations in size and distortions from regularity are either intrinsically negligible or are made to be so by averaging over a protein basis set.

Secondary structure analyses of the MHC antigens could also be affected by two unique features. Processed antigens retained in the antigen-binding site (Bjorkman et al., 1987b) can contribute to the observed spectroscopic properties. It has recently been reported (Buus et al., 1988) that purified murine class II antigens appear to have low-molecular-weight material associated with them. In addition, the sites of papain cleavage of the DR1 alpha and beta chains are not known. Since DR1$_{pap}$ behaves like a soluble molecule and each chain is 4,000-6,000 daltons smaller than the corresponding chain in the detergent-soluble form (Gorga et al., 1987), it is assumed that papain cleaves the DR1 chains somewhere in the connecting peptide regions, that is, in the peptide segments connecting the α2 or β2 domains with the transmembrane regions. Therefore, papain probably removes the cytoplasmic tail, the transmembrane segment, and part of the connecting peptide from each chain. Although papain cleaves the connecting peptide of HLA-A2 at the C-terminus of the α3 domain, removing approximately 9,000 daltons from HLA-A2, it may cleave the connecting peptides of one or both of the DR1 chains closer to the transmembrane region. Thus, the remaining segments of connecting peptide would contribute to the secondary structure. If the additional connecting peptide in DR1 contributes to the contents of β-sheet, turns, or random structure, then the proportion of α-helix in the protein would be less than that in A2$_{pap}$, and conversely, if the connecting peptide contributes to the α-helix content, the proportion of α-helix would be greater.

There are some specific limitations of FTIR spectroscopy (Braiman and Rothschild, 1988). The very strong water band in the Amide I region requires very short pathlengths (6 μm), highly sensitive detectors, and extremely careful difference spectroscopy. Earlier studies have generally used D$_2$O solutions to avoid interference from the H$_2$O band, thereby severely limiting the H$_2$O data available for reference. Second, a few amino acid residues (e.g., Gln and Asn) exhibit bands in the Amide I region. However, such bands are too weak to contribute very much to the overall spectrum, and the estimated contents of these residues are only slightly different for A2$_{pap}$, B7$_{pap}$ and DR1$_{pap}$. Thus the observed FTIR spectral differences between the MHC antigens are unlikely to be due to differences in amounts of the amino acid residues that are infrared active in the Amide I region. Finally, there are problems associated with resolving the various components, quantitating the bands from second derivatives, and possible variations in the intensities of Amide I bands from various types of secondary structure. The results for lysozyme, IgG, and A2$_{pap}$, which are presented here, along with further details to be published (Dong, et al., in preparation), give us confidence that our results provide a satisfactory characterization of the unknown B7$_{pap}$ and DR1$_{pap}$ secondary structures.

There are also limitations of CD spectroscopy compared with FTIR, for secondary structure determination. The resolving power of CD is limited by the fact that the UV bands of the α-helix, β-sheet, β turns, and unordered structures strongly overlap. Contributions of chromophoric side chains pose another serious problem. Theoretical studies (Woody, 1978; Woody, 1987; M. Manning and R.W. Woody, in preparation) indicate that Tyr, Trp, and Phe can make substantial contributions to the far-UV CD of proteins. Experimentally, anomalous CD spectra for various proteins have been attributed to aromatic (Green and Melamed, 1966; Day, 1973) or disulfide (Hider et al., 1988) contributions. In addition, in the present study, instrumental limits restrict measurements to wavelengths longer than about 190 nm. Hennessey and Johnson (Hennessey, Jr., and Johnson, Jr., 1981) have pointed out that CD data down to ca. 175 nm can greatly enhance the information content and thus permit more reliable estimates of β sheet and even β turn content.

Both CD and FTIR show that all three MHC proteins studied have small but significant amounts of α-helix. The α-helix contents must be considered comparable for all three proteins, in view of the differences between the CD and the FTIR results.

The FTIR analysis also indicates that all three proteins have comparable amounts of β-sheet. CD supports this in the case of DR1$_{pap}$ and B7$_{pap}$, but indicates a much higher β-sheet content for A2$_{pap}$. In view of the FTIR results and the strong sequence homology (Brown et al., 1988) between HLA-A2 and HLA-B7, the CD for A2$_{pap}$ must be considered anomalous.

The anomaly in the CD spectrum of A2$_{pap}$ may be due in large part to aromatic side chain contributions unique to this protein. HLA-A2 has a Trp at position 107 which is replaced by Gly in HLA-B7 and is also absent in DR1 (Brown et al., 1988). In addition, there are other points in the chain where an aromatic group in HLA-A2 is replaced by a nonchromophoric side chain in HLA-B7 (His 70$\rightarrow$Gln, His 74$\rightarrow$Asp, Phe 109$\rightarrow$Leu, His 114$\rightarrow$Asp, His 145$\rightarrow$Arg, His 151$\rightarrow$Arg) or by a different aromatic group (Phe 9$\rightarrow$Tyr, Tyr 113$\rightarrow$His). In only one case does such a transition occur in the opposite direction (Val 67$\rightarrow$Tyr). If the entire difference in the CD spectrum were attributed to the replacement of a single Trp, that Trp would have $[\Theta]_{220} \approx 10^6$ deg cm^2/dmol. Calculations of the interaction between a Trp side chain and the two adjacent peptides (Woody, 1987) have shown that such interactions can lead to $[\Theta]_{220} \approx 10^5$ deg cm^2/dmol. Avidin has a CD maximum at 228 nm which corresponds to ca. $+8 \times 10^4$ deg cm^2/dmol for each of the five aromatic residues (1 Tyr, 4 Trp). Given that not all of the aromatic side chains contribute equally, that some may make negative contributions, and that the peptide backbone almost certainly makes a significant negative contribution at this wavelength, individual Trp CD bands substantially in excess of 10^5 deg cm^2/dmol are implied. Therefore, aromatic side chain contributions unique to A2$_{pap}$, especially Trp 107, can account for a large part of the anomaly.

Although sequence homology and similarity in secondary structure contents for HLA-A2 and HLA-B7 determined by FTIR argue for overall structural similarity, the FTIR indicates some differences in detailed structure. The band at 1643 cm^{-1} in the second derivative spectrum for B7$_{pap}$ is much weaker than the corresponding band in the A2$_{pap}$ or the DR1$_{pap}$ spectra (Figure 2). In addition, the B7$_{pap}$ spectrum has a much stronger band at 1676 cm^{-1} than do the A2$_{pap}$ and DR1$_{pap}$ spectra. It has also been reported (Bjorkman et al, 1985) that B7$_{pap}$ present as a contaminant of A2$_{pap}$ is excluded from A2$_{pap}$ crystals, indicating some differences in molecular shape.

It will be of interest to see whether other members of the immunoglobulin superfamily, such as the T cell receptor, CD4, CD8, and the class I-like molecules Qa and Tla, have proportions of secondary structure similar to that of the major histocompatibility antigens.

ACKNOWLEDGEMENTS

We thank Don Wiley's laboratory for providing HLA-A2 and HLA-B7, Todd Willis for performing the amino acid analyses, and Russell Middaugh (University of Wyoming) for providing a copy of Provencher's CONTIN program. Joan C. Gorga is a fellow of the Charles A. King Trust. This work was supported by U.S.P.H.S. grants GM-22994 to Robert W. Woody, HL-15980 to Winslow S. Caughey, and AI-10736 to Jack L. Strominger. This paper has previously appeared in *Proc. Natl. Acad. Sci. USA* 86:2321, 1989 and is reprinted here with permission of the authors.

REFERENCES

Alvarez, J., Lee, D.C., Baldwin, S.A. and Chapman, D., 1987, Fourier transform infrared spectroscopic study of the structure and conformational changes of the human erythrocyte glucose transporter, *J. Biol. Chem.* 262:3502.

Amzel, L.M. and Poljak, R.J., 1979, Three-dimensional structure of immunoglobulins, *Ann. Rev. Biochem.* 48:961.

Becker, J.W. and Reeke, G.N., 1985, Three-dimensional structure of β_2-microglobulin, *Proc. Natl. Acad. Sci. USA* 82:4225.

Bjorkman, P.J., Saper, M.A., Samraoui, B., Bennett, W.S., Strominger, J.L. and Wiley, D.C., 1987a, Structure of the human class I histocompatibility antigen, HLA-A2, *Nature* 329:506.

Bjorkman, P.J., Saper, M.A., Samraoui, B., Bennett, W.S., Strominger, J.L. and Wiley, D.C., 1987b, The foreign antigen binding site and T cell recognition regions of class I histocompatibility antigens, *Nature* 329:512.

Bjorkman, P.J., Strominger, J.L. and Wiley, D.C., 1985, Crystallization and x-ray diffraction studies on the histocompatibility antigens HLA-A2 and HLA-A28 from human cell membranes, *J. Mol. Biol.* 186:205.

Bolotina, I.A., Chekhov, V.O., Lugauskas, V.Yu. and Ptitsyn, O.B., 1981, Determination of the secondary structure of proteins from the circular dichroism spectra. II. Consideration of the contribution of β-bends, *Mol. Biol.* (English Translation of *Molekul. Biol.*) 14:709.

Braiman, M.S. and Rothschild, K.J., 1988, Fourier transform infrared techniques for probing membrane protein structure, *Ann. Rev. Biophys. Biophys. Chem.* 17:541.

Brown, J.H., Jardetzky, T., Saper, M.A., Samraoui, B., Bjorkman, P.J. and Wiley, D.C., 1988, A hypothetical model of the foreign antigen binding site of class II histocompatibility molecules, *Nature* 332:845.

Buus, S., Sette, A., Colon, S.M. and Grey, H.M., 1988, Autologous peptides constitutively occupy the antigen binding site on Ia, *Science* 242:1045.

Chen, G.C. and Yang, J.T., 1977, Two-point calibration of circular dichrometer with d-10-camphorsulfonic acid, *Anal. Lett.* 10:1195.

Day, L.A., 1973, Circular dichroism and ultraviolet absorption of a deoxyribonucleic acid binding protein of filamentous bacteriophage, *Biochemistry* 12:5329.

Dong, A., Messerschmidt, R.G., Reffner, J.A. and Caughey, W.S., 1988, Infrared spectroscopy of a single cell--the human erythrocyte, *Biochem. Biophys. Res. Comm.* 156:752.

Green, N.M. and Melamed, M.D., 1966, Optical rotary dispersion, circular dichroism, and far-ultraviolet spectra of avidin and streptavidin, *Biochem. J.* 100:614.

Gorga, J.C., Hazzard, J.H. and Caughey, W.S., 1985, Determination of anesthetic molecule environments by infrared spectroscopy. I. Effects of solvating molecule structure on nitrous oxide spectra, *Arch. Biochem. Biophys.* 240:734.

Gorga, J.C., Horejsí, V., Johnson, D.J., Raghupathy, R. and Strominger, J.L., 1987, Purification and characterization of class II histocompatibility antigens from a homozygous human B cell line, *J. Biol. Chem.* 262:16087.

Hennessey, J.P., Jr. and Johnson, W.C., Jr., 1981, Information content in the circular dichroism of proteins, *Biochemistry* 20:1085.

Hider, R.C., Drake, A.F. and Tamiya, N., 1988, An analysis of the 225-230-nm CD band of elapid toxins, *Biopolymers* 27:113.

Isenman, D.E., Painter, R.H. and Dorrington, K.J., 1975, The structure and function of immunoglobulin domains: studies with β_2-microglobulin on the role of the intrachain disulfide bond, *Proc. Natl. Acad. Sci. USA* 72:548.

Johnson, W.C., Jr., 1988, Secondary structure of proteins through circular dichroism spectroscopy, *Ann. Rev. Biophys. Biophys. Chem.* 17:145.

Kappes, D. and Strominger, J.L., 1988, Human class II major histocompatibility complex genes and proteins, *Ann. Rev. Biochem.* 57:991.

Karlsson, F.A., 1974, Physical-chemical properties of β_2-microglobulin, *Immunochemistry* 11:111.

Kauppinen, J.K., Moffat, D.J., Mantsch, H.H. and Cameron, D.G., 1981, Fourier self-deconvolution: a method for resolving intrinsically overlapped bands, *Appl. Spectrosc.* 35:271.

Krimm, S. and Bandekar, J., 1986, Vibrational spectroscopy and conformation of peptides, polypeptides, and proteins, *Advances in Protein Chemistry* 38:181.

Kwiatkoski, J.M. and Reffner, J.A., 1987, FT-IR microspectrometry advances, *Nature* 328:837.

Lancet, D., Parham, P. and Strominger, J.L., 1979, Heavy chain of HLA-A and HLA-B antigens is conformationally labile: a possible role for β_2-microglobulin, *Proc. Natl. Acad. Sci. USA* 76:3844.

Levitt, M. and Greer, J., 1977, Automatic identification of secondary structure in globular proteins, *J. Mol. Biol.* 114:181.

Marrack, P. and Kappler, J., 1986, The antigen-specific, major histocompatibility complex-restricted receptor on T cells, *Adv. Immunol.* 38:1.

Parham, P., Alpert, B.N., Orr, H.T. and Strominger, J.L., 1977, Carbohydrate moiety of HLA antigens. Antigenic properties and amino acid sequences around the site of glycosylation, *J. Biol. Chem.* 252:7555.

Provencher, S.W. and Glockner, J., 1981, Estimation of globular protein secondary structure from circular dichroism, *Biochemistry* 20:33.

Rupp, F., Acha-Orbea, H., Hengartner, H., Zinkernagel, R. and Joho, R., 1985, Identical Vβ T-cell receptor genes used in alloreactive cytotoxic and antigen plus I-A specific helper T cells, *Nature* 315:425.

Savitsky, A. and Golay, M.J.E., 1964, Smoothing and differentiation of data by simplified least squares procedures, *Analyt. Chem.* 36:1627.

Sears, D.W. and Beychok, S., 1973, Circular dichroism, *in:* "Physical Principles and Techniques of Protein Chemistry," Part C, F.J. Leach, ed., Academic Press, New York, 445.

Surewicz, W.K. and Mantsch, H.H., 1988, New insight into protein secondary structure from resolution-enhanced infrared spectra, *Biochem. Biophys. Acta* 952:115.

Susi, H. and Byler, D.M., 1986, Resolution-enhanced Fourier transform infrared spectroscopy of enzymes, *Meth. Enzymol.* 130:290.

Susi, H., 1972, Infrared spectroscopy--conformation, *Meth. Enzymol.* 26:455.

Trägårdh, L., Curman, B., Wiman, K., Rask, L. and Peterson, P.A., 1979, Chemical, physical chemical, and immunological properties of papain-solubilized human transplantation antigens, *Biochemistry* 18:2218.

Turner, M.J., Cresswell, P., Parham, P., Strominger, J.L., Mann, D.L. and Sanderson, A.R., 1975, Purification of papain-solubilized histocompatibility antigens from a cultured human lymphoblastoid line, RPMI 4265, *J. Biol. Chem.* 250:4512.

Woody, R.W., 1978, Aromatic side-chain contributions to the far ultraviolet circular dichroism of peptides and proteins *Biopolymers* 17:1451.

Woody, R.W., 1987, Contributions of tryptophan side-chains to the far-ultraviolet circular dichroism of proteins, *Proc. 2nd Intl. Conf. Circular Dichroism*, Budapest, 38.

Yang, J.T., Wu, C.-S.C. and Martinez, H.M., 1986, Calculation of protein conformation from circular dichroism, *Method. Enzymol.* 130:208.

SPECIFICITIES OF GERM LINE ANTIBODIES

Thomas P. Theriault, Gordon S. Rule and Harden M. McConnell

Stauffer Laboratory for Physical Chemistry, Department of Chemistry, Stanford University, Stanford, California 94305, USA

The past few years have seen remarkable progress in research on the structure and function of antibodies. In our own work we have shown that it is possible to obtain extensive significant information on the composition and structure of antibody combining sites using NMR, together with nitroxide spin-label haptens (Anglister et al., 1984a, 1985 and 1987; Frey et al., 1984). This derived information includes the amino acid composition of the combining site region, that is, the number of tyrosines, alanines, etc. that are within ~ 20 Å of the odd electron on the paramagnetic hapten. In antibody molecules there are typically 40-50 amino acids in this combining site region. We have shown that NMR titration data can be used to estimate distances between individual protons on amino acid side chains and the odd electron (Anglister et.al., 1984b; Frey et. al., 1988). These measured distances extend out to about 20 Å, and in to distances of the order of 3-5 Å. Shorter distances can sometimes be estimated from nuclear magnetization transfer experiments. The NMR data also provide a powerful and convenient means of obtaining the on-off kinetics of hapten-antibody reactions, using resonance signals from the hapten as well as from the protein.

Our NMR studies are being carried out on 12 monoclonal anti-dinitrophenyl nitroxide spin label antibodies that we have prepared and sequenced (cDNA) (Leahy et al., 1988). Theoretical models for the Fab fragments of all of these antibodies have been made in collaboration with Dr. Michael Levitt (unpublished). A major purpose of our current work is to compare these theoretical structure models and the NMR data. An essential requirement for this work is the production of antibody mutants that can be used to obtain rigorous NMR assignments.

Figures 1 and 2 show schematically the sequence similarities of the Fab fragments of the antibodies AN01-AN12. Some pairs of these antibodies have sequences that are very similar in both heavy and light chains (AN01,2,3), or (AN05,6), some pairs have similarities in only the heavy chains (AN04,11,12), and other pairs have similarities in only the light chains (AN07-AN10). Clearly many sequences can accommodate the same hapten. Since germline antibody genes doubtless evolved long before dinitro-phenyl was first synthesized, we have developed an interest in the question of whether the germline antibodies would also recognize the dinitrophenyl nitroxide spin label hapten. If the germline antibodies did not recognize this hapten, then we would conclude that their specificity/affinity was largely developed through somatic mutation. If the germline antibodies did recognize the dinitrophenyl nitroxide spin label hapten with essentially the same specificity/affinity, then we would conclude that the specificity of these antibodies was essentially accidental, and we might then speculate on the nature of the primordial antigen and how its structure might be related to dinitrophenyl.

```
          -20         -10          1          10          20        27

           |           |           |           |           |        ABC

ANO2    MDFQVQIFSFLLISASVILSRGQIVLTQSPAIMSASPGEKVTMTCSASS
ANO1    ............M.....M............L.................
ANO3    ........M.....M............L.................
ANO9    .............I..VM...EN........I...L......S.R...
ANO5    MRCSLQFLGVLMFWISGVS.D..I..DELSNPVAS..S.SIS.RSTKSLL
ANO6    MRCSLQFLGVLMFWISGVS.D..I..DELSNPVTS..S.SIS.RSTKSLL
ANO4    MR.LAELLG.LLFCFLGV.CD.QMN...SSL...L.DTI.I..H..Q
ANO8    MRF.VQVLG.LLLWISGAQCDVQI....SYLA.....TIIIN.R..K
AN11    MVFTPQILG.MLFWISA...D........TL.VT..DS.SLS.R..Q
AN12    MHHTSMGIKMES..QV.VFVFLWLSGVD.D..M...HKF..T.V.DR.SI..K..Q
ANO7     MAW.SLI.SLL.LSSGAIS.A.V..ES. LTT....T..L..RS.N
AN10     MAW.SLI.SLL.LSSGAIS.A.V..ES. LTT....T..L..RS.T


          30          40          50          60          70          80

    DEF   |           |           |           |           |           |

ANO2     SVYYMYWYQQKPGSSPRLLIYDTSNLASGVPVRFSGSGSGTSYSLTISRMEAEDAA
ANO1     ..S..F......R...KPW..L.........A...............S.......
ANO3     ..S..F......R...KPW.FL........A....R..........S.......
ANO9     ..N..F......SDA..K.W..Y.....P...A.........N.......S.AG....
ANO5     YK DGKT.LN.FL.R..Q..Q....LM.TR....SD.........DFT.E...VK...VG
ANO6     YK DGKT.LN.FL.R..Q..Q....LM.TR....SD.........DFT.E...VK...VG
ANO4     NINVWLS.......NI.K....KA...HT...S.........FT....SLQP..I.
ANO8     SISK.LA...E...KTNK....SG.T.Q..I.S.........DFT....SL.P..F.
AN11     SVSNNLH.F...SHE......KYA.QSI..I.S.........DFT.S.NSV.T..FG
AN12     DVSTAVA.......Q..K....SA.YRYT...D..T......DFTF...SVQ...L.
ANO7     GAVTTSN.AN.V.E..DHLFTG..GG.N.R.P...A.....LI.DKAA...TGAQT..E.
AN10     GAVT.SNSVK.V.E..DHLFTG..GGSN.R.P...A.....LI.DKAA...AGAQT..E.


          90   95        100    106 109

           |    ABCDEF     |     A  |

ANO2    TYYCQQWSSYPP    ITFGVGTKLEL KRA
ANO1    .........N..    ....S.....I ...
ANO3    .......N.I..    ....A...... ...
ANO9    ......FT.S.     S...A...... ...
ANO5    V.....LVEF.     L...A...... ...
ANO6    V.....LVEF.     L...A...... ...
ANO4    ......GQ...     L.L.G.....I ...
ANO8    M.....HNE..     Y...G.....I ...
AN11    M.F...SN.W.     F...G.....I ...
AN12    V...H.HY.S.     Y...G.....I ...
ANO7    I.F.AL.Y.NH     LV..G....TVLGQP
AN10    V.F.AL.Y.NH     LV..G.A..TVLGQP
```

Figure 1. Deduced amino acid sequences of the V regions of the light chains of the anti-DNP
-SL monoclonal antibodies AN01-AN12. Numbering system according to Kabat et
al. (1987). From Leahy et al. (1988).

```
            -15        -5    1         10         20          30   35 36

             |          |    |          |          |          |    AB

ANO2    MRVLILLWLFTAFPGILSDVQLQESGPGLVKPSQSQSLTCTVTGYSITSDYAWN  WI
ANO1    .K..S..Y.L..I.....................L....S........G.Y..  ..
ANO3    .K..S..Y.L..I.....................L....S........G.Y..  ..
ANO7    .......C...................D......L............G.S.H  ..
ANO5    ME.HW.F.F..SVTA.VH.QF.P.Q..AE.A..GA.VKMS.KAS..TF..YWMH  .V
ANO6    ME.HW.F.F..SVTA.VH.Q...P.Q..AE.A..GA.VKMS.KAS...F.RYWMH  .V
ANO4    MGWSW.F.F.LSGTA.VHCQI..KQ...E....GA.VKIS.KAS...F.DY.IN  .V
ANO9    MGWSY.I.F.VATATDVH.Q....QP.AE....GA.VK.S.KAS..TF..YWMH  .V
AN11     MSW.F.LSGTA.VH.E....Q...E..R.GA.VKMS.KAS..TF..YVMH  .V
AN12    MEWNWVV.F.LSLTA.VYAQG.M.Q..AE....GA.VK.S.KTS.FTFR.S.IG  .L
ANO8    MEWLWN..F.MA.AQS.QAQI..VQ...E.K..GETVRIS.KAS..TF.TAGIQ  .V
AN1O    MNFGFS.IF.VLVLK.VQCE.K.V...G.....GG.LK.S.AAS.FTFS.YAMS  .V


            40          52 53    60          70          82 83    90

             |           ABC      |           |           ABC       |

ANO2    RQFPGNKLEWMGYMS    YSGSTRYNPSLRSRISITRDTSKNQFFLQLKSVTTEDTATYF
ANO1    ............IN     .D.RNN.....KN................K..........Y
ANO3    ............IN     .D.NNN.....KN................K.N.........Y
ANO7    .....H.......IH    .....N.....K.................N..........Y
ANO5    K.R..QG...I..INP   NT.Y.V..QKFKDKATL.A.K.SSTAYM..S.L.SD.S.V.Y
ANO6    K.R..QG...I..INP   ST.Y.E..QKFKDKATL.A.K.SSTAYM..S.L.S..S.V.Y
ANO4    K.K..QG...I.WIYP   G..NNK..EKFKGKATL.I...SSTVYI..S.L.S....V..
ANO9    ..R..QG...I.EINP   SN.R.N..EKFK.KATLNV.K.SSTAYM.IS.L.S..S.V.Y
AN11    K.K..QG...I..INP   .NDG.K..EKFKGKATL.S.K.SSTAYIE.S.L.S..S.V.Y
AN12    K.K..QS...IAWIYA   GT.G.S..QKFTGKARL.V...SSTAYM.FS.L....S.I.Y
ANO8    QKM..KG.K.I.WINT   R..VPK.AEDFKG.FAFSLE..ASTAY..ISNLRND...A..
AN1O    ..T.ERR...VASI.    SGYI.Y.PD.VKG.FT.S..NAR.ILY..MS.LRS....M.Y


             95   100        105  110

              |    ABCDEFGHIJK      |    |

ANO2    CARGWP          LAYWGQGTQVSVSE
ANO1    ...EDDGYYI      FD.....STLT..S
ANO3    ...EGYGYF       FD......TLT..S
ANO7    ...VIYYYGSSYV   WF.......L.T..A
ANO5    ..YYGSS         YFD.....TLT..S
ANO6    ..HYGRS         YFD......TLT..S
ANO4    .V.YGYDG        FG.....L.T..A
ANO9    ...R.GSYVGG     F......NM.T..A
AN11    ...FGYYGR       YWYFDV..A..T.T..S
AN12    ...WD.INRG      F.......L.T..A
ANO8    .G.TDYYGST      YYAMD.....SS.T..S
AN1O    ...WGHRYDVL     D......S.T..S
```

Figure 2. Deduced amino acid sequences of the V regions of the heavy chains of the anti-DNP-SL monoclonal antibodies AN01–AN12. Numbering system according to Kabat et al. (1987). From Leahy et al. (1988).

MATERIALS AND METHODS

Southern Blot Analysis

Balb-c genomic DNA was isolated from the livers of Balb-c mice as described in Maniatis et al. (1982). AN02 genomic DNA was similarly isolated from AN02 antibody producing hybridoma cells. Each of these DNAs were digested with a series of restriction enzymes and analyzed by the method of Southern (1975). The probes used in the hybridization were prepared from the variable regions of the light and heavy chain coding sequences of an AN02 containing plasmid. The light chain probe was a 1000 base pair fragment made by digesting the AN02 genomic kappa chain clone with BglII and PpuMI. This probe begins approximately 800bp 5' to the variable region. The heavy chain probe was a 600bp fragment isolated as an XbaI and EcoRI fragment from an AN02 heavy chain clone which extends to within 40 base pairs of the end of the variable region. The probes were labelled using Pharmacia's oligo labelling kit with α-32P dCTP (New England Nuclear) as label. The hybridizations were carried out as described in Maniatis (1982) at 42 degrees Celsius with formamide concentrations ranging from 35% to 50%.

Germline Gene Cloning and Sequencing

Balb-c liver DNA was digested with the appropriate restriction enzyme, size selected on an agarose gel and isolated using an International Biotechnology Unidirectional Electroelutor. Libraries were made by ligating these fragments into either lambda ZAP or lambda NM590. Recombinant phage DNA was packaged as described in Maniatis (1982) and the resultant libraries were screened using either the heavy or light chain probe. Positive clones were purified and subcloned into pUC322 from lambda NM590 or excised from lambda ZAP into its pBluescript phagemid. The plasmid clones were mapped with 15 to 35 combinations of single and double restriction digests. The clones were subcloned again into M13 vectors and sequenced by the dideoxy method with α-35S dATP as label (New England Nuclear).

RESULTS

Rearrangement of Balb-c DNA

Genetic rearrangement as indicated by restriction fragment length polymorphism is evidenced by the results of the Southern blot experiments shown in Figures 3 and 4. In the case of the heavy chain for AN02, the DNA sequence of the variable region was known to share 97.6% sequence identity with a heavy chain Balb-c germline sequence named SB32 and with the variable region of a gene isolated from an IgM producing hybridoma LB8 published by Dzierzak et al. (1986). The hybridization of fragments obtained by digestion with the enzymes EcoRI, HindIII, and XbaI shows a pattern consistent with what had been seen previously, namely hybridization to a 2.4 kb EcoRI fragment, an 7.5 kb HindIII fragment, and a 7.0 kb XbaI fragment. The AN02 genomic heavy chain gene is known to be contained on a 4.6 kb EcoRI and a 1.9 kb XbaI fragment (G. Rule, unpublished data). These fragments are seen to hybridize as would be expected. Our assumption then was that a Balb-c germline variable region gene which was contained on 2.4 kb EcoRI and 7.0 kb XbaI fragments was rearranged by the processes of recombination and somatic mutation such that it rested on 4.6 kb EcoRI and 1.9 kb XbaI fragments in the AN02 genome. The results for digestion with HindIII are less clear, since the size of the restriction fragment which contains the AN02 genomic heavy chain gene is not known. The patterns seen in the double digestion with XbaI/HindIII and XbaI/KpnI, indicate that the single band in the XbaI lane results from more than one germline variable region.

In the case of the light chain a large number of fragments cross hybridize with the AN02 light chain probe indicating that the AN02 kappa variable gene comes from a large family of related germline genes. Most of the hybridizing bands seen in the Balb-c lanes are matched by equivalent bands in AN02. The major difference is that a 3.7 kb HindIII fragment hybridizes with approximately twice as much intensity in Balb-c as in AN02. This pattern would be expected if only one of the allelic light chain containing chromosomes were recombined during the differentiation of the AN02 producing B lymphocyte and if the rearranged germline variable gene were situated at the 3' end of the family of cross reacting variable regions. Further evidence that the AN02 light chain germline is contained on a 3.7 kb HindIII fragment comes from performing the hybridization under more stringent conditions. Only the bands at 3.5, 3.7, and 3.9 kb appear, and the 3.7 kb band is the most intense.

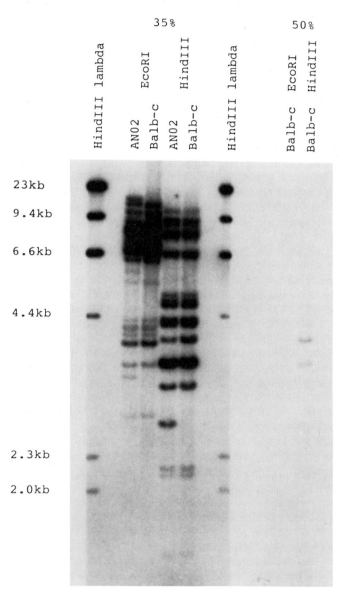

Figure 3. Southern blot results using the light chain probe. AN02 hybridoma and Balb-c genomic DNA were digested with HINDIII and EcoRI. Fragmant sizes were internally referenced by including HINDIII digested lambda DNA in the gel and as a radiolabelled probe. Shown are experiments in which the hybridizations were carried out in 50% and 35% formamide.

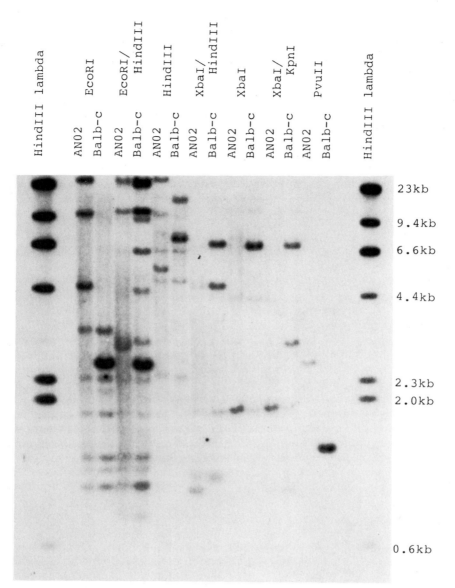

Figure 4. Southern blot results using the heavy chain probe. AN02 hybridoma and Balb-c genomic DNA were digested with the same enzymes. Fragmant sizes were internally referenced by including HINDIII digested lambda DNA in the gel and as a radiolabelled probe.

Cloning and Sequencing of Germline Genes

To determine whether the SB32 gene was indeed the germline gene for AN02 or if a gene with even more sequence identity exists, the 7.0 kb XbaI as well as the 7.5 kb HindIII fragments were ligated into lambda ZAP and lambda NM590 respectively. Multiple positive clones were purified and sequenced from both libraries and in each case a germline gene was obtained that was more than 99% sequence identity to the heavy chain variable region of the antibody AN07 and only 92% sequence identity to AN02. Since the antibody AN07 has a lambda type light chain, the complete germline configuration of AN07 could be deduced. The inability to find the SB32 germline gene or any other AN02-like gene in the XbaI or HindIII libraries upon sequencing 14 independent clones indicated that there may have been a systematic problem with the vectors or hosts used in the cloning process. To better understand the clonability of these variable genes, as well as to obtain partial sequence data for other germline variable regions, a library was constructed which contained EcoRI/XbaI Balb-c DNA fragments of approximately 600 base pairs in length. Two AN02-like germline genes were sequenced from this library in the five clones which have been purified. One of these was the SB32 germline, the other was a gene that was only 87% similar to AN02.

For the light chain germline gene libraries were made in lambda NM590 with HindIII fragments of approximately 3.5, 3.7, and 3.9 kb. Positive clones were purified and sequenced in each case. Two light chain germline genes were found on the 3.5 kb fragments which were 94% similar to the AN02 light chain, one was found on the 3.9 kb fragment that was 96% similar, and one was found on the 3.7 kb fragment which had more than 98% sequence identity. This is assumed to be the germline gene for the light chain variable region of AN02.

Construction of Antibody Genes in the Recombined Germline State.

In order to investigate the binding properties of the IgM class antibody which was found on the surface of the lymphocyte clone that eventually mutated into AN02 or AN07, it is necessary to first make genetic constructs which reflect the sequence of this germline antibody and then express these constructs in a suitable system. Taking the variable regions described above as the true germline variable genes, the remainder of the recombined antibody genes can be found in the germline joining and diversity sequences which have been well characterized (Kabat et al. 1987). For AN07 the diversity segment DFL16.1 and the joining segments JH3 and JL1 were used. For AN02 the diversity segment was too short to determine unambiguously while the joining segments JH3 and JK5 were used. The inferred protein sequence of the heavy and light chain cDNA clones for AN02 and AN07 together with that of the corresponding germline variable regions, diversity segments and joining regions recombined in the proper reading frame are shown in Figures 5 and 6. In each case more than 97% of the protein residues for each

```
          1          10          20        27        30          40          50

          |          |           |        ABCDEF     |           |           |

AN02      QIVLTQSPAIMSASPGEKVTMTCSASS       SVYYMYWYQQKPGSSPRLLIYDTSNL
K5a       . . . . . . . . . . . . . . . . . . . . . . .       . . S . . . . . . . . . . . . . . . . . . . . .

AN07      QAVVTQESA LTTSPGETVTLTCRSSN       GAVTTSNYANWVQEKPDHLFTGLIGGTNNR
Lva       . . . . . . . . .   . . . . . . . . . . . . . . . .T       . . . . . . . . . . . . . . . . . . . . . . . . . . .


          60          70          80        90     95          100        106

          |           |           |         |       ABCDEF      |          A

AN02    ASGVPVRFSGSGSGTSYSLTISRMEAEDAATYYCQQWSSYPP          ITFGVGTKLEL KR
K5a     . . . . . . . . . . . . . . . . . . . . . . . . . . . . . . . . . . . . . . . . . . . . .          . . . . A . . . . . . . . .

AN07    APGVPARFSGSLIGDKAALTITGAQTEDEAIYFCALWYSNH          LVFGGGTKLTVLG
Lva     . . . . . . . . . . . . . . . . . . . . . . . . . . . . . . . . . . . . . . . . . .          . . . . . . . . . . . . .
```

Figure 5. Sequences of AN02 and AN07 light chains compared with their germline gene sequences. Numbering as in Kabat et al. (1987).

```
                1         10        20        30   35        40              52  53      60
                |         |         |         |    AB        |               ABC          |

AN02    DVQLQESGPGLVKPSQSQSLTCTVTGYSITSDYAWN WIRQFPGNKLEWMGYMS    YSGSTRYNP
Gan2    ...............L..................... ................I.    .....S...

AN07    DVQLQESGPDLVKPSQSLSLTCTVTGYSITSGYSWH WIRQFPGHKLEWMGYIH    YSGSTNYNP
G12a    .................................... .......N.........    .........


                70            82  83        90   95   100                    105   110
                |             ABC           |    |    ABCDEFGHIJK            |     |

AN02    SLRSRISITRDTSKNQFFLQLKSVTTEDTATYFCARGWP          LAYWGQGTQVSVSE
Gan2    ..K..................N..........Y...XXX          X.......L.T..A

AN07    SLKSRISITRDTSKNQFFLQLNSVTTEDTATYYCARVIYYYGSSYV   WFAYWGQGTLVTVSA
G12a    ..............................XXX......XX        ..............
```

Figure 6. Sequences of AN02 and AN07 heavy chains compared with their germline gene sequences. Numbering as in Kabat et al. (1987).

antibody can be assigned to their germline states. The ambiguities that exist occur at the junctions between the variable regions, joining, and diversity segments. At splice junctions it is impossible to determine which germline splice partner donated the germline sequence since mutation after recombination could convert one sequence into the other. Compounding this problem in the heavy chains is the mechanism of nucleotide insertion which can place random nucleotides into the V-D and D-J junctions during recombination. However, given that the observed mutation rate in antibody genes during somatic diversification is approximately 0.5% (Gearhart et al. 1983), the small set of ambiguous nucleotides is highly likely to have been the same in the germline as in the mature antibody gene. In the worst case, AN02 with 11 ambiguous nucleotides, the probability is still approximately 95% that the inferred protein sequence for all of these codons correctly represent the germline.

Future Studies

Site specific mutatgenesis has already been used to change the AN02 light chain cDNA clone into the germline gene sequence defined above. Progress is currently underway to construct the other germline genes from the cDNA clones by either site-directed mutagenesis or where suitable restriction sites exist, by replacement with the germline DNA itself. These constructs will then be used in an expression system from which mutant protein can be obtained. Studies will then center on characterizing these proteins in terms of their binding affinities and kinetics for various DNP derivatives.

DISCUSSION

The results of the germline sequencing for AN07 indicate that very little somatic mutation occurred for this antibody. Only two residue changes exist between the known germline genes and the cDNA clones. One of these is removed from the complementary determining regions and is not considered to be a factor. The other occurs in CDR 1 of the light chain. Modelling studies suggest that this residue is removed from the binding pocket and will not be in contact with the bound hapten. Thus we expect the germline antibody and AN07 to have the same affinity for the DNP haptens.

The AN02 germline has a total of 11 residue differences from the sequence of AN02, however only 3 of these occur in the hypervariable loops. Although two of these occur in residues of the second hypervariable loop of the heavy chain they are well removed from our model of the binding pocket and are unlikely to have any significant effect on the binding of the hapten. The residue change in CDR 2 of the light chain, Ser in the germline to Tyr in AN02 at position 31, can also be considered unimportant since NMR studies of tyrosines have implicated only one tyrosine, Tyr34L, as being in contact with the hapten (Rule et al., in preparation). We believe that the antibody in its germline state will have a similar affinity and show similar kinetics to the mature antibody as far as the DNP haptens are concerned.

The plausible assumption that the germline antibodies bind DNP haptens leads us to the conclusion that the DNP specificities of these antibodies are essentially accidental since the germline genes could have only experienced evolutionary pressure in response to naturally occurring antigens. The question then arises as to what the primordial antigen might have been. For example, on the molecular scale, was it big or small? Our NMR studies of the antibodies AN01, AN02, and AN03 show a remarkably large concentration of aromatic residues, especially tyrosines, in the combining site region. In our most studied molecule, AN02, there are 8-10 tyrosine residues in this region (Anglister et al., 1984), many more than could possibly all be in contact with the hapten. Further, as will be discussed elsewhere (Rule et al., in preparation), the sharp proton resonance signals from these residues, with one notable exception, show almost no change on hapten binding, leading us to believe they play no functional role in the binding of DNP haptens. Kinetics studies of antibody-hapten binding also support the view that no significant change in protein structure takes place on hapten binding. We therefore suggest that the primordial antigen recognized by these germline genes was large, with a size of the order of a protein antigen. Studies of other antibody-protein complexes (Amit et al., 1986; Sheriff et al., 1987) typically show that the antibody recognition area is large, on the order of 700 square angstroms of interface. Furthermore, AN02 has been shown to be a cryoglobulin (Theriault et al., in preparation), indicating a possible role for the binding site aromatic residues in idiotype-antiidiotype regulation, the idiotype here being self protein. These ideas will be developed in fuller detail elsewhere.

ACKNOWLEDGEMENTS

We are pleased to thank Daniel Leahy and Dan Denney for providing assistance and guidance in molecular cloning. This work was supported by Office of Naval Research contract N00014-86-k-0388 and American Cancer Society grant NP-631.

REFERENCES

Amit, A. G., Mariuzza, R. A., Phillips, E. V. and Poljak, R. J., 1986, Three-dimentional structure of an antigen-antibody complex at 2.8 Å resolution, *Science* 233:747.

Anglister, J., Frey, T. and McConnell, H. M., 1984, Magnetic resonance of a monoclonal anti-spin-label antibody, *Biochemistry* 23:1138.

Anglister, J., Frey, T. and McConnell, H. M., 1984, Distances of tyrosine residues from a spin-label hapten in the combining site of a specific monoclonal antibody, *Biochemistry* 23:5372.

Anglister, J., Bond, B. W., Frey, T, Leahy, D. J., Levitt, M., McConnell, H. M., Rule, G. S., Tomasello, J. and Whittaker, M. M., 1987, Contribution of tryptophan residues to the combining site of a monoclonal anti-dinitrophenyl spin-label antibody, *Biochemistry* 26:6058.

Anglister, J., Frey, T. and McConnell, H. M., 1985, NMR technique for assessing contributions of heavy and light chains to an antibody combining site, *Nature* 315.6014:65.

Dzierzak, E. A., Janeway, C. A., Jr., Richard, N. and Bothwell, A., 1986, Molecular characterization of antibodies bearing Id-460, *J. Immunol.* 136.5:1864.

Frey, T., Anglister, J. and McConnell, H. M., 1984, Nonaromatic amino acids in the combining site of a monoclonal anti-spin-label antibody, *Biochemistry* 23:6470.

Frey, T., Anglister, J. and McConnell, H. M., 1988, Line-shape analysis of NMR difference spectra of an anti-spin-label antibody, *Biochemistry* 27:5161.

Gearhart, P. J. and Brogenhagen, D. F., 1983, Clusters of point mutations are found exclusively around rearranged antibody variable genes, *Proc. Natl. Acad. Sci. USA* 80:3439.

Leahy, D. J., Rule, G. S., Whittaker, M. M. and McConnell, H. M., 1988, Sequences of 12 monoclonal anti-dinitrophenyl spin-label antibodies for NMR studies, *Proc. Natl. Acad. Sci. USA* 85:3661.

Kabat, E. A., Wu, T. T., Reid-Miller, M., Perry, H. M. and Gottesman, K. S., 1987, "Sequences of Proteins of Immunological Interest", Natl. Inst. of Health, Bethesda, MD.

Maniatis, T., Fritsch, E. F. and Sambrook, J., 1982, "Molecular Cloning : A Laboratory Manual", Cold Spring Harbor Laboratory, Cold Spring Harbor, NY.

Sheriff, S., Silverton, E. W., Padlan, E. A., Cohen, G. H., Smith-Gill, S. J., Finzel, B.C. and Davies, D. R., 1987, Three-dimensional structure of an antibody-antigen complex, *Proc. Natl. Acad. Sci. USA* 84:8075.

Southern, E., 1975, Detection of specific sequences among DNA fragments separated by gel electrophoresis, *J. Mol. Biol.* 98:503.

CONTRIBUTORS

Altman, Russ B. 79
Arseniev, A. S. 111
Bacher, Adelbert 3
Barsukov, I. L. 111
Bayev, A. A. 309
Berry, Alan 333
Beusen, Denise D. 97
Borisova, S. N. 309
Bulatov, A. A. 309
Busch, Robert 61
Bystrov, V. F. 111
Carrara, Enrico 221
Catasti, Paolo 221
Caughey, Winslow S. 355
Champion, Paul M. 347
Chiu, Mark 243
Cubellis, Maria Vittoria 325
Deonarain, Mahendra P. 333
Dobson, Christopher M. 193
Dong, Aichun 355
Emerson, S. Donald 243
Forsén, Sture 291
Golovanov, A. P. 111
Gorbulev, V. G. 309
Gorga, Joan C. 61, 355
Jardetzky, Oleg 79
Jardetzky, Theodore 61
Kantrowitz, Evan R. 35
Karplus, Martin 269
Kieffer, Bruno 139
Kirpichnikov, M. P. 309
Koehl, Patrice 139
La Mar, Gerd N. 243
Ladenstein, Rudolf 3
Lefèvre, Jean-François 139
Lipscomb, William N. 35
Lomize, A. L. 111

Manning, Mark C. 355
Markley, John L. 155
Marshall, Garland R. 97
Maslennikov, I. V. 111
McConnell, Harden M. 367
Moracci, Marco 325
Nicolini, Claudio 221
Nizzari, Mario 221
Pachter, Ruth 79
Parsadanian, A. Sh. 309
Pavlovskii, A. G. 309
Perham, Richard N. 333
Rajarathnam, Krishnakumar 243
Ramakrishnan, V. 49
Rella, Rocco 325
Rigler, Rudolf 257
Roberts, Gordon C. K. 209
Rossi, Mosè 325
Rothbard, Jonathan 61
Rozzo, Carla 325
Rubtsov, P. M. 309
Rule, Gordon S. 367
Schulga, A. A. 309
Scrutton, Nigel S. 333
Skryabin, K. G. 309
Sligar, Stephen A. 243
Sobol, A. G. 111
Stockman, Brian J. 155
Strominger, Jack L. 61, 355
Theriault, Thomas P. 367
Vainstein, B. K. 309
Wiley, Don 61
Woody, Robert W. 355
Wüthrich, Kurt 69
Yu, Liping P. 243

INDEX

1DFT NMR, 195, 229
2D Correlated Spectroscopy *see also COSY*, 73
2D Fourier-Transform *see also Fourier Transform*, 6, 80, 224
2D NMR *see also NMR*, 69, 193, 244, 303
  2DFT NMR, 80, 223, 236
3D Structure Generation *see Structure*
α-Chymotrypsin, 129-131
α-Lactalbumin, 194
ABC Program, 81
Active Site Residues, 40
Active-Site Waters, 282
Allosteric Mechanism, 44
Allosteric Transition, 43, 44
AN02 genomic DNA, 369
Anabaena 7120 flavodoxin, 166
Antibodies, 367
  antibody combining sites, 367
  antibody mutants, 367
  Fab antibody fragments, 159, 367
  heavy chain, 369
  light chain, 370
  specificity, 367
Antigen
  antigen binding site, 62
  antigenic peptide, 145, 159
  class I antigen, 355
  class II antigen, 355
  histocompatibility antigens, 355
  MHC antigens, 61, 358
Archaebacteria 325
  archaebacterial genomic library, 326
  Sulfolobus Solftaricus, 325
Arrhenius Expression, 348
Artificial, Membrane Mimicing Environment, 113
Aspartate Transcarbamylase *see also E. coli*, 36, 41

Aspartate Transcarbamylase (continued)
  PALA, 39
  R-state, 40
  T-state , 40
  quaternary structure, 36
Atomic Fluctuation, 274
Avidin, 62, 63

β-Chiral Deuteration, 156
β-Galactosidase, 326
β-Galactosidase Gene, 327
β_2-Microglobulin (β_2m), 357
β_{60} Capsid, 3, 24
Bacillus subtilis, 3
Bacteriorhodopsin, 111, 128
  left- and right-handed structure, 116
  primary structure, 130
Balb-c Genomic DNA, 369
Balb-c Mice, 369
Biotin, 62
Biotinylated Peptide, 62
BLOCH Program, 89
Blood Clotting, 201
Boltzmann Constant, 271
Boltzmann Distribution, 98
Bovine Immunoglobulin, 355
BPTI *(basic pancreatic trypsin inhibitor)*, 70, 92, 99, 100, 104, 194, 269, 277
Build-up Procedure, 104

Calbindin D_{9k}, 258, 291
Calcium
  calcium affinity, 293
  calcium binding protein, 291+
Calf Liver, 222
Calmodulin, 291
  crystal structure, 292
  secondary structure, 305
Capsid of HRS, 26

Carbon Monoxide, 243
Carbon-13, 155
Cartesian coordinate, 11, 101, 271
Catalysis, 40
Catalytic Chain, 38
Catalytic Mechanism, 334
Catalytic Reaction, 4
CD *see also Circular Dichroism*, 118
CD Spectroscopy, 222, 356
cDNA, 309, 367
CHARMM Program, 270, 282
Circular Dichroism *see also CD*, 42, 222, 355
Cis/Trans Isomerism, 199, 303
Class I Antigen, 355
Class I MHC Proteins, 61
Class II Antigen, 355
Class II MHC Proteins, 61
Cloning, 309, 326, 334, 370
Conformation Changes, 292, 298, 300, 306
Conformational Equilibria, 215
Conformational Heterogenicity, 304
Conformational Substates, 258, 347
CONFORNMR, 122
Contrast Matched, 51
Cooperative Ca^{2+} Binding, 305
Cooperativity, 194, 299
  homotropic cooperativity, 41
Cooperativity Mutant, 283
Correlated Motions, 264
COSY, 73, 160, 224, 244
  DQF-COSY, 224
  RELAYED-COSY, 73
Coupling Constants, 99, 103, 116, 121, 148, 163, 279
  measurements, 181-182
    scalar spin-spin coupling, 73, 99
Crambin, 56, 281
Crystal Packing, 4, 14, 56
Crystal Structure, 62, 72, 92, 99, 117, 151, 199, 209, 228, 244, 269, 292, 316, 334, 355
Crystallographic Refinement Techniques, 79
Crystallographic Symmetries, 7
Crystallography *see also X-ray*, 30, 53, 79, 92, 193, 260
  PROLSQ program, 276
CTP Enzyme, 37
Cyanide, 244
Cyclic Averaging, 17
Cyclosporin, 103-5

Debye-Waller or Temperature Factors, 274
DEPT, 181
Deuterium, 155-6
Deuterium-Assisted Protein NMR Spectroscopy, 159
Differential Scanning Calorimetry *DSC* 222, 223
Dihydrofolate Reductase Binding Site, 209
DIKF *see also Double Iterated Kalman Filter*, 84, 86, 87, 88
Direct Space, 6
Direct-Space Averaging, 14
DISGEO Program *see also Distance Geometry*, 100
DISMAN Program, 102
Dissociation Rates, 300
Distal Effects, 349
Distal Pocket Structure, 349
Distance Constraints, 82, 97, 160
Distance Geometry, 69, 80, 100, 282
  distance distribution function, 86
  DISGEO program, 100
  DISMAN program, 102
Distance Measurements,
  ^1H-^1H distance measurements, 69
  interatomic, 50, 98, 224
  internuclear, 98
  interparticle, 15, 51, 87, 271
  interproton, 70, 98, 145, 175, 279, 367
  long distance geometry, 262
  NOE, 80, 100, 181, 280
  RMS, 91, 272
Disulphide Bond/Bridge, 66, 99, 281, 321, 334, 363
Diversity Sequences, 373
DNA
  AN02 genomic DNA, 369
  Balb-c genomic DNA, 369
  cDNA, 309, 367
  DNA libraries, 326
  mammalian cDNA, 309
Double Iterated Kalman Filter *see also DIKF*, 84
Double Quantum Correlation *DQC*, 157
DR1, 355
  DR1$_{pap}$, 355
Dynamical Simulated Annealing, 103
Dynamics *see also Molecular Dynamics*, 39, 53, 160, 213, 347

E. coli, 53, 160, 211, 243, 292, 315, 325

E. coli (continued)
  aspartate transcarbamylase of *Escherichia coli*, 35
  E. coli dhfr, 211
  E. coli dihydrolipoamide dehydrogenase, 334
  E. coli gor gene, 334
  E. coli thioredoxin, 160
  expression, 325
  lon⁻ and *lon⁺* strains, 295
EcoRI, 295, 326, 370
EGF (epidermal growth factor) domain, 202
Electron Density, 17, 49, 50
  electron density function 7
  electron density map 3, 21
Electron Microscopy, 13, 26
Electrostatic Interactions, 210
Energy Minimization, 72, 101, 271
  energy minimization algorithm, 271
Enrichment Methodologies, 158
Enzymes, 35
  CTP enzyme, 37
Enzymes (continued)
  EcoRI, 295, 326, 370
  HindIII, 370
  mutant enzymes 42, 211, 338
  PALA enzyme, 37
  restriction enzyme, 294, 370
  thermophilic enzyme, 325+
  thermostable enzyme, 325+
  wild-type enzyme, 42, 211, 336
  XbaI, 370

Fab antibody fragments, 367
  and peptide antigens, 159
FAD, 335
Fibrinolysis, 201
Flavodoxin, 23
FlavoProtein, 333
  flavoprotein disulfide oxidoreductases, 333
Folate, 209
Fourier Transform (2D), 6, 224
Fourier Transform Infrared Spectroscopy (FTIR), 355, 356
  FTIR microscopy, 356
  FTIR spectral enhancement, 356

Gaussian Probability Distribution, 347
Gear Predictor-Corrector Algorithm, 272
Gel Chromatography, 111
  gel filtration, 113, 326

Gel Electrophoresis, 63, 295, 335
  Agarose gel, 297, 326, 370
  SDS/polyacrylamide, 63, 222, 295, 316, 335
GENCELL Program, 20
GENERATE Program, 20
Genomic Library, 326
Geometric Redundancy, 10
Germ Line Antibodies, 367
Globular Protein, 69, 75, 112, 193, 221
Glutathione, 333
Glutathione Reductase, 333
GlycoProteins, 61
GOR Method, 228
Gramacidin A, 111
  crystal structure, 117
  primary structure, 114
Growth Hormones, 309+
  3D structure, 311
  bovine, 310
  chicken, 314
  cloning and analysis, 309, 312
  gene expresseed in *E. coli*, 315+
  human, 309, 311
  porcine, 310, 314
Halobacterium Halobium, 128
Heavy Riboflavin Synthase, 3
Heme Proteins, 347
Hemoglobin, 243, 283
HemoProtein, 243, 347
Henderson-Hasselbalch Analysis, 349
HETCOR *see also Heteromuclear Experiments*, 164
Heterologous Hosts, 325
Heteronuclear Multiple-Bond Correlation *HMBC*, 164
Heteronuclear Experiments *see also HETCOR*, 72, 164
Heuristic Refinement, 104
HindIII, 370
Hirudin, 105
Histidine, 340
Histocompatibility Antigens, 355
Histone H1, 221+
  secondary structure, 236
HLA, 355
  HLA-A2 (A2$_{pap}$), 355
  HLA-B7 (B7$_{pap}$), 355
  HLA-DR1, 62
HMQC, 164
Homonuclear ^{1}H NMR, 74

Homonuclear Hartmann-Hahn *HOHAHA*, 160
HPLC, 63
HSBC, 164
Human β2m, 355
Human Chromosome Structure, 222
Human Glutathione Reductase, 334
Human Growth Hormone, 316
Hydrogen Bonding, 56
Hydrogen Exchange, 181
Hyperfine Shifts, 248
Icosahedral β60 Capsid, 3
Immune System, 61
Immunoaffinity Chromatography, 356
Immunoglobulin Constant Regions, 355
Immunoglobulin G (IgG), 357
INADEQUATE, 165
INEPT, 181
Inelastic Scattering *see also Neutron Scattering*, 54, 257
Infrared, 347
Infrared Spectroscopy, 356
Intracellular Ca^{2+}-Binding Proteins, 292
Iron-Porphyrin System, 348
Isoforms, 298
Isotope Labeling, 155, 174
Isotopic Enrichment, 156

Joining Sequences, 373

Kalman Filter, Double Iterated, *see also DIKF*, 84, 86-88
Karplus Equation, 99, 148
Kinetics, 200, 347, 367
  kinetic mechanism, 337, 339
  Ping-Pong, 339
Koshland, Nemethy and Filmer Model, 36
Kringle Domain, 202

Labeled Proteins, 159
Lac-Repressor, 79, 80
  crystal structure, 92
  lac repessor headpiece, 79, 82
  and peptide backbone, 89
  primary & secondary structure, 82
Laue Relations, 7
Lennard-Jones Parameter, 271
Level of Isotopic Enrichment, 156
Ligand Binding, 351
Long Distance Geometry, 262
Low Angle X-Ray Scattering, 44
Lysozyme, 194+, 355

Magnetic Axes, 243
Magnetization Transfer, 160, 195, 367
Main-Chain-Directed Strategy, 74
Major Histocompatability Antigens *(MHC)*, 61, 358
Mammalian cDNA, 309
Mammalian Nucleosome, 221
Maxwellian Distribution, 271
Membrane-Distal Domains, 355
Mercuric Reductase, 333
Mesophilic Hosts, 325
Metabolic Pathway, 35
Metal Binding Proteins, 291
MetalloProteins, 291
Methotrexate, 209
Michaelis Constant, 31, 341
Molecular Dynamics *see also Monte Carlo Simulations*, 72, 101, 257, 269
  molecular dynamics simulations, 271
Molecular Dynamics *(continued)*
  molecular dynamics technique, 72
Molecular Mechanics Calculation, 72
Molecular Modeling, 104
Molecular Packing, 13
Molecular Rotation, 10
Molecular Structure *see Structure*, 257
Molten Globule State, 196
Monod, Wyman and Changeux Model, 36
Monte Carlo Simulations, 104, 278
Mossbauer Effect, 275
Multiple-Bond Correlation *(MBC)*, 158
Multiple Quantum Spectroscopy, 73
Mutagenesis *see Site-Directed, Site Specific and Oligonucleotide-Directed*
Myoglobin, 243, 272, 347
  secondary structure, 272

NADPH, 209, 333
Neutron Diffraction, 49, 57, 243
Neutron Phases, 53
Neutron Scattering, 49, 52, 57, 257
Newton-Raphson Algorithm, 272
Newtonian Equations, 271
Nitrogen-15, 155
Nitroxide Spin-Label Haptens, 367
NMR *(Nuclear Magnetic Resonance)*, 69, 79, 97, 111, 139, 155, 193, 209, 217, 222, 243, 279, 291
  1DFT NMR, 195, 229
  ^{1}H NMR, 297
  2D NMR, 69, 193, 244, 303
    2DFT NMR, 80, 223, 236

NMR (continued)
  CONFORNMR, 122
  COSY, 73, 160
          DQF-COSY, 224
  Coupling Constants, 99, 103, 116, 121,
          148, 163, 279
     measurements, 181-182
     scalar spin-spin coupling, 73, 99
  DEPT, 181
  Deuterium-Assisted Protein NMR
          Spectroscopy, 159
  Double Quantum Correlation *DQC*, 157
  Fourier Transform, 6, 80, 224
  Heteronuclear Experiments *see also*
          HETCOR, 72, 164
  Homonuclear ^{1}H NMR, 74˙
  Homonuclear Hartmann-Hahn
          HOHAHA, 160
  INADEQUATE, 165
  INEPT, 181
  Multiple Quantum Spectroscopy, 73
  NOESY *see below*
  ROESY, 182
  Rotational-Echo Double-Resonance
          (REDOR), 99
  Total Correlation Spectroscopy *(TOCSY)*
          73
NOE *see Nuclear Overhauser Enhancement*
NOESY, 70, 114,160, 196, 205, 224, 244,
          304
NOESY (continued)
  theoretical NOESY spectra, 89
          BLOCH program, 89
Non-Polar Interactions, 213
Nuclear Magnetic Resonance *see NMR*
Nuclear Overhauser Effect 97, 139, 212,
          143, 262, 279
Nuclear Overhauser Enhancement *(NOE)*,
          69, 80, 160, 210
Nucleic Acids, 35, 53, 145, 158, 256, 269
  interactions with proteins, 24-26
Nucleosome sealing, 222+
Nucleotide Insertion, 374

Oligonucleotide, 118, 139, 326
  crystal structure, 151
Oligonucleotide-Directed Mutagenesis, 322
Open Reading Frame *(ORF)*, 327
Optical Absorption Spectra, 347
Optimization Methods, 80, 104
  CHARMM Program, 270, 282
Ordered Sequential Loop, 339

Over-Expression, 334

PALA *see also Aspartate Transcarbamylase*,
          39
  PALA enzyme, 37
Papain, 355, 361
Paramagnetic Enhancement of Relaxation,
          98
Pattern of Enrichment, 156
Pattern Recognition Methods, 72
Patterson Function, 9, 10
Patterson Search Methods, 3
Peptide, 61, 142, 163
  antigens, 145, 159
  backbone, 83, 183, 260, 363
  biotinylated peptide, 62
  bond, 260
  conformation and dynamics, 145
  cyclosporin A, 103
  peptide binding, 62, 355
  peptide-MHC complexes, 61
  sequences, 61, 104
  signal peptide, 314
  synthesis, 145
  transpeptide coupling, 173
  see also Polypeptides
PEPTO, 224
Phase Extension, 17
Ping-Pong, 339
Placental Lactogens, 318
Plasminogen, 201
Point Mutants, 243
Polyacrylamide Gel *see also Gel*
          Electrophoresis, 63, 222
Polypeptides, 111, 139, 230
  antigenic, 142
  membrane spanning polypeptides, 113
  polypeptide chain, 35, 70, 114-7, 201,
          317, 336
  Polypeptide conformation, 70, 103, 111
  polypeptide hormones, 309
     glucogon, 309
  polypeptide structure, 69
     characteristic NOE patterns, 81
     left- and right-handed structure, 116
     3D, 69, 71, 111
     secondary, 72
Potential Surface, 272
Primary Sequence *see also Primary*
          Structure, 79, 102, 224
Primary Structure, 82, 114, 130, 161, 226,
          309, 334

Probabilistic Refinement, 84
Probability Distribution, 347
Prolactins, 318
Proliferins, 318
Proline, 292
PROLSQ Program, 276
PROTEAN Program, 80
  PROTEAN system, 82
Protease Domain, 203
Protein *see also individual proteins*
  binding sites, 292+
  conformation, 53, 111, 155, 347
  crystallography, 274
  design, 79
Protein (continued)
  domains, 91, 336
  dynamics, 160, 193, 209, 274
  engineering, 291, 309, 326, 334
  folding, 101, 139, 193, 258
  globular, 69, 75, 112, 193, 221
  heme, 347
  in solution *see also Solution Structure*,
    79, 111, 155, 356
    PROTEAN program/system, 80, 82
  intracellular Ca^{2+}-binding, 292
  macrostates, 347
  membrane structure, 99, 112, 134
  microstates, 347
  structure determination *see also Structure*,
    69, 70, 79, 97, 193
    symmetric protein assemblies, 3
  structure fluctuations, 53, 181, 212, 269,
    347
  synthesis, 315, 325
  PROTEIN system, 16
  wild-type protein, 297, 323, 336
Proteolysis, 203
Proton Donor, 339
Proximal Pocket Work, 349

Quaternary Structure, 36, 39, 221

R-state *see also Aspartate
    Transcarbamylase*, 40
Raman *see Resonance Raman
    Spectroscopy*, 347
Random Factional Deuteration, 156
Reciprocal Space, 7
Relaxation Matrix, 140, 149
Relaxation Measurements, 97, 111, 119,
    148, 157, 179, 180, 195, 212, 222,
    248, 271, 280

RELAY, 163
Relayed Coherence Transfer Spectroscopy
    (RELAYED-COSY), 73
Resonance Raman Spectroscopy, 349
Restrained Molecular Dynamics *RMD*, 81,
    102
Restriction Enzyme, 370
Restriction Mapping, 327
Ribonuclease A, 282
Ribonucleotide Reductase, 333
Root Mean Square Deviation *(RMSD or
    RMS)*, 71, 91, 86, 105, 272
ROESY, 182
Rotational-Echo Double-Resonance
    (REDOR), 99
Rotational Relaxation, 261

Savitsky-Golay Function, 356
SBC *see Single Bond Correlation*
Scalar Correlations, 171
Scalar Coupling, 73, 99
Scattering Length, 50
  scattering length density, 49
SDS/polyacrylamide *see also gel
    electrophoresis*, 222, 295, 316, 335
Second Derivative Analysis, 356
Secondary Structure, 21, 38, 80, 129, 161,
    172, 194, 221, 272, 305, 355
  ABC program, 81
  secondary structure determination, 72, 81,
    90, 99, 139, 357
Selective Deuteration, 155, 160
Sequence-Specific Resonance Assignment,
    69, 160, 304
Sequencing, 370
Sequential Assignment Strategies, 73, 171,
    304
Sequential Resonance Assignments, 69, 73,
    120, 146, 198, 226 247, 367
Side Chain Motion, 257
Signal/Spectral Assignment, 120, 131, 224,
    244, 291
Simulated Annealing, 277
Single Isomorphous Replacement *(SIR)*, 15
Single-Bond Correlation *(SBC)*, 157, 164
Site Specific Mutagenesis, 43
Site-Directed Mutagenesis, 35, 40, 217,
    291, 315, 326, 334, 347, 355, 374
Site-Directed Structure Modification, 125
Small Angle Scattering *see also X-Ray*, 50,
    53

Solution Structure, 69, 70, 79, 97, 116,
    160, 175, 182, 193, 212, 221, 243,
    262, 356
  of BPTI, 104
  PROTEAN program/system, 80, 82
Somatotropin, 318
Southern Blot Analysis, 369
Spectral Density Function, 144
Spherical Polar Coordinate System, 11
Stable Isotope, 155
Staphylococcal Nuclease, 160, 198
Stereochemical/Stereospecific Assignment,
    100, 120
Steric Tilt Angle, 251
Stochastic Boundary Method, 282
Streptavidin, 62
Streptavidin-Agarose, 63
Streptomyces subtilisin, 160, 165
Structural Heterogeneity, 303
Structural Uncertainty Determination, 79
Structure *see also Solution Structure*, 160
  3D structure generation, 69, 97, 111,
      210, 243, 316
    systematic conformational search, 103
  crystal structure, 62, 72, 92, 99, 117,
      151, 199, 209, 228, 244, 269, 292,
      316, 334, 355
  primary structure, 82, 114, 130, 161,
      226, 309, 334
  secondary structure, 21, 38, 80, 129,
      161, 172, 194, 221, 272, 305, 355
    left- and right-handed structure, 116
    secondary structure determination, 72,
      81, 90, 99, 139, 357
  tertiary structure, 39, 45, 90, 111, 161
  quaternary structure, 36, 39, 221
Structure-Function Relationships, 35, 111,
    139, 209, 222, 235, 291, 309, 347,
    367
Substrate Binding, 26
Subunit Interface, 336
Sulfolobus Solftaricus, 325
Surface Charges, 298
Symmetric Protein Assemblies, 3
Symmetry Averaging Techniques, 10
Systematic Conformational Search, 103

T-cells, 62, 355
T-state *see also Aspartate
    Transcarbamylase*, 40
Taylor Expansion, 51
Tendamistat, 72

Tertiary Structure, 39, 45, 90, 111, 161
Thermal Stability, 31, 336
Thermochemical Properties, 302
Thermodynamics, 31, 194, 283, 303
Thermophilic Bacteria, 325
Thioredoxin, 160, 260
Thioredoxin Reductase, 333
Time Resolved Emission Spectroscopy, 257
Time Resolved Fluorescence Spectroscopy,
    263
Total Correlation Spectroscopy *(TOCSY)*,
    73
TPPI Method, 224
Translation Vector(s), 13
Transmembrane Ion Channel, 113
Trypanothione Reductase, 333
Trypsin *see also BPTI*, 56, 222, 321
Tryptophan, 124, 160, 168, 257

Urokinase, 201

van der Waals, 71, 82, 87, 100, 214, 341
Vector Search Program, 16
Verlet Algorithm, 272
Viral Capsid Structure, 29

Wild-type Enzyme, 42, 211, 336
Wild-type Protein, 297, 336

X-PLOR, 279
X-ray, 44, 49, 61, 358
  low angle x-ray scattering, 44
  reciprocal space, 7
  x-ray crystallography 35, 41, 49, 53, 69,
      79, 97, 112, 116, 193, 221, 244,
      334
  x-ray diffraction 6, 37, 49, 69, 74, 193,
      201, 274, 316
  x-ray refinement, 277
  x-ray solution scattering, 37
  x-ray small angle scattering, 4, 26, 50, 53
  x-ray structure analysis, 3, 41, 116, 243,
      269, 334, 351
  "liquidity" factor, 56
XbaI, 370